MAPPING HISTORICAL LAS VEGAS

MAPPING HISTORICAL LAS VEGAS

A CARTOGRAPHIC JOURNEY

JOE WEBER

UNIVERSITY OF NEVADA PRESS | *Reno & Las Vegas*

University of Nevada Press | Reno, Nevada 89557 USA
www.unpress.nevada.edu

Cover design by Diane McIntosh

LIBRARY OF CONGRESS CATALOGING-IN-PUBLICATION DATA
Names: Weber, Joe, 1970– author.
Title: Mapping historical Las Vegas : a cartographic journey / Joseph Weber.
Description: Reno, Nevada : University of Nevada Press, 2022. | Includes bibliographical references and index. |
Summary: "This book takes readers on a cartographic journey through thousands of years of history in Las Vegas, illustrating the history of the city and surrounding region from the time of the ancient Anasazi farmers to the present. It provides a geographic perspective on the city's growth, showing the influence of water, public land surveys, transportation routes, and casinos on the city's evolution. The visual elements also depict the development of the surrounding region, including public lands, military bases, and also reconstructs the settlement and geography of the canyons and valleys of the Colorado River before the Hoover Dam created Lake Mead."—Provided by publisher.
Identifiers: LCCN 2021041555 | ISBN 9781948908405 (paperback) | ISBN 9781948908399 (ebook)
Subjects: LCSH: Las Vegas (Nev.)—Maps. | Las Vegas (Nev.)—History. | Las Vegas (Nev.)—Description and travel.
Classification: LCC F849.L35 W44 2022 | DDC 979.3/135—dc23
LC record available at https://lccn.loc.gov/2021041555

The paper used in this book meets the requirements of American National Standard for Information Sciences—Permanence of Paper for Printed Library Materials, ANSI/NISO Z39.48-1992 (R2002).

FIRST PRINTING

Manufactured in the United States of America

Frontispiece photo © Hanyun, stock.adobe.com; *page 2*, Basin and Range National Monument, courtesy of the Bureau of Land Management.

Contents

Maps

Preface

My many years of experience with Las Vegas and my even longer fascination with the city spurred me to write this book. I grew up in Death Valley, California, in the 1970s and 1980s and then briefly lived at what was then the western end of Sahara Avenue. Nothing but desert was beyond Hualapai Way at the time; now it's just another intersection. Las Vegas is the closest place I've known to a hometown, and it has always been a source of wonder for me.

This book is my attempt to better convey the history of the greatest city in the world—and to encourage others to seek out that history. My own experiences inspired many of the maps. The 1975 Caesars Palace flood, the 1980 MGM Grand fire, the 1982 Thunderbirds crash, and the 1988 PEPCON explosion are vivid memories for me. So was seeing water go over the Arizona spillway of Hoover Dam in the summer of 1983, a sight that may very well never be seen again because of long-term drought and falling lake levels. I also remember waiting at the Spring Mountain Road railroad crossing for a seemingly endless freight train to pass; visiting McCarran International Airport when there were no D gates or people-movers; and driving the Strip before it was jammed with traffic. My parents told me the first words I spoke as a child were "Aku Aku," the name of a Polynesian-themed lounge at the Stardust hotel, promoted by Easter Island statues out front. I grew up afraid of showgirls the way children in other cities grow up afraid of clowns. I remember seeing movies at the Huntridge Theater, Cinerama dome, Parkway Theatres 1-2-3, Boulevard Twin Theaters, and especially the amazing Red Rock 11 on West Charleston Boulevard, which I still consider to be the greatest movie theater ever built (chapter 8). All of these memories made it into the book.

I learned a tremendous amount about the city while making these maps. I never actually knew where Las Vegas Creek flowed, had never heard of Duck Creek, didn't know Henderson had a downtown, or that there had been two McCarran airports. I certainly had no idea what was underneath Lake Mead. And I was very happy to discover that those statues from the Stardust survive and are on an island at Sunset Park. I hope you learn something new about the city as well.

This book is also an attempt to write about the growth of Las Vegas without resorting to the usual clichés about casinos, the Mafia, or "Sin City" (a term that appears nowhere else in this book). It is also a rebuttal to frequent Las Vegas bashing and myths about the city that manage to persist. The Flamingo was not the beginning of the Strip or even the city; the Strip does not receive all of its power from Hoover Dam; and, yes, the city not only has history, but that history survives and is increasingly valued. It is my hope that this book will encourage others to explore more of the city's past and perhaps even to help preserve more of it.

A special thanks to my wonderful wife, Amy, and our dogs, Teddy and Lady, for their patience during this never-ending process, to Rex Rowley at Illinois State University for inspiring me to finally get around to writing a Las Vegas book, and to Justin Race, Jinni Fontana, and Paul Szydelko at the University of Nevada Press for their efforts on this project. And I'd like to also thank my father-in-law, Hassell Smith (Teamsters Local 391), whose pension funds helped build this wonderful city.

MAPPING HISTORICAL LAS VEGAS

1

Introduction

Las Vegas is the greatest city in the world. It has a long and fascinating history, but unfortunately the many books that have been written about it rarely provide maps. This book differs by taking a geographic perspective, emphasizing the importance of understanding locations and placing events and features in that context. This book illustrates the growth of the city, its evolution within the region, and its changing relationships to the environment. It includes not just the city of Las Vegas but the entire valley and surrounding hinterland, outlying towns such as Pahrump, federal recreational lands such as Lake Mead and the western end of the Grand Canyon, and the vast military-nuclear complex of the Nevada Test and Training Range and the Nevada National Security Site.

More than just depicting earlier versions of the city or how it has grown, the book also reveals what isn't there anymore. Much of what was essential to the Las Vegas area's birth and growth has already been lost. The original lush meadows, woodlands, and flowing springs that gave the valley its name are long gone. More water was here than we probably realize today. While Lake Mead continues to be a prominent aquatic playground, other valley floors and canyons that were once inundated were largely unknown. No vegetation, faunal, archaeological, or geological surveys were performed in this region before it disappeared.

The book does not need to be read sequentially, and readers can skip around to maps that interest them. Cross references within the text to other maps and chapters aid those interested in specific subjects. The chapters are, however, arranged in a general chronological order. Those reading it this way will encounter a certain unavoidable amount of repetition when the same topics are introduced in several different maps.

The next chapter begins the story of Las Vegas by examining its natural setting. Everyone knows Las Vegas is in the desert, and anyone can see it lies within a large valley. But there is much more to the story than this. Elevations matter: the towering heights of the Spring Mountains made the springs and meadows of the valley possible, and they are closely related to slopes, rainfall, and environmental patterns of plants and animals throughout the region.

The human history of the Las Vegas area before the town was founded in 1905 is covered in chapter 3. These maps depict the several Native American cultures, including Pueblos and Paiutes, that have occupied the region for at least ten thousand years. The Las Vegas Valley was the westernmost outpost of the Anasazi or Ancestral Pueblo culture, better known for their spectacular cliff cities in national parks in Arizona, Utah, Colorado, and New Mexico. Several major archaeological sites exist from this era, including the Lost City complex along the Muddy River. The Paiutes lived throughout the valley and mountains, and they had a well-developed way of life with trade routes and place-names. They are still here, though their way of life, settlements, and place-names have largely been erased.

In addition to being a home for indigenous peoples for thousands of years, the valley was also a resting place for long-distance travelers. The springs received the name "Las Vegas" in January 1830 when Antonio Armijo led an expedition through the south end of the valley on his way from Santa Fe to Los Angeles. It first showed up on a map in 1844 when the explorer John Frémont included the word "Vegas" on his map of his travels. By then, travelers on the Old Spanish Trail, a seven-hundred-mile route between New Mexico and California, regularly stopped at the springs. Others were involved in a slave trade that exploited the local Paiutes. Newcomers arrived in the 1850s when Church of Jesus Christ of Latter-day Saints pioneers established an agricultural outpost in the valley as well as in the Muddy Valley to the east. They built the Old Mormon Fort, parts of which still stand as Southern Nevada's oldest building. Las Vegas was not always in Nevada; when it was first named, it was part of Mexico and later became part of New Mexico and Arizona Territories, as well as having been claimed by Mormons as part of a proposed state of Deseret. The area that includes Las Vegas didn't join Nevada until several years after the new state was created.

Chapter 4 begins with the founding of the railroad town of Las Vegas in 1905. This little town was created by the San Pedro, Los Angeles and Salt Lake Railroad, which purchased the life-giving springs and harnessed them to provide water for the railroad as well as to supply the residents. It remained a little town dependent on the railroad for jobs and water until the 1930s, when the first of several massive government projects arrived in Southern Nevada. World War II was the next transformative event, necessitating an air base with thousands of jobs as well as an industrial center for war materials (and the new town of Henderson). Through it all, the city of Las Vegas remained quite small. Growth accelerated after the war when hotel-casinos were built south of town along US Highway 91. These eventually became the Strip, the Las Vegas Valley's economic heart and one of the world's best-known urban landscapes. The maps show the rapid growth of the city and outlying areas after World War II.

The construction of Hoover Dam and Lake Mead transformed Las Vegas and Southern Nevada in a variety of ways shown in chapter 5. It caused a population surge, established Boulder City, and created a world-famous attraction. But the lake created by the dam flooded canyons and wide basins along the Colorado and Virgin Rivers. Towns, roads, railroads, mines, and farms were abandoned. At the time, few were concerned about this, and, with a few notable exceptions, the history and geography of these places have largely been forgotten. This chapter re-creates features that are now underwater, including the routes of its explorers and the locations of settlements and mines. Among these was the town of St. Thomas, a salt mine along the Virgin River that had been mined for hundreds, if not thousands, of years, and the short-lived river port town of Callville.

The construction of Hoover Dam required a vast infrastructure of railroads, roads, industrial plants, and communities, most of which have long been submerged. The filling of Lake Mead produced many changes to the region, including earthquakes and sinking land caused by the sudden weight of water. Planners assumed the lake would last hundreds of years before it slowly filled with mud from the Colorado River, but since the late 1990s, drought has been the norm. The lake has shrunk, exposing many areas long submerged, and will likely continue to shrink. The town of St. Thomas reappeared from the lake, and more places will emerge as the water level diminishes. If the drought persists, the lake will reach an elevation known as "dead pool," below the level of the hydroelectric intakes on Hoover Dam. But even if the water drops below that level, much of the old canyons and basins have already been buried permanently under river mud. Recent surveys of sediment in the lake provide a glimpse of how much has accumulated. Some places, such as Callville, will likely never be seen again.

Water is the most important natural resource in the valley and the underpinning of the economy. The presence of large springs and creeks attracted Native Americans, Mormon pioneers, and the railroad to what became Las Vegas. Chapter 6 maps out the development of these water resources and some of the problems that have accompanied it. The reliance on a few springs was reduced when hundreds of wells were drilled throughout the valley. But dependence on this groundwater reached its limits by the middle of the twentieth century, and during this time the city literally sank as the water was pumped out of the ground. Tapping the Colorado River was the obvious next step, but developing these water resources has entangled the city in a web of regional, national, and international legal issues, including the ever-increasing reliance on Colorado River water and the controversial plan to construct a pipeline from northern Nevada. While less glamorous, the disposal of wastewater has also been an important story. Las Vegas Wash has become the city's sewer, and Lake Mead the destination for much that goes down the drain. The Las Vegas Valley is subject to flash flooding, and tremendous accomplishments in flood control have been essential to its growth since the 1990s.

Second only to water as the lifeblood of the community, transportation put Las Vegas on the map and has sustained it ever since. This is covered in chapter 7. It was named by those who used it as a welcome watering place on an almost thousand-mile trade route between California and New Mexico in the nineteenth century, founded as a rail station between California and points east, kept alive in its early decades by railroad employment, and sustained by road and air connections bringing in visitors and their money from the rest of the West, the entire country, and an increasingly globalized world. This chapter maps the evolving transport systems of Las Vegas, beginning with the area's railroads. Many of these lines are long gone, and it is surprising just how many existed. The development of highways in the Las Vegas area was fundamental to its growth as a gambling center and is the subject of several maps showing the region's place within larger highway networks as well as how each route has changed. Las Vegas has a long experience with aviation, including one of the nation's first airmail routes, and the valley's many airports and airline connections to the outside world are covered.

Chapter 8 examines a number of historical topics thematically, exploring individual elements of the city's history: boundaries; the influence of land surveys on the city's street patterns; street names; labor unions; visitors; neighborhoods; casinos; movie theaters; fires; population; electricity; mining; and even waste disposal. These are generally topics of little concern to visitors, and even locals may know little about them. Las Vegas gets very little electricity from Hoover Dam; almost all of it instead comes from power plants burning natural gas from underground pipelines.

Las Vegas is surrounded by vast public lands, many of which have been protected almost as long as the town has existed and have become part of the town's cultural landscape. Chapter 9 examines the evolution of these places beginning with national forests and wildlife refuges, the development of a state park system, the protection of the spectacular Red Rock Canyon west of the city, and new national monuments around the city. Some of these areas were protected soon after the city was founded, but many are only a few years old. Basin and Range, Tule Springs Fossil Beds, and Fossil Butte National Monuments are the newest of these protected areas. Las Vegas is near the western end of the Grand Canyon, and one map explores the Grand Canyon West area within the Hualapai Indian Reservation. There are even wilderness areas around the city, some in view of the Strip. Some lands, most of which are off limits to the public, are used by the US Air Force or the Department of Energy for a variety of training purposes. Although the ground no longer shakes from atomic explosions, the skies still echo with the sound of military jets; the city is home to premier air combat training facilities.

It is easy to assume all of this was inevitable. Chapter 10 provides ample evidence that it wasn't and shows the future that never was (or at least, has not yet arrived). Countless dreams have gone unrealized in Las Vegas, including those of casino developers, railroad promoters, city planners, and even federal agencies such as the US Bureau of Reclamation and the Department of Energy. This chapter looks at the Las Vegas that never was: casinos that were never built; railroad projects that never materialized; failed transportation plans; the long-delayed second Las Vegas airport; and the politically stalled Yucca Mountain nuclear waste repository. Some of these represent entirely different futures for Las Vegas, such as schemes to make the small town of the 1930s into an irrigated agricultural region like today's Imperial Valley in California. While everyone knows that Hoover Dam was once known as Boulder Dam, the fact that it was originally planned to be built in Boulder Canyon is less known. What might the area have looked like if the dam had been built there rather than in Black Canyon? In the postwar era, a dam even larger than Hoover was planned for the western Grand Canyon. This Bridge Canyon dam was eventually canceled, but its location and the roads and other developments can be mapped. Interstate 11 is already familiar to drivers because of the Boulder City Bypass, but there may be far more to this highway in the future. Other projects are less likely, such as the Ivanpah Valley airport and the northern Nevada pipeline.

Finally, the history of Las Vegas is further explored in chapter 11 by looking at the stories of individuals and places associated with the city's history. Frank Sinatra, Elvis Presley, Howard Hughes, Clint Eastwood, and even the mythical James Bond all spent time in the city and became closely associated with many casinos and other locations, though many of the places with which these people are associated are long gone. The Sands was one of these. Elvis learned how to swing his hips and found one of his hits there in 1956. It was the location of the legendary Rat Pack, where crowds including presidential candidate John F. Kennedy watched Sinatra, Dean Martin, Sammy Davis Jr, and others perform in 1960.

What remains of early Las Vegas history? Quite a lot, actually. Three maps explore these places, the first indicating the many casinos that have disappeared in the city's endless cycles of renewal. The area is home to several dozen entries on the National Register of Historic Places, a list of more than one million properties nationwide that are of local or regional significance. Shown on a map, these buildings and neighborhoods make up an important part of the city's history. Few are open to the public, but there are other opportunities to explore the area's rich history at museums such as the incomparable Neon Museum in downtown Las Vegas or the Clark County Museum in Henderson. The last map shows the many remnants of the city's history located in out-of-the-way places or even in plain sight, such as the original McCarran airport entrance pillars or a statue from the long-lost Stardust relocated to Sunset Park.

Many other topics and phenomena could be mapped but were not because of limited space, missing data, or other difficulties. For that reason there are no maps of commuting, individual casinos, foreclosures, housing values, bus routes, gated communities, schools, traffic flow, crime, Death Valley National Park, Mojave National Preserve, Laughlin/Bullhead City, Searchlight, all the movies filmed in the Las Vegas area, or other various aspects of the political and social life of the city.

Readers will also note similar patterns emerging on different maps. One is the city's location on routes connecting Southern California with Utah. Most of the town's links to the outside world were to Southern California, including trail, road, air, rail, pipeline, and electricity. Connections to the rest of Nevada have always been much slower to develop; these have been much less important to the city. For that reason, many maps depict much of California and Arizona, but the entire state of Nevada appears on only a few maps.

Another facet of the city's history that is revealed in these maps is the persistence of several early decisions. It may have seemed a minor event when Western Air Express Field was built far to the northeast of town in 1929, but that tiny airfield remains in use as Nellis Air Force Base, one of the city's economic pillars. Imagine if that field had been built elsewhere in the valley; it would likely be where Nellis is today. Similar outcomes existed for the decision to build a dam in Black Canyon instead of Boulder Canyon; the decision to build a wartime magnesium plant outside of Las Vegas instead of Needles, California; and even the decision by a handful of Mormon pioneers to locate their settlement below Las Vegas Springs rather than at the springs or farther south along Duck Creek. If any of these decisions had been made differently (or never made), the map of the city today would be quite different.

Several additional notes about this cartographic journey are necessary. The maps were constructed using a wide range of geographic information system (GIS) data, much of it from the US Geological Survey (USGS); the US Census Bureau; and Clark County, described in more detail in chapter 12, along with references to all sources. This is a work of historical GIS, some based on existing data, but much of it created from various sources. In many cases, the map data were created by the author in Google Earth in 2017 and 2018 using a variety of current and historic imagery to display objects such as lakes at different points in time. The University of Nevada, Las Vegas (UNLV), has many early maps of Las Vegas in its digital collections available online, and these were used to identify early boundaries, vegetation, and other features. Census data comes from 2020 where available. All of the data used are available over the internet free of charge.

It will be evident that the name Las Vegas is used to refer to the entire valley and urban region, not just the city of Las Vegas. Many names have changed over the years. Casinos are usually referred to by the name they had at the time of the map. Likewise, McCarran International Airport was transitioning to become Harry Reid International Airport in 2021; it is referred to here as McCarran in the past and as Harry Reid in the present. Finally, the word gambling is used here rather than the insipid euphemism of "gaming."

The Natural Setting

2

The terrain, climate, and natural vegetation provide the stunning backdrop for Las Vegas, not only making it possible but also shaping the city's growth. Las Vegas is in the Mojave Desert, but the tall mountains that ring the valley are high enough to generate orographic precipitation, which occurs when air must rise over a high mountain. The rainfall soaks into the ground and flows down the mountain to emerge at numerous springs. These springs attracted Native Americans, Mormons, and later the railroad that created modern Las Vegas.

The rain that falls in the mountains results in a pattern of different ecosystems at different elevations, with each ecosystem having different plants and animals. These have been disrupted at lower elevations, and the valley has changed tremendously in the last century, with much of the original springs and vegetation having disappeared. For that reason, this chapter can be read as a description of what Las Vegas was like before humans, or at least Europeans, arrived.

It all begins with the mountains. They surround Las Vegas and create an impressive natural skyline. The tallest are the Spring Mountains, to the west, which run more than sixty miles from US 95 in the north to Potosi Mountain in the south. The range reaches an elevation of 11,916 feet above sea level at Charleston Peak, the highest point for more than a hundred miles in any direction. The tall, often snow-covered mountain can be seen from many locations in Southern Nevada and even from the summit of Mount Whitney in the Sierra Nevada Mountains, 146 miles to the west.

The second tallest range near Las Vegas is the Sheep Range, though it is not visible from much of the city because it is hidden behind the lower Gass Peak in the Las Vegas Range. Even from US 95 on the way north to Indian Springs, only the southern end of the range is visible; it extends fifty miles northward. The highpoint of this range is the 9,912-foot Hayford Peak, toward the southern end of the mountains.

To the south of the city are the lower McCullough and River mountains, while to the east Frenchman and Sunrise Mountains provide a dramatic backdrop to the city. These two low rugged mountains are often confused: Sunrise is to the north, and the slightly taller Frenchman Mountain is on the south. Lake Mead is surrounded by several rugged ranges, among them the Muddy Mountains on the west, Virgin Mountains to the east, and the Black Mountains east of Hoover Dam.

Charleston Peak. Southern Nevada's highest point at 11,916 feet above sea level, 2008. Photo by Stan Shebs.

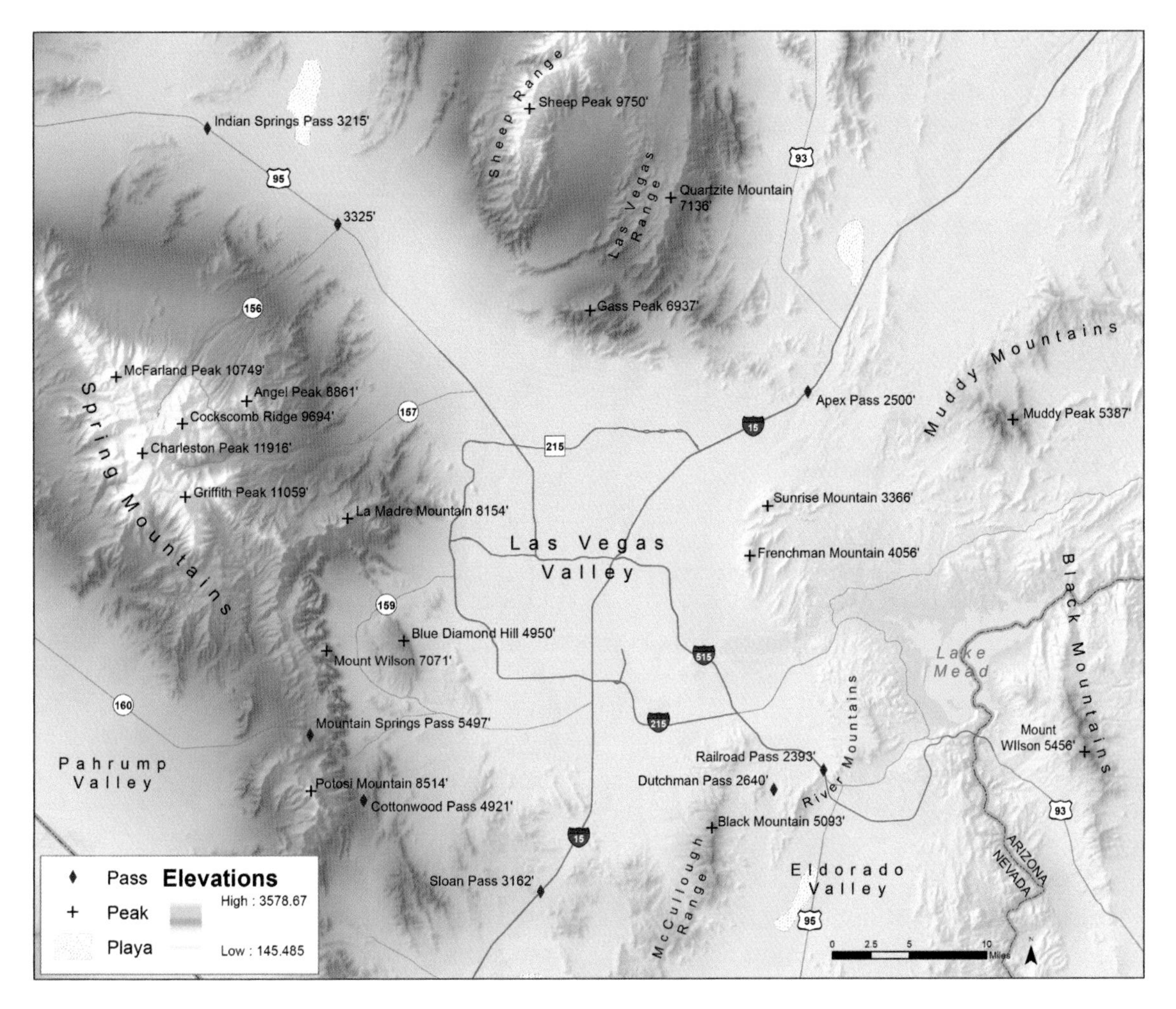

Map 2.1. Elevations. This map is based on a Digital Elevation Model (DEM) from the US Geological Survey, and employs a hypsometric color scheme to depict elevations. This is a common method for showing elevations using a sequence of colors, but care must be taken not to interpret the colors as vegetation or snow-covered mountains.

The mountain walls surrounding the Las Vegas Valley are breached by passes at many locations, though unlike the mountains few of these are named. To the northeast, Interstate 15 and the Union Pacific Railroad cross over a low pass with an elevation of 2,500 feet near Apex. Another higher, but less obvious, pass is south of Sloan. US 95 leaves the valley on its way to Indian Springs over an almost imperceptible divide between the Las Vegas Valley and the Great Basin watershed (map 2.5) a few miles south of Indian Springs (though the highest point on the road is at the Lee Canyon Road intersection). To the southeast, US 93 and 95 cross between the Las Vegas and Eldorado valleys via Railroad Pass, the lowest entrance to the valley on a main highway. This name is familiar to many thanks to the Railroad Pass casino located there. But among several "Railroad Passes" in Nevada, it is the only one through which a railroad ever passed.

The most spectacular pass and highest route out of the city is at Mountain Springs between Las Vegas and Pahrump. This road reaches almost 5,500 feet in elevation and may have snow during winter storms. This route was once a part of the Old Spanish Trail (map 3.5), for which the springs were an important water source. To the east, Lake Mead Boulevard and Lake Mead Parkway connect the Las Vegas Valley to Lake Mead, the former through an unnamed gap between Frenchman and Sunrise Mountains and the latter by Las Vegas Wash. This latter route was the first by which people of European ancestry entered the Las Vegas Valley and is also the lowest elevation in valley, about 1,450 feet at the upper end of Lake Las Vegas (another much smaller manmade lake developed in the 1990s).

The valley floor is not flat; elevations generally increase farther west in the valley. Downtown Las Vegas is 2,026 feet above sea level, the Meadows Mall is 2,170 feet, and Downtown Summerlin is 3,030 feet. The Stratosphere tower, the valley's tallest building, is 1,149 feet tall. It stands on ground slightly lower than downtown, giving the 869-foot-high observation deck an elevation of 2,909 feet above sea level. Observers on the tower therefore look up at Downtown Summerlin and the neighborhoods beyond, the highest in Las Vegas.

Valley floors in Southern Nevada are higher than many in the Mojave Desert. This elevation helps limit Las Vegas's summertime heat; the average July high temperature is 104 degrees, much more pleasant than that found in the lower elevations of Death Valley to the west. Las Vegas also has a higher elevation than many other large cities. When the Oakland Raiders relocated to Las Vegas in 2020, Allegiant Stadium along Interstate 15 became the second-highest home field of any NFL team. Only Denver's Empire Field at Mile High is at a higher elevation.

The range of elevations in the Las Vegas Valley is important for several reasons. One is that temperatures decrease 3.5 degrees Fahrenheit with every 1,000-foot increase in elevation. Charleston Peak is therefore about 35 degrees cooler than downtown Las Vegas, which anyone driving from the city up to Kyle or Lee Canyons has experienced. This applies at lower elevations as well. Boulder City is about 2,500 feet in elevation, only slightly higher than Las Vegas, but its location was chosen because it was 1,800 feet higher than the Colorado River, where the first worker's camp and a shantytown had been located (map 5.7). The extra elevation and exposure to breezes made the site much more tolerable than the oven-like canyon bottom.

Another reason that elevation is important is that precipitation is greater at higher elevations. Rainfall and winter snow occur where air rises over the mountains; the higher the peaks, the greater the effect (map 2.4). This snow and rain not only support forests and ski resorts, but runoff trickles down into the valleys and creates the plentiful groundwater that once made Las Vegas's many springs flow with abundant water. Before the

1940s, when the Colorado River was tapped as a water source, these springs made Las Vegas's existence possible.

Not all of the precipitation created by the mountains is beneficial. Differences in elevations and rainfall between the high mountains and valley floors also cause flooding in the valley, though an elaborate system of flood control basins has reduced this in recent decades (map 6.8).

Cooler temperatures and greater precipitation also mean that different plants and animals will live at different elevations; the pine forests high on Charleston Peak are evidence of this. Although protected today, these plants and animals were important resources to Paiutes and early settlers (map 2.6).

Elevations in Las Vegas have changed. The heavy pumping of groundwater over many decades caused the ground to drop, an effect called subsidence (map 6.5). This was first noticed in Las Vegas in 1948 because of precise surveys done in 1935 by the Coast and Geodetic Survey for the Hoover Dam project. These inadvertently revealed that Las Vegas had sunk as much as three inches since 1915. The sinking has continued at even greater rates since then as groundwater pumping has increased.

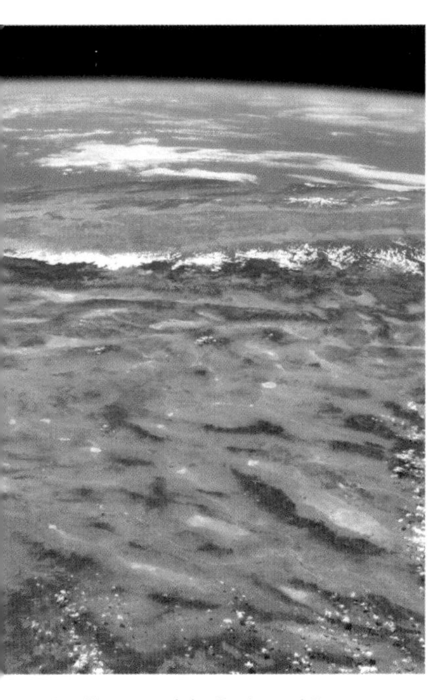

Las Vegas and the Basin and Range country as seen from the International Space Station, 2010. The view is to the southwest; Las Vegas is in the lower left center, just to the left of the white clouds over the Spring Mountains and Sheep Range. The darker areas are mountain ranges and the yellow-tan are valley bottoms. Portions of the Grand Canyon can be seen in the lower left. Photo courtesy of NASA.

Las Vegas lies within the Basin and Range region of the western United States with alternating mountain ranges and valleys. This region extends from Oregon and Idaho south through almost all of Nevada into California, southern Arizona, parts of New Mexico and west Texas, and into Mexico. This region began forming about seventeen million years ago when Earth's crust began pulling apart, causing the surface to become fractured into large blocks. Some blocks dropped, forming valleys, while others rose, forming mountains. The result is a landscape marked by three common landforms: north-south running mountain ranges, valley floors, and alluvial slopes between them.

All three landforms are found in the Las Vegas area. The mountains stand high above Las Vegas and other valleys (map 2.1) and are easily visible on the map by having the steepest terrain. The steepest slopes in the Las Vegas area are the magnificent red and white cliffs in the mountains south of Red Rock Canyon, where near-vertical cliffs provide a spectacular backdrop. But few of the valley's mountain passes are associated with steep slopes, and the low divide between Las Vegas and Indian Springs is imperceptible on the ground.

Most of the Las Vegas Valley bottom is within a few degrees of being completely flat, but there is a definite decrease in elevation from the west to the east side of the valley (map 2.1). Few features break up this pattern, notably Whitney Mesa in Henderson, northwest of the Galleria Mall. The valley differs from most of the surrounding ones in not having a playa or dry lake

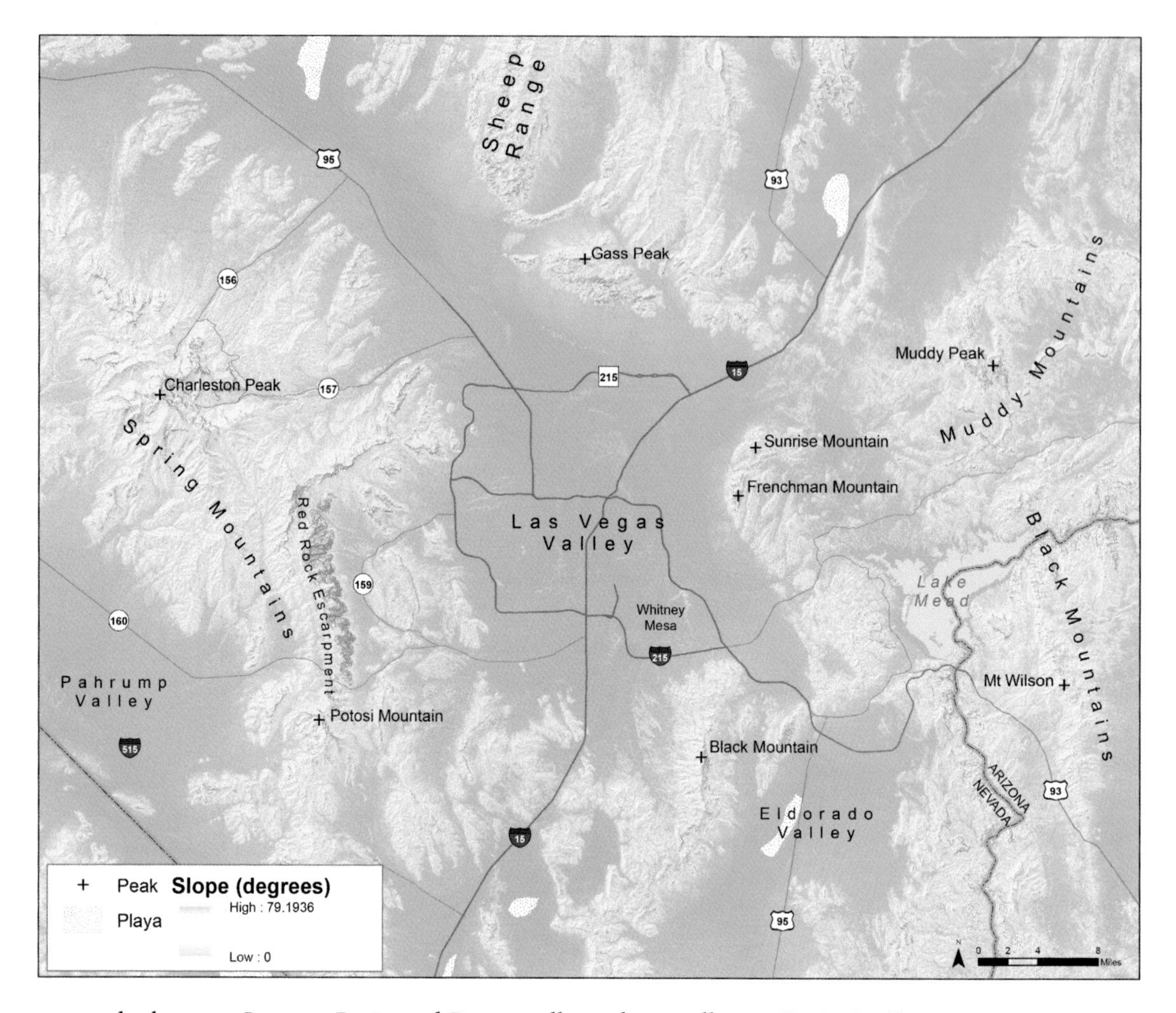

Map 2.2. Landforms. Redder colors denote steeper slopes, and green areas are flatter. The slope surface was created from the DEM used for elevations and hillshaded terrain.

at the bottom. In most Basin and Range valleys, these collect all the runoff from the mountains above, but in Las Vegas any runoff will flow down Las Vegas Wash to the Colorado River.

The third landform component of the Basin and Range region are alluvial fans. These are most evident around the east side of the Spring Mountains, where separate fans from individual canyons have merged to create a continuous bajada. This is crossed by US 95 between Las Vegas and Indian Springs. Smaller fans have developed around the valley's shorter ranges. The transition between valley bottom and fan is not always obvious, especially after the city's growth has expanded onto lower alluvial slopes to the south and west. The steeper slopes of the mountains have so far seen little development.

The elevation change and heavier precipitation in the mountains produce flash flooding, especially during summer months. Although the steeper slopes are found in the mountains, it is in the flatter slopes where the floods do the most damage to the city and its roads (map 6.8). Fortunately, larger rocks carried

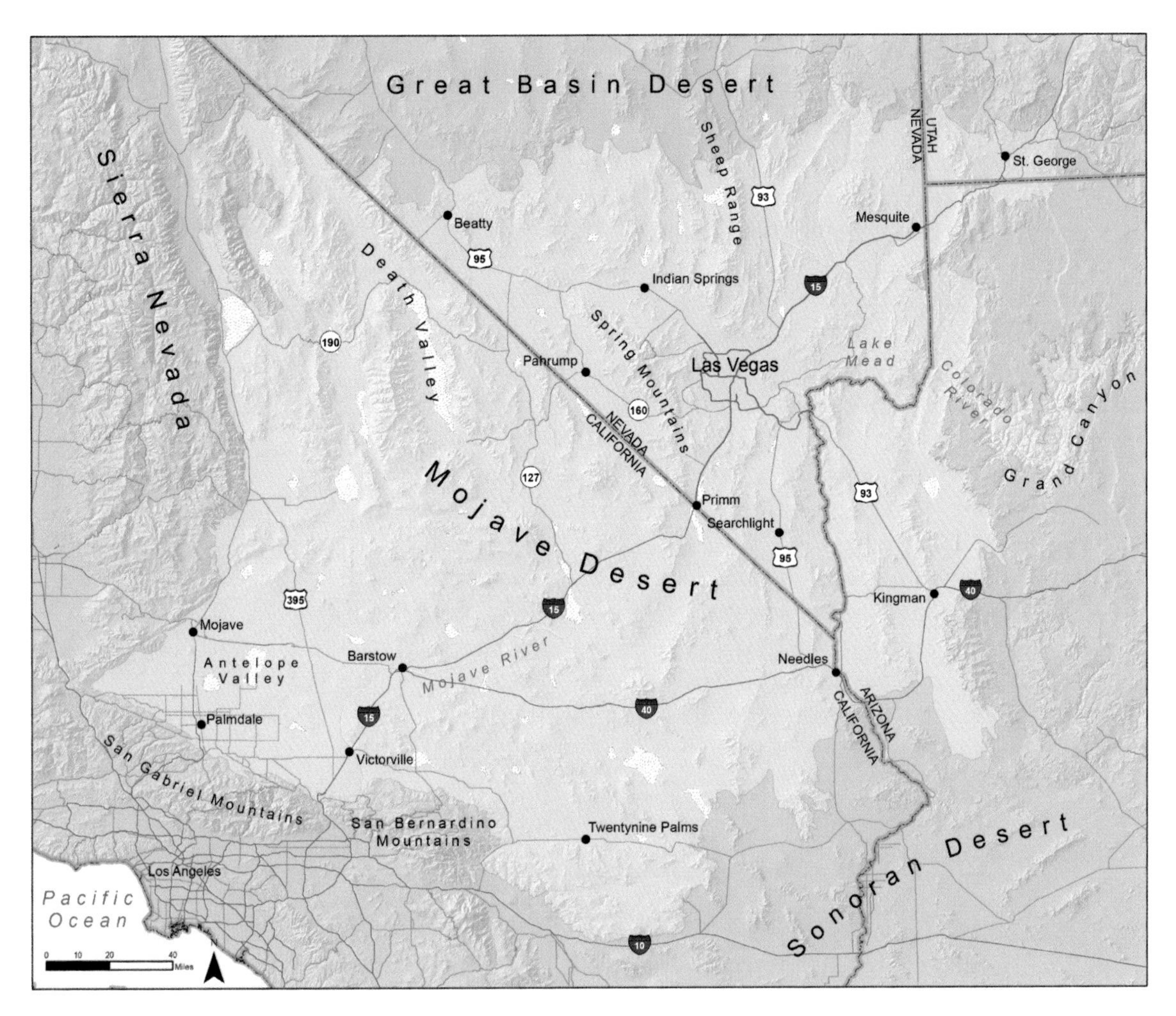

Map 2.3. Mojave Desert. While much of the desert is in California, it includes southern Nevada and even portions of Arizona and Utah. This map employs the EPA's Level IV Ecoregions.

by floodwaters are deposited high on an alluvial fan, with only smaller rocks and mud moving farther down the fan. For this reason, the floods that have swept through Las Vegas do not carry boulders.

Las Vegas is a desert city, but exactly which desert? There are usually said to be four deserts in the United States, but the boundaries between them are not always clear. One is quite far from Las Vegas: the Chihuahuan Desert is found in west Texas and northern Mexico. The Sonoran Desert includes southern and western Arizona, parts of Southern California, and northwest Mexico. To the north of Las Vegas is the Great Basin Desert, which occupies a similar area to the Great Basin landform region (map 2.5). It has the hot summer characteristic of a desert, but with fairly high elevations also has cold, snowy winters. Sagebrush is one of the most common vegetation types in the desert's valleys.

Farther south, the Mojave Desert is the smallest of the American deserts and the home of Las Vegas. It lies mostly in California but also includes all of Clark County and parts of Nye and Lincoln Counties in Nevada. It even extends up the Colorado River into the Grand Canyon and includes a small portion of southeast Utah around St. George. It is bounded by the Sierra Nevada and other mountains to the west, the Sonoran Desert to the south, high plateaus to the east, but the northern boundary is a less distinct transition into the Great Basin Desert.

The Sierra Nevada Mountains are responsible for the Mojave Desert's existence. Prevailing winds from the west must rise over these mountains to reach the Mojave. As they rise, the moist oceanic air cools, and precipitation is created on the windward (west) slope of the mountains, an effect known as orographic precipitation. On the lee (east) side of these mountains, little moisture is left in the air. The Mojave, and all of Nevada, is in a rain shadow produced by these mountains. The dry air downwind of the mountains will also heat and cool more quickly than more humid air, which also explains the extremes of temperature found in Las Vegas during the day.

Despite sometimes intense summer storms, there is little moisture during the summer growing season to support trees. Because of relatively high elevations in the Mojave, winter temperatures can also be quite cold, limiting the growth of cactus common in the Sonoran Desert. One of the few large plants that can survive these extremes is the Joshua tree. These can grow to almost fifty feet tall and live for up to one thousand years. They live at moderate elevations and can be seen near several of the mountain passes near Las Vegas. The largest stand of Joshuas is not far south of Las Vegas, on Cima Dome, a large hill in California's Mojave National Preserve; it can easily be reached from Interstate 15 using the Cima Road exit.

The largest river in the Mojave is the Colorado, which originates in the Rocky Mountains but flows through the Mojave before entering the Sonoran Desert. The Colorado receives little water while passing through the Mojave Desert, with the Virgin River the only substantial tributary (map 5.1). The only other significant river in the desert is the Mojave River, which flows out of the San Bernardino Mountains past Victorville and Barstow in California before ending in Soda Lake. Those driving Interstate 15 between Las Vegas and Los Angeles can glimpse it from bridges in Victorville and Barstow.

Soda Lake is one of many dry lake beds or playas in the Mojave, each occupying the lowest part of a valley. These receive what little precipitation falls in the valley, and the water does not soak into the ground. Water in these dry lakes is a rare sight, usually occurring only after winter storms, and does not last long. Because any surface flow drains into the Colorado River, the Las

Vegas Valley does not have a dry lake, which makes it unusual in the Mojave Desert. The closest dry lakes to Las Vegas are Ivanpah and Roach Lakes, both near Primm south of the city, and south of Boulder City in the Eldorado Valley.

Despite the popular perception, the Mojave Desert has few sand dunes. Several exist around Death Valley and near Kelso, but the largest in North America are in the Sonoran Desert, in southeast California near Yuma, Arizona. Small sand dunes can be found in the Las Vegas Valley, with the largest in Paradise Valley between Warm Springs Ranch and Whitney, though these have long been erased by urban growth. Sandhill Road was named for these dunes, though only the south end of the modern road was in the dune area. Other dunes are found north of Corn Creek Springs and northeast of Nellis Air Force Base.

Las Vegas gets little rain. The annual average precipitation of only 4.17 inches makes it the driest big city in the country. Nearby Phoenix has almost twice the annual rainfall at just more than eight inches, while Los Angeles is drenched by almost fifteen inches of rain each year. But the higher the elevation in Las Vegas, the greater the rainfall. Kyle Canyon in the Spring Mountains, with an elevation of 4,500 feet, matches Phoenix's rainfall; at the 6,000-foot level, the annual rain total is equivalent to that of Los Angeles. Charleston Peak receives 27.5 inches annually, making it the rainiest place in Southern Nevada, and considerably wetter than San Francisco or Honolulu (but with a very different climate).

The Spring Mountains and Sheep Range are also the only areas around Las Vegas that receive any significant snow, because of both higher precipitation and cooler temperatures at their highest elevations. Many Las Vegans are familiar with Lee Canyon's ski resort, the closest opportunity to play in the snow (map 9.2). During the Ice Age, enough snow fell high in Kyle Canyon to create a small glacier. It disappeared at least thirty-seven thousand years ago. The nearest glaciers to Las Vegas today are in the Snake Range in northeast Nevada and California's Sierra Nevada Mountains, though those are shrinking rapidly in the West's warming climate. The Sheep Range receives less snow than would be expected from its height, as wind patterns are slightly different in the winter and the Spring Mountains block precipitation from the downwind Sheep Range during snowstorms.

The Spring Mountains receive considerable precipitation, but where does it go? Rain and snowmelt runoff flow down canyons and sink into the desert ground of the Las Vegas and Pahrump Valleys, with winter precipitation providing the greatest contribution because of lower evaporation during cooler weather. This groundwater is the source of the life-giving springs that

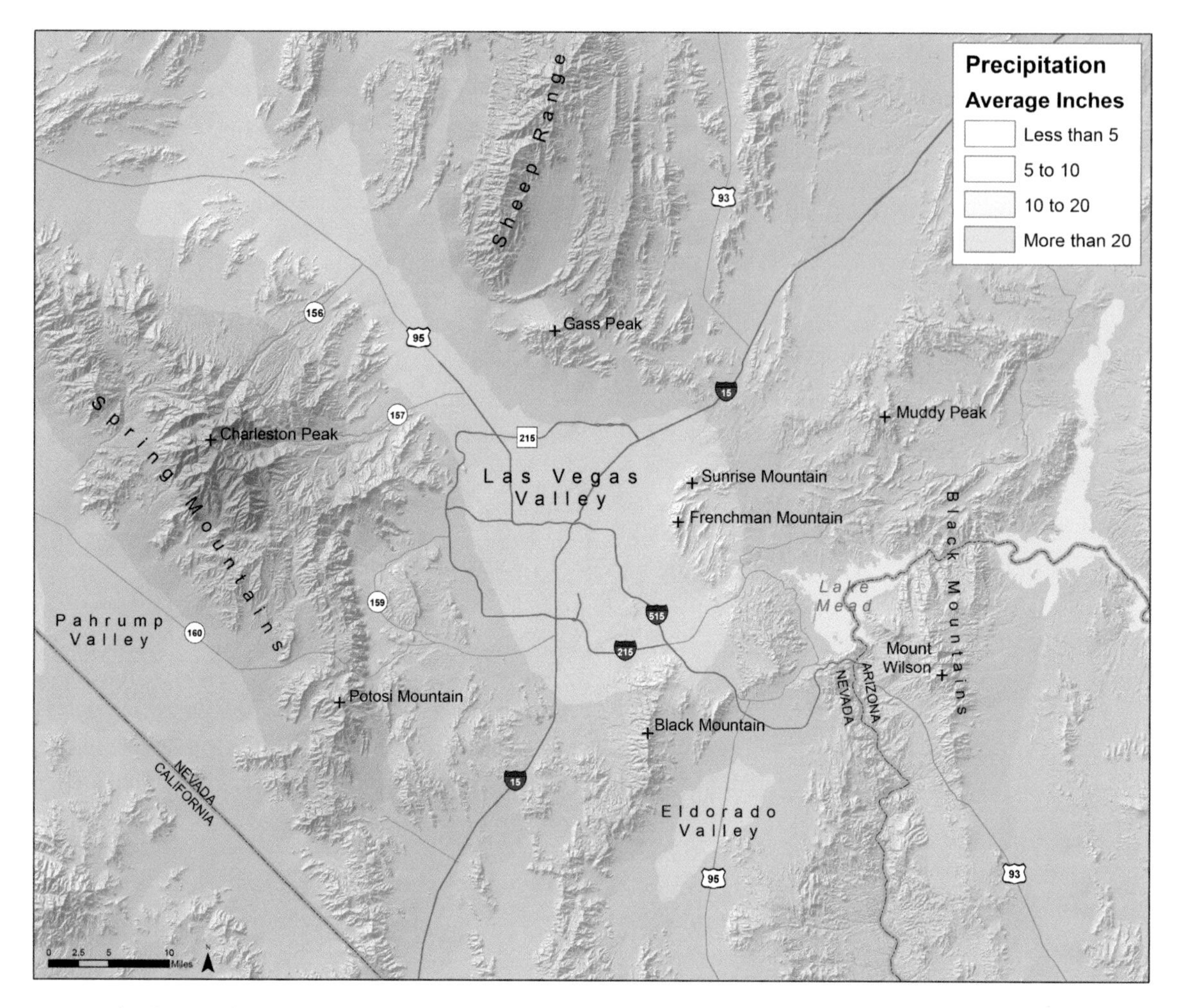

made these valleys so attractive to early settlers, although most springs in both valleys have long ago stopped flowing because of excessive groundwater pumping (map 6.5). Virtually none of this water directly flows into Lake Mead.

Map 2.4. Rainfall. Southern Nevada's rain varies tremendously from mountaintops to valleys. Based on the National Oceanic and Atmospheric Administration's (NOAA's) Gridded Precipitation Data, which was then interpolated into a continuous surface using the Inverse Distance Weighting method.

Sometimes so much water falls in the mountains that the ground can't absorb it. This is especially common during summer storms, with flash floods the result. These have devastated the road and railroad lines that connected the Las Vegas Valley to the outside world as well as causing widespread damage within the valley (map 6.8). Perhaps the worst flood in Clark County's history occurred on July 3, 1975, commonly known as the Caesars Palace flood. Afternoon thunderstorms dumped up to three inches of rain on the desert west of the city, although only a few areas in the city received any precipitation. The floodwaters from this deluge flowed down washes and became concentrated in the Flamingo and Tropicana Washes, which were overwhelmed by their highest-ever flows. The Flamingo Wash crossed the Strip just north of Caesars Palace. Unfortunately, lulled into thinking of the wash as simply vacant land after several dry years, the casino had extended its parking lot into the wash bottom. The floodwaters submerged seven hundred cars, damaged other property, and killed two across the valley. A flood control program was created in 1985 to build dams and detention basins to catch these floodwaters throughout the Las Vegas Valley. Without these, the growth of Las Vegas would have been quite different.

Precipitation levels vary annually. Nevada's weather is subject to the same drought and flood conditions caused by El Niño temperature variations in the Pacific Ocean that plague California. During an El Niño winter, greater precipitation can be expected in Southern Nevada; in La Niña, it is lower. Rainfall in Las Vegas will also be different as the average annual temperature increases because of a warming global climate. Precipitation will decrease in the summer but may increase slightly in the winter; with warmer conditions, less will fall as snow. These conditions will lead to more flooding and less groundwater infiltration, stressing both the region's water supplies and flood control systems.

Most of Nevada lies within the Great Basin, a vast area of internal drainage where runoff flows into playas or dry lakes at the bottom of valleys and then evaporates. The Great Basin also includes parts of Oregon and Idaho, western Utah, and most of the Mojave and Colorado Deserts in California, totaling about 209,000 square miles. Pahrump Valley and Eldorado Valley south of Boulder City are perfect examples of Great Basin valleys, with playas collecting whatever runoff exists.

But the Las Vegas Valley is different. It lies within the Colorado River watershed, and any surface runoff flows out of the valley into that river. The Colorado River, its headwaters in the mountains of Wyoming and Colorado, flows southwest toward Las Vegas and then abruptly turns south to the Gulf of California (map 5.6). During its passage through Nevada, several small streams flow into it, including the Las Vegas Wash.

Although rarely noted by residents, the Las Vegas Wash is the valley's longest watercourse. Its usually dry channel can be traced from the Corn Creek area southeast through the Las Vegas Valley and down to Lake Mead. Much of it has been diverted into concrete channels or underground pipes in residential areas; the remainder east to Lake Mead has been buried in pipes beneath Lake Las Vegas. If you do catch a glimpse of it downstream of the city, you may notice water flowing; this is not from springs or rainfall but rather the city's treated sewage and runoff from sprinklers that went into storm drains (map 6.9).

The boundary between the Colorado River and Great Basin watersheds circles the valley, running along the crest of the mountains which enclose it. Almost every ridge you can see on the skyline to the south and west of Las Vegas is the boundary

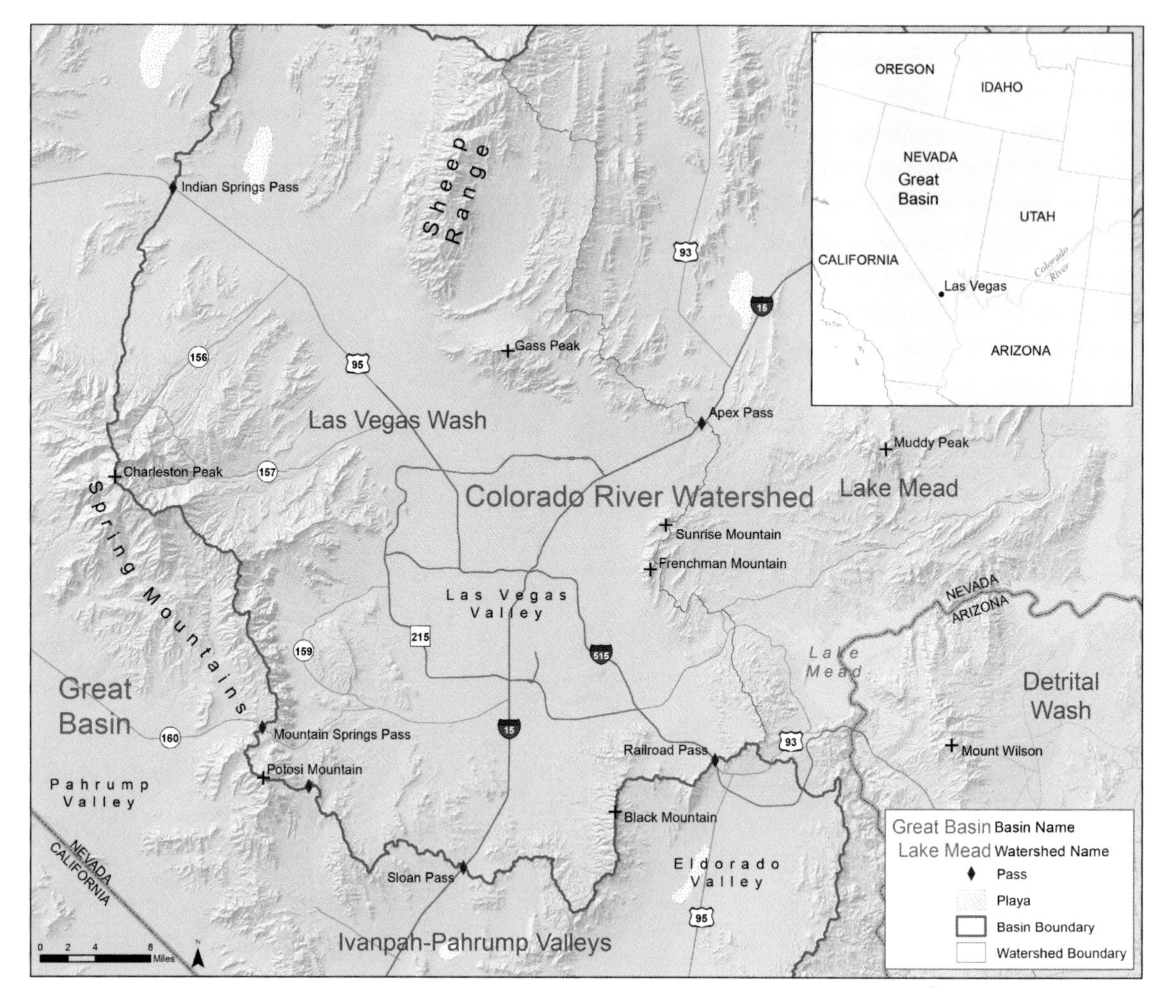

Map 2.5. Watersheds and the Great Basin. While much of Nevada is in the Great Basin, Las Vegas is not. Watershed boundaries from the National Hydrography Dataset, from the USGS National Map.gov website.

between these watersheds. Charleston Peak is one of many peaks on this watershed divide. North of that peak, the divide descends to cross a wide desert basin east of Indian Springs. Farther east, the Sheep Range is unusual in that its crest is not a major watershed boundary; the south half of the range lies in the Colorado watershed, but the north end is in the Great Basin.

Northeast of Las Vegas are the Muddy and Virgin Rivers. The Virgin's source is in the mountains of southwest Utah, and it in turn flows into the Colorado. The Muddy River is unusual for a desert stream in that its headwaters are not in distant mountains but only a short distance upriver in an area of springs. These produced a nearly constant amount of water year-round that could be easily diverted for agricultural use (map 5.3). Farther north, a series of springs also waters the Pahranagat Valley. A long tributary of the Muddy, the Meadow Valley Wash occasionally brought destructive floodwaters from distant mountains to the north. It caused particular troubles for the railroad that connected Las Vegas to the outside world (map 7.1).

Several watershed oddities can be seen on the map. A small

playa northeast of Las Vegas is in the aptly named Dry Lake Valley, which has a very small drainage basin. It should be part of the Great Basin but is classified by the United States Geological Survey (USGS) as part of the Muddy River watershed for convenience. Another dry lake is found west of the Sheep Range within the Colorado River drainage. It is possible that if enough rain were to fall, these lakes would overflow into nearby streams that do reach the Colorado, but this has not happened in recorded history.

Although it is only a few miles from Lake Mead, most of Boulder City is within the Great Basin watershed. The city is in Eldorado Valley, a classic desert basin with a playa at its bottom, though it has always been supplied with water from the Colorado River. When driving from Las Vegas south to Hoover Dam, you must leave the Colorado River watershed and enter the Great Basin at Railroad Pass; the road only reenters the river's watershed after passing Boulder City.

An ecoregion or ecosystem is an area with similar climate, vegetation, and animals. The Environmental Protection Agency EPA has defined 967 of these across the country using its most detailed classifications, called Level IV Ecoregions. Only a few of these are in Southern Nevada, defined largely by elevation, the principal determinant of temperature and precipitation (maps 2.1 and 2.4).

At the highest elevations, above eight thousand feet in the Spring Mountains and Sheep Range, is the Eastern Mojave High Mountains environment. It is heavily forested with ponderosa and limber pine and white fir common. The top of the Spring Mountains is above the treeline, where conditions are so cold and windy that nothing but grasses and small shrubs can grow. Below this, the Great Basin bristlecone pine may be found; this is famous as the longest-lived life form on Earth. The oldest bristlecone known, 4,852 years old in 2021, is on California's White Mountains; other long-lived examples are found on Nevada's Snake Range in Great Basin National Park. The oldest in the Spring Mountains is only about three thousand years old, but the range is home to the largest stand of these trees. These bristlecones and other trees are closely related to those on widely scattered mountains in California, Nevada, and Utah, but could never survive just a few miles away on the hot dry valley bottoms that surround each mountain. For that reason these mountain forests are sometimes referred to as sky islands, with forests scattered amid a sea of desert.

These mountain forests were important resources in the early years of Las Vegas, because the timber needed for construction, framing mine tunnels, and firewood was always in short supply

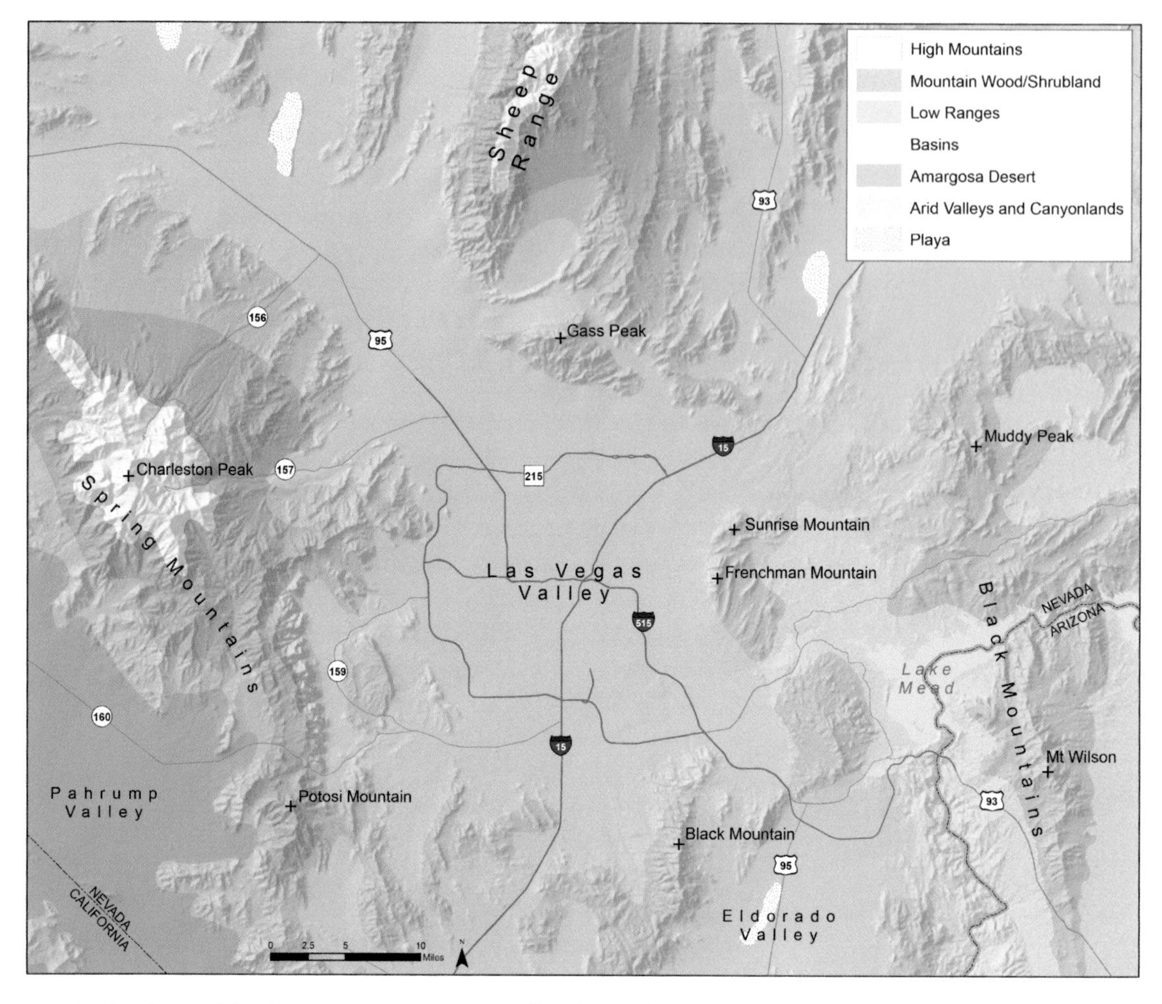

Map 2.6. Ecoregions. While Las Vegas has vegetation typical of Mojave Desert basins, the surrounding mountains and canyons are quite different. Based on the EPA Level IV Ecoregions to map the Mojave Desert.

in the desert. The Spring Mountains were the closest forests to the city, and several sawmills were active in Kyle and Lee Canyons beginning in the nineteenth century. Mining in Goodsprings, south of Las Vegas (map 7.3), required considerable lumber in the early 1900s, and a stand of yellow pine near Charleston Peak was almost wiped out because of this demand.

Below eight thousand feet and down to about six thousand feet is the Eastern Mojave Mountain Wood and Shrubland ecoregion. This is forested with pinyon pine and juniper trees as well as a variety of large shrubs. It makes up much of the lower forested areas in the Spring Mountains and Sheep Ranges and can also be found at the highest elevations of the McCullough Range south of Las Vegas. It also appears east of the city at the tops of the Mormon Mountains and a few ridges within Gold Butte National Monument (map 9.11). The McCullough Range is unusual in that the three major subspecies of pinyon, living in the Great Basin, Arizona, and California deserts, can all be found together. Although this tree is not commercially logged like those at higher elevations, it was particularly important in Nevada and adjacent states during nineteenth-century mining booms. Mines, smelters, and towns needed wood for cooking, heating, and especially for making charcoal with which to smelt ore, for which pinyon was excellent. The pines in several ranges in central Nevada were completely cut over and have only gradually returned. The cooler, wetter, and forested Spring Mountains and Sheep Ranges are also home to many animals not usually found at lower elevations. Deer can be seen, and a herd of Rocky Mountain Elk was introduced to the Spring Mountains. Mountain lions are rarely seen.

The Eastern Mojave Low Ranges are found below these woodlands on the Spring Mountains and Sheep Range and include the entirety of many lower desert ranges, such as the River Mountains between Las Vegas and Lake Mead. This desert environment has sparse vegetation with small cactus and sometimes Joshua trees. In addition to the Joshua trees on Cima Dome in California noted earlier, others can be seen when driving north along US 95 to Indian Springs and along Blue Diamond Highway headed to Pahrump.

Desert bighorn sheep live in the rugged mountains surrounding the city and can be found in this ecoregion as well as the Arid Canyonlands below. Hemenway Park in Boulder City, at the foot of the River Mountains, is a good place to see them. Bighorn do not mix well with busy roads, and highway overpasses or underpasses were built on US 93 south of Hoover Dam and on the new Interstate 11 bypassing Boulder City to allow bighorns to cross the road safely.

Joshua Tree in Red Rock Canyon, 2005. Photo by Stan Shebs.

The Las Vegas Valley and the bottoms of most neighboring basins are within the Eastern Mojave Basins ecoregions, with Creosote bush, white bursage, and galleta grass the most common plants, though mesquite and willow can be found near water sources. Native animals of this ecoregion include desert pupfish and desert tortoises. Unfortunately for both animals, the city's growth and the disappearance of the valley's springs, creeks, and woodlands have transformed this ecoregion. The pupfish that once lived in the Las Vegas springs and creek are gone, and the desert tortoise habitat is shrinking as the city expands.

The Arid Valleys and Canyonlands ecoregion includes the lowest, and therefore hottest and driest, elevations in the Las Vegas region, though much of this area is now under the cold waters of Lake Mead. The natural vegetation that existed in these canyons before the lake was built is little known today (map 5.1). Early accounts make clear that riparian vegetation was very sparse, with mesquite likely to have been the only tree common along the river. Reported vegetation along the Muddy

River included mesquite, cottonwood, willow, creosote, arrowweed, and an abundance of grasses making for an ideal pasture. There are even fewer accounts of the animal life in the region, but travelers reported ducks, quail, and coyotes along the river, and beaver were also known to live in it.

Even in areas not flooded, the settlement of Las Vegas and surrounding communities caused many changes to the area's natural vegetation. Not only did formerly common species disappear, but exotic plant and animal species arrived from other parts of the country and around the world. Tumbleweed and salt cedar are two of the best known among these in the American West, though only the second is common around Las Vegas, found in the Basins, Canyonlands, and Amargosa Desert ecoregions. Salt cedar is native to North Africa and grows as a large shrub or tree. It can grow in thick stands that crowd out other species and has no value to wildlife. It was introduced into the United States in the early nineteenth century but was not reported in the Lake Mead area until after the lake filled in the late 1930s.

Many plants were deliberately brought to Las Vegas for landscaping purposes. The city hosts the tallest eucalyptus, Italian cyprus, and African sumac trees in Nevada, among others. Many cacti can be seen in the city, including the magnificent multiarmed saguaro, but this is native to the Sonoran Desert of southern Arizona and not naturally found in Las Vegas. Along Northshore Drive near Lake Mead, several natural oases feature abundant palm trees. These are California fan palms, a widely distributed species in lower elevation deserts. They are not however native to Southern Nevada; exactly when and how palms were introduced to the Lake Mead area are not known.

The ecoregions described here have existed for thousands of years, but they have also been subject to elevational changes as the climate warms or cools. During the Ice Age, each ecoregion was about one thousand feet lower and the area above the treeline greater. As the climate warms and dries during this century, they will move upward in elevation. The open meadows above the treeline on Charleston Peak will shrink, and some mountaintop plant species may be lost altogether as their habitat disappears. Joshua trees are also on the move because of climate change, and Pahute Mesa in the Nevada National Security Site (map 9.14) could be one of the best places to see Joshuas in the future.

3

Before Las Vegas

Las Vegas has thousands of years of human history, but little of it is known today, gleaned mostly from fragmentary evidence at a handful of archaeological sites. Only the past two hundred years is known with much detail, beginning when Southern Nevada was still the homeland of the Paiute Indians. They did not know it, but their home was also claimed by Spain, Mexico, and the Church of Jesus Christ of Latter-day Saints (Mormons) before being transferred to the United States. Three new counties were created for Las Vegas, one in Utah, one in Arizona, and one in Nevada. Trade routes were opened across Southern Nevada, and decades later a group of Mormon pioneers arrived to found a mission and farm at one of the valley's major springs. A stable but limited economy was established based on ranching, farming, and mining that lasted decades before being disrupted by the arrival of a railroad in 1905.

This is the first of several chapters outlining the history of Las Vegas, with maps depicting different eras in time or particular events. The founding of the town of Las Vegas in 1905 is used as an inflection point in the area's history; events from that point on will be covered in chapter 4.

The first inhabitants of the Las Vegas region were Native American big-game hunters who followed the region's megafauna, such as mastodons, horses, and camels, as many as ten thousand years ago. They are known today from fragmentary archaeological evidence at a few locations throughout the region. Gypsum Cave, Tule Springs Wash, Pintwater Cave, and Stuart Rockshelter are important archaeological sites for understanding Las Vegas's early human history. Along with their prey, the big-game hunters disappeared, leaving few clues as to their existence. The second wave of people in what would become the Las Vegas area were farmers who lived in small villages or pueblos. They are known today as the Anasazi or Ancestral Puebloans. The Anasazi realm extended from Southern Nevada east through southern Utah, northern Arizona, Colorado, and New Mexico. Many national park units have been created around Anasazi ruins in the Four Corners states, most spectacularly Mesa Verde in Colorado, Canyon de Chelly and Navajo in Arizona, and Chaco Canyon in New Mexico. In contrast, the Nevada Anasazi have been given little attention and don't even show up in many books on the topic.

Why did the Anasazi expand into this area? Population growth and the need for more farmland are the primary reasons. They

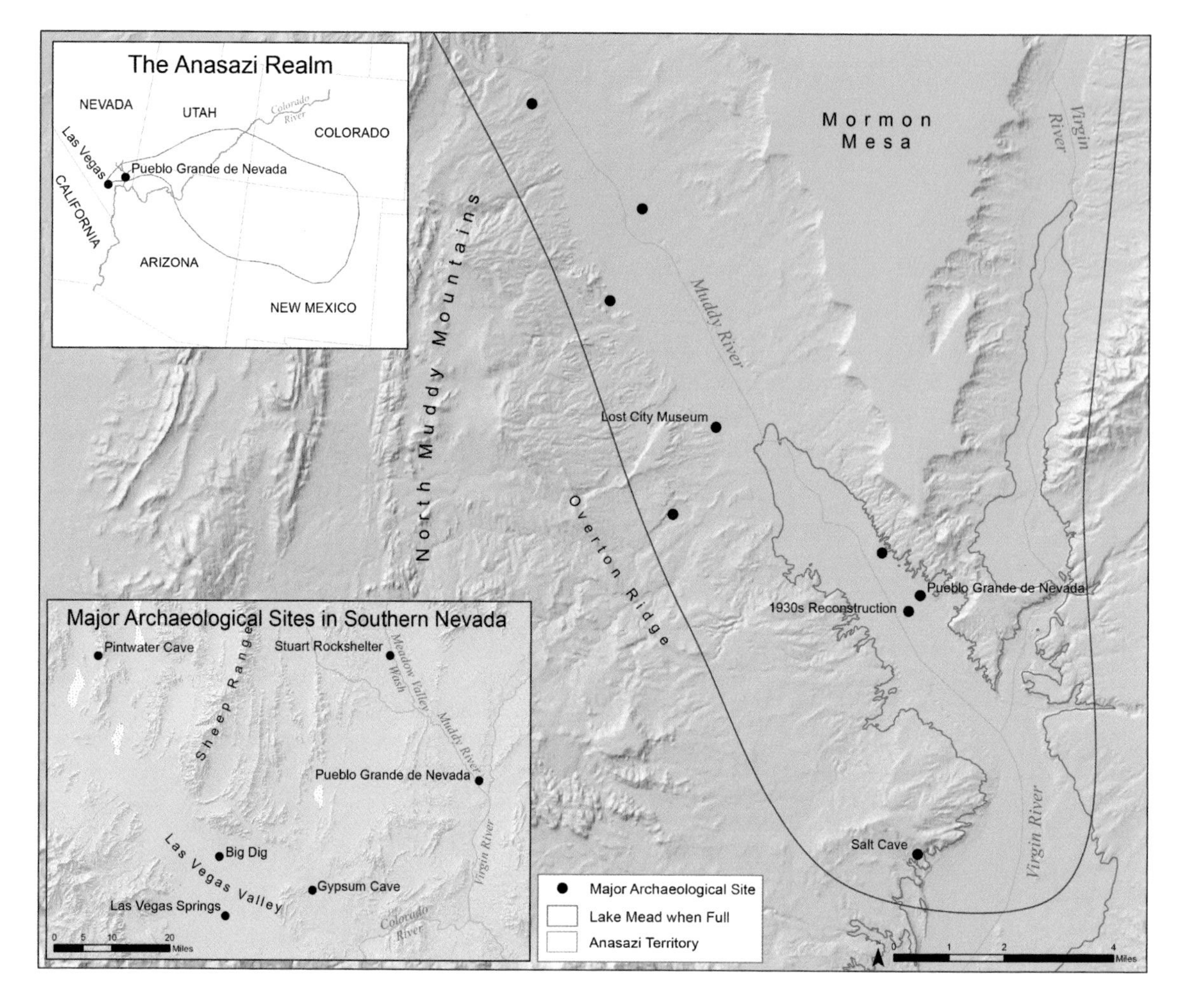

Map 3.1. Anasazi Las Vegas. Archaeological sites in the Muddy River Valley include Pueblo Grande de Nevada. *Lower inset*: major archaeological sites throughout Southern Nevada; *upper inset*: the place of Las Vegas in the vast Anasazi realm.

lived in permanent settlements along the Muddy and Virgin Rivers, where they farmed corn, beans, and squash, and gathered other available food, such as pinyon nuts and mesquite beans. But the Anasazi seem to have abandoned the area by 1150, probably moving back to their core in northern Arizona. We don't know why they left these settlements, but drought may have been part of the answer.

The largest of the Anasazi settlements was known as the Lost City, Pueblo Grande de Nevada, or Main Ridge site. This area was excavated in the 1920s and 1930s, but the record of what was found is not clear. Several groups of archaeologists carried out excavations with little coordination; records and artifacts were distributed among several museums and sometimes lost; and those records that do survive are not consistent. Even the name "Lost City" is confusing; it may refer to the main Pueblo Grande de Nevada ruins, a 1930s reconstruction of a pueblo where outdoor pageants were held, a museum in Overton, or to all archaeological sites found throughout the Muddy River Valley. The record is further confused because many photographs that show the Lost

City ruins being inundated by Lake Mead show the 1930s reconstruction being flooded; the Native American ruins are on a hilltop east of the Muddy River and were never flooded.

A few tantalizing hints from travelers in the 1920s suggest that more pueblo ruins, and perhaps irrigation canals, were to be found where the Virgin and Colorado Rivers joined. Unfortunately, archaeologists never investigated these sites before Lake Mead flooded them. Accounts also exist of a Pueblo ruin at Las Vegas Springs investigated by an amateur archaeologist in the late 1930s, but development work at the springs destroyed this site. Much of the Anasazi history of Las Vegas has been lost before it was ever known.

The Anasazi were not the only group living in the area at the time. Paiutes arrived in Southern Nevada about one thousand years ago and remain in Las Vegas today. They relied on hunting and gathering during most of that time, often making seasonal movements between mountains and valley floors, and also engaged in farming near the valley's springs (map 3.2).

The Paiutes, also known as Nuwuvi, called the Las Vegas area Yiwaganti. They developed a stable way of life based on farming, hunting small game, and gathering wild plants. They made full use of the region's elevation range, and, like later settlers, congregated near the abundant springs on the valley floor.

Paiutes often camped at Las Vegas Springs (Parampaiya) as well as other springs and in Las Vegas Wash. Kiel Springs, or Kwaintomanti, was one of their largest camps, and as many as fifty-two people lived here before the arrival of the Mormons in 1855. The Paiutes lived at these springs in brush huts, with the doorways always facing east to avoid the winds Las Vegas experiences much of the year.

They farmed by using irrigation ditches at most springs in the valley, growing crops such as corn, beans, squash, and melons. They also gathered mesquite beans and hunted small game, though roadrunners and snakes were never eaten. The Paiutes also farmed several areas along the Colorado River, with camps at the mouth of Callville Wash, and Las Vegas Wash. Here they relied on the annual flood of the river to irrigate their crops rather than digging ditches. They hunted beaver, and families went to the mountain in the fall to gather pinyon nuts and hunt deer and bighorn sheep.

They tended to live in small groups of ten to fifty people, made up of several families. Each camp group circulated among a number of springs and used different areas in the mountains. They were quite mobile and carried out trade with others living around them, trading tobacco or bows for turquoise and rabbit-skin blankets, among other items. They also traveled great

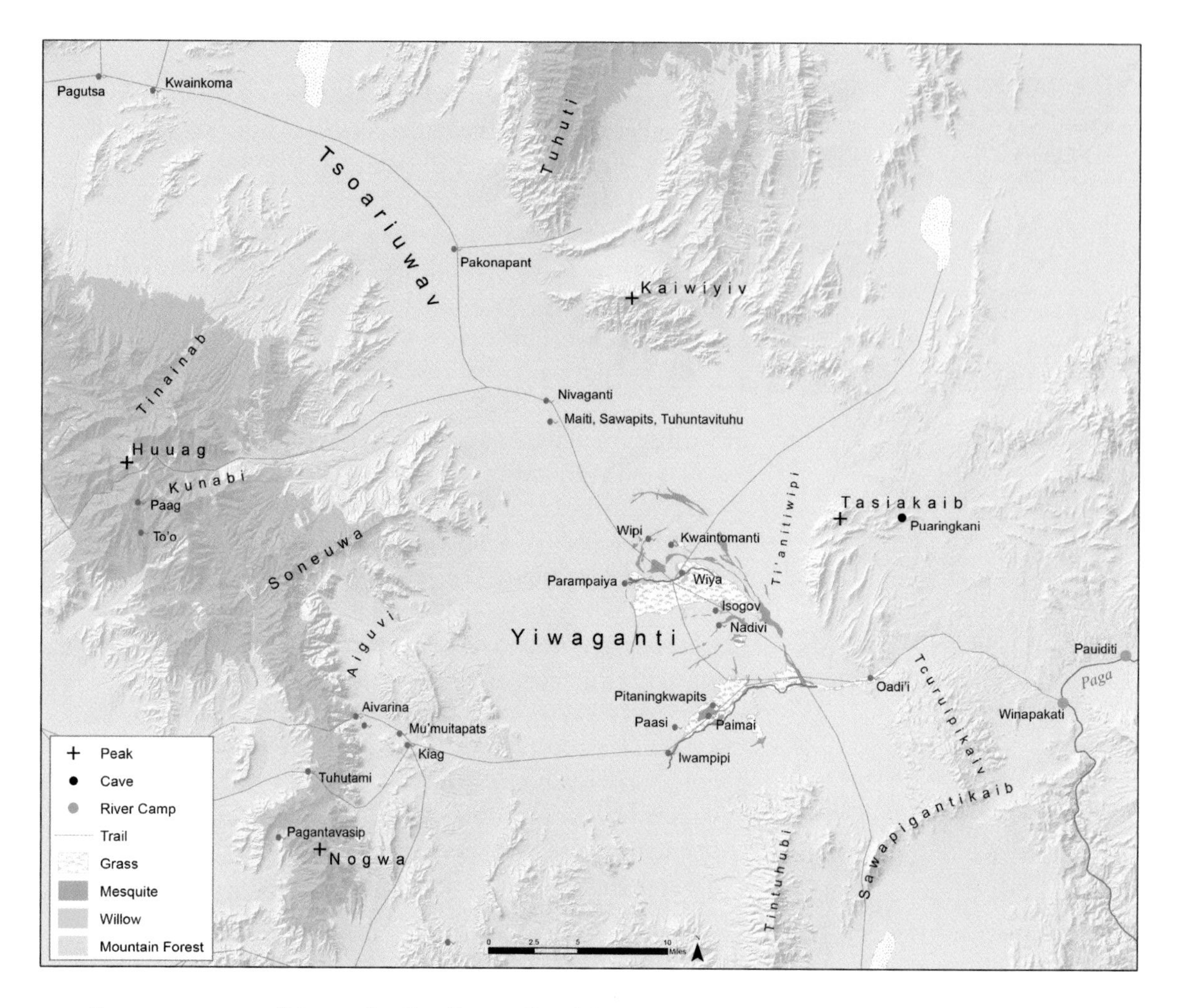

Map 3.2. Yiwaganti. Much of what is known about Paiute place-names and how they lived is through the efforts of Isabel Kelly, an anthropologist who interviewed Paiute elders in 1933 and 1934. Kelly also purchased many Paiute baskets, household items, and pieces of clothing, now stored in the Peabody Museum at Harvard University, the American Museum of Natural History in New York City, and the Museum of New Mexico in Santa Fe.

distances, even walking to the San Bernardino Mountains near Los Angeles to hunt deer. Trails linked these areas as well as provided easy routes between the valley and nearby mountains and the Colorado River.

The Paiutes suffered considerably during the nineteenth century from a number of events. While early Euro-American exploring parties passed through peacefully, the arrival of thousands of hungry and thirsty animals during the days of the Old Spanish Trail (map 3.5) must have been devastating to the Paiutes' use of small desert springs and farms. With the Old Spanish Trail also came slave traders to prey on the Paiutes. The arrival of Mormon farmers in 1855 was the beginning of their permanent loss of springs and land in the valley (map 3.7). The Paiutes now only lost their water sources and farms but also suffered the indignity of the Mormons attempting to teach them how to farm.

As their land and water resources diminished after white settlers arrived, they adapted by working for wages in towns and on farms. Towns such as Las Vegas typically had a Paiute village on the outskirts; in Las Vegas, this was north of town. Residents

Group of Paiutes at Cottonwood Springs (today's Blue Diamond), 1875. Courtesy of the National Archives and Records Administration. Photo by Timothy O'Sullivan.

sought paid work in the town or on Muddy River farms at least part of the year, and they might also engage in traditional hunting or gathering at other times. Their labor was much in demand in the early years of Las Vegas when there were never enough workers for construction and other jobs.

Many Paiute words remain on the map as place-names. Among the most common are those involving water, or pah. Ivanpah, Mizpah, Nopah, Tonopah, the Pahroc and Timpahute Ranges, and the Pahranagat Valley are all examples, though they were all bestowed by white settlers or explorers and their original meaning is unknown. Pahrump is the only place-name used by Paiutes that remains on the map in the Las Vegas region. But people from Pahrump are no longer called Parimpaniwi, as they were by the Paiutes.

The Paiutes of the Las Vegas Valley did not live in isolation. In the late nineteenth and early twentieth centuries, there were thought to be fifteen distinct Paiute bands. Those in the Las Vegas area occupied the valley, Spring Mountains, and several valleys to the west and south. The Moapa band lived along the Muddy and Virgin Rivers to the east, while the Chemehuevis were to the south. Other bands lived to the northeast in Utah and Arizona.

The Paiutes were surrounded by many other tribes, including the Navajo, Yavapai, and Havasupai. The Utes were once culturally indistinguishable from Paiutes, but their territory was more suitable to horses, and after the arrival of these animals in the 1600s the horse-riding Utes and horseless Paiutes slowly diverged. To the south, the Paiutes and Chemehuevi bordered the lands occupied by the Mohaves, who lived and farmed along the east side of the Colorado River. The Chemehuevi farmed along the west side of the Colorado River, especially around Cottonwood Island (now under Lake Mohave) and in Chemehuevi Valley (now partly filled by Lake Havasu).

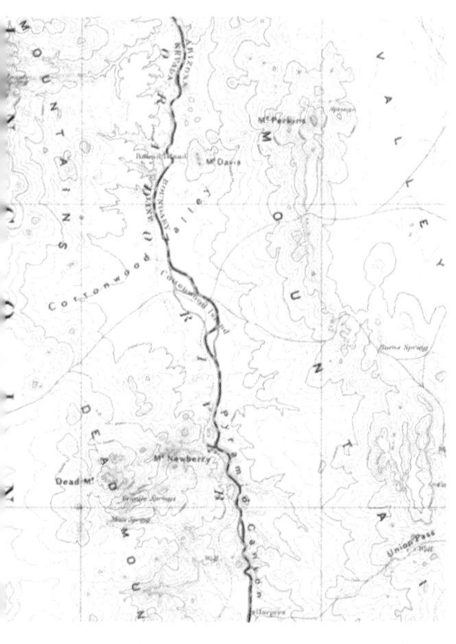

Detail of 1892 Fort Mohave topographic quadrangle showing Cottonwood Island. Courtesy of USGS.

The Chemehuevi and Mohave did not always coexist peacefully. A conflict between them broke out in 1865, with scattered battles north of Cottonwood Island, Chemehuevi Valley, at Topock, near Camp Mohave, and near Parker. After a battle or raid, the Chemehuevi always disappeared into mountains where Mohaves would not venture, leading to a stalemate. After several years of fighting, peace was eventually restored. Fort Mohave was created in 1859 as a military outpost to protect emigrants from the Mohave, and, except for a short absence during the Civil War, was garrisoned until 1890. No army operations were ever carried out against the Paiutes.

The Paiutes never sold their lands or gave any up in a treaty; newcomers simply moved in and occupied the farmland, took over the water, and began claiming land. With the loss of land

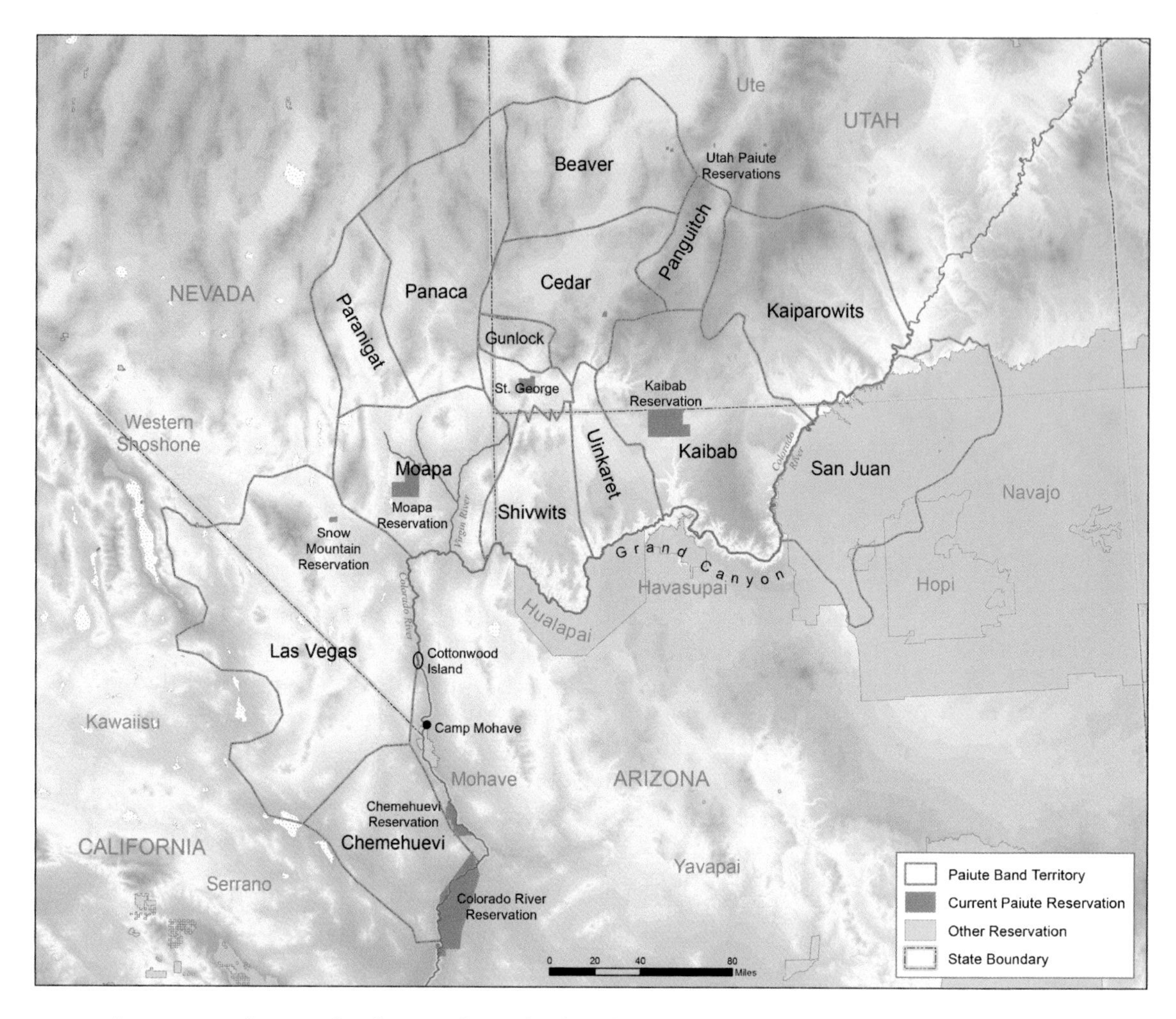

Map 3.3. Paiute Bands. Reservations occupied by Paiute groups in the nineteenth century and today.

and resources, disease, the slave trade, and other depredations, Paiute numbers fell during the nineteenth century. The government created a reservation on the Muddy River where the surviving Paiutes were to be relocated, but few moved there (map 3.11). Ten more small Paiute reservations were later created in Utah, Arizona, and California, but by that time the damage had been done.

The Uinkaret band had disappeared by the 1930s, and only one individual from the Panguitch, St. George, and Gunlock bands remained. The need for reservations seemed much reduced, and four Paiute reservations in Utah were eventually abolished. But the government policy of eliminating reservations slowly changed, and in 1980 a new Utah reservation for the Cedar, Indian Peaks, Kanosh, Koosharem, and Shivwits bands was created as a replacement for those lost. This reservation had ten separate pieces of land scattered in four counties, the largest near St. George. Four thousand acres were returned to the Paiute tribe in 1983 in the form of the Snow Mountain reservation along US 95 northwest of Las Vegas, which can be accessed from the Paiute

Group of Chemehuevis along the Colorado River in 1900. Photo by C. C. Pierce.

Drive exit (US 95 passes through the reservation). This includes the Las Vegas Paiute Golf Resort and a considerable amount of vacant land, but none of the springs that had been the focus of Paiute settlement. They are more fortunate than the San Juan Paiute band, whose land has been absorbed by the vast Navajo Reservation in Arizona. The San Juan still live in several communities within Navajo territory and are still working to regain their ancestral land.

In 1535 King Charles V of Spain established the Viceroyalty of New Spain, which contained Spanish possessions and claims in North and Central America, Cuba, and even the Philippines. Included within this vast territory were the desert basins and mountains that would later be home to Las Vegas.

The Viceroyalty of New Spain became part of Mexico when it became independent in 1821. Map 3.4 shows this new country circa 1830, at the time of the first European forays into what would become Las Vegas, but before the map was reshaped by the independence of Texas in 1836 and the Treaty of Guadalupe Hidalgo in 1848. Like the American West, Mexico's north was a frontier, and the government sent expeditions, promoted settlement, and established forts to protect those settlements. Two areas that had been settled early on were the Rio Grande river valley, known as Nuevo Mexico, and Alta California, which included a series of missions along the coast and several towns, among them Los Angeles. Between them were hundreds of miles of deserts, canyons, mountains, and high plateaus. Nuevo Mexico produced textiles, pottery, and other goods, while Alta California had an abundance of horses and livestock.

Interest naturally arose in connecting these two areas, and expeditions sought to find a feasible route between them. The first people of European ancestry to see Las Vegas were Mexicans on a trading expedition under the command of Antonio Armijo. The group headed northwest from Abiquiu in what would become New Mexico in November 1829, and passed through Colorado and Utah before turning southwest toward Los Angeles. In what would become Nevada, the party of sixty-one men descended the Virgin River to the Muddy River (which they called the Rio de Los Angeles) and followed it south to the Colorado River (which they named the Rio Grande). Armijo then turned west and followed the Colorado downstream to Las Vegas Wash (or the Arroyo de Yerba Buena). Here the group decided the best route was to leave the river rather than continue downriver through deep canyons. They ascended Las Vegas Wash and then headed west across the valley to a group of wooded springs, which they named Las Vegas (the meadows). These were not the later springs of that name west of downtown but instead what would be known as

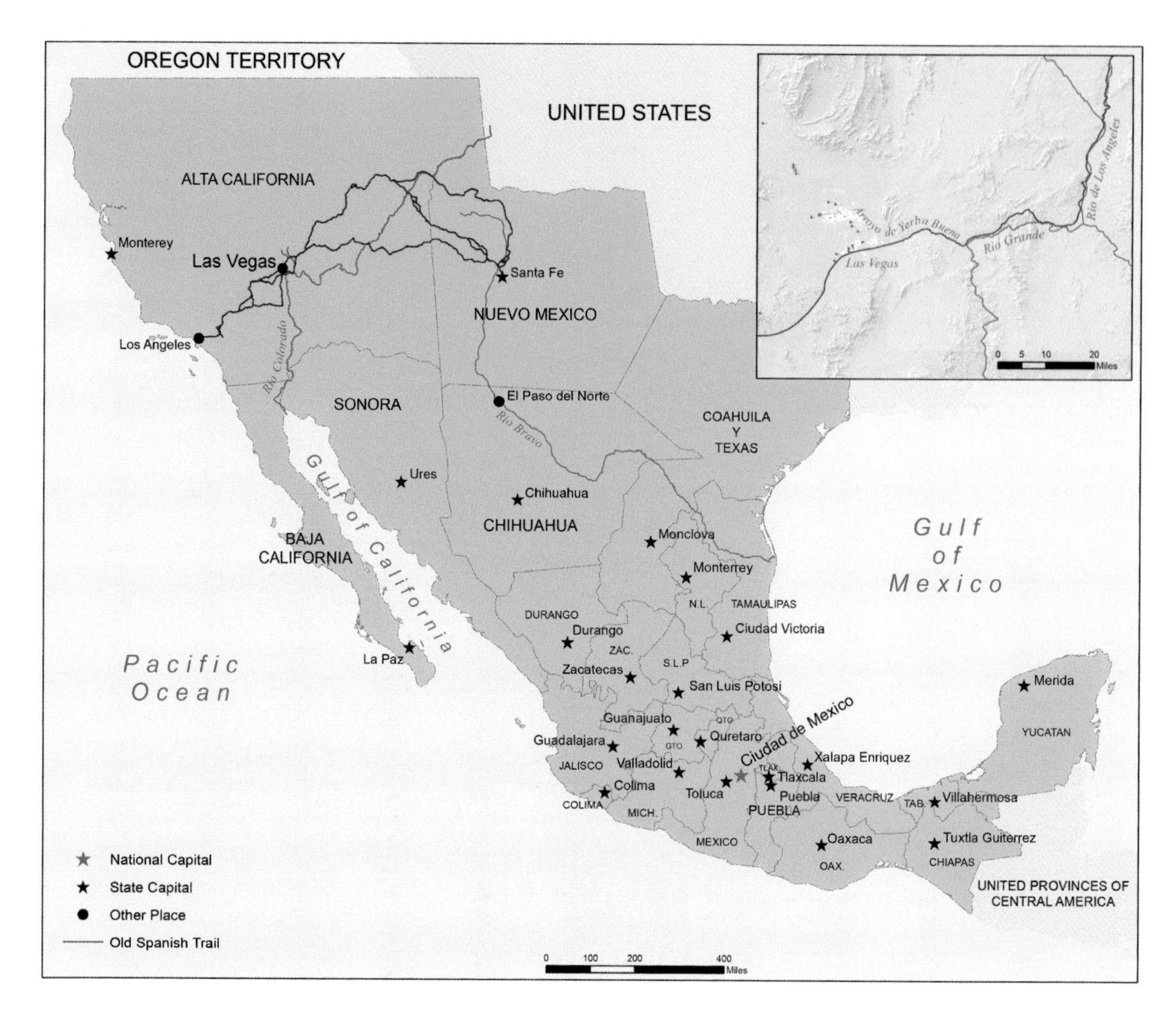

Map 3.4. Mexican Las Vegas. The 1830 location of what would become Las Vegas was in the vast province of Alta California. Inset: the route of Armijo's expedition through the Las Vegas region.

Duck Creek or Paradise Valley (map 6.3). After resting there in early January 1830, Armijo's group continued to the southwest, eventually reaching the Mojave River and ascending it to Cajon Pass, the gateway to Southern California.

Armijo's route down the Virgin and Colorado Rivers was soon replaced by another that made use of other springs within the Las Vegas Valley and crossed the Muddy River to the north of the future Lake Mead. It was probably first used in 1831 or 1832 as an annual trading expedition between New Mexico and Southern California. This became a popular horse and mule route between Southern California and Santa Fe and known as the Old Spanish Trail, though it was developed by Mexicans and was only fourteen years old when it was first named (map 3.5).

Aside from Paiute trading paths, the seven-hundred-mile Old Spanish Trail was Las Vegas's first connection to the outside world (map 3.4). Traders from New Mexico would carry trade goods to California and exchange them for several thousand horses to be herded back to New Mexico. Although it was not a

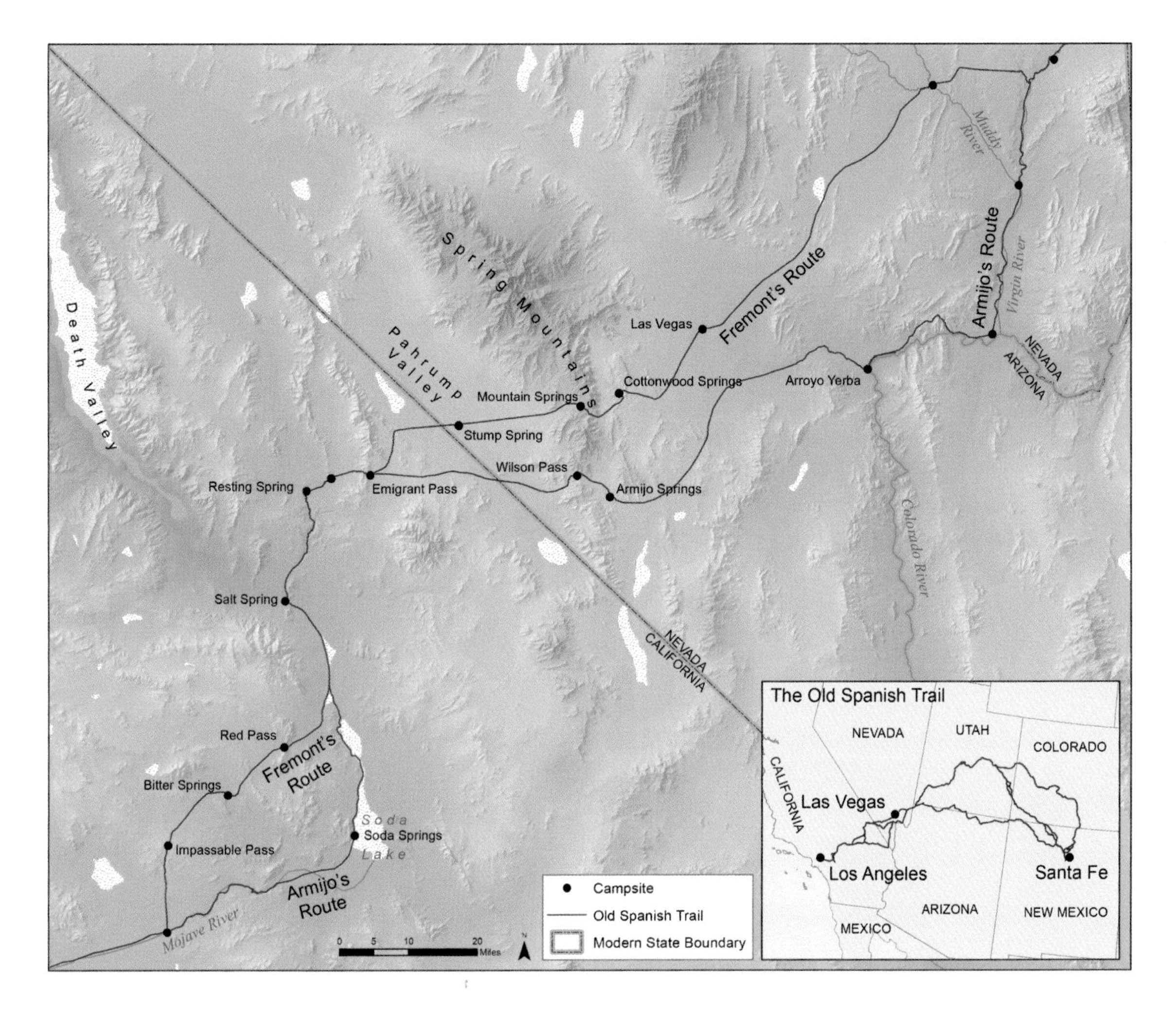

Map 3.5. The Old Spanish Trail. This route connected New Mexico and California, with several different trails used by different parties. The campsites were usually at springs.

direct route between its endpoints, the Old Spanish Trail avoided Native American groups in Arizona that the Mexican officials considered hostile. It was usable only from late fall through early spring because of the dryness of much of the route. The springs in the Las Vegas Valley were a vital stopping point in both directions along this trade route.

From the west, the Old Spanish Trail entered the Mojave Desert through Cajon Pass between the San Gabriel and San Bernardino Mountains and followed the Mojave River downstream until a place known as Forks of the Roads. The Old Spanish Trail then branched off to the north, and another road, the Mojave Trail, continued east down the river and eventually to the Colorado River. The Old Spanish Trail ran north to Bitter Springs (also known as Agua de Tomaso) and then east through Red Pass. The next water was at Salt Spring at the southern end of Death Valley. Heading north, travelers ascended the Amargosa River Canyon, where water flowed year round, to the site of modern Tecopa. The next water was at Resting Springs. From there the trail climbed Emigrant Pass, then across California Valley to Stump Spring in Pahrump Valley. Mountain Springs was the next destination, after which the trail descended into the Las Vegas Valley to Cottonwood Springs (now the town of Blue Diamond). The trail then crossed the valley bottom to Las Vegas Springs.

This was an important stop because the fifty-seven-mile section from Las Vegas Springs to the Muddy River was one of the hardest along the trail, with no water to be found anywhere between these locations. Travelers were required to make a dry camp along this stretch and often set out only after dark. From the Muddy River the trail climbed over Mormon Mesa, down to the Virgin River, then up this watercourse to modern Littlefield, where it left the river and headed north into Utah.

Armijo named Las Vegas Springs in 1830, but many other names appeared over the years, including Ojo de Las Vegas, Ojo del Gaetan, Ojo de Cayetana, Ojo de Galleta, Ojo de Quintana, and Vega Quintana. It was John C. Frémont who fixed the name as Las Vegas in 1844. He journeyed along this route returning from Southern California on his second western expedition in 1844. He noted the name Las Vegas was already in use for the area and described two streams flowing east from the springs. He included the site on his map as 'Vegas,' the first time the Las Vegas area appeared on a map. Traveling farther east, he gave the Virgin River its name, after Thomas Virgin, a member of Jedediah Smith's 1826 expedition down that river (map 5.2). Twelve years after passing through Las Vegas, Frémont became the Republican Party's first presidential candidate; he lost to James Buchanan and died in 1890. But in 1905, Frémont's name was memorialized on the map of Las Vegas as the city's first main street (map 8.6).

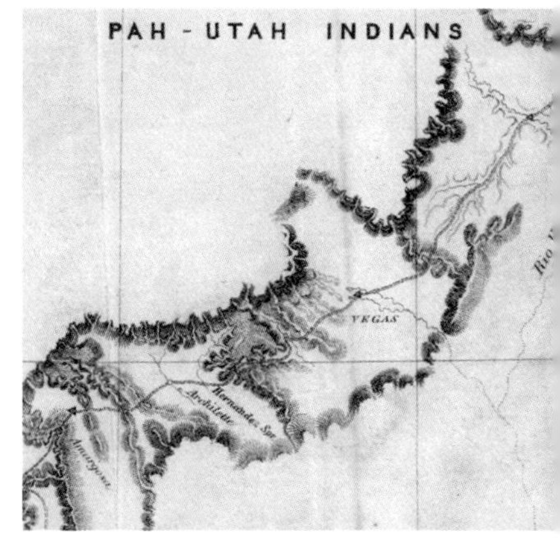

Detail of John Frémont's 1844 map, showing "Vegas," the first appearance of the locality on a map. The dashed line is the Old Spanish Trail. Courtesy of the Library of Congress.

The trade in goods and horses was soon accompanied by one in people. A slave trade flourished in what would become Utah and Nevada in the 1830s and 1840s, perpetuated by Utes and Mexicans against the Paiutes. Slavers especially prized women and children, who were sold in Southern California or New Mexico where labor was needed. This slave trade became an integral part of the annual trading caravans along the Old Spanish Trail across Paiute territory. The numbers of Paiutes enslaved will never be known, but there is reason to believe it took a horrendous demographic toll on the small band. It also helps explain why travelers saw few Paiutes: they were either hiding or had already been taken away.

The days of the Old Spanish Trail ended in the late 1840s. The end of the Mexican-American War brought this territory into the United States. Traffic between Southern California and Utah switched to more direct roads, leaving the horse and mule trail to disappear. The new road was used heavily by emigrants heading to California, especially after news of the Donner Party's disastrous 1846 attempted crossing of the snow-covered Sierra Nevada Mountains. Taking the Mormon Road south from Salt Lake City avoided these mountains, but the Manly party left the road in 1849 to take a shortcut to California. After struggling across Nevada, they ended up in Death Valley. Those who stayed on the Mormon Road did not always fare better; in 1857, every member of an emigrant party was murdered or taken to be wives or children by Mormons at the Mountain Meadows Massacre in Utah.

Most of the Old Spanish Trail around Las Vegas has long since been erased by time and the growth of the city. Efforts have been made to find, map, and mark surviving sections, especially on Mormon Mesa and between Blue Diamond and Mountain Pass. A pristine surviving section of trail can be seen at Emigrant Pass near Tecopa, California. The Old Spanish Trail Association now works to preserve the trail and promote its history.

In 1847, the first Mormons reached the shores of the Great Salt Lake far to the north of the springs of Las Vegas. They had fled the United States and the religious persecution they had experienced there and sought to build their society in the vast western frontier. Unfortunately for them, the United States followed them as it expanded west.

The Mormons reacted to the growth of the United States by asking Congress in 1849 to create the state of Deseret, which would encompass what became Utah as well as most of Arizona, Nevada, and Southern California from Mexico to north of Los Angeles. With the exception of California, the boundaries of Deseret were set by watershed boundaries; including California guaranteed the Mormons continued control of the trade route from Utah down the Mormon Road to Los Angeles. Congress quickly rejected the state of Deseret and instead made Utah a territory on September 9, 1850. This Utah Territory was reduced several times before Utah became a state in 1896 to give Nevada more land (map 3.10).

Regardless of the shifting political boundaries, the Mormons expanded their cultural domain, founding more than five hundred settlements throughout the West as well as in Mexico and Canada. Many of these were agricultural colonies necessary to support the growing society, including St. Thomas along the Muddy River in 1865. Some outposts also served strategic purposes; a small colony founded at Las Vegas in 1855 was one of these

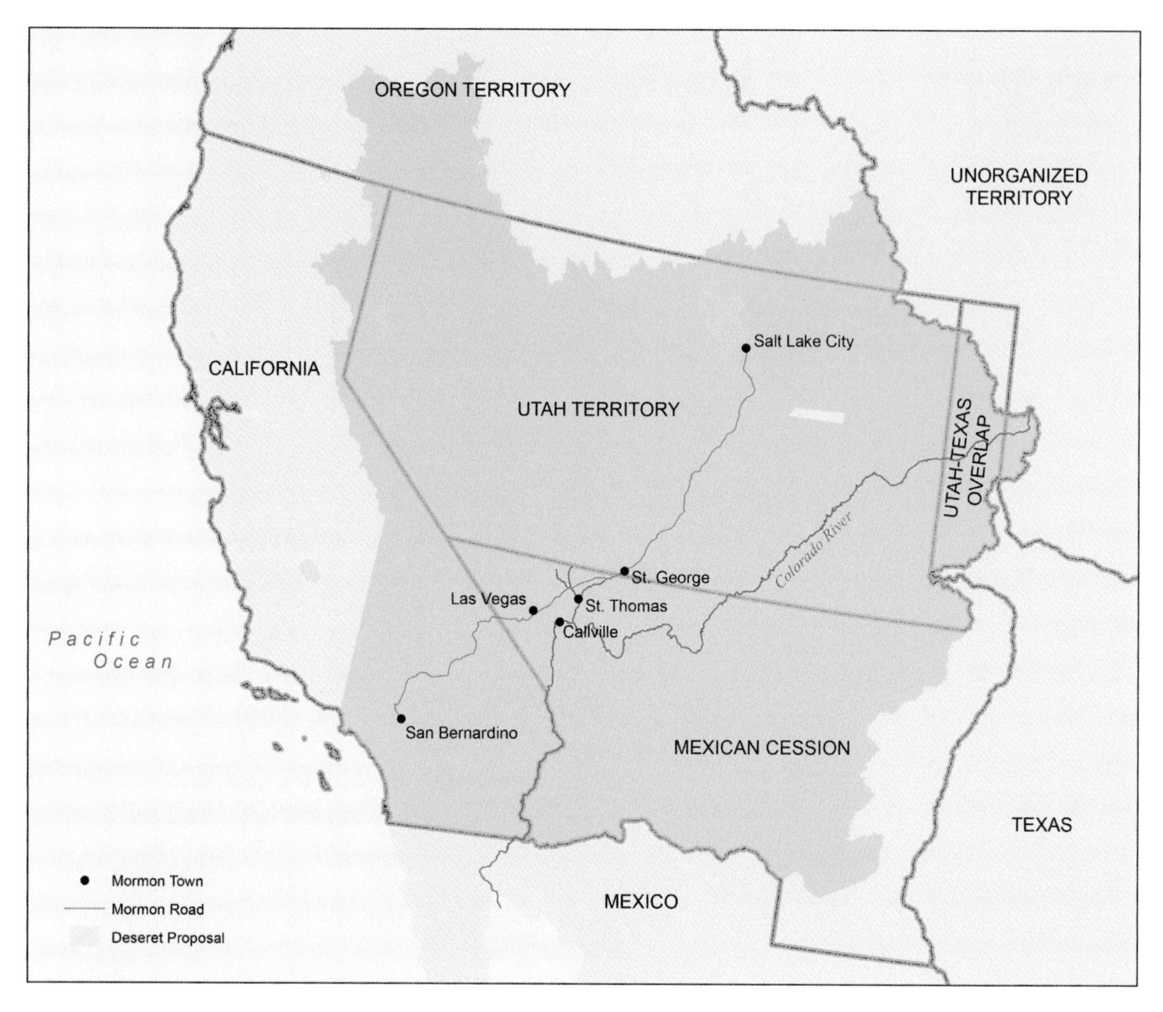

Map 3.6. Mormon Las Vegas. The proposed new state of Deseret and the eventual boundaries of California and Utah Territory. At that time, Las Vegas was a Mormon outpost between Salt Lake City and San Bernardino.

strategic locations (map 3.7), and along with San Bernardino, California, served to facilitate travel along an important trade route between the Mormon realm and elsewhere. Originally part of the Old Spanish Trail (map 3.5), it later became the Mormon Trail, with Interstate 15 as its eventual successor.

Travel by pack mule and wagon trains was slow and tedious, and more efficient options for travel were investigated. Mormons founded the town of Callville in 1864 to establish a port along the Colorado River, which offered the possibility of easier and less expensive travel connections (map 5.3). It did not succeed because steamboat traffic on the Colorado River was efficient only along the southern part of the river, and the arrival of the transcontinental railroad in Utah in 1869 caused the Church of Jesus Christ of Latter-day Saints to abandon the river trade idea.

Not all Nevadans were happy with the changing Utah-Nevada state line, among them the farmers in St. Thomas and along the Muddy River (map 5.3). The Muddy River region had never been surveyed, but locals considered themselves to be residents of Utah Territory and paid their taxes accordingly. They even had their own county when the territorial legislature of Utah created Rio Virgen County in 1869.

However, the realization slowly grew in the Nevada state government that the Muddy Valley might be in Nevada. Surveys confirmed that this was the case. Residents of St. Thomas were outraged when told they owed the state of Nevada several years of back taxes, which had to be paid in gold. The majority responded by packing up and heading back to Utah, abandoning St. Thomas and other towns. Rio Virgen County was eliminated in 1872, though very little of it had been in Utah.

The Mormons eventually learned to accept Nevada and returned to St. Thomas and the Muddy River, at least until Lake Mead forced a second abandonment. The Mormon presence in Las Vegas and Southern Nevada remains strong today; Mormons are the third-largest religious group in the city.

After Mormons began settling what would become Utah, they sought routes to the outside world. Among these was the Old Spanish Trail, soon to be known as the Mormon Road (map 3.5). Maintaining this wagon route to California was important to the Church of Jesus Christ of Latter-day Saints in the 1850s, and for that reason thirty men were sent south from Salt Lake City to Las Vegas Springs to create an agricultural community, proselytize among the Paiutes, and serve as a way station for travelers between Utah and Southern California. Mormons created the settlement of San Bernardino, California, for similar reasons several years earlier.

The Las Vegas settlers arrived in June 1855 and decided to build their colony not at the springs themselves but three miles east along Las Vegas Creek. Paiute Indians occupied the springs, and the settlers thought it best to avoid any confrontation over land and water. The Mormons chose the location for their settlement very carefully. Although mostly erased by urban growth, several terraces or scarps ran north-south in the Las Vegas Valley; along these scarps, the ground dropped several feet to the east. The selected site was on level ground between two of these scarps. This location allowed water to be easily diverted from the creek upstream to irrigate fields around the fort.

The first task was to build an adobe fort to house the settlers. This was not a building but a compound 150 feet to a side, with walls about fourteen feet tall. The interior walls were lined by houses for the missionaries as well as storerooms. Despite the defensive appearance and name, the fort was never attacked, and the Mormons never finished the structure while they occupied it. A post office was briefly established at the Mormon colony, but since the valley was then in New Mexico Territory it could not be named Las Vegas because a town of that name already

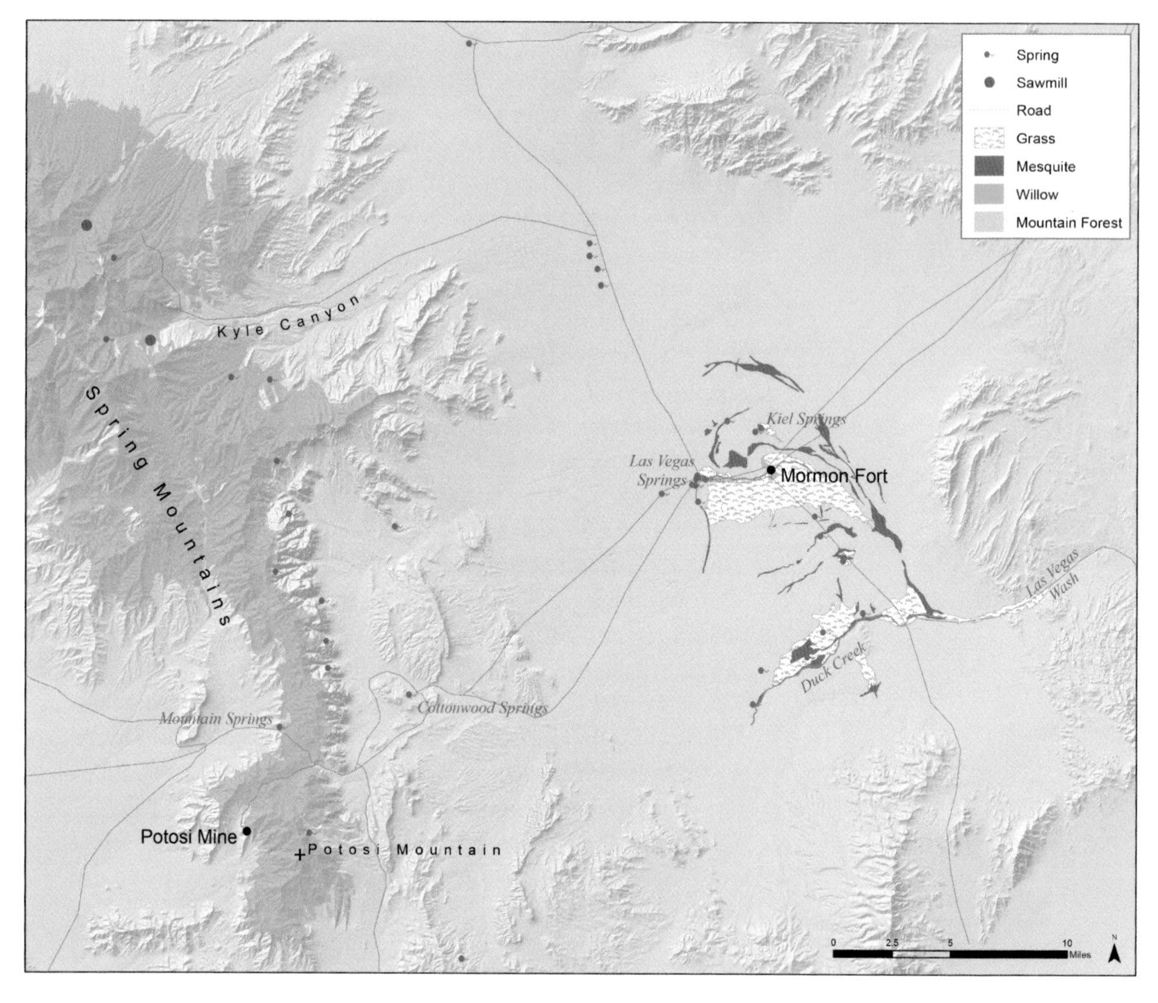

Map 3.7. The Mormon Fort and Potosi. The Mormon fort was a walled building built in 1855 as an agricultural settlement. Many settlers instead spent their time many miles away at the Potosi Mine.

existed east of Santa Fe. It was instead named Bringhurst after the man in charge of the Mormon mission.

Each of the men in the group was assigned 2.5 acres of agricultural land along the creek on which they planted a total of forty acres of corn. Although water was sufficient, the alkali soil was not good for farming, a fact that later residents would continue to rediscover (map 10.5). One of the men, John Steele, drew a map of the fort and surrounding land in 1855, the first detailed map made of Las Vegas. It shows the fort and several fields on the north side of the creek, with the Mormon Road (labeled as the California Road) passing nearby. He depicted Sunrise Mountain in the background.

The Mormons established another farm (called the Indian Farm) at what is now Kiel Springs, which they used to teach Paiutes farming. The Paiutes had been farming at this same spring for countless generations, but the Mormons were unaware of this. In addition to farming, the settlers were instructed to investigate a lead outcropping reported southwest of Potosi Mountain. The outcrop was found, and in August 1856 a mining camp was

Reconstruction of Old Mormon Fort, 2012. Photo by David Stanley.

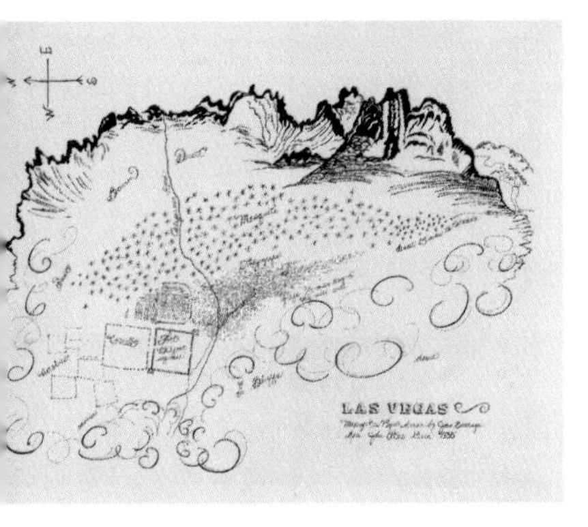

First map of Las Vegas, drawn in 1855 by John Biddinger, based on a map by John Steele. North is to the left, and the fields around the Mormon Fort, mesquite woodlands, and Sunrise Mountain are to the east. Courtesy of UNLV.

founded and tunnels begun to mine the ore from the mountain and smelt it. The Mormons simply referred to this as the lead mine, but the camp would later be known as Potosi. Despite difficult working conditions, nine thousand pounds of lead were shipped out to Utah before the camp was abandoned in 1857. Considerable tension had been created over the mine in the small Mormon community; many of them were strongly opposed to the pursuit of mineral wealth, and the small group had spent so much time at the Potosi mine that their agricultural efforts suffered. This marked the first time in Las Vegas history when the pursuit of quick wealth caused people's livelihoods to suffer.

The settlers were called back to Salt Lake City in 1858, and the fort and fields were abandoned. Dissension among the men over mining was a contributing factor, as was ongoing resistance by the Paiutes against the loss of their land and water. It was a rare setback for Mormon colonization in the West. But the fort and fields they left behind would not stay empty for long.

The Mormon pioneers' colonization efforts were not successful, but others saw an opportunity with their abrupt departure. Albert Knapp occupied the abandoned fort and farmed there in 1861, but he soon gave up and left. Another pioneer, Octavius Gass, visited the fort while traveling the Mormon Road in 1864 and moved into the fort with two partners the next year. He later bought them out along with an adjacent ranch to be the sole landowner of what he called the Las Vegas Rancho. He added more buildings and remodeled the fort, demolishing much of it in the process. He continued farming, trying wheat, oats, barley, beets, cabbage, corn, onions, potatoes, watermelons, and beans, along with apples, apricots, peaches, and figs. He also bought more land, including the springs, and became involved in local politics.

A worker on his ranch, Conrad Kiel (or Kyle), started his own 240-acre ranch several miles to the north at what is now known as Kiel Springs. He also opened a sawmill in what is now Kyle Canyon in the Spring Mountains to supply his ranch and others with lumber and firewood. Another ranch started up about 1867 near Red Rock Canyon (map 9.6), bringing more neighbors to the ranch.

Gass sold his Las Vegas Rancho to Archibald Stewart in 1882 and moved to Southern California. He lived until December 10, 1924, and considerably outlived Stewart, who died soon after purchasing the ranch and was buried just to the west in what would later be known as the Stewart burial ground, Las Vegas's first cemetery. His widow, Helen, continued operating Stewart's ranch and in 1893 opened a post office next to the old fort. Since by then Las Vegas was no longer in New Mexico, the postal

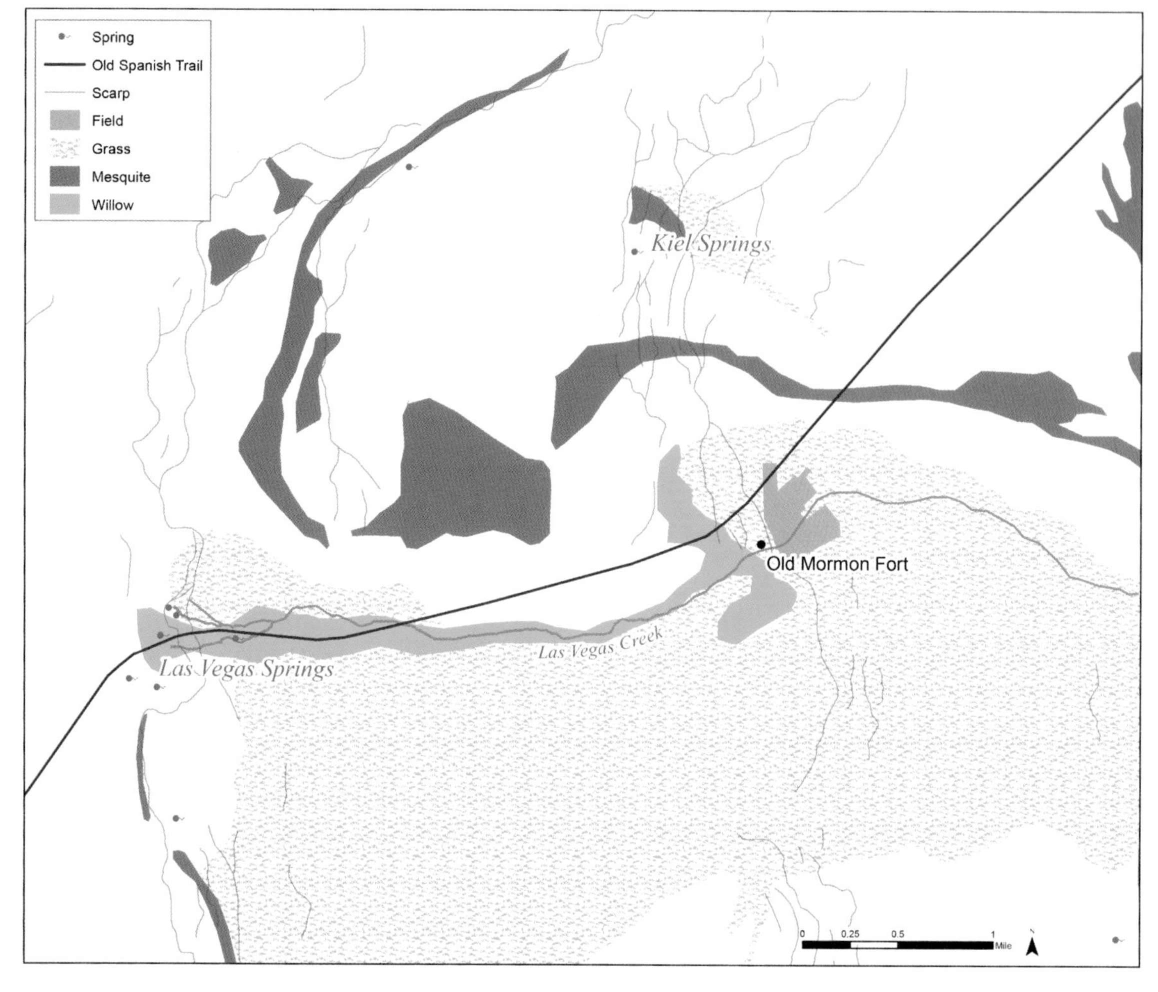

Map 3.8. The Las Vegas Rancho. The Las Vegas Springs created a creek and associated vegetation in the nineteenth century.

service finally approved that name for a post office, though it was spelled Los Vegas.

Helen Stewart kept the ranch until she sold the land to the San Pedro, Los Angeles, and Salt Lake Railroad in 1903 (the railroad purchased the Kiel Ranch as well). At this time the ranch had 557 peach trees, 114 apple trees, 40 acres of hay, 6.5 acres of wheat, 2.5 acres of sweet potatoes, and more than a thousand grapevines, as well as cattle. She retained ownership of the Stewart burial ground. She also retained a Paiute encampment until 1911, when she sold the ten acres it occupied to the government to become the Las Vegas Indian Colony.

The railroad was being built to connect Salt Lake City with the port of San Pedro, south of Los Angeles, and was completed in 1905 (map 7.1). On May 15, 1905, the railroad founded the new town of Las Vegas on the east side of the tracks south of the old Rancho, while a speculator founded a rival townsite on the west side of the tracks outside railroad lands (map 4.1). The old Rancho remained an agricultural area as well as a public park, with flowing water and tall trees. The area faded

Las Vegas Creek in 1895. Photographer unknown.

in importance as Las Vegas grew; by the 1940s, the old Stewart burial ground disappeared off maps and is now the site of an empty lot on the south side of Washington Avenue. The end of what remained of the old Rancho finally came in 1948 when the Las Vegas Water District shut off water to the area. The land was sold for development, and streets were built through the Rancho lands. Only the southeast corner of the original 1855 fort survives and is now part of Old Las Vegas Mormon Fort State Historic Park.

Kiel Ranch was also sold and was known as the Park Ranch, the Taylor Ranch, and finally the Boulderado Ranch before eventually becoming a city park. It is now Kiel Ranch Historic Park, at West Carey Avenue and Commerce Street. Several early buildings survive and have recently opened to the public.

The Las Vegas Indian Colony still exists, though it suffered from years of neglect. The government sought without success to sell it off in the 1950s and did not provide electric or telephone service until 1965. Eventually it was expanded to thirty-one acres, and it includes a number of homes, a cemetery, and a store.

The conclusion of the Mexican-American War with the Treaty of Guadalupe Hidalgo in 1848 transferred most of Mexico's northern frontier to the United States. The area that would become Las Vegas became part of New Mexico Territory. The area became part of Mohave County in the new Arizona Territory in 1863, created out of the west half of New Mexico Territory. Bringhurst then became Las Vegas.

Gass, who took over the Mormon Fort in 1865 (map 3.8), was a strong supporter of the Las Vegas region and was elected to the Arizona territorial legislature. He used his influence to create a new Pah-Ute County that year from the northern end of Mohave County. This included Las Vegas, St. Thomas, and the river port of Callville, the first county seat; the failure of river traffic to materialize led to the county seat being moved to St. Thomas in 1867.

This change was moot because in the previous year Congress passed a law authorizing the transfer of Pah-Ute County to Nevada if the state would accept it, which it did on January 18, 1867. That part of Pah-Ute and Mohave Counties west of the Colorado River then became part of Nevada and split between Nye and Lincoln Counties. Congress also moved the eastern boundary of Nevada east several miles, which set off a series of events culminating in the abandonment of St. Thomas and nearby Mormon settlements.

Although he was now living in Nevada, Gass continued to serve out his term in the Arizona territorial legislature until 1869. He also attempted to create a Las Vegas County out of the

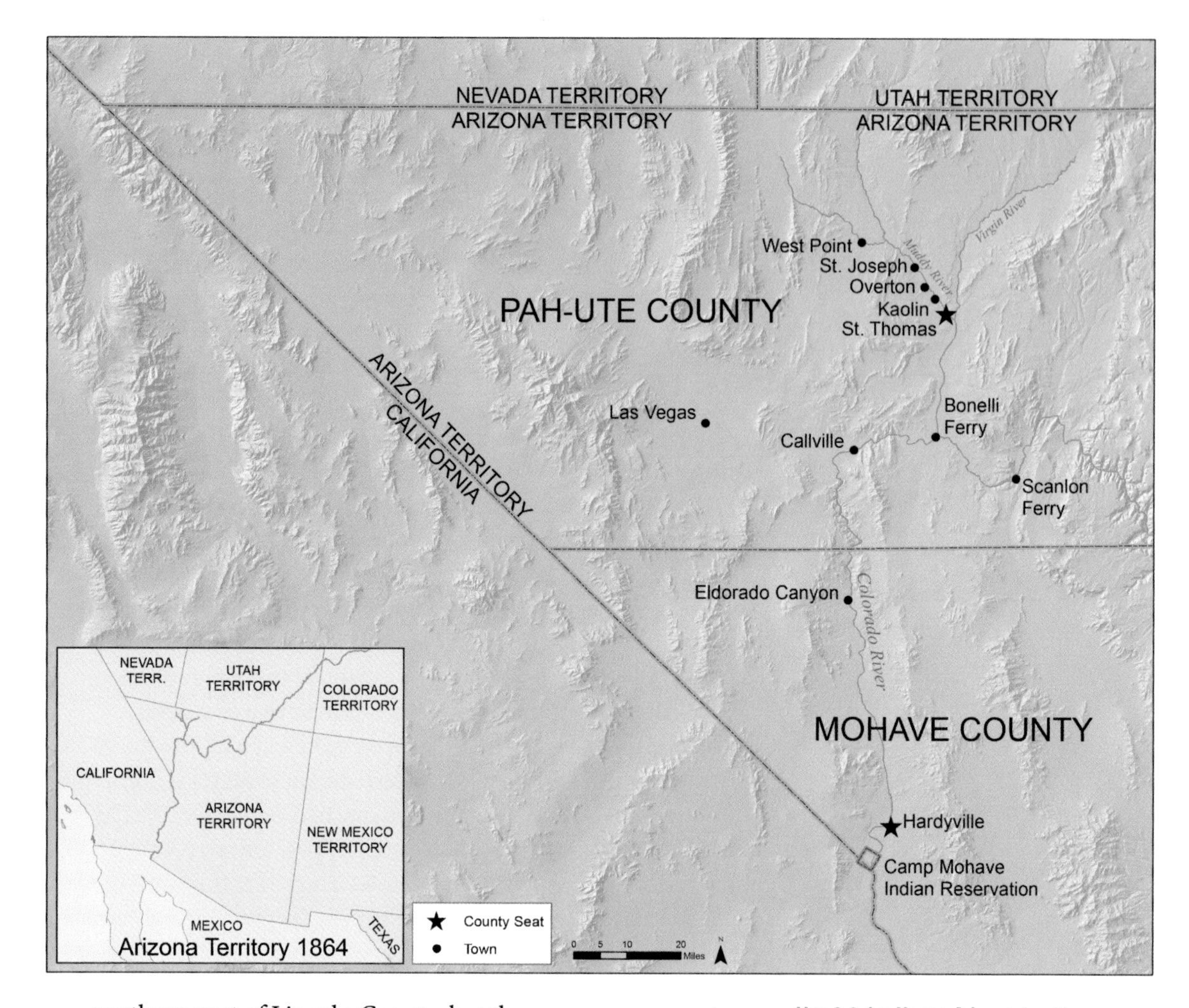

Map 3.9. Las Vegas, Arizona. Las Vegas was part of an Arizona county in 1864.

southern part of Lincoln County, but the move was premature. The abandonment of St. Thomas and consequent decline in population eliminated his political influence, which in Nevada was extremely concentrated in the northern part of the state, and hastened his retirement from politics.

The northern border of Pah-Ute County was specified as the 37th parallel, which today remains the border between Arizona and Utah as well as that between New Mexico and Colorado. The use of this latitude to divide territories dates to the Compromise of 1850, an attempt to divide the territory acquired by the Treaty of Guadalupe Hidalgo. Southern congressmen wanted to create New Mexico Territory south of 36 degrees 30 minutes latitude, allowing them to claim that New Mexico Territory was covered by the Missouri Compromise and therefore allowed slavery. Northerners preferred 38 degrees as it would prevent slavery from spreading west, and 37 degrees emerged as a compromise between these groups. The west end of Pah-Ute County, where it intersected the state of California, was therefore the westernmost relic of struggles over the extension of slavery in the United

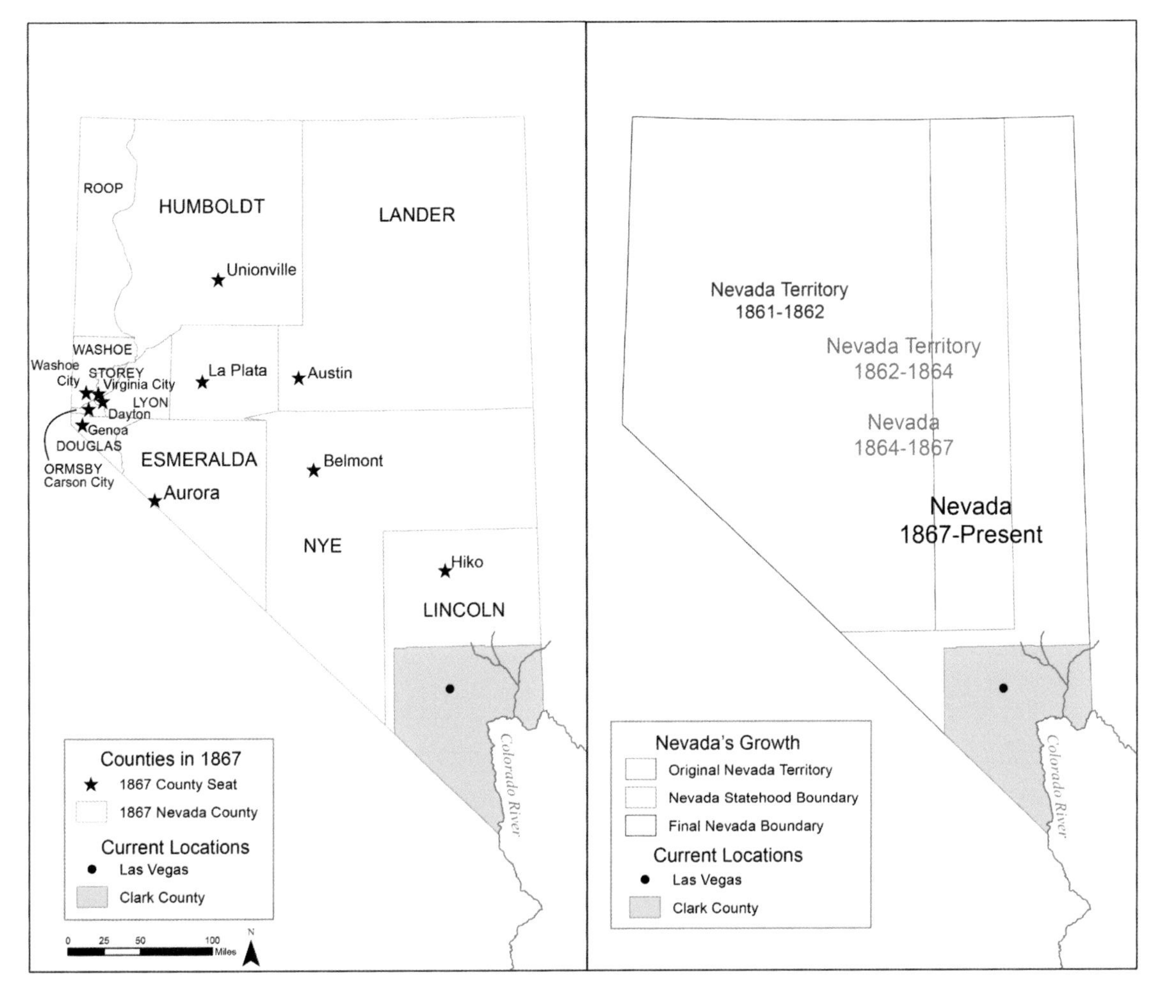

Map 3.10. Las Vegas, Lincoln County. *Left*: Nevada counties in 1867, when Las Vegas was still part of Lincoln County. *Right*: Nevada's expansion from its territorial days in 1862 to its modern boundary in 1867.

States. A roadside historical marker commemorates this line north of Beatty on US 95.

Nevada Territory was created March 2, 1861, out of the western end of Utah Territory. It was enlarged in 1862 and became a state on October 31, 1864, when President Abraham Lincoln signed the legislation to admit it to the Union. But the new state was still much smaller than the one we know today; it was again extended east in 1866, and received part of Pah-Ute and Mohave Counties from Arizona Territory in 1867.

Although Nevada now had its final boundaries, its counties were quite different from today. Southern Nevada was made up of Nye and Lincoln Counties. The former was named after James Nye, the first governor of Nevada Territory. The latter, the westernmost of the state's sixteen counties, was named after the president. In 1867, Lincoln County had fewer than three thousand people. Hiko was the county seat, but the booming mining town of Pioche replaced it in 1871. When the town of Las Vegas was founded in 1905, it was within Lincoln County; residents had to travel 150 miles to Pioche to visit the courthouse for legal

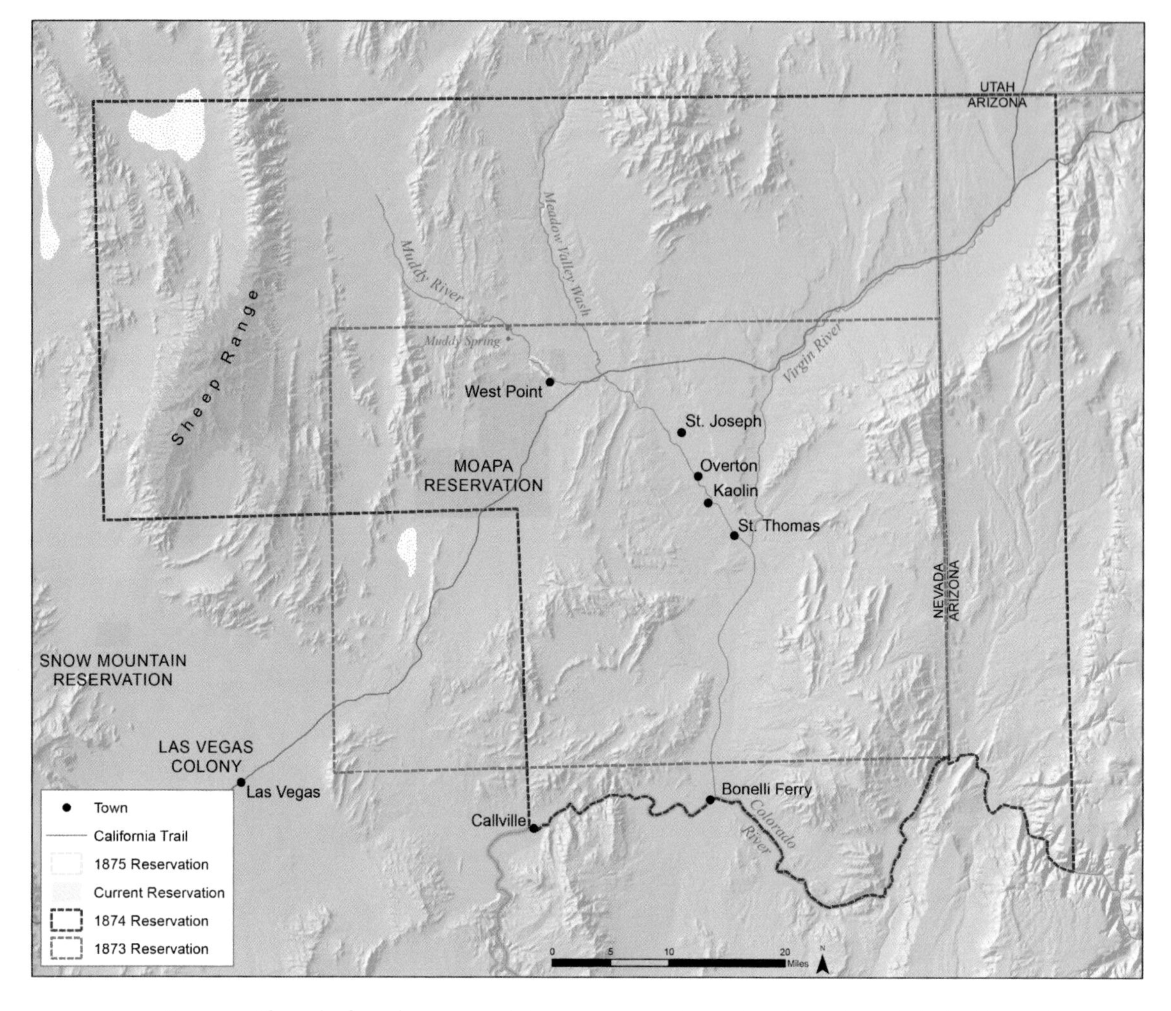

matters. It was not long before they wanted their own county, and they won it on February 5, 1909, when Clark County was created.

The new county was named after William A. Clark (1839–1925), a Montana senator who had become one of the wealthiest men in America by shrewd investments in copper mining and other ventures. How did his name end up on a Southern Nevada county? One of his many business schemes was to build a railroad between Los Angeles and Salt Lake City (map 7.1), along which his agents founded the new town of Las Vegas. The new county's name honored the man ultimately responsible for creating the modern city of Las Vegas (the small town of Clarkdale, Arizona, part of his copper mining empire, was also named for him). Clark County has more than two million residents today, while Lincoln County has little more than five thousand.

The Mexicans, Mormons, and Americans who arrived in the nineteenth century did not recognize the Paiute occupancy of their land (maps 3.2 and 3.3). The Paiutes did not fit into their view of how the land should be used, and one solution was to create reservations in which the Paiutes were expected to live.

Map 3.11. The Enduring Moapa. The Paiutes living along the Muddy River have had a series of reservations with four different boundaries.

The Moapa band received one of these in 1873 when President Ulysses S. Grant created a 3,900-square-mile reservation along the Muddy and Virgin Rivers, extending from the Arizona state line westward to the Las Vegas Valley. The assumption was that all Paiutes would live there, though relatively few ever did; they were accustomed to ranging widely among springs and between mountains and valleys, and the reservation allowed few opportunities for this. The reservation achieved little.

Though this reservation encompassed much of southeastern Nevada, a survey by the explorer John Wesley Powell in 1873 showed only six thousand acres were fit for farming, which had been planned to be the main activity of the Paiutes. He recommended expanding it north to include more farmland along the Virgin and Muddy Rivers and west to include the forested slopes of the Sheep Range. Grant duly changed the reservation boundaries in 1874, creating, in theory, a vast territory for the Paiutes and one of the largest reservations ever. Local farmers, ranchers, and prospectors, with the help of Nevada congressmen, strongly objected to this large reservation and the next year persuaded the president to again change the boundaries, this time reducing it to a mere one thousand acres along the Muddy River in 1875.

By the 1880s, both the Moapa and the government had all but abandoned the reservation. The coming of the Public Land Survey (map 3.12) showed the reservation boundaries had not been surveyed correctly, leading to a prolonged land dispute with settlers not resolved until 1903. This slowly brought the reservation back to the government's attention, and small sections of land were added in 1903 and 1912. The tiny reservation was expanded to its present size of almost seventy-two thousand acres in 1980.

The reservation is currently home to 238 Moapa and has tried to develop economically with mixed results. Most recently, a 250-megawatt solar power plant opened on the reservation, part of the growth of solar power around Las Vegas (maps 8.16 and 8.17). The Moapa continue to work to regain their ancestral homeland. The odds are long; the land around Las Vegas is controlled by a variety of government agencies loath to give up even empty desert (map 8.1).

While the Paiutes lived peacefully among the springs and mountain forests, a new country won its independence on the opposite side of the continent and set about expanding west. The government of the new United States wanted a simple method for transforming the western frontier into surveyed land that could be sold for farmland and other uses. This method was invented by a future president of that country, Thomas Jefferson, in 1785.

This is the Public Land Survey System (PLSS), also known as the Rectangular Survey System or Township and Range System.

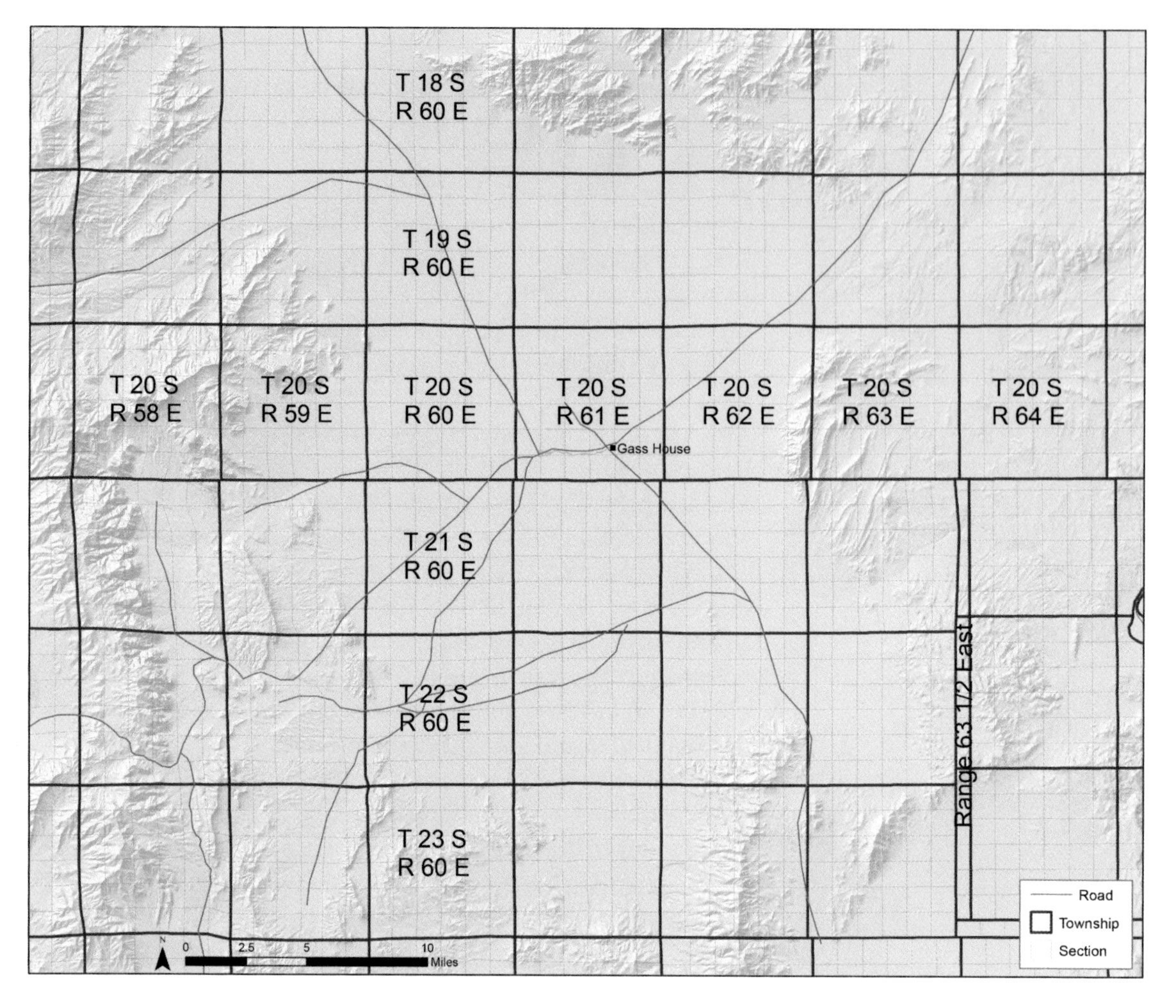

Map 3.12. The Public Land Survey. The roads shown are from the original plats (or maps) drawn by nineteenth-century land surveyors as they worked their way across Las Vegas Valley. Based on the Bureau of Land Management's mapping data for the survey.

The basis of the system is the township, which is a square six miles to a side. Townships in turn are divided into thirty-six sections, each of which is one square mile or 640 acres. These in turn can be subdivided into quarter sections of 160 acres or even smaller squares and rectangles as desired. A quarter-quarter section, or forty acres, was usually the smallest parcel homesteaders could claim. The location of a township is based on distance from a meridian and baseline; they are numbered north and south from the baseline and east and west from the meridian.

The system was first applied in Ohio in 1785 and spread west from there, being used in every state west of the Mississippi except Texas. It is perhaps most famous for its application in the Great Plains, where it resulted in a checkerboard of fields and roads laid out from horizon to horizon.

The PLSS arrived in Southern Nevada in the 1880s when government surveyors mapped out lands along the Virgin River and the Las Vegas, Pahrump, and Eldorado Valleys. While most states had their own initial survey point that defined the meridian and baseline, Nevada was mapped using the Mount Diablo baseline and meridian established near San Francisco in 1851. For that reason, all townships in Las Vegas are south and all ranges east.

The survey clearly did not go smoothly. A number of irregularities occur in section lines, especially west of the city where many section lines are not straight north-south lines. A narrow Range 63½ was also required to the southeast of the city. These sorts of irregularities are common with the PLSS. An offset of the north-south lines is apparent between Townships 20 and 21 South, but this was not an error. Rather, it was an adjustment necessary to take into account the curvature of the earth, which would otherwise cause the north-south lines to get closer together the farther north they ran.

The original 1880s survey maps can be found online today from the Bureau of Land Management website but show very few details around Las Vegas. The surveyors had instructions to note agricultural potential and timber, but there was little of either around Las Vegas. The map for Township 20 South, Range 61 East, which includes the Las Vegas Rancho and later townsite, shows isolated sections of roads and washes, Las Vegas Springs and Creek, fields at Kiel Ranch, and the Gass House at the Las Vegas Rancho (map 3.8).

The future site of Las Vegas was surveyed January 18–24, 1882. During those seven days, the surveyors, listed only as Woods and Myrick, laid the foundation for the future city. They surveyed the land and made possible the homesteading of farm and ranch lands as well as subdivisions added to the growing town. The section lines they laid out, based on a survey marker placed on a remote mountaintop near San Francisco in 1851, would later become the city's main streets (map 8.5). The adjustment required between Township 20 and 21 South because of the earth's curvature would later become a nuisance to Las Vegas drivers crossing Charleston Boulevard.

The Founding of Las Vegas and Its Growth

4

The arrival of surveyors working for the San Pedro, Los Angeles and Salt Lake Railroad in 1905 set in motion the transformation of Las Vegas from a sleepy desert rancho to a big city. The surveyors laid out the railroad line and then marked out streets and blocks along the tracks. The lots were sold and the town was born. But growth was very slow in the first two decades. The town suffered setbacks such as floods and fires. Residents also faced a long trek to the county seat at Pioche and endured the loss of railroad jobs to Caliente.

But Las Vegas survived into the 1930s when the electrifying news of a massive new dam on the Colorado River arrived. That meant another railroad line, another fledgling town in the form of Boulder City, an influx of people and even a visit by the president of the United States. Soon after this project was completed, another growth boom ensued when the start of World War II brought an airbase, an enormous industrial plant, and another new town (Henderson) to the region.

To the tremendous relief of locals, these economic assets survived the end of the war and growth continued. The airbase and industrial plant are still pillars of the Las Vegas economy. But little by little, sprawling hotel-casinos popping up along the highway south of town began to define the future. It also began an era when events in the larger world brought rapid change to the region, including a seemingly permanent cold war, atomic testing, and racial integration. Despite rapid urban growth, the Las Vegas Valley remained very low density, with a vast acreage of empty desert even next to the Las Vegas Strip, until the 1990s when its greatest growth boom occurred.

The San Pedro, Los Angeles, and Salt Lake Railroad was completed between those cities in 1905 (map 7.1). The railroad crossed the Las Vegas Valley because the large springs offered a reliable water source for thirsty steam engines. The railroad purchased Helen Stewart's Las Vegas Rancho (map 3.8) to gain control of the springs as well as land for locomotive maintenance facilities nearby. A town was necessary for workers to reside, and a location south of Las Vegas Creek was chosen for it. The railroad decided to create a town with a grid street pattern, a style used when founding towns since the days of the Roman Empire.

Surveyors laid out a grid of streets running parallel or perpendicular to the tracks on the dusty ground. Streets were eighty

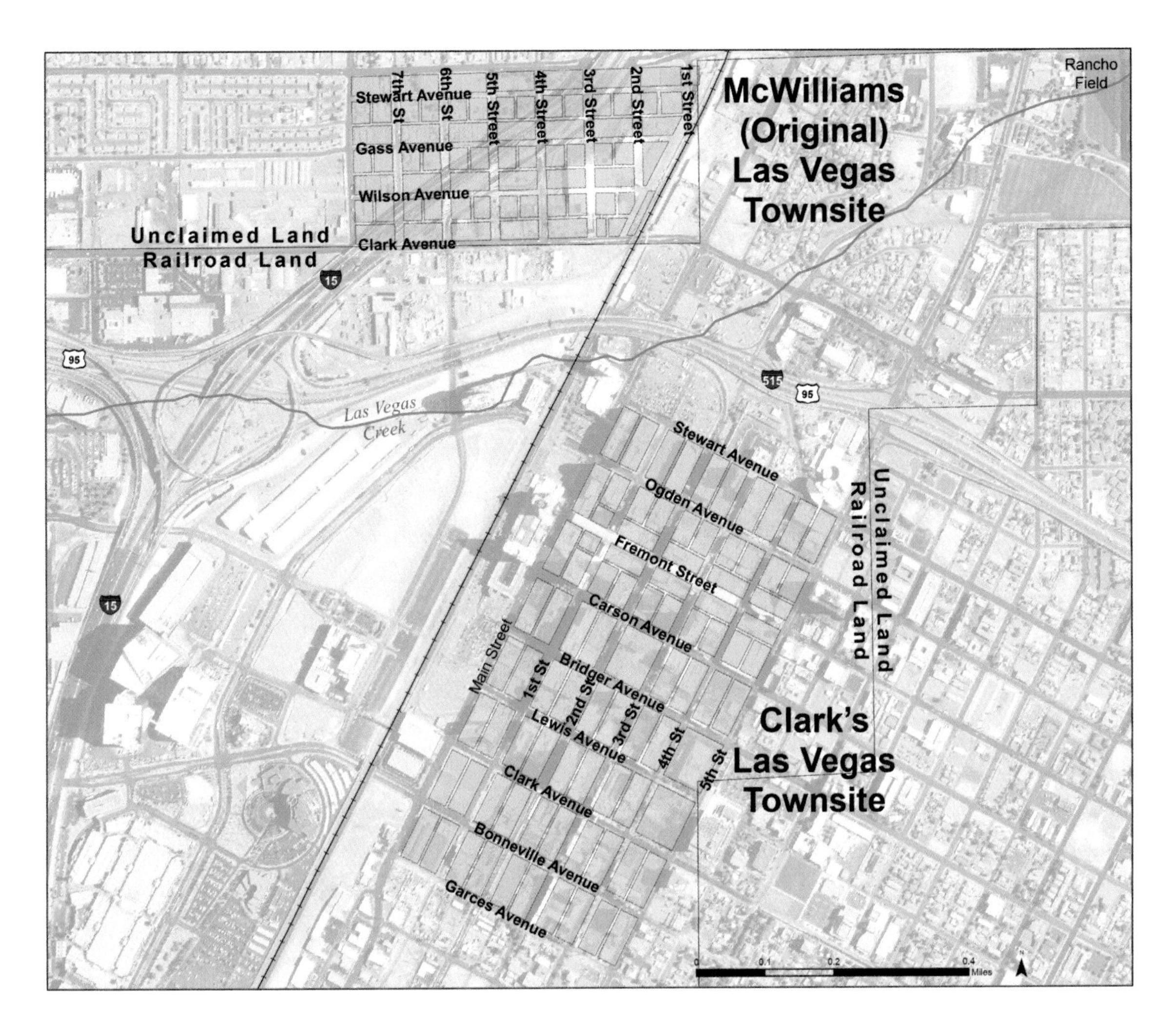

Map 4.1. The Las Vegas Townsites. Two competing townsites were surveyed for Las Vegas in 1905. Clark's became downtown, while the McWilliams became the Westside. Block boundaries and original street names are shown over a modern aerial photo.

feet wide and alleys twenty feet, with lots usually having twenty feet of street frontage. Fremont Street was planned as the main commercial street, with lots along this road oriented toward it rather than the tracks, as with every other street. One block was reserved for a courthouse, should Las Vegas ever become a county seat, and Block 16, on First Street between Ogden and Stewart Avenues, was designated as the city's red-light district. No land was reserved for parks. A railroad yard and maintenance facilities were built west of the tracks.

Las Vegas came into being on May 15, 1905, with a land auction where downtown's Plaza Hotel is now located. The bidders were merchants, railroad workers, investors, and speculators looking for a place to move their business or live, all gambling that the new town would survive and prosper. Some lots in good locations sold for as much as $850, but the odds on success must have seemed long as only 176 of the two thousand lots sold that day.

Las Vegas had two original townsites. The nation's land laws not only allowed individuals to homestead farmland or stake out mining claims but to survey towns on public land to be sold off to potential residents. Countless towns were surveyed and touted by promoters, but few survived long enough to become actual communities. J. T. McWilliams, hired by Stewart to survey her land for sale, saw an opportunity to create his own town just off the railroad land but adjacent to the tracks. He called this the "Original Las Vegas townsite" as he started selling lots in 1904, before the railroad auction of what became known as Clark's Las Vegas townsite, after the railroad's owner, William Clark.

The Original Las Vegas townsite (also called the McWilliams Townsite) was also a grid, but one that ran in cardinal directions, with the streets laid out in a north-south or east-west direction. Streets here were seventy feet wide, and the lots varied by direction and size, with most being fifty- by one-hundred-fifty feet. Although the area saw some early growth, Clark's townsite became the more successful of the Las Vegases. It was expanded many times and is now the heart of downtown Las Vegas. The McWilliams Townsite later became known as the Westside and would appear again in Las Vegas's history (maps 4.9 and 4.11).

The arrival of the San Pedro, Los Angeles, and Salt Lake Railroad and the creation of Las Vegas opened Southern Nevada to settlement, ranching, and mining. The town quickly became the heart of a road and railroad network radiating into the surrounding desert. The Las Vegas and Tonopah (LV&T) Railroad was built from Las Vegas to the booming mine towns of Rhyolite and Goldfield in 1906 (map 7.2), and a narrow gauge line started operating from Arden west to mines in the Blue Diamond Hills in 1907 (map 7.3). Rough roads extended to other mines and ranches in the area.

Early maps showed many railroad stations along the valley's several tracks, but most of these were simply telegraph stations with water tanks for locomotives. Arden, the only one that might be considered a town, remained on maps of Las Vegas for many years. It thrived as the junction between branch lines to gypsum mines to the west and the main railroad line. Today it is the home of a Union Pacific yard that replaced the original one downtown in the 1990s. An old water tower, one of the oldest surviving structures in Las Vegas, still stands next to the tracks.

The valley had many other springs that attracted settlement. Tule Springs became a water stop on stagecoach lines between Las Vegas and Rhyolite; it was homesteaded as a ranch circa 1916. It was changed in 1941 into a divorce ranch, where couples could live and wait out their six weeks of Nevada residency before gaining a divorce. It also served as a dude ranch where city slickers could have a Wild West vacation. To the north, Corn Creek Springs became a stagecoach water stop, and a station

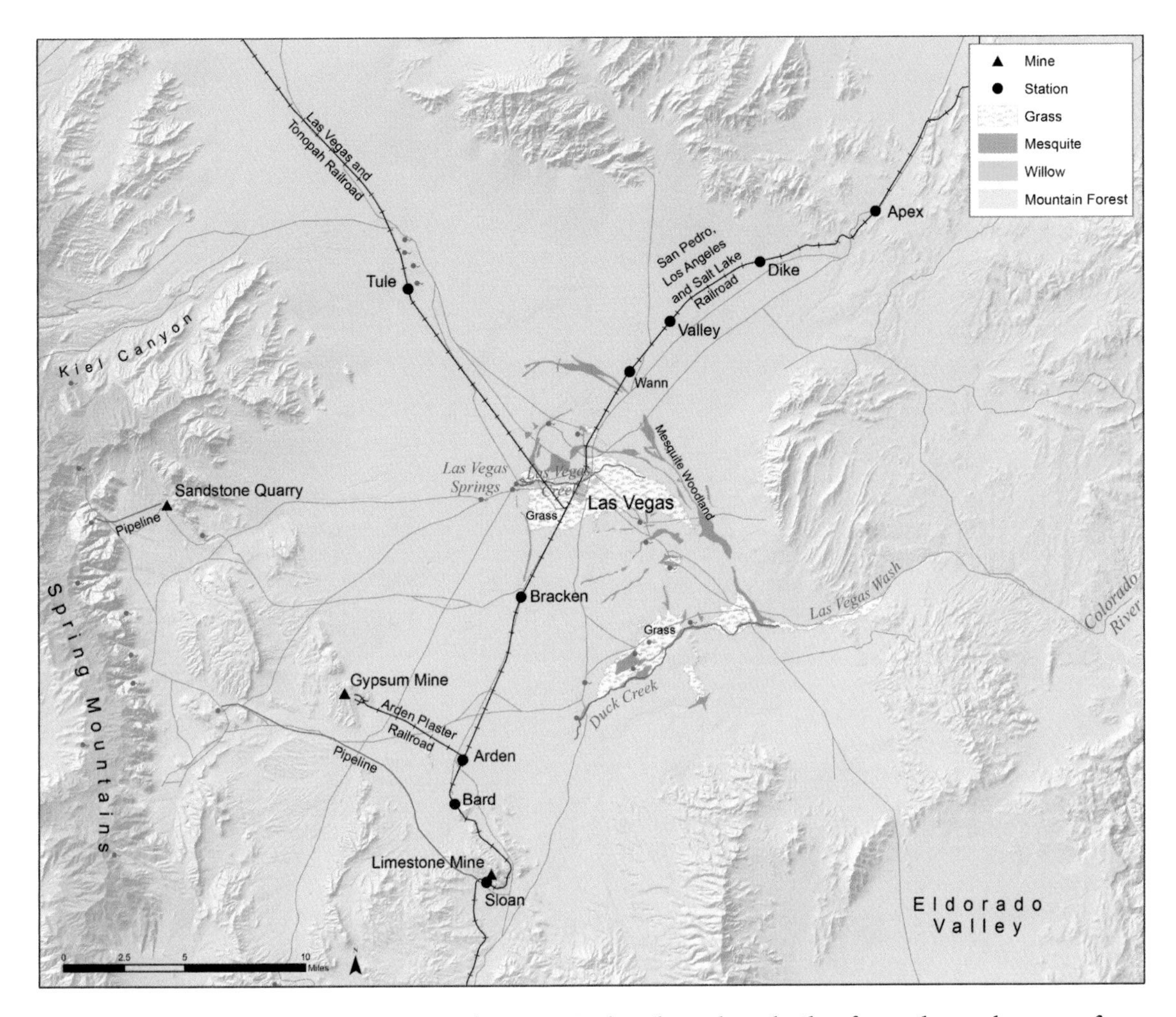

Map 4.2. Early Las Vegas. The small railroad town was situated within a well-watered area of the valley, with mesquite woodlands nearby. Based on the 1907 Las Vegas USGS topographic map.

on the LV&T Railroad was later built a few miles to the west of the springs. A ranch started up at the springs circa 1915, and it is today the headquarters for the vast Desert National Wildlife Refuge (map 9.3). Ranches also developed along Duck Creek in the southern end of the valley (map 6.3), and along the base of Wilson Cliffs west of town (map 9.6).

The high, forested mountains with pinyon pine, juniper, ponderosa pine, white fir, limber pine, and bristlecone pine west of Las Vegas are recreational resources today, but in the town's early years they were essential to its growth. Endless timber was needed for constructing buildings, railroad ties, and framing mine tunnels, while firewood was always in short supply in the desert. Settlers naturally turned to the closest mountain forests to supply these needs. Conrad Kiel was the first to do this when he built a sawmill in the Spring Mountains in the 1870s. His sawmill provided the name for Kyle Canyon (a variant spelling of his last name), one of the oldest place-names still on the area's map. Several other sawmills operated near the head of Cold Creek, one

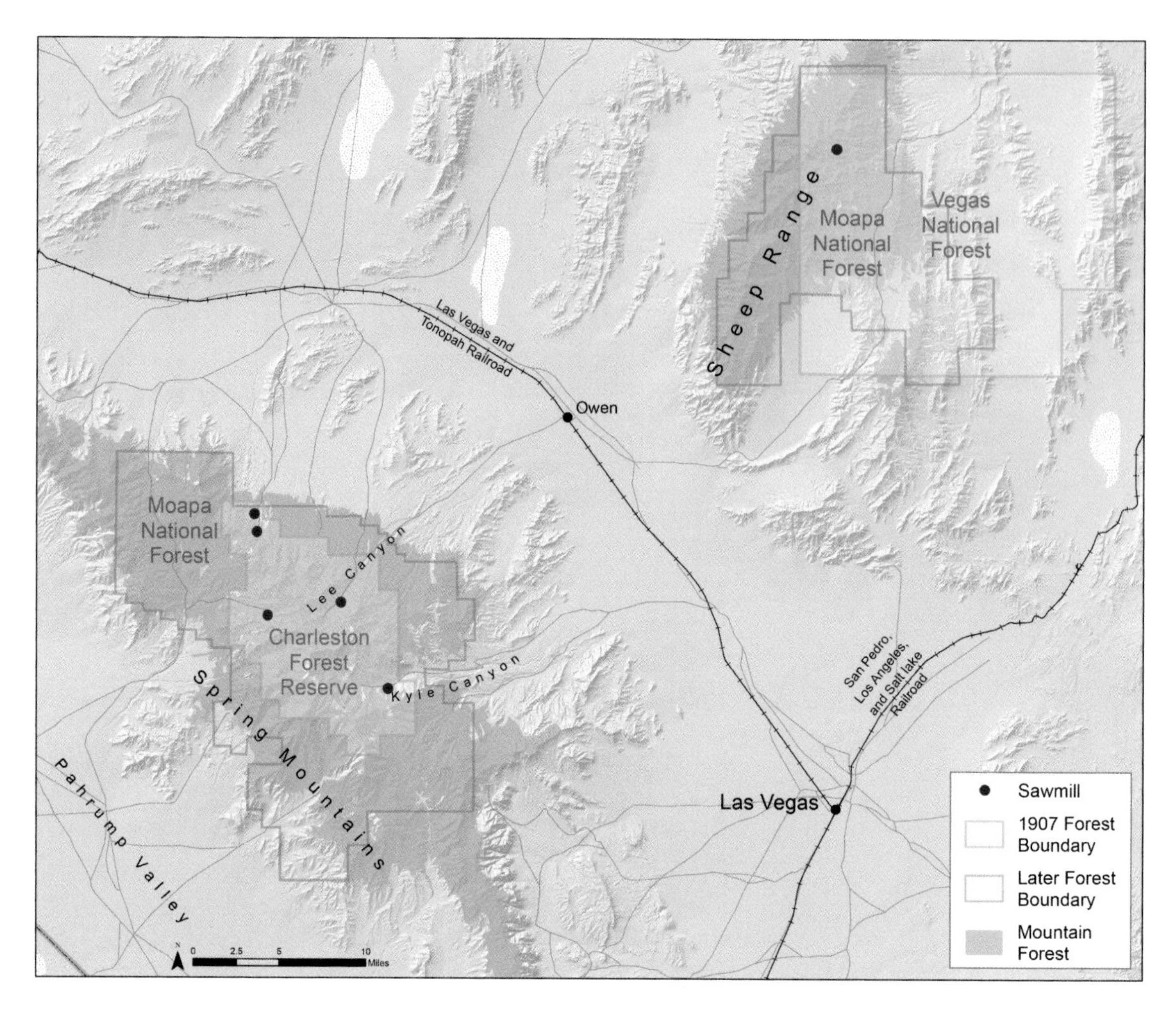

Map 4.3. Logging. National forest boundaries have changed in the mountains outside Las Vegas. Sawmills, supplying the desert town with lumber and firewood, were once common.

on the western slopes of the Spring Mountains, and at least one could be found in the Sheep Range.

These sawmills were so important that the LV&T Railroad established a station, Owens, where timber hauled down from the various sawmills could be loaded and shipped to the mining towns of Rhyolite and Goldfield. Lumber was therefore perhaps the first commodity exported from the Las Vegas area. But this demand for lumber had consequences for these small mountain forests, and the canyons around sawmills were nearly clear-cut. A stand of yellow pine near Charleston Peak was almost wiped out to supply the needs of the mines in Goodsprings, south of Las Vegas. Pinyon pine, a particularly useful tree with which to make charcoal for mine smelters, was completely cutover in several central Nevada ranges; the forests on the Spring Mountains and Sheep Ranges were lucky to have survived the era.

This widespread logging and deforestation across the West provoked government attempts to protect these resources. An 1891 law allowed the president to create forest reserves on public land, known today as national forests. President Theodore Roosevelt directed his attention at Las Vegas when, on November 5, 1906, he created the Charleston Forest Reserve in the Spring Mountains to protect the remaining forests on this range. Roosevelt proclaimed the Vegas National Forest on the Sheep Range the following year, and in 1908 these two forests were combined under the name of the Moapa National Forest. The United States Forest Service (created only a few months before Las Vegas was founded) was now in charge of Las Vegas's mountain forests, and had the power to regulate logging, grazing, and other activities, with the goal of ensuring that the forests survived.

The boundaries of these new forests were based on the straight lines of the Public Land Survey System PLSS, which allowed boundaries to be specified quickly and easily (map 3.12). These boundaries were sometimes also inaccurate, and the boundaries of the Vegas National Forest were noteworthy for missing most of the range's forest and including a large area of desert. This was adjusted in 1911, and further changes began in 1915 when the Moapa National Forest was merged into the Toiyabe National Forest in northern Nevada. It was withdrawn and instead transferred to Utah's Dixie National Forest the following year, but in 1918 the Sheep Range unit was eliminated from the forest and the Spring Mountain unit's size was reduced. Even more changes came in 1937 when the Spring Mountains were transferred to a new Nevada National Forest, before once again being reassigned, this time back to the Toiyabe National Forest. It remains with that name today (map 9.1).

Las Vegas grew slowly in its early decades. Little attracted new residents besides railroad jobs, and the hot desert town must have seemed an unlikely prospect for growth. Fortunately, more than a few pioneers were willing to take a chance, and the population increased from 800 recorded by the 1910 census to 2,304 in 1920.

What was it like to be one of those residents? The 1920 census unfortunately tells us little about the population of the town of Las Vegas, but at that time 93 percent of Clark County's 4,859 residents were white; 710 of those were foreign born, with about half listed as being from Mexico. England was the next largest country of origin, followed by Italy, Canada, Germany, Scotland, Ireland, and Greece. There were also 214 Native Americans, nine Chinese, and 62 Japanese reported in the area. The county was also mostly (60 percent) male, reflecting its pioneer status, though Las Vegas was probably more balanced than the small desert mining camps found throughout the county. The county was however one of the state's smallest counties by population.

Sanborn fire insurance maps provide a fascinating glimpse into what the town was like in the 1920s. The Sanborn Map Company was founded in 1866 and for many decades created

Map 4.4. Las Vegas in the 1920s. Insurance maps produced by the Sanborn Map Company include every building in Las Vegas in 1923. Buildings are color-coded by their function, and some are labeled by name.

amazingly detailed maps of more than thirteen thousand towns in the United States showing every building, its function, construction materials, and number of floors. The company surveyed Las Vegas in 1923, providing a snapshot of the town. Fremont Street was well established as the main commercial area, with many substantial two-story buildings. Businesses along Fremont included hotels, drug stores, and even two movie theaters. But beyond Third Street, homes and vacant lots prevailed. The town boasted two lumber yards, both located off Fremont Street where land values were lower. One of these was run by Ed Von Tobel, a name that would be well known to Las Vegans until the 1980s because of his giant hardware store on Maryland Parkway. Several warehouses and tanks owned by Standard Oil were found along the railroad tracks; it would be many years before gasoline would arrive by pipeline (map 8.18).

Other needs were provided in different areas of the town. Religion has always been a large part of life in Las Vegas, and it was well established from the earliest days. Five churches (two

Catholic, one Episcopal, a Methodist, and a Baptist congregation serving the town's African American population) were scattered amid homes and businesses. In Block 16, north of Fremont Street at the edge of town, the red-light district, several brothels facing First Street were mapped by the Sanborn Company. These remained in business until 1942 when they were shut down at the insistence of the US Army, which did not want off-duty airmen from the town's new air base patronizing them. The remaining brothel buildings were demolished, and the block redeveloped in 1946; it is parking for nearby casinos today.

Clark County was created with Las Vegas as the new county seat in 1909. The first courthouse was on the south side of Bridger Avenue between Second and Third Streets. Las Vegas became an incorporated city in 1911, and the courthouse became city hall and a library when a new courthouse opened in 1914. This was built on a block reserved for the purpose in the original townsite (map 4.1). The growing community also needed schools; the railroad, which owned unsold lots, donated two blocks of vacant land between Bridger and Lewis Avenues from Fourth to Fifth Streets for a school in 1909. That use prevailed until 1963, when a federal government building was built on the north half and a courthouse annex on the south half. That federal building remains in use today.

Most of the town was residential, with a few rooming houses and many small houses. The first planned community in the city was started in 1909 when the San Pedro, Los Angeles and Salt Lake Railroad built sixty-four stone houses for its employees on Second, Third, and Fourth Streets. Only one of these houses survives on its original location (map 11.9). Aside from these substantial buildings, most of the buildings were made of wood (though a few residents still lived in tents), with most commercial buildings built from stone or concrete. A few small houses were also stone or adobe.

Other improvements were out of sight but very important. Water lines had been laid throughout the city by the railroad, which functioned as the city's water department until 1954 (map 6.4). The town's first sewer system was constructed in 1912, a much-needed improvement. When the Sanborn Company surveyed the town, the streets were all dirt, but Fremont was paved from Main to Fifth Street in 1925.

Although it was still a small railroad town, indications of changing times are apparent; several gas stations were indicated on the map, though they were quite different from modern stations in that they operated out of storefronts fronting directly on the street. One building was listed as being used for aircraft storage; Las Vegas's first airport opened in 1920, when the first airplane arrived and landed at the new Anderson Field (map 7.15). This was far outside town, at the southwest corner of what would later be Sahara Avenue and Paradise Road. It had no facilities, so an aircraft owner must have had to find a suitable hangar in town.

The insurance maps show nothing beyond the built-up area, but a number of new businesses had been created outside town. The Las Vegas Rancho had become the Las Vegas Park Resort, with nightly dances and a concrete swimming pool. Ladd's Resort opened in 1911 with a swimming pool east of town, at what would later be Fremont and Twelfth Streets. Development of Lorenzi Park west of town began in the early 1920s, with lakes, a pool, and a dance hall. Kiel Ranch north of town began to be developed as a dude ranch in 1924, and much farther out of town a road up Kyle Canyon was built and the Mount Charleston summer resort opened in 1915. Other needs were met when Woodlawn Cemetery opened in 1914 north of the old Rancho on land donated by the railroad. It remains in use.

It was several years before the original townsite (map 4.1) filled up, but not before speculators saw opportunities to cash in on future growth by surveying additions to the townsite and selling off lots. Real estate development on the periphery of the town became a new industry, one familiar to Las Vegans today.

The early additions continued the street pattern of Clark's townsite to the east on lands not owned by the railroad. Buck's Addition, Ladd's Addition, and the Fairview Tract were laid out in 1911, and Wardie Addition in 1913. Although the streets ran in the same direction as Clark's townsite, their boundaries often followed the PLSS grid, an indication of the decreasing importance of the railroad on the early town (map 3.12).

The South Addition was laid out in 1926 and was one of the last to continue the city's original street patterns. Several additions to the south and later additions to the east did not follow this practice, and instead streets were laid out to the points of the compass. By this time, much of the valley had been claimed by homesteaders using the PLSS, and the promotors of these additions clearly believed that this grid was more important than the original street pattern based on the railroad. Additions to the McWilliams Townsite were also based on the PLSS grid, which meant that as the city grew the two original townsites (and later Henderson) would eventually be able to grow together seamlessly.

But the transition between the two early street grids resulted in many awkward intersections, still visible where the downtown street pattern meets newer streets. It is especially apparent at the five-way intersection where Fremont Street crosses Charleston Boulevard at a sharp angle. To accommodate automobile traffic, some of these intersections had to be redesigned; in the late 1960s,

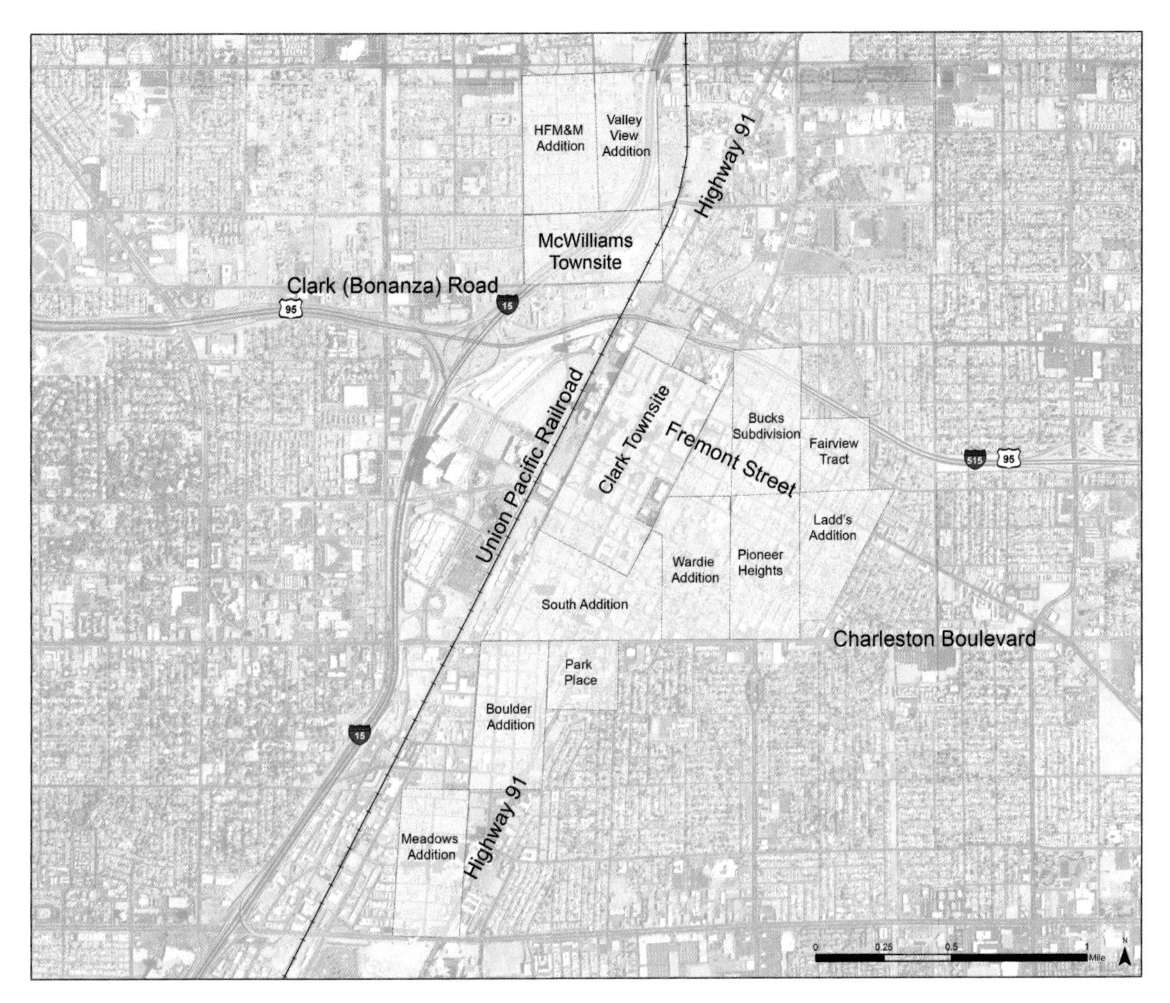

Map 4.5. Early Additions. The small town grew when new neighborhoods were surveyed and lots sold on the edge of town. A modern photo provides the background to early additions to both Clark's and the McWilliams townsites.

Twenty-fifth Street was rerouted to connect to Eastern Avenue and carry north-south traffic around this intersection (map 8.5).

In 1930 Clark County had 8,532 people, of whom 5,165 lived within the city of Las Vegas. This must have seemed nearly miraculous because the town faced bleak times following the loss of many railroad jobs in 1922 after a prolonged strike, and the country was entering an economic depression with no end in sight. As encouraging as the growth during the 1920s were, the 1930s turned out to be the most momentous decade yet for the small town, setting it on a path to growth beyond the wildest imaginings of its early promotors.

This new path began with two big events in 1931. The first was the legalization of gambling in Nevada; the town government acted quickly to limit casinos to Fremont Street, the town's commercial area, between First and Third Streets, though this was later expanded to Fifth Street. The Northern Club was the first casino in town when it opened March 20, 1931, but locals probably thought of casinos as little more than local diversions.

The second event was the beginning of the Boulder (now

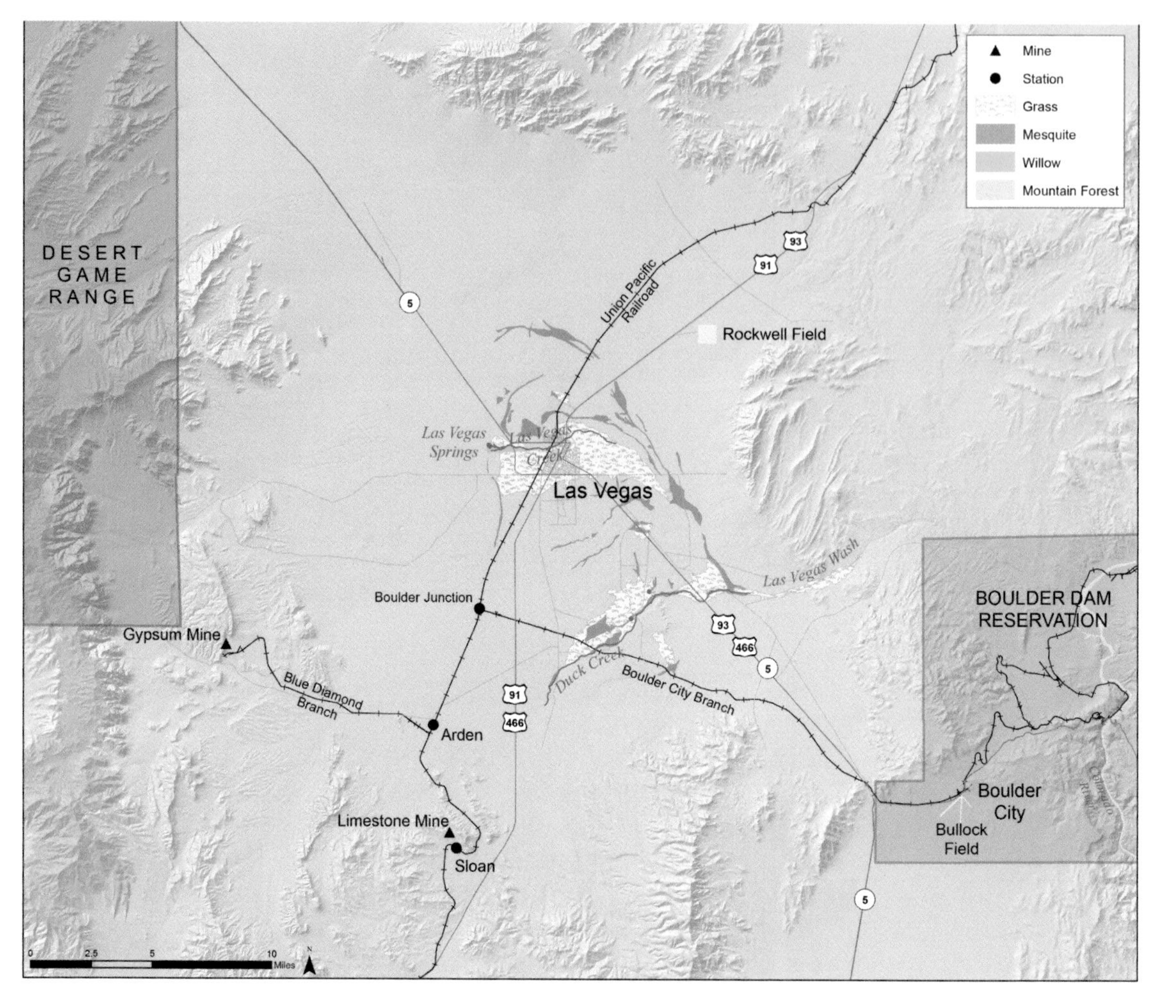

Map 4.6. Las Vegas in the 1930s. New boundaries and railroad lines appear, but much of the valley's vegetation and water were still intact.

Hoover) Dam construction project. This had been discussed with various levels of seriousness for several decades and had been approved back in 1928. But nothing happened until 1930, when a new railroad line was built from the Union Pacific Railroad south of town southeast through Railroad Pass toward the Colorado River. Thousands of workers and massive quantities of building supplies would be funneled through Las Vegas and Railroad Pass in ensuing years, swelling the small town even as the Great Depression took hold across the country.

The government created the vast Boulder Dam Reservation, with room for the new dam, reservoir, and a construction camp. Another government reserve that spanned several mountains and valleys was north of town; this was the huge Desert Game Range created in 1936 to protect the bighorn sheep. Little did Las Vegans know that in following decades these remote mountains and empty valleys would be transformed in ways unimaginable by even the most optimistic town booster, and perhaps even more profoundly than in the Boulder Dam Reservation.

There were many changes in town as well. The two original

Las Vegas townsites (map 4.1) were still distinct areas separated by a busy railroad line in the early 1930s. But a railroad underpass was built in 1937 to allow people to cross safely without delays (map 7.6). Other new developments included the city's first golf course built where the Las Vegas Convention Center is today and the opening of the Clark County Indigent Hospital in 1931. This became Clark County General Hospital, and is now the University Medical Center of Southern Nevada.

The valley's network of roads was improved, bolstering connections with neighboring states. US 91 appeared in 1926, US 93 and 466 in 1935. US 95 appeared later, replacing Nevada Highway 5 (map 7.10). Anderson Field, Las Vegas's first airport, closed in 1929, but Rockwell Field opened up eight miles north of town along US 91. This became Western Air Express Field after that company began airline service there, and was later renamed McCarran Airport after one of the state's senators (map 7.18). On April 29, 1930, biplanes from the First Tactical Fighter Group landed at the tiny airfield while passing through Southern Nevada; ten years later, the military would return permanently.

The Great Depression began in 1929, but the small town of Las Vegas was not as strongly affected as the rest of the country. Unlike the Panic of 1907, which devastated or killed booming mining towns that funneled their traffic through Las Vegas, the town prospered during the Great Depression because of the arrival of several big government projects. The first and biggest of these was the Boulder Dam project (map 5.7). This was approved in 1928, before the Great Depression, but brought thousands of prospective workers and millions in construction dollars to Las Vegas and the new town of Boulder City in 1930 when work began.

For many Americans, the Great Depression continued to worsen until the new presidency of Franklin D. Roosevelt established a set of programs collectively known as the New Deal in 1933. Las Vegas received a number of New Deal projects, and these not only put people to work but provided much-needed infrastructure. The Works Progress Administration (WPA), a new agency helping the country survive the Great Depression, built the Clark Avenue underpass, Las Vegas Grammar School, and War Memorial Building (demolished in 1971).

An important component of the New Deal in Southern Nevada was the Civilian Conservation Corps (CCC). More than three million young men served in the CCC from 1933 to 1942 working on a variety of rural conservation projects, including developing parks and planting trees. (Although he worked in California, one CCC enrollee was future Las Vegas hotel-casino developer Kirk Kerkorian.)

Road sign near Boulder City, 1930s. Courtesy of the National Archives and Records Administration. Photograph by Bureau of Reclamation.

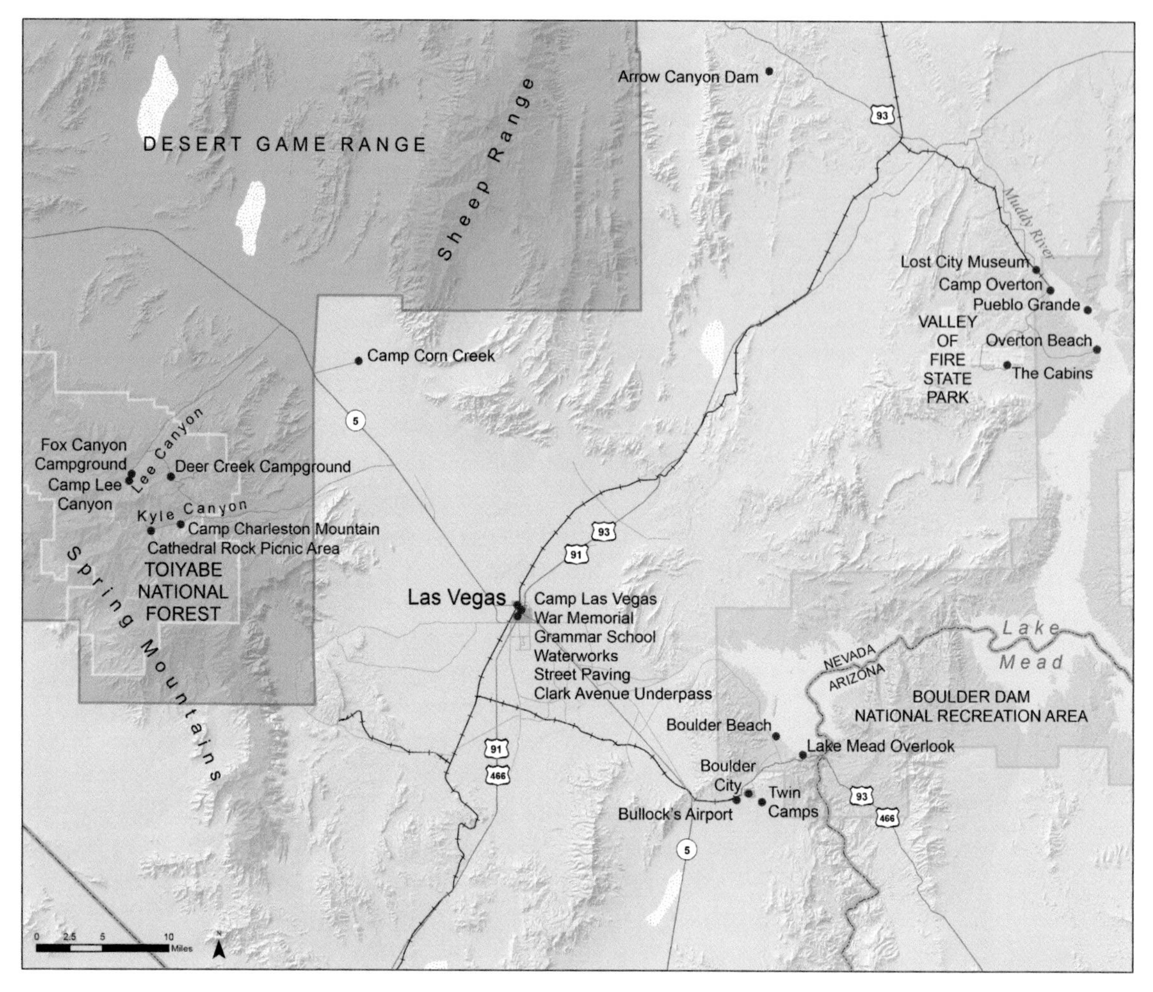

Map 4.7. The New Deal in Southern Nevada. A variety of government programs were initiated in the 1930s to put people to work. The red dots show many of these projects, including campgrounds, street paving, and beaches for Lake Mead.

Two CCC camps were established at Boulder City, in former dam worker housing in a complex called Twin Camps. These workers helped develop Lake Mead for recreation by building Boulder Beach, campgrounds, and an overlook near the dam. They also improved the Bullock Field airport. Another CCC camp was established near Overton. Crews there built the Lost City Museum, took part in archaeological digs at the Lost City, and built recreation facilities in Valley of Fire State Park. The stone cabins they built there for campers (but now a picnic area) are among the more picturesque of their projects in the state. Another CCC camp was established at Pearce Ferry in Arizona near the eastern end of Lake Mead, and the remains of the camp can still be seen there. Located in Kyle Canyon next to the current ranger station, Camp Charleston Mountain operated from 1933 to 1942. This camp operated during summers only, with the workers moving to a camp outside of Las Vegas for the winters. They built trails, a ranger station, water system, and several campgrounds (one where the Cathedral Rock Picnic Area is, another at what is now the Mahogany Grove Group picnic area, and a third now occupied by the Foxtail snow play area). The CCC was also involved with agricultural endeavors. Crews began building the Arrow Canyon Dam and reservoir in 1935 on the Muddy River, but the project was abandoned when the foundation failed (map 10.5). The CCC went on to build small dams along the Muddy River and a diversion dam on the Moapa reservation in 1937.

The completion of the dam brought Roosevelt to Southern Nevada for the dedication. On September 30, 1935, his train arrived at the Boulder City station and he was driven to the dam and across it into Arizona, where the ceremony took place. That afternoon he returned to his train, which traveled the short distance to Las Vegas. He toured the War Memorial Building to see the handiwork of the WPA, drove down Fremont Street, and then took a drive up to visit the CCC enrollees at Camp Charleston Mountain. But he never made it. The president's driver got lost and attempted to turn around on a narrow road along a steep hillside, and barely avoided tumbling down the slope with the president in the back seat. Roosevelt arrived safely back at the Las Vegas train station two hours late and left the city that evening.

World War II's outbreak ended both the Great Depression and the New Deal. Although distant from Southern Nevada, the war would have enormous consequences for Las Vegas, even more than the momentous 1930s.

Even before the attack on Pearl Harbor spurred the United States into the war, preparations for war were underway in Southern Nevada. The army was given permission earlier in 1941 to operate an aerial gunnery school at McCarran Field, to be known as Las Vegas Army Air Field. Since the United States was not yet in the war, the army facility would operate alongside the regular airport operations. To accommodate flight training needs, a vast area was withdrawn from public use northwest of Las Vegas; the southern two-thirds was called the Las Vegas Range, and the northern end the Tonopah Range. Indian Springs Army Airfield was established in Indian Springs to support this, and several auxiliary airfields were set up throughout the range. Tonopah Army Air Field was located farther north and used the northern part of the range, with a number of auxiliary airfields established for it as well.

Once the war began, operations at Las Vegas Army Air Field accelerated. There were 9,117 people stationed there in early 1942, with thousands of trainees rotating through every six weeks. Housing construction was an early priority for the base, but never kept up with needs; by 1945, employment had swollen to 12,955 and housing was still scarce. In addition to the gunners being

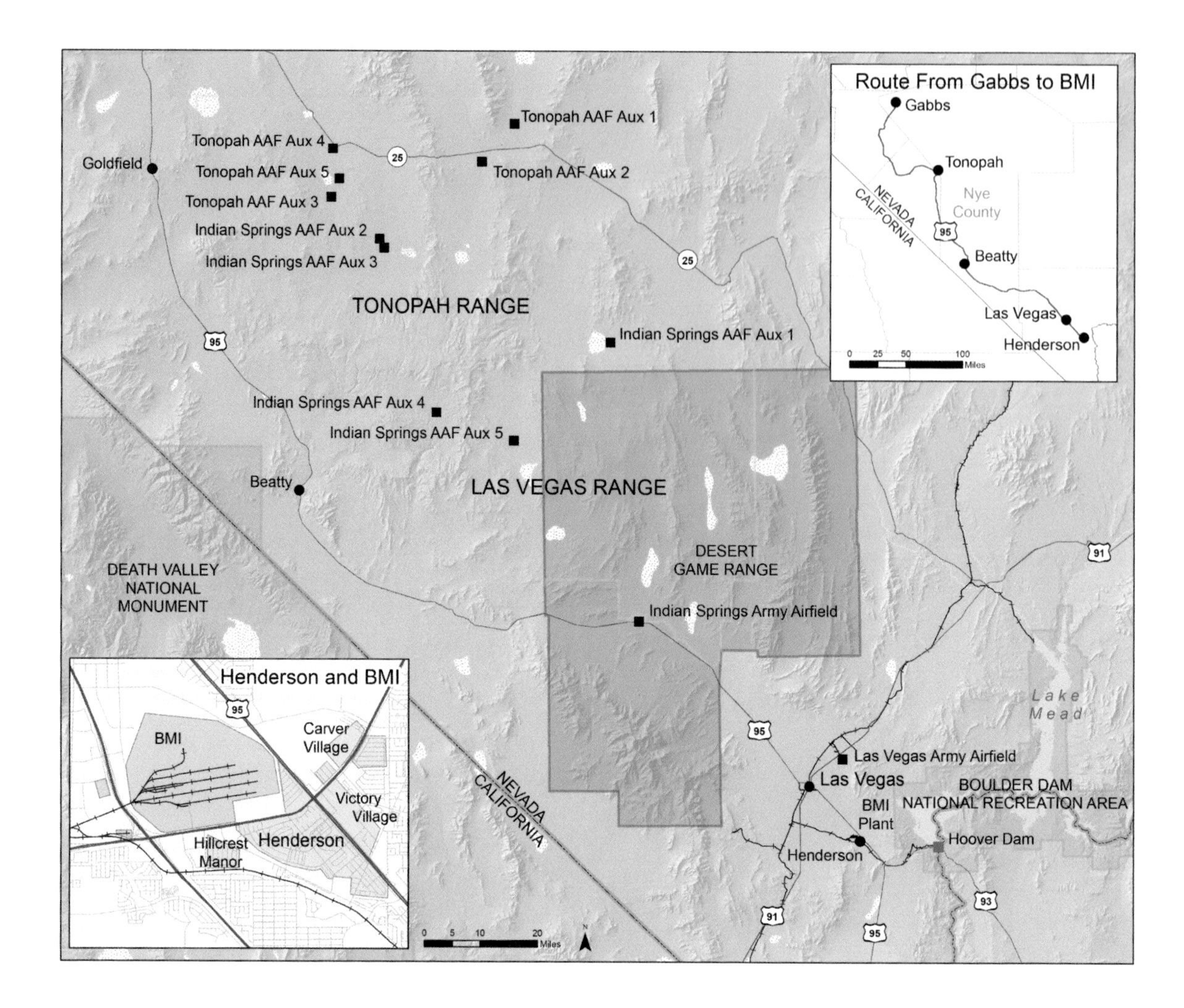

Map 4.8. World War II. A vast military training base with many new airfields was created north of Las Vegas. *Insets*: Ore was shipped south from Gabbs to the new town of Henderson, where the BMI industrial facility was constructed to refine magnesium for industrial use.

trained at the airfield, more pilots were needed, and Alamo Airport opened south of town in 1942 as a privately operated pilot training operation under contract to the military.

The same year, the Las Vegas Valley was selected over Needles, California, to be the site for a massive magnesium-processing plant. This light but strong metal was in high demand for use in aluminum alloys for aircraft parts and munitions. The plant was built along the Union Pacific's Boulder City branch, near Boulder Highway. The location was considered ideal because obtaining pure magnesium from ore is based on electrolysis of magnesium compounds in water; this requires large amounts of both water and electricity; these were supplied by Lake Mead and Hoover Dam, respectively. The Union Pacific Railroad would be used to ship the finished products to war plants in California and elsewhere, and Las Vegas was far enough inland to be considered safe from Japanese attack. A new company, Basic Magnesium Incorporated (BMI), was created to undertake the construction and operation of the plant for the government.

Work on the plant started on September 15, 1941, and was completed in July 1943. Twice as many construction workers were employed as for the Hoover Dam project, and it employed fourteen thousand workers once it was operating. A pipeline and water intake on Saddle Island were built to bring water from Lake Mead to the plant, and powerlines from Hoover Dam were built and electricity diverted from Hoover Dam's regular customers (map 8.17). The processing of magnesium at the BMI plant was joined by that of similar sounding but very different manganese ore, mined at the Three Kids Mine east of Henderson in 1943.

As with Hoover Dam, a new town was needed to house construction workers and plant employees. This was laid out south of the plant along Boulder Highway with Water Street planned to be the commercial district (though gambling was prohibited). The new town was originally called Basic after the plant before being renamed Henderson after former senator Charles Henderson. Many simple houses were quickly built for war workers, and one of the originals can be seen at the Clark County Museum. The huge plant required far more labor than was available locally, and recruiters sought out African Americans in Mississippi, Louisiana, and Arkansas, bringing a substantial new demographic presence to the area. Carver Village was constructed as a segregated housing area for these workers and their families across Boulder Highway, but many preferred to live in the Westside of Las Vegas where there was more of a social life (map 4.9). A hospital was also needed, and this opened in 1942 as the Basic Magnesium Hospital. It still operates as St. Rose Dominican Hospital (Rose de Lima campus). Schools were also needed. Railroad Pass was the closest school district and was quickly expanded to include Henderson and renamed after the town.

Where did the magnesium come from? A magnesium mine had been opened in central Nevada at the same time as the BMI plant was being planned, and the town of Gabbs was founded for its workers. By this time there were no longer any railroad connections between central Nevada and Las Vegas (map 7.2), so the magnesium was trucked down US 95 to Henderson. This heavy truck traffic put great demands on city streets in Las Vegas. To cope with this, a bypass was built around the west side of the city; trucks from Gabbs would continue south on Rancho Drive to a sweeping curve east on Charleston Boulevard and cross the railroad tracks south of town. Boulder Highway was also widened to four lanes from Charleston Boulevard to Henderson, the first four-lane highway in Nevada.

Hoover Dam was vital to the war effort as a source of electricity and irrigation water. It was also a well-known and easily found landmark. Throughout the war, there were persistent concerns that German saboteurs might strike at the dam to shut down electrical production or panic downstream populations. A heavy guard was placed on the dam at the beginning of the war, and three observation bunkers were built. One of these can still be seen on the Arizona side overlooking the spillway. An attempt at camouflaging the dam from the air was also carried out, using netting extended across the reservoir adjacent to the dam. From above, any attacking pilot would see what appeared to be the outline of the dam several hundred feet upstream, and hopefully aim at this fake dam instead. To house soldiers defending the dam, the army set up Camp Sibert in Boulder City, renamed Camp Williston in 1942. It was located where Boulder City High School is now. The camp closed in 1944 as the threat diminished.

World War II bunker overlooking Arizona side of Hoover Dam. Photo by author.

While thousands of workers descended on Las Vegas for war work, others left to join the fighting. A total of sixty-nine people from Clark County died in World War II, including William H. Nellis, a P-47 fighter pilot killed when he was shot down in Europe in December 1944. His name would later be added to the map of Las Vegas and remains known to fighter pilots worldwide (map 9.13). Las Vegas was honored in 1944 by the naming of the USS *Las Vegas Victory*, a navy cargo ship built in California. The freighter hauled ammunition from forward bases in the Pacific Ocean to resupply navy ships, and earned a battle star for combat service during the Okinawa campaign against Japan. The ship was decommissioned in 1946 and remains to this day the only navy ship named after Las Vegas.

Although Germans never attacked Hoover Dam and the BMI plant was considered safe from attack, the Las Vegas area was bombed by the Japanese. On July 20, 1945, the remains of a Japanese balloon bomb were found near Indian Springs. This was one of more than 9,300 similar weapons the Japanese had built and launched against the United States in the last year of the war. Each balloon was about thirty feet in diameter, made of silk or paper, and filled with hydrogen gas. The Japanese had discovered the existence of the jet stream and realized balloons released into it would cross the Pacific eastbound in only three days, quicker than the hydrogen gas could leak out. Each balloon carried an incendiary bomb that was released automatically after three days when the balloon should have reached the United States. The goal was to start forest fires in the Pacific Northwest to reduce lumber supplies and panic the population. About 350 bombs made it to the United States, Canada, and even Mexico, and several small forest fires were started, but lumber production was never in danger and the government kept the balloons secret to prevent panic. No damage or panic seems to have occurred from Southern Nevada's Japanese attack.

The end of the war prompted celebration but also worry about the town's future. A mass exodus of war workers was inevitable when the Army shut down its operations at the airfield, while the BMI plant had produced so much magnesium that it had shut down before the end of the war. The plant, and the town of Henderson, appeared to have no future. With the massive boom of the 1930s and World War II over, Las Vegas looked as if it would again become a sleepy desert town.

The bright lights of Las Vegas have attracted many through the years, and one of the first groups to come here looking for a better life were African Americans from small towns and farms in the Deep South. This migration was instigated by the decision of the Japanese government to attack the United States, launching the country into World War II.

The first African Americans arrived in Las Vegas as railroad workers in the town's early days. These workers primarily lived on the north side of Las Vegas, and later some Blacks owned businesses there. Only sixteen African Americans lived in Las Vegas in 1910, mainly near Block 16 (map 4.4). Their numbers remained small despite the huge amount of workers needed for Hoover Dam's construction project, which was considered a whites-only job. The dam was built by an industrial consortium called Six Companies Inc., which won a contract from the US government to build the dam. This included a clause barring the hiring of "Mongolians," meaning anyone of Asian descent, and noncitizens. Although allowed to hire Blacks, the company chose only white workers; after complaints were voiced, forty-four Blacks were hired out of the twenty thousand workers employed during the life of the project. They were put to work in the Arizona gravel pits, an unpleasant and demanding place to work with few prospects for upward mobility. The new worker town of Boulder City was whites only, so these few Black workers had to commute all the way from Las Vegas each day.

World War II changed everything. As world events drew the United States closer to war, President Roosevelt issued Proclamation 8802 on June 25, 1941, banning racial discrimination in war-related industries. Its effects first became evident a few months later when the BMI project started up in the new town of Henderson (map 4.8). This needed enormous numbers of workers, and labor recruiters scoured the country to find them. Blacks in the South were eager to seek out a better life elsewhere. Two towns in particular, Tallulah, Louisiana, and Fordyce, Arkansas, sent many young workers to help create Henderson and win the war.

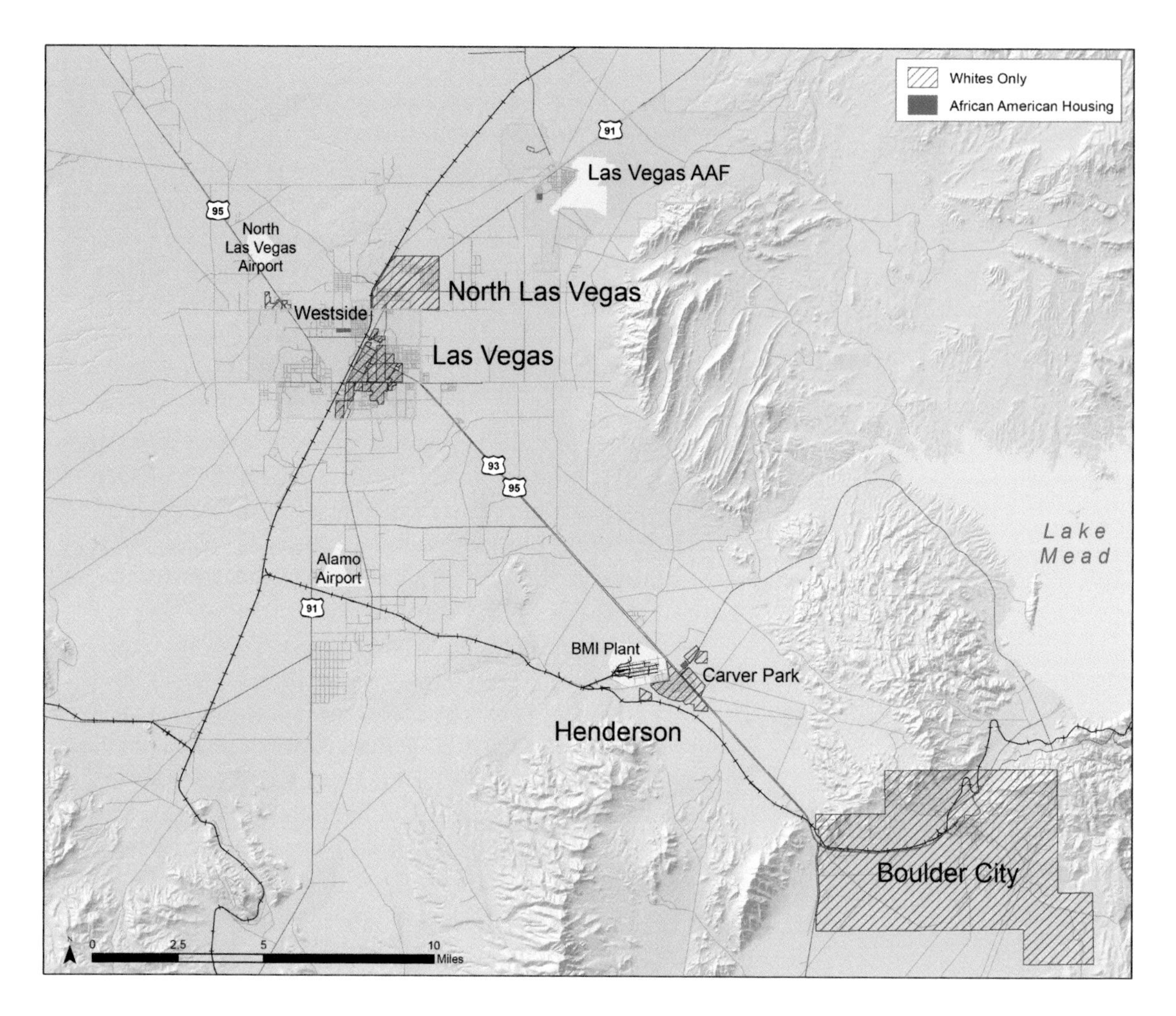

Map 4.9. Segregated Las Vegas. In the 1940s, most of Las Vegas and its suburbs were strictly segregated, with only a few neighborhoods welcoming African-American defense workers.

But while no discrimination was allowed in hiring, Roosevelt's proclamation said nothing about housing. Henderson was strictly segregated, with Blacks allowed only in the Carver Park community built on the east side, across busy Boulder Highway. This development had 64 units for single men, 104 one-bedroom units, 104 two-bedroom units, 52 three-bedroom units, a school and recreation hall when it opened in October 1943. What it didn't offer was much of a social life or sense of community.

While Black workers were arriving in Henderson, racial attitudes in Las Vegas changed rapidly for the worse. Whites pushed the small numbers of Blacks living in Las Vegas into the Westside area, essentially making it the city's Black neighborhood. Black business owners were also told they had to relocate to the Westside to receive a new business license. The wartime boom led to several new housing subdivisions to be built around Las Vegas; all excluded Blacks. As many as three thousand Blacks may have lived in the Westside in the 1940s, enduring severe housing shortages and relatively primitive living conditions.

Although housing, roads, and other services in Westside were far below the standards of the rest of the city, many residents of Carver Park moved there because it offered more social life than segregated Henderson.

Wearing a uniform did not bring better treatment. Nellis Air Force Base was segregated, as was the entire military at the time. Two "colored" Army units were stationed there and had their own barracks, dining hall, and pool. The military was officially desegregated by presidential proclamation in 1948, but the process moved slowly; in 1949, the first winners of the Air Force Gunnery Meet at Nellis was the 332nd Fighter Group, better known as the Tuskegee Airmen, still an all-Black squadron.

The substandard living conditions in the Westside remained after the war was over; racial attitudes did not improve. But improvements finally arrived in the 1950s with paved streets, a public swimming pool (allowing Blacks to be completely excluded from all other pools in Las Vegas), the Marble Manor housing development, and even the Berkley Square subdivision. Schools were never segregated in the fashion of housing; the Westside had three elementary schools serving local residents while junior and high school students attended outside the neighborhood. In 1972, bussing began to send elementary school students out of the area and bring in sixth graders from other schools to achieve racial balance. This lasted until 1993, the last vestige of segregated Las Vegas.

Las Vegas was a small city in 1950 with only 24,624 people, most of them living in an area not much larger than present-day downtown and the Westside. The shutdown of the BMI plant in Henderson (map 4.8) had almost killed that town, reducing it to 3,643 people. Boulder City was only slightly larger with 3,903 people, a small fraction of its peak during the 1930s. But there were signs of growth; two new small communities even appeared on the map after the war. Blue Diamond was founded in the 1940s at Cottonwood Springs to house workers for the nearby mine (map 8.19), and North Las Vegas was founded in 1946, largely to eliminate the possibility of the area being annexed into Las Vegas. But the prospects for growth of these towns were not promising.

Fortunately, several big developments occurred not long after the end of the war. The air force had been split from the Army in 1947 and decided to reopen the military airfield in Las Vegas the following year, but this time the military wanted total control of it. Local leaders were happy to hand over the city's airport to anyone promising jobs. To replace it, the city began looking for a new one; the solution it found was to purchase Alamo Airport south of town and build the facility into a modern airport. On December 19, 1948, the new airport was dedicated as the second

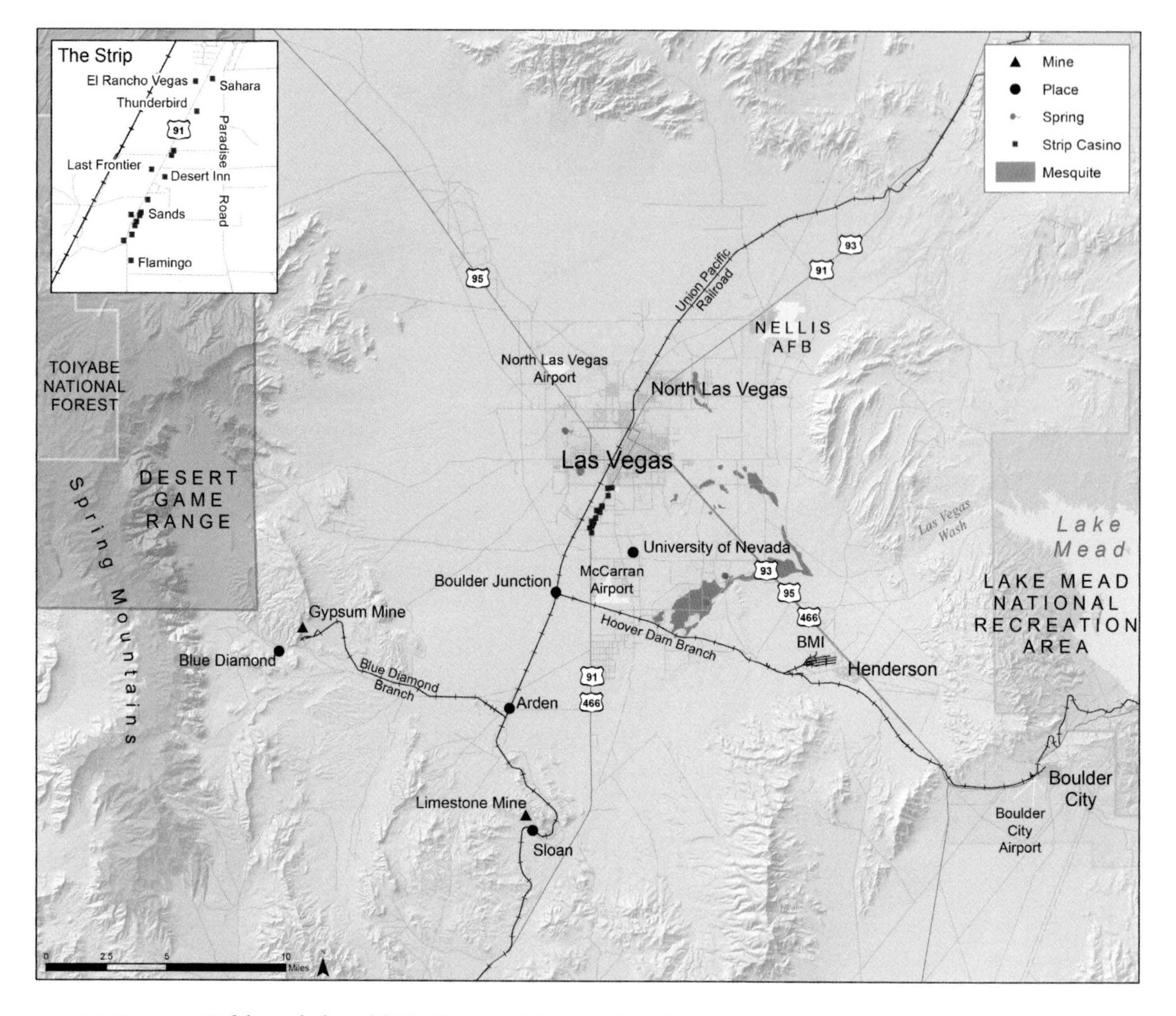

Map 4.10. Postwar Las Vegas. The Strip had begun to develop by 1952, but the town was still very small. Mesquite groves were still common and the Las Vegas springs still flowed.

McCarran Field, and the old McCarran Airport then became Nellis Air Force Base.

A second development was the sale of the dormant BMI plant in Henderson to the Colorado River Commission in 1948. The commission, responsible for overseeing water usage from the Colorado River (map 5.6), sought new tenants for the plant and was gradually able to build up a varied industrial base for the town. A chemical plant began producing chlorine; another plant processed a new metal called titanium for high-performance airplanes; while another began producing ammonium perchlorate, a solid rocket propellant, for the US Navy. There would be unfortunate consequences for the city because of these plants in later decades, but for now the population was happy to save their town and attract new workers.

A third important development was the opening of casinos along Highway 91. This had started before the war when El Rancho Vegas opened in 1941; it flourished after the war before burning to the ground in 1960. It was never rebuilt, but others

El Rancho Vegas Hotel, 1930s-40s. Tichnor Brothers Postcard.

followed. The Last Frontier opened in 1942, followed by the Flamingo in 1946. This last was considered a significant step in marketing the city as more than just a dusty desert town, and was the first to involve theming that went beyond Old West styles. New casinos followed quickly in the 1950s. The Desert Inn opened in 1950, the Sahara and Sands opened their doors in 1952, followed by the Dunes and Riviera in 1955.

Sometime during these years, Guy McAfee, a casino operator from Los Angeles, nicknamed this section of Highway 91 "The Strip" after the Sunset Strip back home. He meant it as a joke, because the dusty rural highway was nothing like his beloved urban corridor, but in time Las Vegas's Strip outshone the Sunset Strip of Los Angeles.

A growing Las Vegas required improved roads, and the entire Boulder Highway had been made into a four-lane highway by 1950. A railroad underpass was built on Charleston Boulevard by 1950 to improve traffic flow on the Las Vegas bypass (map 7.6). It was provided with good drainage so no pumps were considered necessary in case of rain, as was the case with the old Clark Avenue underpass. However, the new Charleston underpass filled with water after a thunderstorm in June 1955—the first of many times this would happen (map 6.8). The town also began installing stoplights to handle traffic flow, but there were few outside downtown.

The growing town sought other amenities, including higher education opportunities. One consequence was the opening of the University of Nevada, Southern Branch, in 1957 at a dusty site along Maryland Parkway. Maude Frazier Hall was the first building on this new college campus.

After the end of the war, Blacks who had labored for the war effort in the BMI plant, as well as returning veterans, found that segregation in Las Vegas had not changed (map 4.9). Many

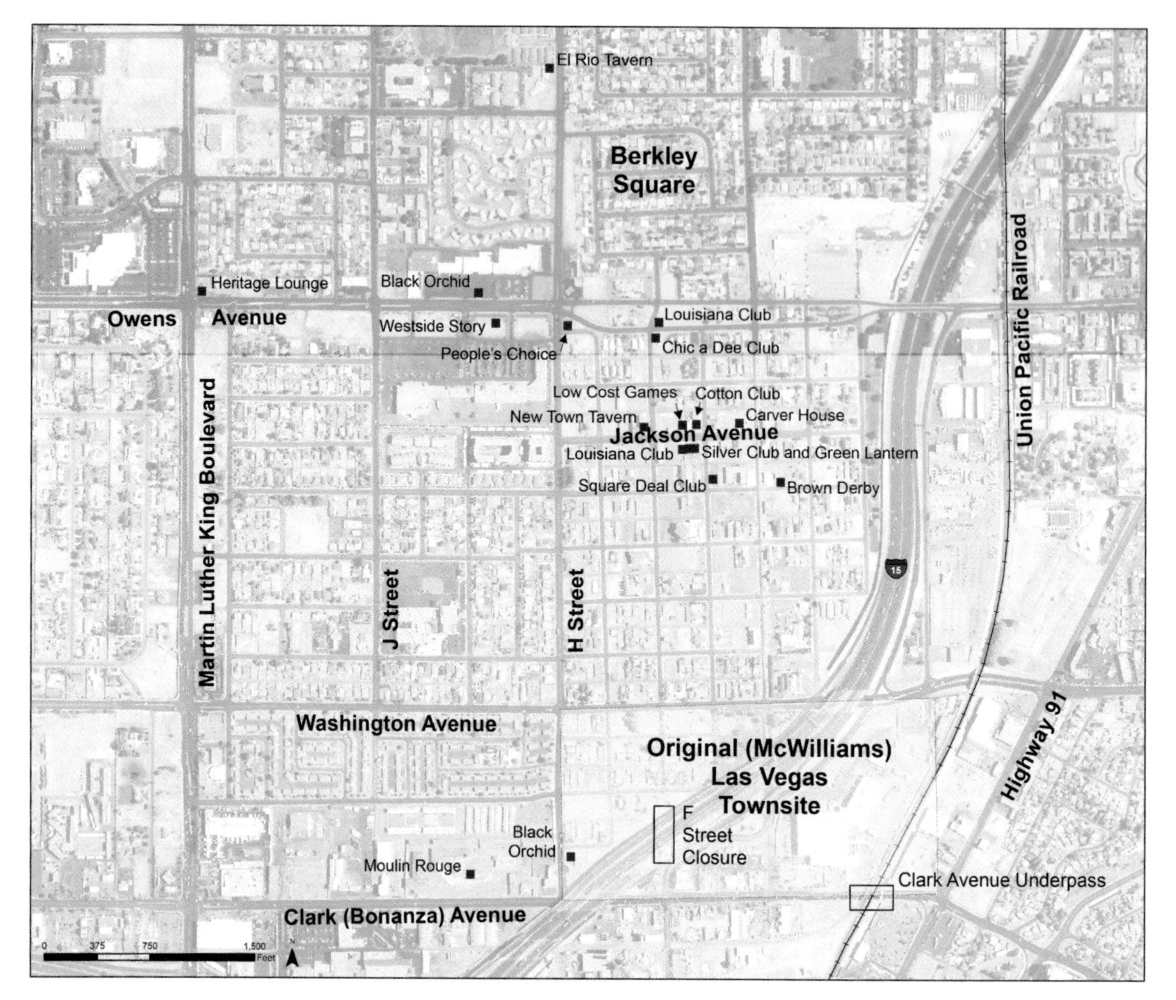

Map 4.11. The Mississippi of the West. The Westside was once a thriving African-American neighborhood with its own casino strip along Jackson Avenue. Berkley Square was one of the first suburban neighborhoods built for African-Americans. The red squares denote former casinos in this area. A modern aerial photo provides the backdrop.

whites had expected that their presence was a temporary necessity until the war emergency was over; the lack of interest shown by the city in improving the Westside may have been a result of this belief. Blacks were also told they were not welcome in casinos, except as menial workers. Black performers had appeared in Las Vegas casinos since the Deep River Boys played El Cortez in 1944, but even Black entertainers such as Sammy Davis Jr. (map 11.1) were not allowed to gamble, eat, or sleep in the Strip casinos in which they performed. Only at Foxy's Deli, across from the Sahara, were Blacks allowed on the Strip as customers, but only at a few tables in the back. Movie theaters were also segregated, and several major taxi companies wouldn't pick up Black riders. These policies earned the town the nickname of the "Mississippi of the West."

Jackson Avenue emerged as the main business street in the Westside in the 1950s and 1960s with a number of small gaming halls, a few hotels, and rooming houses. Among these was the Harrison's Guest House, where Davis and other entertainers stayed while performing in Las Vegas. The first racially integrated hotel-casino in Las Vegas was the Moulin Rouge, which opened in the Westside in 1955. This attracted tremendous attention from the national media for its biracial crowd but only remained in business six months; its fame did not earn the owners enough money to pay the bills. It reopened on several occasions in later years but never achieved the prominence of its brief 1955 incarnation.

Frustration over the racial discrimination practiced in Strip casinos led to plans for a protest march down the Strip in 1960. Representatives of casino management and local politicians were finally moved to action to prevent this, and agreed in a meeting at the closed Moulin Rouge hotel to allow Blacks into Strip casinos. Not all owners went along with this: the Sal Sagev (Las Vegas spelled backward, and now the Golden Gate) and Binion's Horseshoe Club downtown still refused to serve Blacks (though, ironically, Benny Binion lived in the Westside). But the Strip had finally been desegregated. Unfortunately, as was the case in many southern cities, this desegregation was fatal to Jackson Avenue and Black-owned businesses in the Westside, and the community lost much of its vitality. When Martin Luther King Jr. made his only visit to Las Vegas in April 1964, he stayed not in the Black Westside but at the integrated Sands on the Strip.

While the Moulin Rouge meeting opened up most Las Vegas casinos to Black gamblers, it did little to increase hiring of more Black workers in casinos, especially for higher-level positions. Complaints about this to the US Justice Department led to a consent decree in 1971, in which eighteen casinos and four labor unions agreed to change their hiring and promotion practices. The government would monitor them for progress until sufficient gains had been made. A number of casinos reached that point later in the decade, while newly built ones were not subject to the decree. By 1986, only the Tropicana and Riviera were still subject to its terms.

The Westside is one of the few places in Las Vegas to be impacted by freeway construction; much of the original McWilliams Townsite and the southeast corner of the Westside were destroyed by the construction of Interstate 15 in the 1960s. F Street was temporarily closed in September 2008 for the reconstruction of the highway. The city asked the state department of transportation to make the closure permanent, and the highway bridges over F Street were demolished. Westsiders strongly opposed this closure, and at a cost of $13.6 million, Interstate 15 was torn up again to allow new overpasses to be built, and F Street reopened in 2014.

When World War II ended, the airbase and BMI plant were shuttered, and wartime Las Vegas quickly lost its war footing (map 4.8). But rising international tensions brought the need

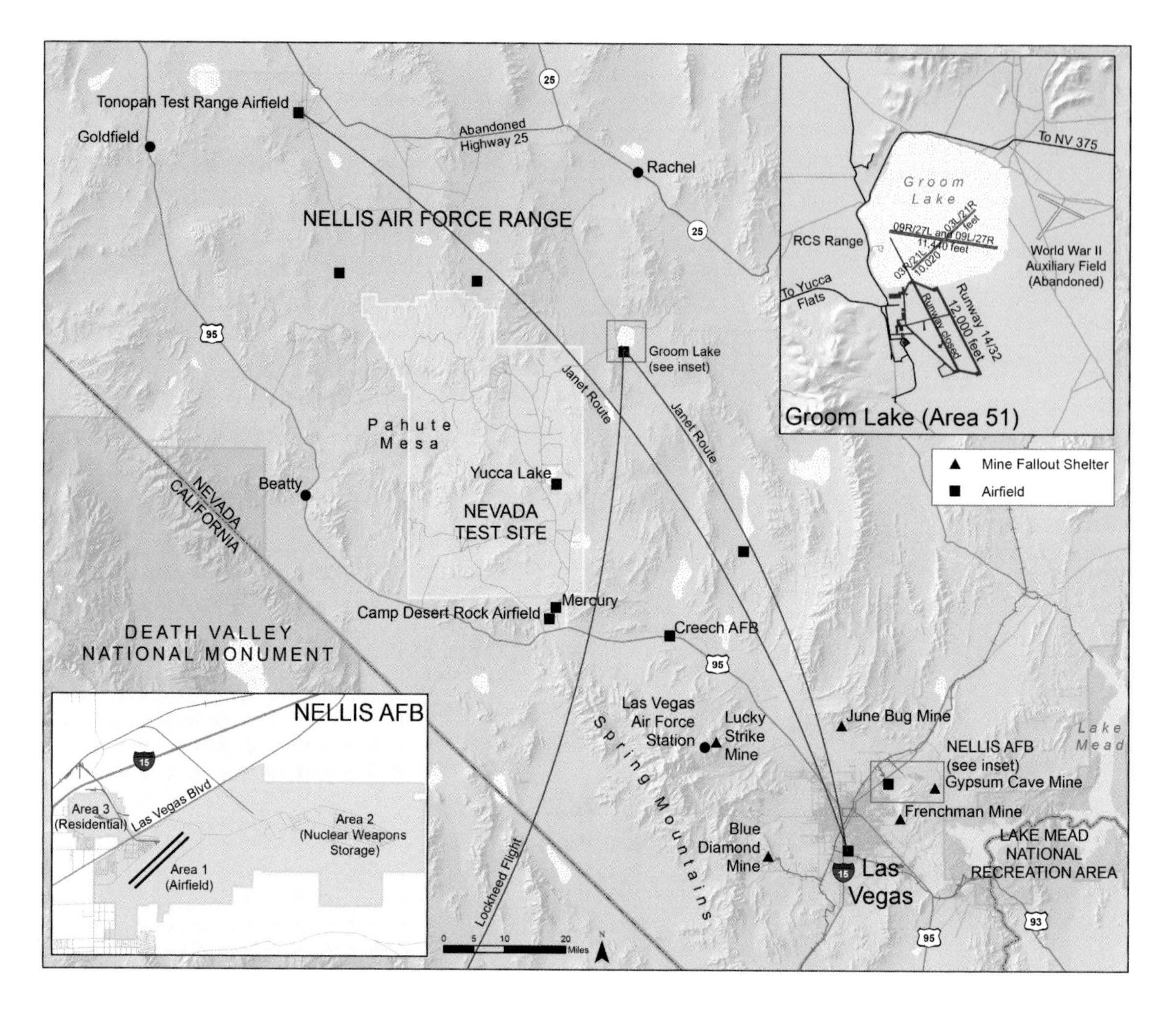

to reactivate the region's military facilities, and even build some new ones. The newly developing Cold War, lasting from about 1947 to 1991, made a lasting impact on Las Vegas.

This began in 1948 when the Las Vegas Army Air Field reopened as Nellis Air Force Base. The base inherited the vast Las Vegas and Tonopah Ranges, which were made permanent and expanded several times, permanently closing off a huge part of central Nevada to the public. This required the construction of a new Nevada State Route 25 to replace the old road through the north end of the range. Indian Springs Army Air Field was reopened and became Indian Springs Air Force Base (but Tonopah Army Air Field farther north remained closed).

Nellis specialized in training fighter pilots; the outbreak of the Korean War in 1950 made this vital. These pilots were taught well; those who learned to fight in the F-86 Sabre at Nellis shot down more than five hundred enemy jet fighters in that war. Nellis became "the home of the fighter pilot," and in the 1950s was the busiest airbase in the world. After the Korean War, the

Map 4.12. The Cold War. The airbase created for World War II remained in use, with the addition of the Nevada Test Site for atomic testing. *Lower inset*: subdivisions in Nellis Air Force Base; *upper inset*: the new top-secret aircraft testing base at Groom Lake, better known as Area 51. Google Earth was used for the details of Groom Lake and paved roads in the Nellis Range and Nevada Test Site.

base transitioned to the F-100 Super Sabre jet; the Thunderbirds aerial demonstration team had been flying that plane at an Arizona base but were transferred to Nellis in 1956. They have had their home here ever since.

Several new facilities were added to this vast airbase and range complex. The US Navy opened the Lake Mead Base as a nuclear weapons storage site in 1953, and it became Area 2 of Nellis in 1969. It remains in use as one of three air force nuclear weapons depots, with more than one thousand nuclear bombs and missiles stored in bunkers. At night, Area 2 is plainly visible east of the city because of its grid of orange streetlights. These lights are utterly unremarkable compared to those of the Strip, but once you realize what they represent all the golden dreams the Strip become insignificant.

Other new facilities opened north of town. The Las Vegas Air Force Station opened on 8,799-foot-high Angel Peak in the Spring Mountains in 1952, serving as part of the country's air defense radar system. The facility closed in 1969, but the Federal Aviation Administration (FAA) still has a radar on Angel Peak. The former base is now the Spring Mountain Youth Camp, a juvenile correctional facility. An area larger than Rhode Island within the Nellis Range was selected in 1951 as a testing ground for atomic weapons (map 4.13). This was the Nevada Test Site, now called the Nevada National Security Site (map 9.14). Almost one thousand nuclear bombs were detonated there, along with an amazing variety of other nuclear testing.

Two important new airbases were opened within the vast Nellis Range. An airfield was built south of Groom Lake for secret flight tests of the U-2 spy plane in 1955. This base, off limits even to the air force pilots flying at Nellis, was called Paradise Ranch (or just the Ranch), Watertown, and later Area 51. It was eventually used for the SR-71 spy plane and for the development of stealth fighters. Employees from Lockheed, the builder of both planes, commuted to work at Area 51 weekly by plane from Burbank, California. (One such flight went off course and crashed into Charleston Peak, discussed in map 7.20.) Other workers came to work on the "Janet" airline from Las Vegas (map 7.17). Area 51 attracted attention in the 1990s for a supposed high-performance spy plane named *Aurora* thought to be undergoing tests there and for allegedly housing alien spacecraft. Whatever is housed there, the base remains active and has been continually expanded.

The Tonopah Test Range airfield opened in 1957 before being abandoned and then reactivated in the 1970s as a base for carrying out secret flight tests using captured Russian fighter planes. It later became the base for F-117 stealth fighters before they

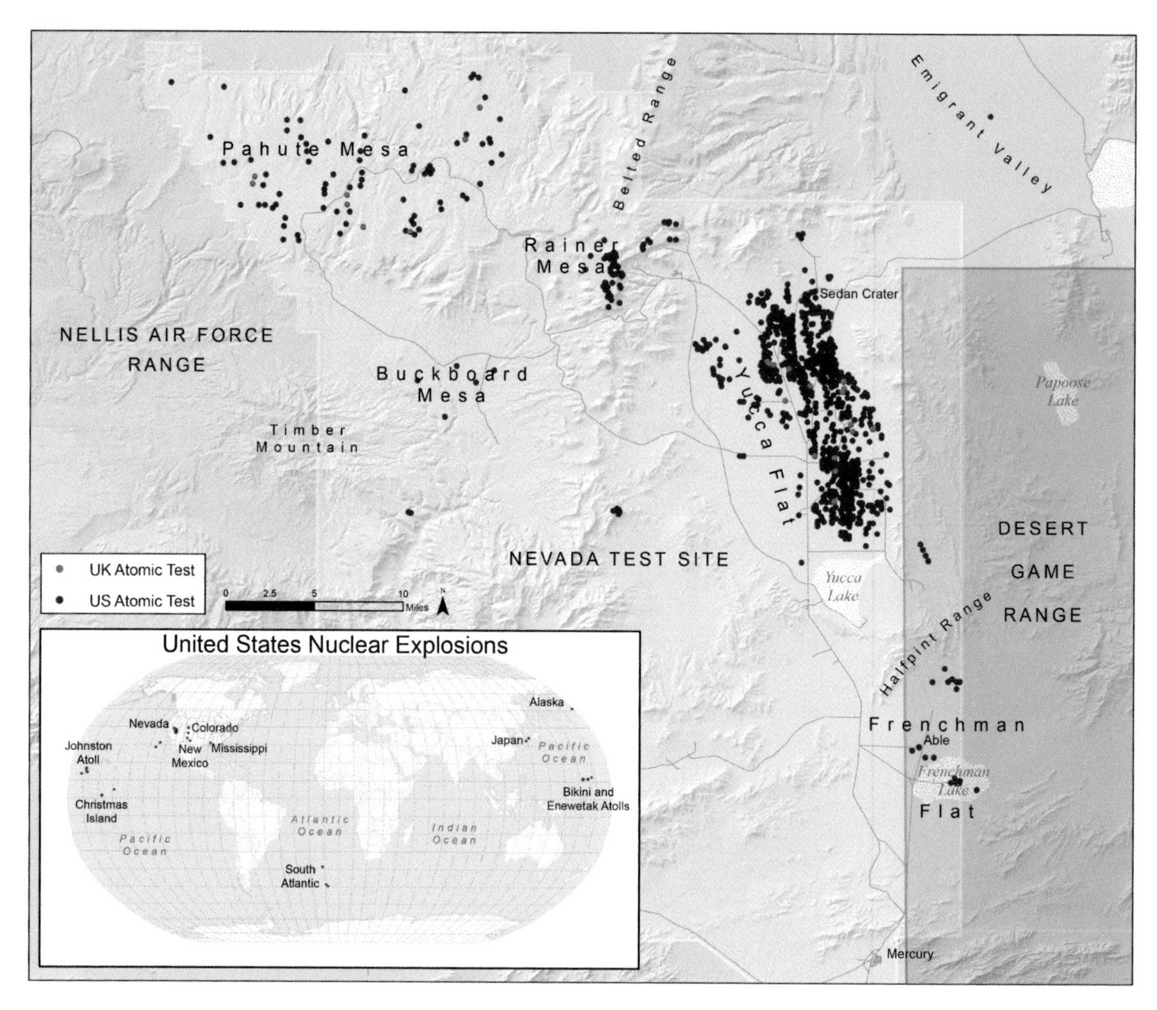

Map 4.13. Atomic Las Vegas. Every atomic test at the Nevada Test Site is depicted. They started at Frenchman Flat before moving north to Yucca Flat and the mountains beyond. *Inset*: every nuclear test the United States has conducted worldwide, along with the two bombings of Japanese cities during World War II.

deployed operationally. When the entire fleet of stealth fighters was taken out of service in 2008, they were brought back here for storage. The base currently operates several drone aircraft.

What if the Cold War had turned hot and the radars on Angel Peak warned of incoming Soviet bombers? The military facilities and nuclear weapon storage would have made Las Vegas a likely target for the Soviet Union. Several abandoned underground mines in the Las Vegas area were selected to serve as national fallout shelters. Las Vegans would presumably be sent to hide inside them before the bombers arrived, though little planning seems to have ever been carried out for this. By the 1960s, the Soviet Union had developed missiles capable of hitting Las Vegas (and any other place) within a few minutes, making these shelters useless. Fortunately for Las Vegas, the Soviet bombs never fell, though many others did.

In the early-morning darkness of January 27, 1951, a B-50 bomber took off from the Albuquerque, New Mexico, airport with a one-kiloton Mark 4 atomic bomb on board. The bomber slowly made its way west across New Mexico and Arizona before turning to the northwest. The crew headed for the lights of Las Vegas and armed the bomb. It was dropped a few minutes later over the dry lake at Frenchman Flat, northwest of the town, and exploded 1,060 feet above the ground at 7:44 a.m. Las Vegas had entered the atomic age.

This bombing mission started in New Mexico because that was where the atomic bomb was developed and the first one tested in July 1945. After the end of World War II, continued atomic testing took place in several south Pacific atolls, but the cost and difficulty of those operations made a domestic testing area necessary. A site within the continental United States was needed, but where was a good place to set off nuclear bombs?

Potential locations on North Carolina's Outer Banks, Texas barrier islands, the White Sands bombing range in New Mexico (site of the first atomic bomb test), Utah's Dugway Proving Ground, and the Las Vegas Gunnery Range were all examined. The western sites had the advantage of already being on government land (map 8.1) and remote from population, but also the disadvantage that radioactive fallout from the blasts would blow east across the United States. Government planners decided this was not a major problem and selected the Las Vegas Gunnery Range as an atomic testing area in 1951. This became the Nevada Test Site, now called the Nevada National Security Site (map 9.14). It was under the control of the Atomic Energy Commission and later the Department of Energy.

Atomic Cannon at Frenchman Flat. Courtesy of National Nuclear Security Administration.

The first atomic tests were conducted at Frenchman Flat, a small valley with a playa near the southern end of the test site. Other mushroom clouds followed quickly as the military tested new designs and tried to fit atomic warheads into every kind of weapon it had; bombs, rockets, anti-aircraft missiles, and even bazookas. One test in 1953 fired an atomic warhead from a giant cannon named Atomic Annie. Every other blast was triggered remotely from a safe distance; this one required a soldier standing next to the cannon to fire it. To his great relief, the warhead did not explode in the barrel but over Frenchman Lake several miles away. This was the first and only use of a nuclear cannon.

After the first fourteen atomic explosions at Frenchman Flat, tests were moved farther north to Yucca Flat, a much larger valley. A total of 739 nuclear tests were carried out here, the last on September 23, 1992, the last US nuclear test. A total of 928 tests took place at the Nevada Test Site. Not all of these blasts were American; the United Kingdom exploded twenty-four underground bombs there from 1962 to 1991.

Although watching mushroom clouds has become a part of Las Vegas lore, most nuclear tests took place underground to prevent nuclear fallout. The last of 219 US above-ground tests took place July 17, 1962. Later blasts took place in a shaft excavated in the valley bottom before lowering the bomb into it and sealing it shut. An underground nuclear explosion created a cavity that usually later collapsed, creating a crater on the surface; Yucca Flat is today littered with these craters. The biggest of these was from the Sedan test on July 6, 1962, created by a 104-kiloton bomb. This was not the largest ever set off at the site, but it was deliberately placed in a shallow borehole to create the largest possible crater, measuring 1,280 feet across and 320 feet deep. This was part of Project Plowshare, a program to investigate peaceful uses of nuclear bombs. Large craters were thought useful for building harbors or excavating canals. A harbor in northwest Alaska and a route through the Bristol Mountains for Interstate 40 in California were two of the planned applications, but the program ended in 1977 with the Sedan crater as one of the few relics of its existence.

Sedan Crater. Courtesy of National Nuclear Security Administration.

Another large group of explosions took place underneath Rainier Mesa. Many of these were at the ends of horizontal tunnels bored into solid rock from the side of the mesa. Others, much farther west on Pahute Mesa, were bored vertically.

Activities at Nevada Test Site required a large construction workforce to build roads, buildings, and excavate tunnels and shafts as well as technicians to assemble the bombs and run the blasts. The solution, by now familiar to Southern Nevadans, was to create a new town to house them. This was named Mercury, taking its name from an abandoned mine in the area. It was founded in 1950 and served as the base of operations for nuclear tests, with housing, food services, and administrative facilities for the workers. By the 1960s, as many as ten thousand workers and their families lived there, with a school, movie theater, pool, and other recreational amenities provided. President John F. Kennedy even visited the small town on December 8, 1962.

After nuclear testing ended in 1992, the workers moved away, and only limited facilities now exist for the handful of remaining staff. The Nevada Test Site became the Nevada National Security Site and is still used for a variety of training and research programs (map 9.14). The National Atomic Testing Museum in Las Vegas provides a glimpse of the Nevada Test Site during its atomic heyday (map 11.9).

Although US nuclear testing has ended, the nuclear era is far from over. The country still has 5,500 nuclear weapons (many of them stored in Las Vegas, not far from the Las Vegas Motor Speedway), and the need to make sure the complex weapons still work has led to calls to begin testing again. Other countries' recent tests have all been underground; thanks to monitoring

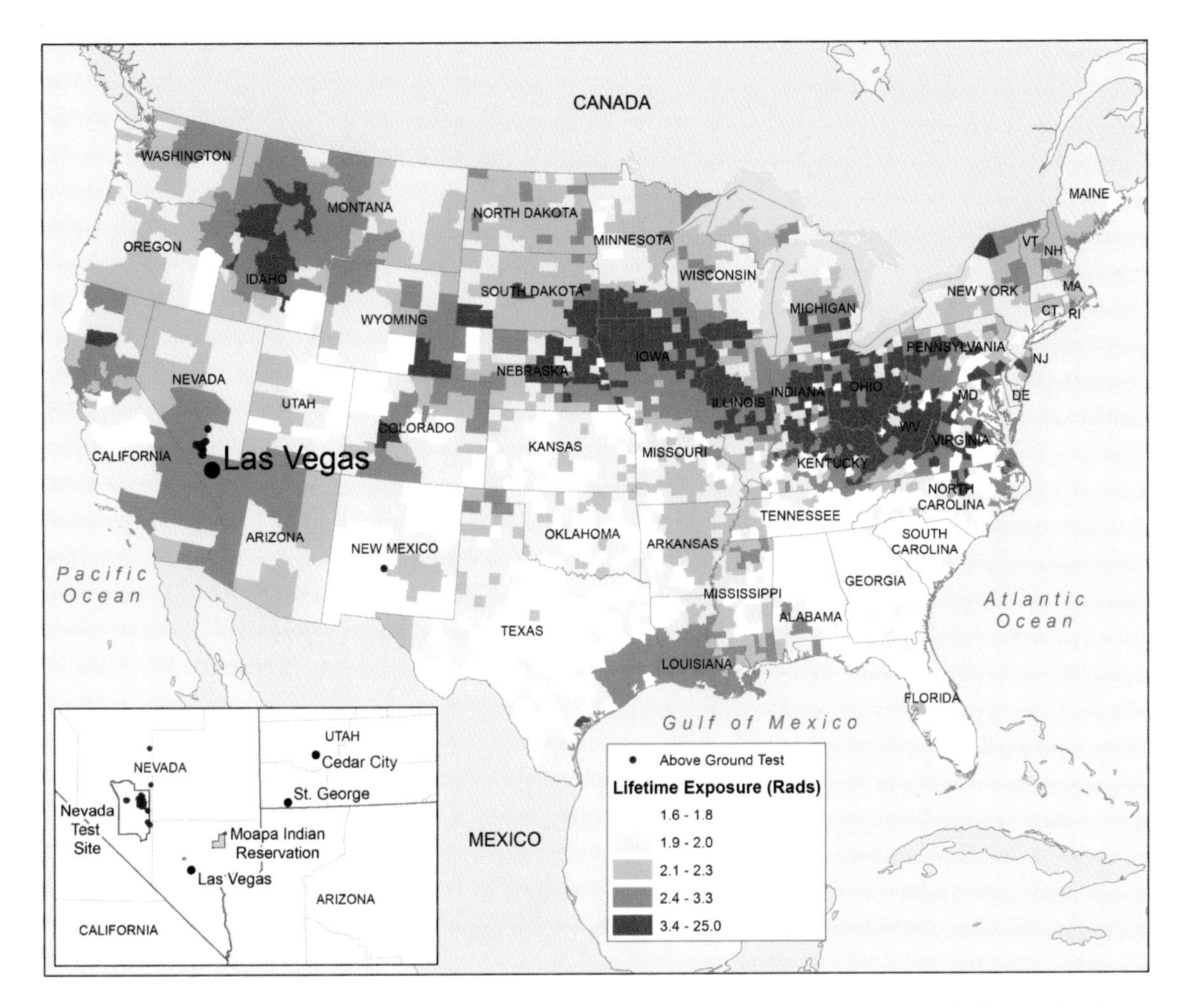

Map 4.14. Downwinders. Above-ground atomic testing at the Nevada Test Site dispersed radiation throughout the country. Darker shades indicate greater exposure to radioactive fallout. *Inset*: Areas immediately downwind of the Nevada Test Site.

methods pioneered in Nevada, they were instantly detected and the size of the blast calculated.

If you were born before July 17, 1962, when the last above-ground test was conducted, you are a downwinder. No matter where you lived in the United States, you received a dose of radiation from fallout blown downwind from at least some of the 216 above-ground or atmospheric nuclear explosions carried out by the United States.

None of this was unexpected; the government had considered testing atomic bombs on North Carolina's Outer Banks to prevent Americans from being exposed to this fallout, but decided on Nevada since the land there was already under government control. An increased number of birth defects and mutations because of radiation exposure was a worry in the early years of the atomic age. But aside from stimulating a large number of bad horror movies in the 1950s (one of them, *The Amazing Colossal Man,* was even filmed in Las Vegas), this has not been a problem. Instead, the main risk from this fallout has been an increased risk of cancer. When fallout settles out of the air, it can coat the ground and buildings and irradiate people and animals, especially if it is inhaled. It may also be eaten if fallout settles on farm crops or enters the body in the form of milk that has been produced after cows ate contaminated hay. Once inside the body, the fallout can create cancers. There are many different radioactive particles, but one that has attracted the most attention is Iodine-131 (I-131) because it remains radioactive for up to eight days and if eaten or drank will become concentrated in a person's thyroid. This greatly increases the chance of thyroid cancer, an otherwise rare form of cancer.

Not all Americans were affected equally by this fallout. Map 4.4 shows the amount of radiation received by each county from all US above-ground nuclear tests. This radiation included radioactive soil and isotopes sent high into the atmosphere by the nuclear fireball and then spread downwind, slowly settling to the ground. The radioactivity of the fallout slowly dissipates, so the closer it is deposited to the explosion the more radioactive it is. Wind patterns and weather conditions are fickle, giving some areas far from Nevada, such as Idaho and Montana, a heavier cumulative dose than closer places. The Moapa Indian Reservation and St. George, Utah, areas were particularly strongly affected, but fallout even reached the East Coast.

On May 19, 1953, a 32-kiloton explosion produced the most radioactive fallout ever recorded from a test in Nevada, and this fell on 3,046 of the nation's 3,142 counties. The blast had been code-named Harry and was often known as "Dirty Harry" because of its tremendous fallout. Among the victims of this fallout were John Wayne and other cast and crew of *The Conqueror*, being filmed outside of St. George. Wayne and many of the cast and crew of the Genghis Khan epic later died of cancer, though no link to Dirty Harry was ever definitively established.

Radiation sometimes escaped from underground blasts if they had not been dug deep enough or the blast was more powerful than expected. Venting of radiation into the atmosphere was noted at 484 of the 709 underground tests in Nevada, and at least fourteen underground bombs created substantial atmospheric radiation. The Baneberry test in December 1970 was one of the worst; the leakage afterward led to a six-month suspension of nuclear tests while engineers tried to figure out what happened and how to prevent it from recurring.

The harm done to Americans by American bombs has been substantial; atomic fallout may have increased the number of thyroid cancer cases in the United States by 10 percent, and leukemia has also likely increased, though to a lesser extent. The Radiation Exposure Compensation Act of 1990 allows claims to be filed by former Nevada Test Site workers and downwinders. About $2.4 billion has so far been distributed to these people. But despite the hazards of fallout from Nevada tests, the greatest spread of radiation across the world came not from nuclear bombs but from the Chernobyl nuclear accident in the former Soviet Union (now Ukraine) in 1986. Radiation from this disaster circled the globe, creating a new generation of downwinders.

Even though the United States ended nuclear testing in 1991, the risk of fallout has not gone away. One area of the Nevada Test Site is known as Plutonium Valley after being deliberately contaminated with that substance during a 1956 test. It is still lethally radioactive to anyone venturing in without protective clothing. A planned massive nonnuclear explosion in 2007 was canceled after the public became aware that the blast would spread this radioactive dust throughout the region. The problem will not be resolved anytime soon; the plutonium coating the Nevada desert will remain radioactive for 250,000 years.

Outsiders have driven most of Las Vegas's growth. The Bureau of Reclamation, the US Army, the US Air Force, Department of Energy, and many other organizations have funneled money and jobs to the area. Another of these organizations was the Mafia. Although this Italian-American organized crime syndicate is most commonly associated with eastern cities and remains most active there, it could not resist the opportunities for easy money offered by Las Vegas. But it didn't just make money off the city, it helped it grow.

The Flamingo was the first mob casino, though started by legitimate businessmen. California gangster Bugsy Siegel was approached by the builders as a source of funding to finish the hotel after its construction stalled, and he bought a large stake in the hotel. It opened under mob control in 1946, but poor management limited profits; in disgust, the mob had Siegel murdered at his Los Angeles home in 1947. The Flamingo remained a mob-run casino until 1967 when it was bought by Kirk Kerkorian. Many other hotels were secretly financed or controlled by organized crime following the Flamingo, including the Tropicana, Desert Inn, Thunderbird, Sands, Stardust, Riviera, and Dunes. Money was skimmed off casino earnings and sent back east to several waiting Mafia families in addition to being invested in the growing city. The mob also helped build Sunrise Hospital, the Boulevard Mall, and the Las Vegas Country Club.

A new era for the Mafia in Las Vegas began in the late 1960s when Lefty Rosenthal arrived in town. Although not a member of the Mafia, he worked with them to skim money from several casinos. He got a job at the Stardust in 1971 and became the casino's food and beverage director, but he was running the casino

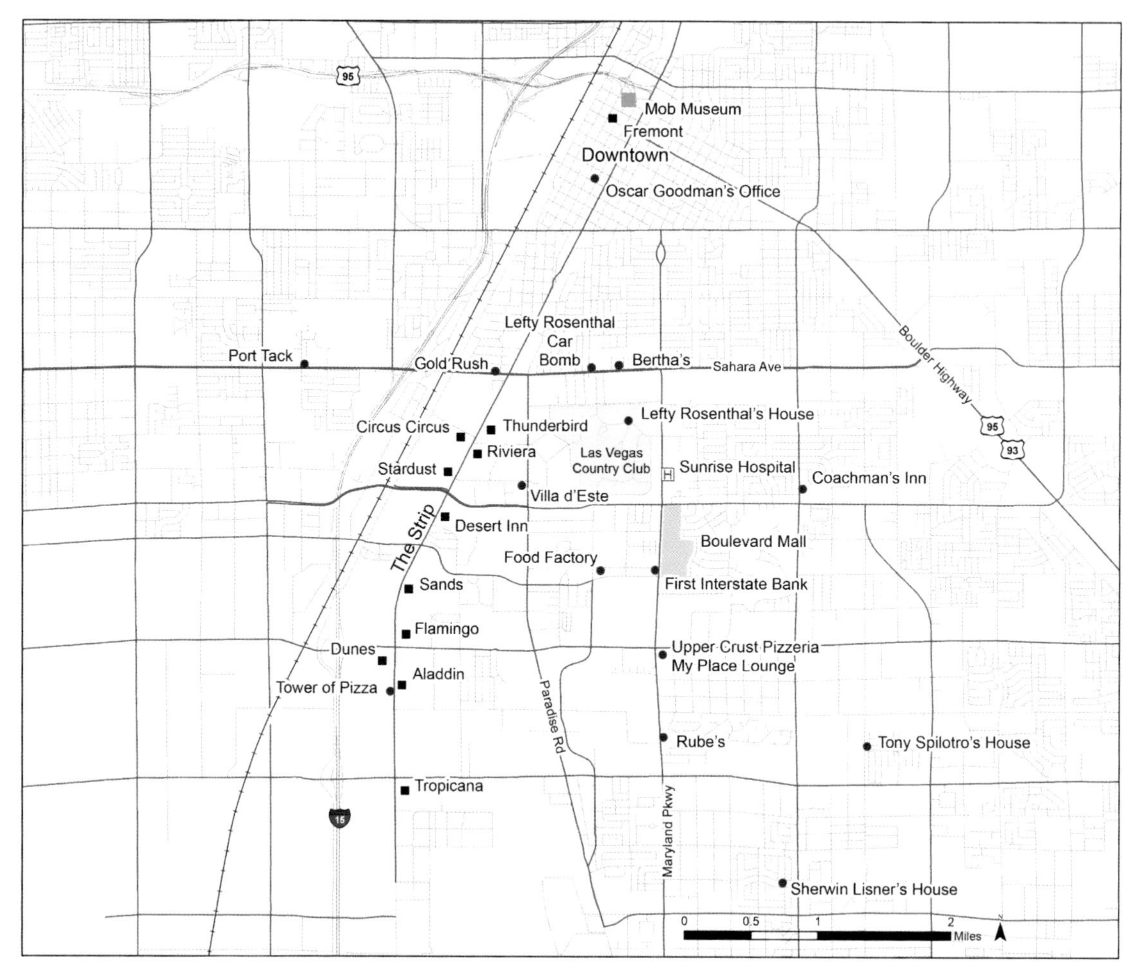

Map 4.15. The Mafia. The black squares are casinos that were financed in part by the Mafia, while the red dots are other places associated with the Mafia.

in place of the official owner, Allen Glick. Not long after Rosenthal settled into his new role, Tony Spilotro arrived from Chicago. Las Vegas had been an open city; any Mafia family could come in and set up operations without having to seek permission from any other family. There were also to be no killings in town, since these would draw attention to the mob's presence. Spilotro changed this when he arrived, requiring a "street tax" on any loan shark, pimp, dealer, or other criminal activity, and killing those who refused. The murder rate went up after he arrived, and he was suspected to be involved in several of the killings. One of these was the murder of Sherwin Jerry Lisner in his home at 2303 Rawhide Avenue on October 11, 1979. Spilotro gave the order for the hit, but Frank Cullotta, one of Spilotro's enforcers who followed him from Chicago, carried it out. Spilotro's group started burglarizing homes and businesses, often smashing through the wall to deal with alarms, earning them the name "Hole in the Wall Gang." The gang used their earnings to get involved in local businesses: Spilotro opened a gift shop in Circus Circus called Anthony Stuart Ltd., before moving on to owning the Gold Rush jewelry store on Sahara Avenue. Cullotta opened the Upper Crust pizzeria with his earnings from burglaries.

The authorities were well aware of Spilotro's and Rosenthal's backgrounds and paid close attention to them. Rosenthal was seen exercising considerable power despite his job title and was forced to apply for a gaming license; he was turned down in early 1976 and forced to leave the casino. This changed nothing, as he simply ran the skim from home. The heat intensified in May 1976, when a government audit revealed the biggest skim operation ever found in Las Vegas at the Stardust. Instead of laying low, Rosenthal launched a legal and public relations campaign to get a gaming license with the help of his lawyer, Oscar Goodman. In 1977, he was allowed to work in the casino again and returned to his role as food and beverage director. He also started a newspaper column and even his own TV show. *The Frank Rosenthal Show* ran on Channel 13 on Saturday nights for about two years. Most episodes were filmed at the Stardust and even featured celebrities such as Frank Sinatra.

Back in Chicago, the Mafia was becoming increasingly annoyed with both Spilotro's and Rosenthal's high profiles. The two friends also grew to dislike each other; Spilotro began to hate him for his arrogance and started sleeping with his wife. In turn, Rosenthal thought Spilotro was a joke; the Federal Bureau of Investigation (FBI) agreed with this assessment and viewed him as a messenger boy and leader of a barely competent band of burglars. The end began when Spilotro and his gang were caught robbing Bertha's Gifts and Home Furnishings. They had planned to dig through the ceiling to bypass alarms, with the job planned for Independence Day so fireworks would cover the sound of their work. But it was a trap: the authorities had been tipped off and were hiding nearby. Spilotro got out on bail and was called back to Chicago; he and his brother were beaten to death by the Mafia and buried in a shallow grave in Indiana.

Back in Las Vegas, Rosenthal's car was bombed on October 4, 1982, outside Tony Roma's restaurant, a favorite eatery of his, on East Sahara Avenue. He narrowly survived because of an extra steel plate under the driver's seat installed on his car; nobody was ever charged with the bombing. It had already been a bad year for him; his wife, Geri, left him and took jewelry worth $1 million from their safe deposit box at First Interstate Bank. She died of an overdose in early November in Los Angeles, by which time the jewelry had disappeared. Rosenthal left Las Vegas for Florida in 1983, a year that marked the end of the mob era in the city.

The Stardust and several other casinos associated with the Mafia are long gone. Tony's Gold Rush store later became a Kawasaki dealership and then demolished. Bertha's Gifts and

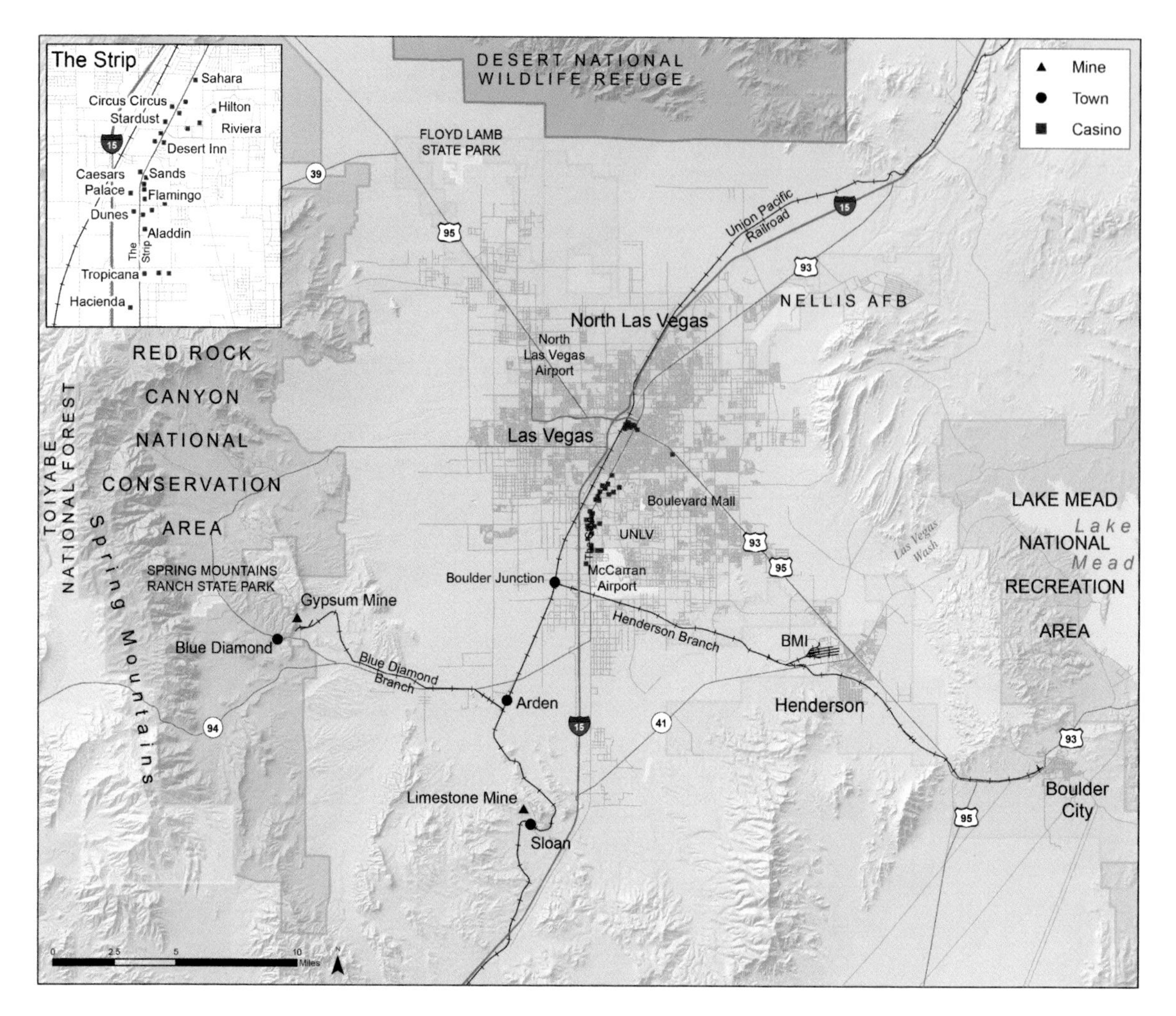

Home Furnishings later became Tower of Jewels but closed several years ago. Other Mafia sites were scattered around the city. Spilotro liked to eat at Coachman's Inn, Food Factory, Upper Crust, Port Tack, Rube's, My Place Lounge (where he ordered Cullotta to kill Lisner), and Frank Cullotta's Tower of Pizza. The Villa D'este restaurant was also popular with the Mafia as well as Sinatra and other celebrities.

The favored Mafia lawyer, Oscar Goodman, had (and still has) an office downtown at 520 South Fourth Street. Oscar later had another office in City Hall, serving as the mayor of Las Vegas from 1999 to 2011. His wife, Carolyn, succeeded him. During his time in office, he was instrumental in getting a museum created to commemorate the mob era in the city. This opened in 2012 as the National Museum of Organized Crime and Law Enforcement, better known simply as the Mob Museum. This is located in a former federal courthouse and post office on Stewart Avenue downtown, where the United States Senate had once held hearings on organized crime in 1950.

Map 4.16. Las Vegas in the 1970s. Las Vegas filled much of the valley and had several freeways, but there was still ample vacant land near the strip, and Henderson was still a separate town.

Las Vegas had become the biggest city in the state by 1970 with 125,787 people, and the US Census Bureau classified Clark County that year as a Metropolitan Statistical Area (a large city together with suburbs) with 273,288 people. Las Vegas had finally become a big city, though with a tremendous amount of vacant land scattered within it. One feature that did not grow was the city limits; much of this population growth took place either in Henderson, North Las Vegas, or outside the city limits.

One of the biggest changes to the map since the 1950s was that Interstate 15 had replaced US 91 as the main road through town. The freeway opened in the city in stages, with the downtown section not finished until 1974. The city's railroad heritage remained evident in the form of the freeway's broad curve around the railroad yards on the west side of downtown. (The curve still exists, though the railroad yards are long gone.) The city's second freeway was being built west from downtown, but would not be finished for more than a decade.

The Strip continued to grow rapidly. The Hacienda opened in 1956 and for many years marked the south end of the Strip. The Stardust opened in 1958, Aladdin in 1963, Caesars Palace in 1966, Circus Circus in 1968, the International (later Hilton, LVH, and Westgate) and Landmark in 1969, Castaways in 1970, and the first MGM Grand (now Bally's) in 1973. But not all casinos survived; El Rancho Vegas burned to the ground in 1960 and was never rebuilt.

The Strip and Interstate 15 grew together. In 1963 Interstate 15 was open to the south end of the Strip, where a crossover carried traffic to the Strip, and in 1967 it was extended to Sahara Avenue, where a new railroad overpass allowed easy connections between the two highways (map 7.13). Visitors from Southern California could use the new highway to reach the Strip, but those heading on to downtown Las Vegas still had to use Highway 91. Many Strip casinos were built or expanded after Interstate 15 was constructed and their back property lines follow the highway and ramp curves. The freeway not only provided easy access but a western boundary to Strip development.

Hotelier Jay Sarno first came to Las Vegas in 1963. He was hooked from the first day but felt that the Flamingo and other hotels along the casino corridor were too plain or even downright shabby. He made the decision to abandon plans to build a new hotel in Birmingham, Alabama, and instead build one in Las Vegas, which he vowed to make the classiest in town. He expanded on a vaguely Roman theme he had used in his hotels in Atlanta, Dallas, and Palo Alto, California, and was particularly inspired by a visit to St. Peter's Square in the Vatican. He named his new creation Caesars Palace, deliberately dropping the apostrophe as he wanted every (male) visitor to feel it was theirs. At a time when the Strip was still a dusty rural highway, the massive fountains built in front of the hotel became the city's leading landmark and main attraction. Among those attracted was stuntman Evel Knievel, who sought to jump it on his motorcycle to raise money for an attempt to jump the Grand Canyon. His attempt was spectacularly unsuccessful, and was famously filmed by actress Linda Evans.

A convention center opened in 1959 east of the Strip, between it and the new Maryland Parkway corridor. The futuristic dome was once a Las Vegas landmark, but it was removed during one of many expansions that has given it more than fifty times the exhibit space it had when it opened. The center's construction also marked the beginning of the growth of Paradise Road as a major street; the International and Landmark Hotels opened on adjacent land in 1969. A new airport terminal (now Terminal 1) opened on Paradise Road on the eastside of McCarran Airport in 1963, but the street was also cut short by a runway extension (map 7.18).

Maryland Parkway replaced Fremont Street as the region's commercial core by the 1970s; the Boulevard Mall opened in 1968 and drew department stores out of downtown. A massive new Von Tobel indoor lumber store opened nearby, while Sunrise Hospital opened between them. Farther south on Maryland Parkway, the University of Nevada campus, newly renamed the University of Nevada, Las Vegas (UNLV), continued to grow, though well behind the growth in enrollment of college campuses in Phoenix, Tucson, or Albuquerque.

One feature that did not grow much was the city limits. Casino owners resisted annexation by the city, and Sahara Avenue would remain the southernmost extent of the city of Las Vegas. The Strip, airport, UNLV and growing residential and commercial areas south of Sahara Avenue would instead become part of various towns (map 8.2). Despite urban growth, much of the valley remained vacant land, very apparent in several movies made in the city during the 1970s (maps 11.4 and 11.6). Large empty parcels existed along the Strip, Maryland Parkway, Paradise Road, Boulder Highway, and other major streets. Henderson was still a distinct town separated by miles of open land from Las Vegas. It had allowed gambling in 1956 and several small casinos opened in its downtown, but these offered no competition to the ones in Las Vegas.

The 1990 census indicated that 741,459 people called Clark County home, of whom 258,295 people lived within the city limits of Las Vegas. It had spread considerably, nearly filling the valley from mountain to mountain. Although Henderson was

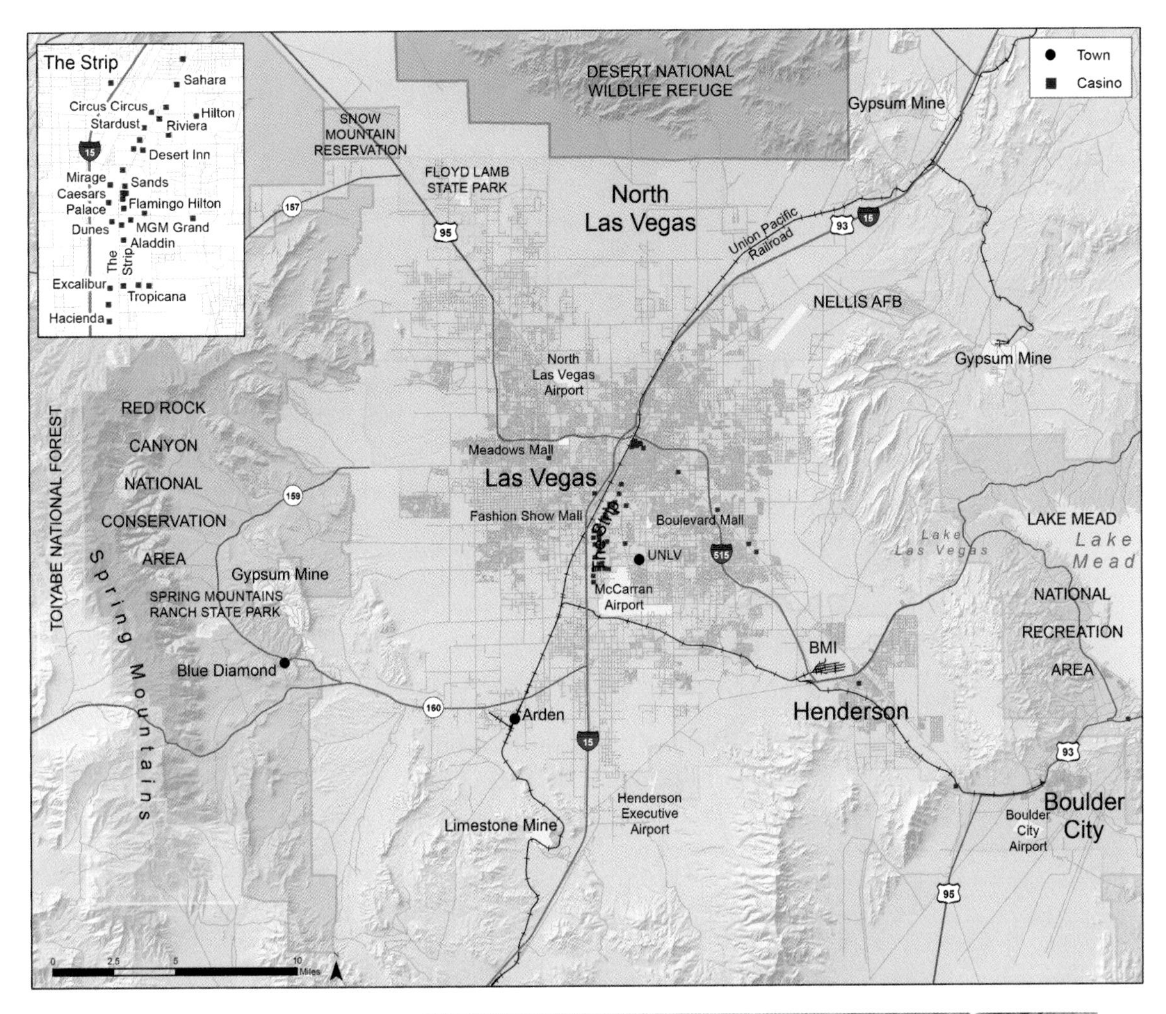

Map 4.17. Las Vegas in 1990. The city was poised to begin the 1990s building boom. The Snow Mountain Paiute reservation, I-515, and the removal of the Blue Diamond Railroad were some of the biggest changes to the map.

Satellite view of Las Vegas, 1972. Courtesy of USGS.

still clustered along Boulder Highway, it began to grow westward with the new Green Valley master-planned community, which would alter the character of the former industrial town. Maryland Parkway became only one of several commercial cores for the city, with lots of new competition. The Meadows Mall opened in 1978, part of a large retail area west of the Strip that became the city's premier shopping destination throughout the 1980s.

The city, and especially the Strip, was poised to begin its greatest era of growth. The transformation of the Strip began when the Mirage opened in 1989, but other new themed mega-resorts soon followed: the Rio and Excalibur appeared in 1990, the second MGM Grand hotel, Luxor, and Treasure Island in 1993. Many old casinos would disappear from the city's skyline that decade, many through spectacular implosions, and for the first time a casino would finally surpass Nellis Air Force Base as the largest employer in town.

A new geography of casinos emerged in the form of scattered neighborhood casinos catering to locals in the 1990s. These were designed not to appeal to visitors from distant places but to residents. The first of these was the Showboat at the north end of Boulder Highway; the Moulin Rouge was another early example. Another early example was the Bingo Palace (later Palace Station). From the late 1980s, many new ones were built, including all of the Station Casino properties.

The map is noteworthy for how few freeways are visible compared to today. But the rapid growth of the city stimulated demand for more (map 7.13). As the freeway system began to expand, the railroad network contracted. The Blue Diamond branch serving gypsum mines west of town shut down in the late 1980s, and the Boulder City branch tracks crossing Boulder Highway at Railroad Pass were paved over, cutting the line that helped build Hoover Dam (map 7.5). There were many fewer railroad overpasses than today (map 7.6), and getting across town was often much harder.

On the outskirts of the city, other changes were evident. The spectacular Red Rock Canyon sandstone area west of the city was protected as a national conservation area operated by the Bureau of Land Management BLM (map 9.6). To the north, the Paiute tribe received a small reservation straddling US 93, known as the Snow Mountain Reservation. On the east side of Henderson, Lake Las Vegas was being constructed in Las Vegas Wash and became the valley's newest water body and high-end retail development.

Las Vegas owes its existence to the Spring Mountains, whose height is responsible for orographic precipitation that brought water to the desert. Much of this water soaked in and became part

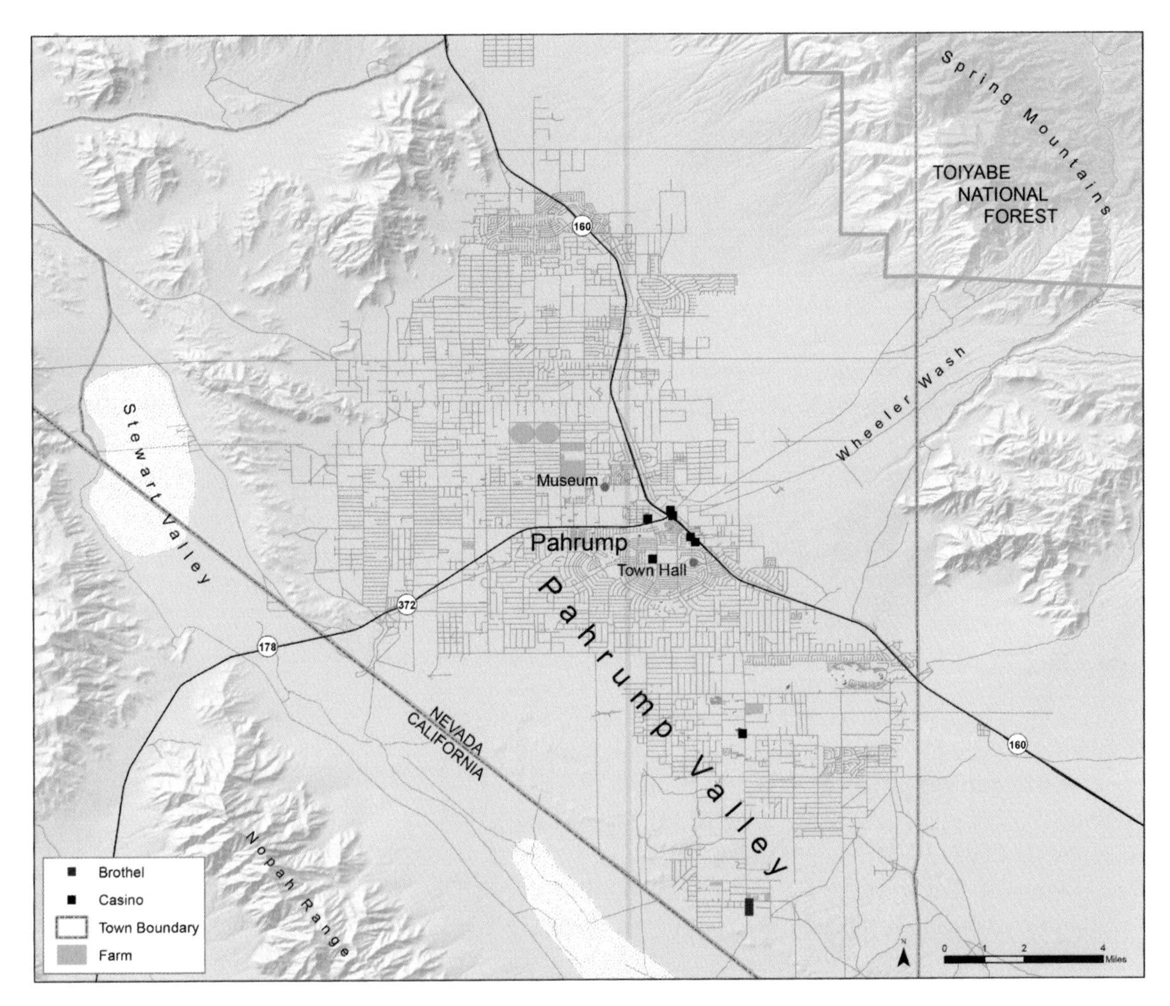

Map 4.18. Pahrump. The rapid growth of this farm town turned suburb has been remarkable, but like Las Vegas, it is hemmed in by public lands and limited water.

Satellite view of Las Vegas, 1992. Courtesy of USGS.

of the groundwater that emerged at the valley's springs. These springs in turn became vital to Paiutes, Mormons, and the railroad that produced modern Las Vegas.

The same process that brought water and people to the Las Vegas Valley also operates on the west side of the Spring Mountains, making Pahrump a kind of mirror image of the city, with its own springs, creeks, and meadows. As in Las Vegas, Paiutes camped and farmed at the valley's springs before people of European ancestry arrived. These newcomers founded a ranch in the valley at Manse Springs in the 1860s, followed by one at Pahrump Springs in 1875.

Unlike Las Vegas, the Pahrump Valley also had soil suitable for agriculture. The ranches founded here had a mix of crops and cattle and profited from selling food to nearby mining camps. Agriculture was firmly established in the valley at eight ranches by 1945. Some farms raised cattle and alfalfa as feed, others were dairy farms and feed, and others were orchards and specialty crops. The first cotton known to be planted in the valley was in 1936, but this failed. A second attempt succeeded in 1948, and cotton soon became a major crop in the valley, though mainly at the foot of the broad alluvial fans from the Spring Mountains. A cotton gin opened in Pahrump in 1959, making it even more profitable.

Access to the town was slower to develop. Although a state highway existed to Las Vegas from 1947, it was a rough dirt road and little used (map 7.12). Travelers to Las Vegas instead went north to US 95. The road west to Shoshone was also an important road because it had been the main route for shipping cotton for ginning in California. A paved road to Las Vegas through Mountain Springs pass was only built in the early 1950s, and became NV 160 in 1954.

Thirty-three ranches were found in the valley by 1965, and just more than ten thousand acres were being farmed in the peak year of 1968. But cotton growing began to decline in the 1970s because of weak prices and the sale of the largest farm, the Pahrump Ranch, to a landholding company in 1970. Without this farm, the cotton gin could no longer stay in business; and without that, cotton could no longer be profitably produced in the valley. Many farmers switched back to alfalfa, but prices were also weak and the valley's wells began to show signs of groundwater depletion.

Land for purely residential purposes was sold beginning about 1960, with unfarmed areas in the western part of the valley being subdivided and sold. The sale of the Pahrump Ranch was the beginning of the end of agriculture and the valley's transformation into the low-density city it is today.

Nevada law allows for the creation of limited town governments instead of cities (map 8.2). Pahrump became a town in 1962, giving it a local government to provide services to the small but growing community. Outside electricity arrived in 1963 when a precursor to the Valley Electric Association (VEA) completed a power line from Henderson to Pahrump (map 8.17). The VEA) obtained power from Hoover Dam. Power lines were extended to Amargosa Valley, Fish Lake Valley, Death Valley, and Shoshone and Tecopa. Outside telephone service arrived in the valley in 1965.

Pahrump's population has skyrocketed since the 1990s, growing from several thousand to almost forty thousand people. New powerlines have been needed, and NV 160 has been slowly rebuilt as a four-lane highway to handle the huge increases in traffic between Pahrump and Las Vegas. Great Basin College opened a campus here in 2006, bringing higher education to the city.

Unlike Las Vegas and surrounding cities, Pahrump never had a townsite from which it grew, and it has never had a downtown. The intersection of Highways 160 and 372 instead mark the center of town and the heart of its commercial strip. Most of the casinos in town are clustered near this intersection, though the town's two brothels are far to the south on the edge of the city (Pahrump is in one of ten Nevada counties where prostitution is allowed). Aside from these highways, most of the town's development is based on the same Public Land Survey System grid as Las Vegas (map 3.12). The city is very low density and scattered, much as Las Vegas was as late as the 1980s.

Will Pahrump ever grow to be a second Las Vegas? The city's population of thirty-six thousand will continue to grow but is already reaching the limits of available water. The city is dependent on groundwater pumped by a large number of wells, 11,280 of them for domestic use. Several water companies provide water and/or sewer service to selected areas while many residents have their own wells and septic tanks. These wells draw more water than is being naturally recharged because of rainfall, resulting in water table declines since the 1950s. At the end of 2017, the Nevada state water engineer put limits on new residential well drilling. Much of the problem stems from excessive water rights claims rather than pumping, and the engineer suggested the city's water supplies could support eighty thousand people if wisely used.

New sources of water will eventually be needed if the city is to continue growing beyond this level. The only feasible option would likely be groundwater pumping from other desert valleys, much as Las Vegas has sought to do in northern Nevada (map 10.11). This was attempted in the 1970s when wells were drilled in Ash Meadows to the northwest as part of a large residential development. This was ended in 1976 by the US Supreme Court, which ruled in favor the National Park Service's right to regulate groundwater pumping to protect the tiny pupfish in Devils Hole. Pahrump would have to look much farther away for groundwater, perhaps as far as the Fish Lake Valley (already served by Pahrump's Valley Electric Association utility). Or perhaps growth will eventually require a pipeline tunneling through the Spring Mountains from Las Vegas.

Another limit on future growth is land. Like Las Vegas, Pahrump sits on a large block of private land surrounded by government land administered by the Bureau of Land Management (BLM), with several private outliers to the south. Pahrump is not part of the Las Vegas Valley Disposal Boundary that surrounds Las Vegas (map 8.1), so once this private land has been developed the only way for the city to expand is for Congress to pass a law authorizing the BLM to sell more land. Future population growth will therefore require infill development and higher densities or an act of Congress; neither of these are appealing in extremely libertarian Pahrump.

Did you know Las Vegas is at war? Thousands of air strikes have been flown from the city in the last two decades, killing both enemy soldiers and civilians. Those enemies are shooting back, and dozens of American aircraft have been shot down during these missions. You've seen the video footage on the news, and you might even live or gamble next to one of the pilots.

The aircraft involved are drones flying out of bases located in several countries in the Middle East and Africa; the targets are in Iraq, Afghanistan, Pakistan, Yemen, and other countries, while the pilots are housed inside windowless buildings at Creech Air Force Base in Indian Springs, commuting daily from Las Vegas. Creech is the only US airbase dedicated to drone warfare and the air force's only drone pilot training base. But don't call them drones: remotely piloted vehicle or even unmanned aerial vehicle are preferred terms.

The first of these was the MQ-1 Predator, entering service in 1995. Originally just for observation, the use of armed drones was pioneered here in 2001. When the United States invaded Afghanistan in 2001, armed Predators were on the scene. They flew in Iraq beginning in 2003 and helped capture Saddam Hussein. The larger MQ-9 Reaper entered service in 2007. It can carry several missiles or bombs and fly missions of up to twelve hours. They also fly reconnaissance missions, cruising at high altitudes that often make them undetectable from the ground. They can identify targets at night as well as day and communicate what

An armed Predator drone flies over southern Afghanistan, controlled by an operator in Las Vegas, 2008. Courtesy of US Air Force. Photo by Leslie Pratt.

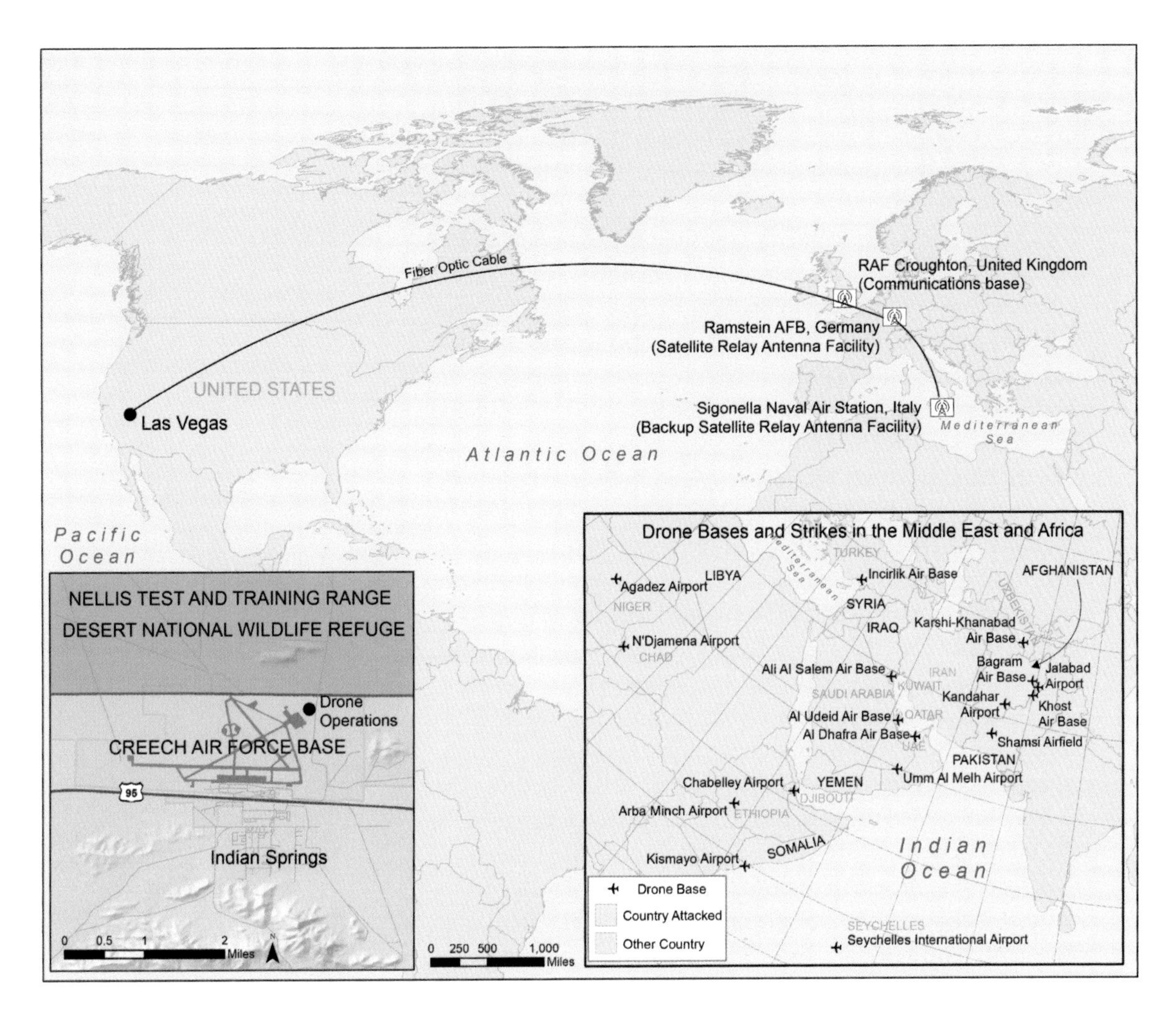

Map 4.19. The Drone Wars. Communication links between Las Vegas and military bases in Europe connect to drones launched from a variety of bases in Africa and the Middle East. *Right inset*: major drone bases and countries where attacks have been carried out; *left inset*: the drone control center at Creech Air Force Base in Indian Springs.

they see to friendly forces; and they can attack targets themselves if needed.

Where have Las Vegas drones carried out attacks? The government is not forthcoming about these, but they are known to have operated in Afghanistan and Pakistan against the Taliban, in Iraq, Yemen, Libya, and Somalia against ISIS, and in Syria against government forces. The results of these attacks are even less certain; thousands of terrorists and enemy forces have been killed in these countries, perhaps more than eleven thousand people. Unfortunately, the view on computer screens is not particularly clear, and more than fifteen hundred civilians have been among those killed, including hundreds of children.

Las Vegas's drones have operated out of a number of bases in the Middle East and Africa. Once in the air, the drones are controlled from Indian Springs; commands are passed through a fiber optic cable from Indian Springs to Ramstein Air Force Base in Germany and then beamed to a satellite in geosynchronous orbit, which relays commands to the drones. Sigonella Naval Air Station in Sicily is being developed as a backup to Ramstein. It is not just American pilots operating out of Creech; the British Royal Air Force No. 39 squadron has been based here since 2005 and has also flown combat drone missions in Afghanistan. At least 429 drone strikes have taken place in Pakistan; ironically, the Pakistan Air Force regularly sends planes and pilots to Las Vegas to participate in Red Flag military exercises (map 9.13).

More drone communication facilities are being built to cover the Pacific and Asia. Those Las Vegans operating drones will soon be able to fire a missile against anyone, anytime, anywhere in the world, and make a brief stop to play some slots on the way home. Las Vegas is a perfect home for the 24/7 drone wars.

Clark County, with 1,951,269 people in the 2010 census and 72 percent of Nevada's population, is the state's largest county by population. Of that total, more than 583,000 people lived within the city of Las Vegas. Las Vegas has become a metropolitan area of more than 2.1 million people (counting Henderson, North Las Vegas, Boulder City, Pahrump, Mesquite, Searchlight, Laughlin, Kingman, Lake Havasu City, and even Tonopah), the twenty-eighth largest in the country. Almost the entire Las Vegas Valley has been filled with streets and subdivisions. Many of the peripheral areas not yet covered with roads never will be because the city has nearly filled the land Disposal Boundary, a congressionally mandated urban growth boundary for the city (map 8.1).

The aforementioned transformation of the Strip that began with the opening of the Mirage in 1989 and others continued with Monte Carlo in 1996, New York–New York in 1997, Bellagio in 1998, and Mandalay Bay, Venetian, and Paris Las Vegas in 1999, each introducing a new theme but with similar buildings. The newest are the Wynn in 2005, Palazzo in 2007, Encore in 2008, CityCenter in 2009, Cosmopolitan in 2010, and Resorts World in 2021. These new casinos replaced many long-familiar names: the Castaways, Silver Slipper, Landmark, Dunes, Sands, Hacienda, Desert Inn, Stardust, Frontier, Aladdin, and the Riviera are all gone (map 11.7).

Elsewhere in the city, the Fremont Street Experience was completed in 1995 as an attempt to reinvigorate tourism downtown. The project closed off to motorists five blocks of the street, once the commercial heart of town and where President Franklin D. Roosevelt toured the small town, Elvis raced, and James Bond outran the police. A massive barrel-vault canopy was installed, its Viva Vision light show shining nightly.

Retail activity continued to spread throughout the city as it grew. The Galleria Mall opened in Henderson in 1996, marking the transition of the gritty industrial town into an upscale suburb, while the opening of Downtown Summerlin in 2014 reflects the continued westward growth of the city. The city has seen changes

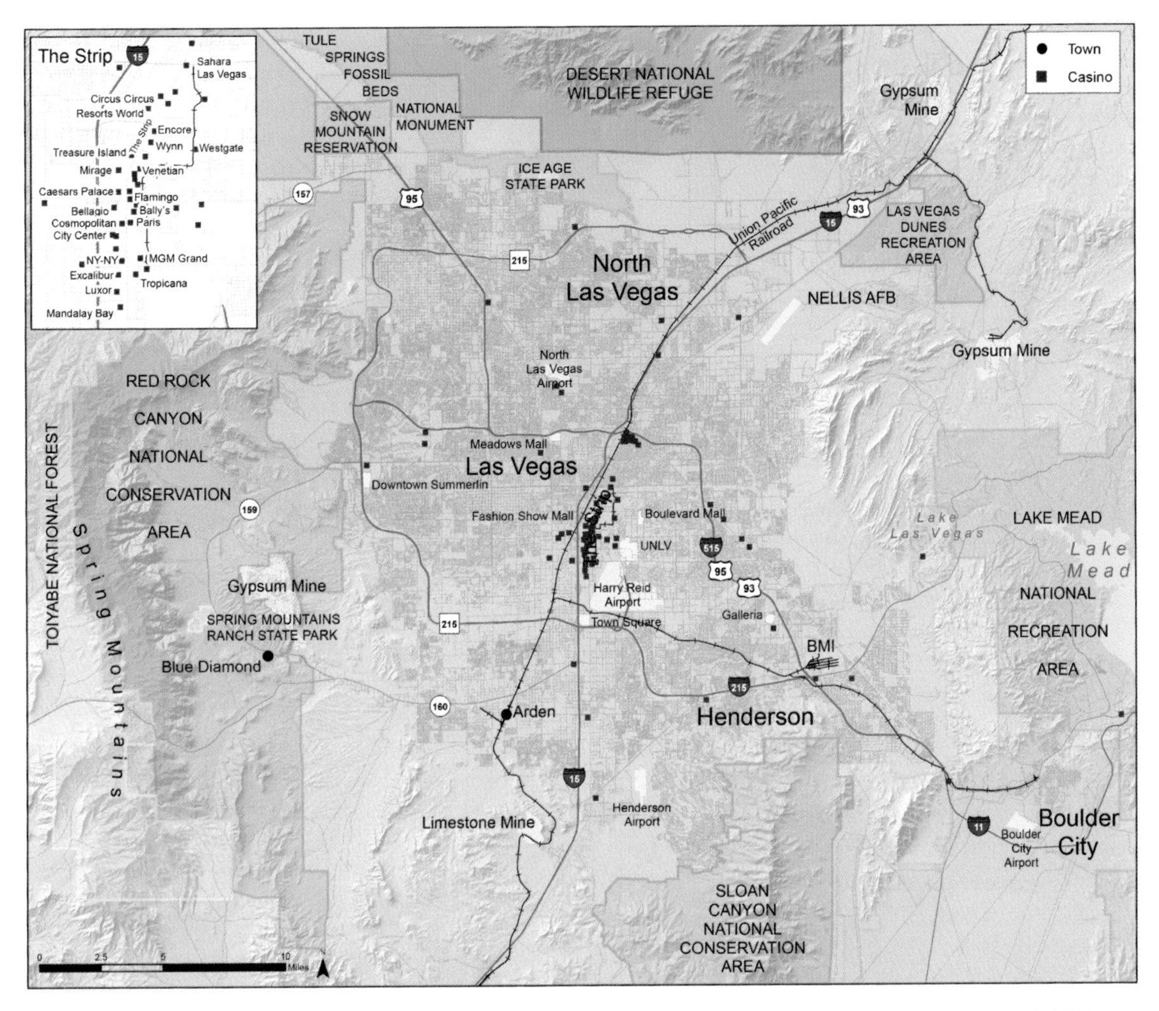

Map 4.20. Las Vegas Today. A metropolitan area of two million people, with an extensive freeway system, several distinct clusters of casinos, and a dizzying array of parklands in the surrounding mountains.

Satellite view of Las Vegas, 2017.
Courtesy of USGS.

in its demographic landscape as well. Las Vegas's Chinatown is along Spring Mountain Road, mostly between Decatur and Valley View Boulevards, though unlike other Chinatowns it is not a residential area. Instead, it is centered in a shopping center, Chinatown Plaza, which opened in 1995.

One of the biggest changes apparent on the map is the proliferation of freeways (map 7.13). Interstate 515 was extended south to Railroad Pass, and the county began building a beltway. The first section, south of the airport, opened in 1994 in conjunction with the Airport Connector tunnel under McCarran International Airport. The entire route was open by 2003, but will not be built out to freeway standards until about 2025. The railroad, which created Las Vegas in 1905, further reduced its presence in the city by moving its downtown yards to Arden. The area has since been redeveloped and now contains county offices, a shopping center, and the World Market, a five-million-square-foot exposition center for home and hotel furnishings. The curve built in Interstate 15 when it was constructed around these yard remains as the only indication of their former importance.

In Henderson the BMI plant was renamed the Black Mountain Industrial Center and remains in use. Unfortunately, decades of industrial operations there resulted in soil and groundwater contamination that will take years to remedy; much of the site and evaporation ponds to the east have already been cleaned up at a cost of tens of millions of dollars. One corner of the plant has been redeveloped as a shopping center, and it is likely that over time the Black Mountain Industrial Center will gradually be erased. But other industries remain in Henderson along the railroad line; at the far north end of the valley, North Las Vegas has sought tenants for its new Apex Industrial Park.

The population of the former dusty railroad stop swells by 40 million more people who visit the city each year. Las Vegas's former rivals of Pioche and Caliente still survive, but with populations of 1,002 and 1,130 people, respectively, have been all but forgotten by Las Vegans. Clark County is forecast to exceed 3 million people by 2060. It will fill up its urban growth boundary (map 8.1) as it grows, leading either to higher-density infill development or to growth in outlying communities such as Pahrump or Mesquite. A new airport will eventually be needed, and one has already been planned for Roach Lake south of the city in Ivanpah Valley (map 10.9). The source of the city's water supply for that year has yet to be identified.

5

Creating Lake Mead

The construction of what became known as Hoover Dam and the resulting Lake Mead represents one of the most important events in the history of Southern Nevada—and the country. The dam's creation filled Lake Mead, but the rivers and valleys flooded by the lake have largely been forgotten. It was one of the few areas of Southern Nevada and adjacent states not surveyed by scientists intent on documenting vegetation, geology, or archaeological sites. We know surprisingly little about this region. Several maps on the following pages show the towns, roads, mines, and landmarks that existed before they were submerged beneath the rising waters.

When the lake filled, one story ended but another began. The weight of the water caused earthquakes and made the land sink. The water level of Lake Mead has been dropping in recent decades, and this is likely to continue because of a crippling drought throughout the Southwest. Boat ramps and other marinas have been moved, and long-forgotten places have emerged from the waters. The last few maps will examine the future of the reservoir and what remains underneath it.

Before the construction of Hoover Dam and Lake Mead, the Colorado River flowed through a series of valleys and narrow canyons in Southern Nevada. Descending from the Grand Canyon, a river traveler would encounter Iceberg Canyon, Hualapai or Virgin Canyon, Boulder Canyon, and Black Canyon. Black Canyon received the greatest attention from explorers because of its great length and shadowed depths. Early explorers and settlers thought of these canyons as obstacles, but dam builders would eventually consider them to be an important resource. The Grand Wash Valley, Gregg Basin, the Virgin-Detrital Wash Valley, and the Callville Basin were valleys that accommodated early efforts at agriculture as well as being among the few areas where the river could be crossed.

The river changed tremendously throughout the year. Before dams regulated its flow, the Colorado River was highly seasonal, with water levels peaking in late spring. At other times of the year, many small islands and sandbars were present. Within the canyons, numerous rapids formed where tributary streams dumped boulders and gravel into the main river channel. Among these were Hualapai rapids in Virgin Canyon and Reverse Rapids in Black Canyon.

The best-known feature along this stretch of the Colorado was its Great Bend, where the westward-flowing river abruptly

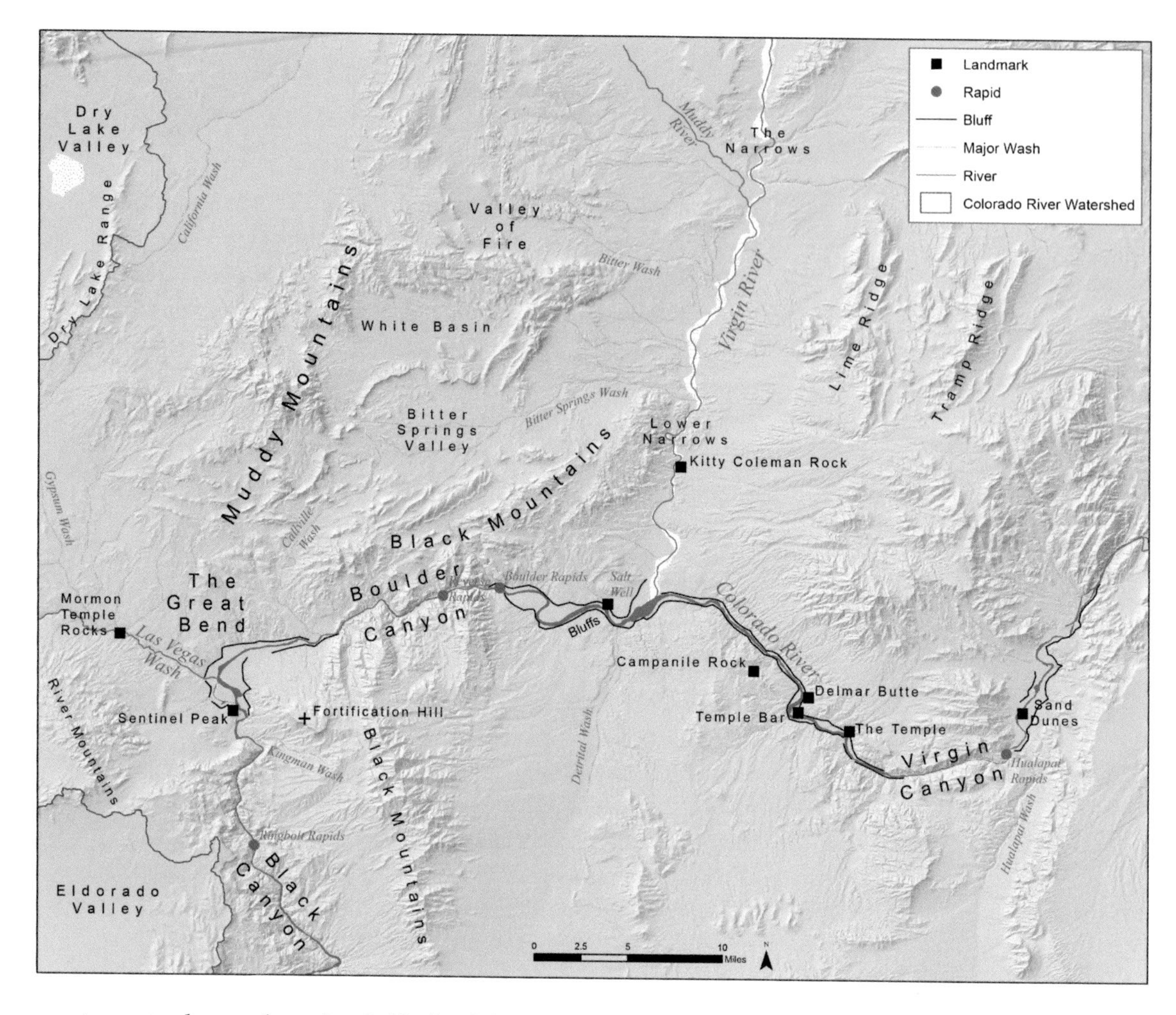

Map 5.1. The Great Bend of the Colorado River. Many features along the Colorado and Virgin Rivers were flooded when Hoover Dam was built. Map based on Smith and WESTEC Services, Inc., early USGS topographic maps, and Longwell's surveys of the area's geology.

turns to the south to the Gulf of California. (The Big Bend of the Colorado State Park near Laughlin refers to a different and much smaller bend on the river.) This curve, in the broad valley between Boulder and Black Canyons, was just above the location chosen for Hoover Dam. Photographs of this valley show a varied landscape quite different from most Southern Nevada valleys. A prominent hill in the center of the valley, capped with dark lava, was called Fortification Hill by Joseph Ives during his expedition in 1857 (map 5.2), with sand dunes near its base. Today, the hill is Sentinel Island; the name Fortification Hill was later transferred to the large volcanic mesa to the east of the river. An area with colored soil below the modern Fortification Hill was known as the Painted Desert but today as the Paint Pots.

Throughout Southern Nevada river canyons and valleys, terraces or bluffs paralleled the river. They ranged from a few dozen feet to 750 feet high. These ran along one or both sides of the river and often confined it to a narrow channel, even within a broad valley. The bluffs along these rivers often contained spectacular erosional features in the form of cliffs or buttes. Napoleon's Tomb, the Campanile, Delmar Butte, and The Temple were prominent formations between Virgin Canyon and Bonelli's Ferry, and they are still visible today above the lake's water level. Other named rocks included Kitty Coleman Rock at the Lower Narrows of the Virgin River and the Mormon Temple in Las Vegas Wash. These were inundated by Lake Mead, though the eroded bluffs of the Mormon Temple are now above water.

These bluffs were important to early explorers and settlers. Callville and Bonelli's Ferry (map 5.3) were built on top of these terraces, providing some protection from floods. The terraces across from Callville provided all of the gravel used to make concrete for Hoover Dam, which in turn submerged them.

Several tributaries flowed into the Colorado from the north. The Virgin River, sourced in the mountains of southwest Utah, usually flowed through a broad valley but passed through the Upper Narrows (where the river cut through a gravel deposit) and the Lower Narrows before reaching the Colorado. In the Lower Narrows the river cut a gorge through volcanic rock, confined to a valley no wider than three hundred feet, with one-hundred-foot cliffs in places. Although the river appears quite wide on early maps it was described as sluggish stream, usually dry in its lower section, within a broad sandy bed. Sand dunes and windblown sand were common along the river. It could be forded anywhere except at times of high flow.

The Muddy River is an even smaller stream flowing into the Virgin below the Upper Narrows. It is unusual for a desert stream in that its headwaters are not in distant mountains but only a short distance upriver in an area of springs. These produced a nearly constant amount of water year round that could be easily diverted for agricultural use. However, a small tributary, the Meadow Valley Wash, occasionally brought floodwaters from distant mountains that were very destructive in this valley.

The vegetation along the Virgin and Colorado Rivers before Lake Mead is little known today because no surveys of vegetation were done before the lake filled. Early accounts make clear that riparian vegetation was sparse, with mesquite likely to have been the only tree common along the river. However, the Colorado carried a sizable amount of driftwood, which provided a valuable source of firewood for those along the river. Vegetation along the Muddy River reportedly included mesquite, cottonwood, willow, creosote, arrowweed, and an abundance of grasses making for an ideal pasture. There are even fewer accounts of what the animal life in the region was like, but the 1923 US Geological Survey (USGS) expedition along the Colorado River (map 5.4) reported many ducks, quail, and coyotes along the river.

Mineral deposits drew more attention from explorers. A

number of salt deposits were found in the bluffs along the west bank of the Virgin River. Native Americans had mined these for hundreds, if not thousands, of years; later residents continued the practice into the twentieth century. The northernmost of these deposits was a particularly well-known landmark. Jedediah Smith provided the first account of it in 1826, calling it the Big Cliff salt mine, and in 1869 George Wheeler referred to it as Salt Mountain. This deposit extended along the bluffs for several hundred feet, with tunnels dug into the cliff. The salt deposits were large and of good quality, and were intermittently mined commercially in the twentieth century. Several other salt mines were found farther south along the river, including the Calico, Fairview, Black, and Bonelli Mines. Near the Colorado River was the Salt Well, a large circular steep-sided forty-three-foot-deep depression with a pool of salty water 118 feet across at its bottom. Many travelers noted this as a curiosity. The USGS investigated these salt deposits in the 1920s out of concern that a reservoir built in the area would become a salt lake once these deposits eroded into the lake water. They concluded this concern was groundless: the salt deposits were tiny in relation to a vast reservoir, and much of the salt was soon be buried under sediment.

Fifteen fur trappers led by Jedediah Smith in 1826 provided the first written account of this area. This group, on its way to California, came down the Virgin River (which Smith called the Adams River). They camped near the confluence with the Muddy River and continued down the Virgin River (with a stop at the salt mine) to camp at the Colorado, which Smith referred to as the Seedskeedee (Padre Eusebio Kino had first applied the name Colorado to the mouth of the river in 1701, but it took many years before it was applied everywhere.) Here they rested a day before crossing the river on a raft. Because Boulder Canyon blocked travel to the west, they continued south up Detrital Wash before turning west and reaching the Colorado River at Willow Beach. After resting here, they departed southward up Jumbo Wash. Smith repeated the journey in 1827.

Spanish traders seeking a route between Santa Fe and Los Angeles were the next to pass through this area. Antonio Armijo's expedition of 1829-30 also descended the Virgin River, reaching the mouth of the Muddy River, where they rested a day. They continued down the Virgin and reached its confluence with the Colorado River (which they called the Rio Grande). Here Armijo broke from Smith's route and turned west to follow the Colorado downriver, though he detoured around Boulder Canyon. Arriving at Las Vegas Wash (which he called Yerba del Manso Arroyo), his group left the Colorado to ascend the wash into Las Vegas Valley. They crossed the valley and discovered a profusion

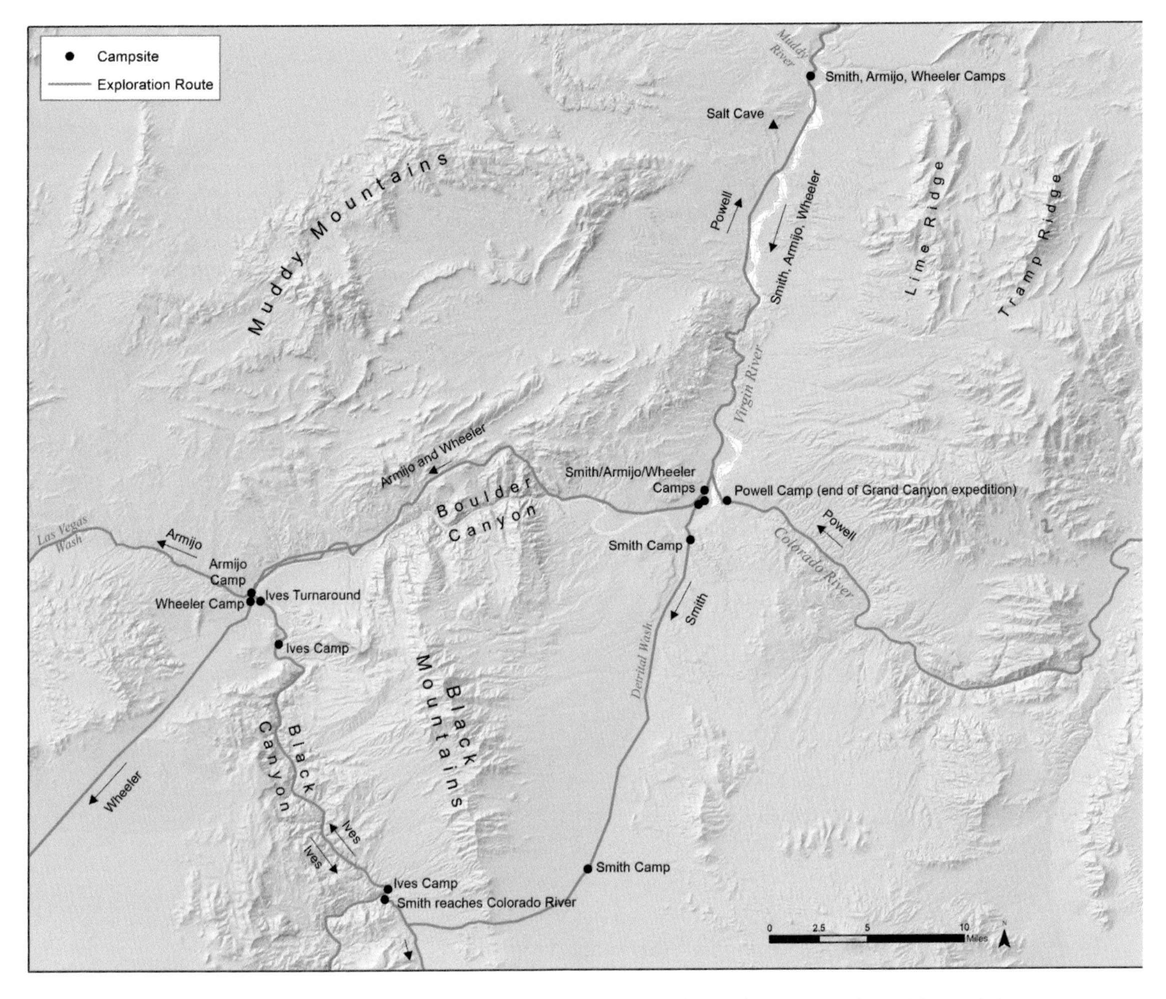

Map 5.2. Early Explorations. The first explorers followed rivers to what would become the Las Vegas region. Jedediah Smith and Antonio Armijo led parties down the Virgin river in the 1820s, Joseph Ives rowed up the Colorado River to the Great Bend in 1858. John Wesley Powell completed his epic 1869 descent of the Colorado River through Grand Canyon at Bonelli's Ferry, weeks before George Wheeler passed through. Map based on the routes recorded in accounts of several early explorers.

of springs and lush meadows they named Las Vegas. These were not the Las Vegas springs that Mormon pioneers later settled near, but those along Duck Creek at the south end of the valley (near today's Sunset Park) (map 6.3).

The route Smith and Armijo followed down the Virgin and Colorado Rivers was soon replaced by the Old Spanish Trail, which made use of the springs of Las Vegas and crossed the Muddy River to the north of the future Lake Mead. John C. Frémont was one of the first to use this route, passing northeast from Southern California on his second expedition in 1844. He was also the first to put Las Vegas on a map. This became a popular route between California and New Mexico in the nineteenth century (map 3.5), but the Colorado River canyons were still little known.

The next group of explorers to venture into this area sought to explore these canyons. Lieutenant Joseph Ives, traveling by steamboat, led an expedition. He commanded an 1857–58 army expedition in with the goal of traveling up the Colorado River from Yuma, Arizona, to reach the Great Bend of the Colorado, the Big (now Grand) Canyon, and determine where the head of navigation on the river was. After a long and difficult journey upriver, Ives was tremendously impressed with Black Canyon's depth, sheer cliffs, varied rock formations, and darkness at the river. His first attempt to ascend it was ended by a boulder protruding from the river that severely damaged the steamboat. Rather than risk further damage to their means of transportation, Ives decided to continue upriver in a rowboat. They encountered many rapids, the most difficult of which they named Roaring Rapids for its sound. They ended their first day of rowing by camping on a small patch of gravel in the vicinity of present-day Willow Beach. They continued laboring upriver on the second day, and just when they thought they would be spending another night deep inside Black Canyon, they reached its head.

They camped for the night in a ravine at the base of a peak on the west side of the river they named Fortification Hill, as noted earlier not today's landmark by that name. They climbed to the top of this hill (now known as Sentinel Island) before sundown and looked north to see the Great Bend of the Colorado River. To the east they saw blood-red hills, today's Paint Pots at the base of Fortification Hill.

The next day they rowed a short distance upriver to a broad shallow salty stream, which they assumed to be the Virgin River. They were expecting a much larger river, and it was a tremendous disappointment to them. In fact they were at the mouth of Las Vegas Wash. But because they had now reached the Great Bend and believed that the river would not be navigable beyond this point, they turned back. They descended the river to their waiting steamboat below Black Canyon and then set off overland toward the Grand Canyon. They eventually descended Diamond Creek to become the first Americans of European ancestry at the bottom of that canyon.

The Ives expedition showed that steamboats could make it as far as the mouth of Black Canyon during low water and preserved hopes that the ascent would be easier with higher water levels. The expedition included the artist Balduin Möllhausen and the mapmaker and artist Frederick W. von Egloffstein, who created several woodcut illustrations depicting the trip. Among these is one of Black Canyon, showing towering vertical walls extending out of sight above. This was one the first images of the Las Vegas area to be recorded.

At least a dozen years would pass before another expedition to this region, and then two passed by within weeks of each other. John Wesley Powell's legendary 1869 expedition down the Colorado through the Grand Canyon ended at the mouth of the Virgin River on August 30. Several Mormons had been sent from

View of Black Canyon from the 1858 Ives Expedition. This is the first published image of the Colorado River near Las Vegas. Courtesy of David Rumsey Historical Map Collection. Sketch by F. W. Egloffstein, 1861.

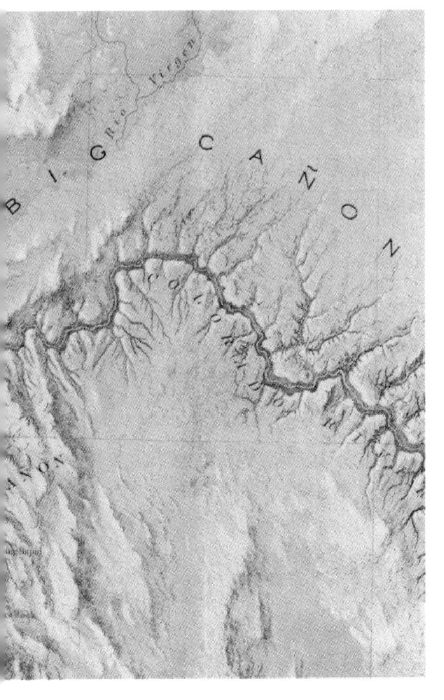

Detail of 1858 Ives Expedition map showing Black Canyon and "Los Vegas." Courtesy of David Rumsey Historical Map Collection. Map by Joseph Ives.

St. Thomas, Utah Territory, to camp there to spot any sign of the expedition and were happy to find them still alive. Powell's group rested there for several days before Powell traveled up the Virgin River by wagon on his way to Salt Lake City, while others in his group continued downriver to Fort Mohave, Arizona (above modern Needles, California).

Several weeks later, an army reconnaissance led by Lieutenant George Wheeler journeyed south through the Muddy River Valley. They camped at the confluence of Muddy and Virgin and, like Ives, were disappointed by the Virgin River's small size. The next day, they passed Salt Mountain and reached the Colorado River. They headed downriver from there, but, like Armijo before them, bypassed Boulder Canyon (which they referred to as Virgin Canyon, not to be confused with today's Virgin Canyon upriver). They continued downriver to the mouth of Las Vegas Wash, where the group split. One group went up the wash to Las Vegas, while another left the river farther south to bypass Black Canyon and reach El Dorado, a mining town below Black Canyon. Like Ives, Wheeler was impressed by the scenery of Black Canyon. He would get to experience it as Ives did several years later (map 5.4).

In 1847 the first Mormons reached the shores of the Great Salt Lake in present-day Utah, having fled persecution in the United States. In time the Mormons would expand throughout the region, establishing agricultural colonies wherever possible. One such colony was established at Las Vegas Springs along the Old Spanish Trail in 1855 (map 3.6).

River navigation on the Colorado River also became an interest to the Mormon leadership in the 1850s. If it could be developed, it would reduce overland travel distance to and from Utah as well as help struggling Mormon outposts in southern Utah. A formal plan to establish a settlement on the Colorado River near the mouth of the Virgin River was enacted in 1864. Anson Call, an early Mormon convert who traveled across the plains in 1847 and now living in St. George, Utah, was selected to lead this mission. He followed the route of earlier travelers down the Virgin River, but bypassed Boulder Canyon by pioneering a new route up Echo Wash, past Bitter Springs, and down another wash to the Colorado River. This location was selected as the site of Call's Landing or Callville on December 2, 1864. A townsite was surveyed with streets lined up parallel and perpendicular to the Colorado River, and a sizable warehouse for the expected river trade was built out of stone in early 1865.

Callville was reached by a barge rowed upriver on January 14, 1865, and by steamboat the next year, proving the possibility of river trade. A post office, stamp mill, stone corral, quarry,

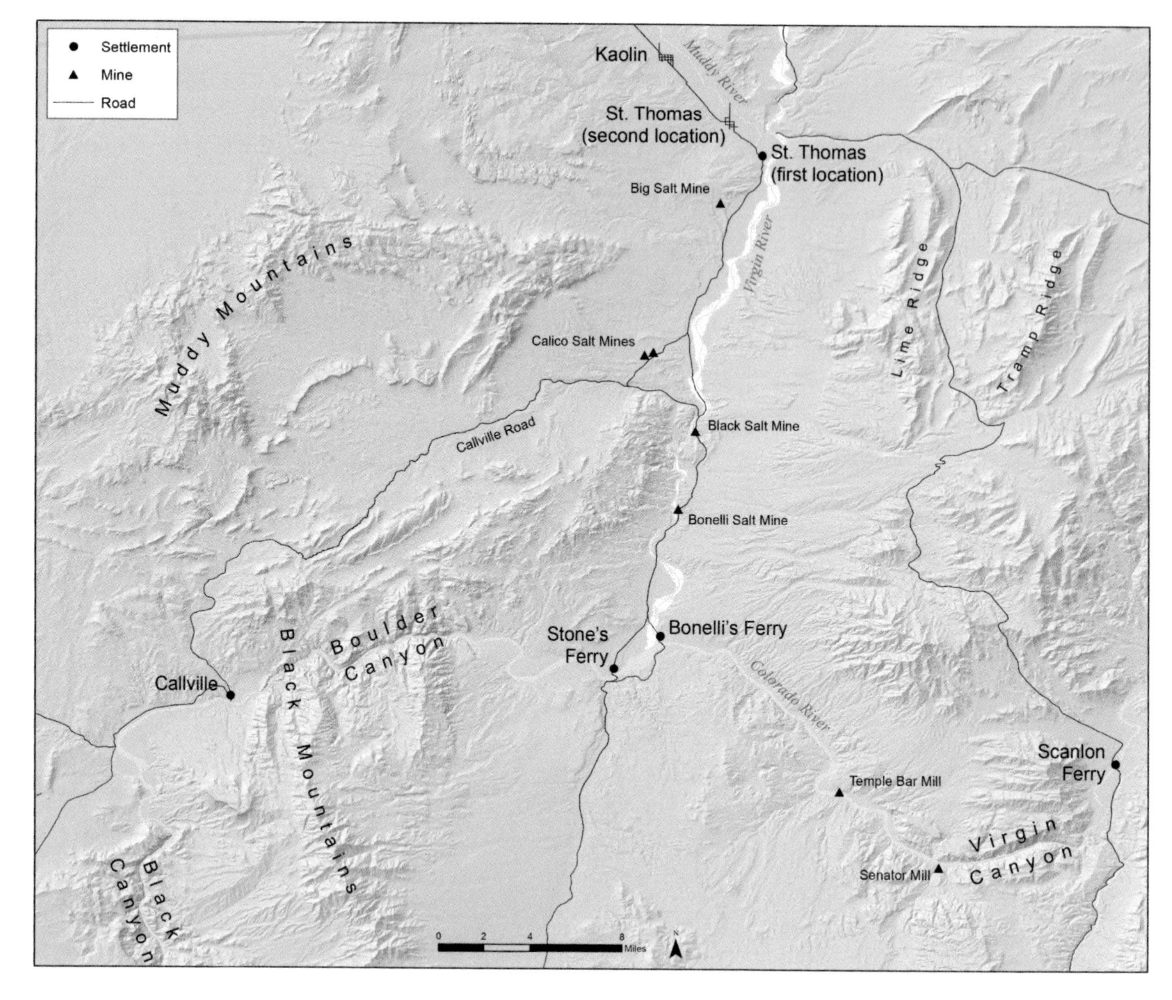

and other buildings had been built by 1866, and a small army detachment was even stationed there in 1867 and 1868. However, the river trade did not materialize, and the settlement was short lived. The post office finally closed in June 1869 when the town was abandoned. Many photographs were taken in the 1930s of the remains of Callville, usually showing the massive stone walls of the warehouse and corrals (for which reason it was sometimes called Fort Callville). Many of the photographs showed these walls sinking beneath the waters of Lake Mead. The site is now buried under mud at the bottom of the lake and will never be seen again (map 5.13).

Not long after Callville was founded, Mormon leadership initiated the Muddy Mission to support trade routes southward from Utah and to produce cotton growing for other Mormon settlements. The farming community of St. Thomas was founded January 8, 1865, at the confluence of the Muddy and Virgin Rivers, though soon relocated two miles upriver. This was eventually followed by St. Joseph (now called Logandale) in May 1865, West Point (later Moapa) in 1868, and Overton in 1869. Life was hard

Map 5.3. Early Settlement. The Colorado and Virgin Rivers held many attractions to those willing to settle in this remote region in the nineteenth century. Some farmed along the Muddy river, others mined or operated ferry services across the Colorado River. Locations of towns, mines, and roads in the late nineteenth century are based on early USGS topographic maps and Smith and Westec Services Inc.

but the colonists were able to keep their settlements alive, if never thriving. A number of crops were tried, but cantaloupes soon became one of the best.

Other early settlements appeared along the Colorado River. Stone's Ferry was established at the confluence of the Virgin and Colorado in 1870, providing one of the few crossings of the Colorado River in the region. Brigham Young visited this Mormon outpost on March 17 of that year but was reportedly not impressed. The ferry was later operated by Daniel Bonelli at a new location upriver adjacent to a small farming community (variously known as Junction City, Rioville, or just Bonelli's Ferry) with a post office. This settlement was reached by the steamboat *Gila* on July 8, 1879, and perhaps twenty more times during the next few years, the farthest any steamboat made it up the Colorado.

Farther up the Colorado, people lived and worked along the river at a few favorable locations. The Smith Ranch and a ferry crossing were in the valley between Virgin and Iceberg Canyons. The ranch, established in 1881, irrigated one hundred acres. The ferry crossing was known as Scanlon Ferry from 1881 to 1900 and then Gregg Ferry, and the area was still being ranched well into the 1920s.

Gold mining arrived in the area in the form of placer mining operations along the Colorado River above Bonelli's Ferry in the late nineteenth century. The Senator Mill began operating in 1895, and mining near Temple Bar started in 1897. These mining efforts were not based on tunnels or shafts dug into the ground but finding free gold in river mud. At Temple Bar, claims were staked out for two miles along the river and a half-mile inland on both sides. A boom was extended across the river to catch driftwood coming down the river, which provided the main source of firewood. The mining supported seventy-five workers and a ferry crossing in 1898, and it was served by a newly built wagon road and telephone line from Kingman, Arizona. A railroad was planned from Kingman, and some of the route may even have been graded, but the project was soon abandoned when mining shut down.

The completion of the transcontinental railroad in 1869 ended remaining Mormon interest in the Colorado as a trade route, but others were increasingly attracted to it. A number of expeditions ascended or descended the river to better understand the area's geology and the river's potential uses in an industrializing country. These river trips created modern maps of the region and also established a number of modern place-names.

Wheeler returned to the area in 1871 as part of a series of annual army expeditions, this time traveling upriver in rowboats

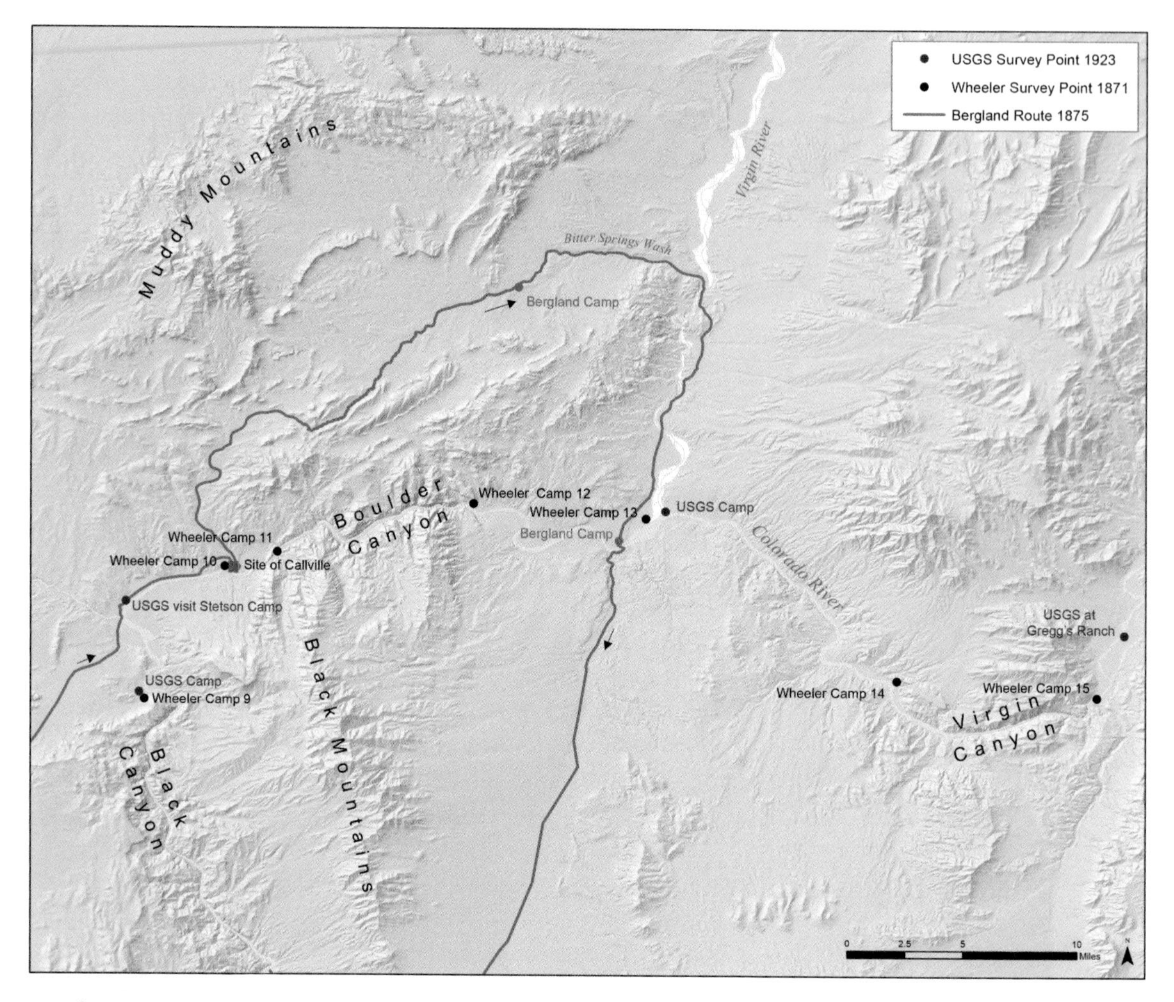

Map 5.4. Colorado River Surveys. After the first explorers came later parties along the river. Wheeler went upriver, while a 1923 USGS survey came down the river looking for places to build a dam. Map based on accounts of the various expeditions.

from Camp Mohave, Arizona, to the Grand Canyon. Because of their slow travel against the current, they made many camps and took four days just to climb Black Canyon. They created the first widely used maps of the region and put Fortification Hill in its present position. (They referred to Ives' Fortification Hill as Black Butte.) Photographer Timothy O'Sullivan, a member of the expedition, also captured the first photographic images of this area.

Wheeler continued his annual expeditions in later years but did not return to the Lake Mead country. However, one detachment of his 1875 expedition, led by Lieutenant Eric Bergland, was assigned to carry out a reconnaissance in this area. Bergland traveled overland upriver from Fort Mojave, bypassing Black Canyon to the west and reaching the Colorado at the mouth of Las Vegas Wash. From there his group went upriver to the site of old Callville, then went up the old road to the Virgin River before heading down that river back to the Colorado River at Stone's Ferry. They camped here for almost two weeks before crossing the Colorado and heading south up Detrital Wash. One

Entrance to Black Canyon, Colorado River From Above [upstream], 1871, photographed during the Wheeler Expedition up the Colorado. Hoover Dam would be built sixty years later, not far around the bend in the canyon; this is now the deepest part of Lake Mead. Courtesy of the Library of Congress. Photograph by Timothy O'Sullivan.

accomplishment of their stay was a detailed map of the Stone's Ferry region, including the Salt Well.

Several later expeditions descended the river to investigate it as a resource to be put to use. Robert Stanton led a survey through the Grand Canyon in 1890 to investigate the possibility of using the river for a railroad line from Grand Junction, Colorado, to Yuma, Arizona (map 10.3). A US Geological Survey (USGS) expedition in 1923 rafted the Grand Canyon, surveying potential damsites in support of the federal government's ambitions to harness the Colorado. They stopped at Pearce's Ferry and likely changed Pearce to Pierce with a misspelling before camping at the mouth of Grand Wash. They descended Iceberg Canyon the next day and stopped at Gregg's ranch and ferry, where they purchased canned goods and homegrown fruits from the Smith family (a photograph by famed Grand Canyon photographer Emery Kolb recorded the meeting). They noted Temple Butte, Napoleon's Tomb, and abandoned mining works before stopping to make camp at the mouth of the Virgin River. Bonelli's Ferry was abandoned by this time, and the ruin of a stone house was the only structure visible at this deserted location. It was the end of an era for this location, which had once served as a crossroads of a vast desert region.

By the 1920s, the Virgin and Muddy River Valleys were rapidly modernizing and becoming integrated into the larger world. The Union Pacific Railroad completed a branch line to St. Thomas in 1912, connecting the region to the outside world. Modern highways soon followed. The Arrowhead Trail, the first automobile highway between Southern Nevada and Utah, was completed through St. Thomas in 1915. This road soon became part of the new Nevada state highway system (map 7.8) as Nevada Route 6. The old road south to Bonelli's Ferry was Nevada Route 12 by 1926, but it had been rerouted east toward the mines at Gold Butte by 1929.

The town of St. Thomas had a small commercial street with a garage, store, post office, and even an ice cream parlor. However, Overton had long since eclipsed it as the main commercial center of the Muddy Valley, and St. Thomas declined after the Arrowhead Trail was rerouted to the north in the early 1920s. Much of the land along the Muddy and on the Virgin to the north and south of the Narrows was being farmed or used as pasture. About five thousand acres were being farmed, with cattle being grazed in the desert beyond. Alfalfa and grains now made up the primary crops, with cantaloupe production having decreased considerably.

General Land Office records show that very few individuals

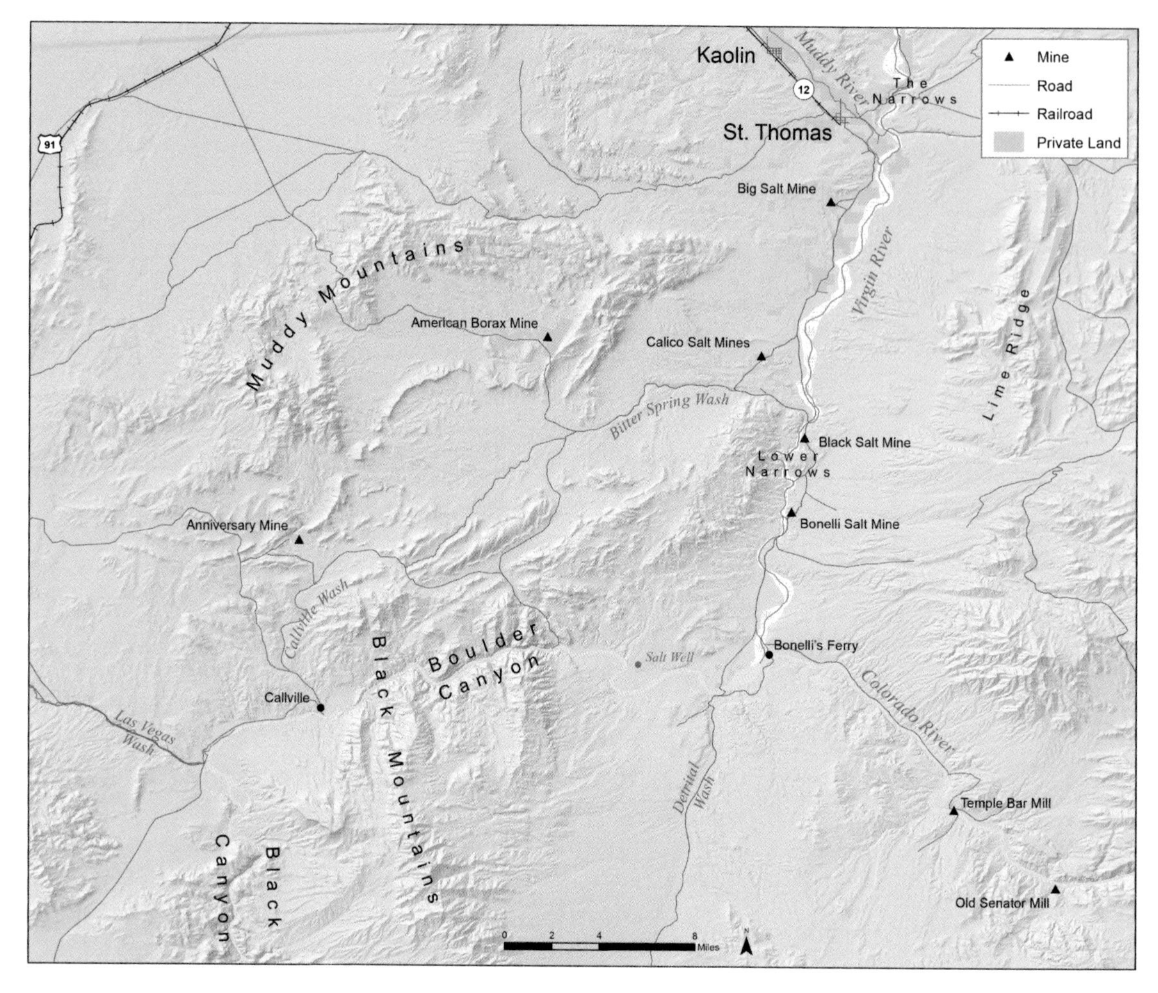

Map 5.5. Before Lake Mead. Locations of towns, mines, roads, and private lands circa 1930 are based on USGS topographic maps, Nevada Highway Department road maps, and Smith and WESTEC Services Inc.

in the area had bothered to homestead or patent the land they were using before the Boulder (now Hoover) Dam project began; there was simply no need to. The initial settlers functioned under the guidance and control of the Church of Jesus Christ of Latter-day Saints, which allocated land to families. The absence of patented land was a common feature of Mormon settlements. Only when government agents began discussing purchase of these lands before they were flooded by Lake Mead was holding title important.

Salt and gypsum mining had been carried out in the area using the St. Thomas branch railway to haul out ore. The Big Salt Cliff, Black, and Bonelli mines were active at this time. Some copper mining had been undertaken to the east of the Virgin Valley in the 1910s, for which St. Thomas provided supplies and was the point at which ore was loaded for shipment. The road to the mines crossed the Virgin River and ran fifty-four miles to the mines using horse-drawn wagons, supplemented briefly with a truck in 1913. Borax prospectors located two extensive depots to the west of the Virgin Valley in 1920, but these were accessed by roads from the west. All of these activities would soon end because of events far to the south and then in the nation's capital that would transform the region beyond anything the explorers and settlers could imagine.

The origin of Hoover Dam and Lake Mead had nothing to do with Las Vegas, but was instead a result of agricultural development in California's Imperial Valley and along the lower Colorado River. It was evident that enormous areas of the desert near the river could be farmed if water were obtained, but the Colorado had a highly variable flow during the year and was very inconsistent between years. It also carried a tremendous silt load, which would clog irrigation ditches and stifle farm fields. Taming the Colorado was beyond the capabilities of farmers.

The transformation of the Colorado River came slowly in a series of steps in laws passed in Washington, DC. The passage of the Newlands Reclamation Act in 1902 authorized the federal government to build dams for irrigation, and the Reclamation Service (Later the Bureau of Reclamation) was created to carry out its provisions. Coincidently, taming the Colorado became the focus of national attention following the accidental diversion of the entire river into California's Salton Sink from March 1905 to February 1907, creating the Salton Sea. The need for one or more dams along the river to control floods and store water for irrigation was inescapable, but the scope of the project made such a dam a difficult undertaking.

Two important steps were taken in 1922. The Fall-Davis report on the irrigation problems of the Imperial Valley called for a huge dam on the Colorado River in or near Boulder Canyon. This was necessary to control floods and to store water for irrigation. The Colorado River Compact signed that year divided the water of the Colorado River watershed into upper and lower basins at Lees Ferry, Arizona. The upper basin, where most of the river's flow came from, consists of Wyoming, Colorado, New Mexico, and Utah. The lower basin is made up of Arizona, Nevada, and California. Each basin was allocated 7.5 million acre-feet of water. The amount to be given to each state was not settled until 1964 when the Supreme Court settled a case between California and Arizona. California was to receive 58.7 percent of the 7.5 million acre-feet of water in the lower basin, Arizona received 37.3 percent, and Nevada only 4 percent. (Were the river flow to exceed 7.5 million acre-feet, Arizona and California would split the surplus evenly.) One million more acre-feet was later guaranteed to Mexico. Unfortunately, the total flow of the river was assumed to be much greater than existed, leading to persistent shortfalls in flow. Today California, Nevada, and Arizona are

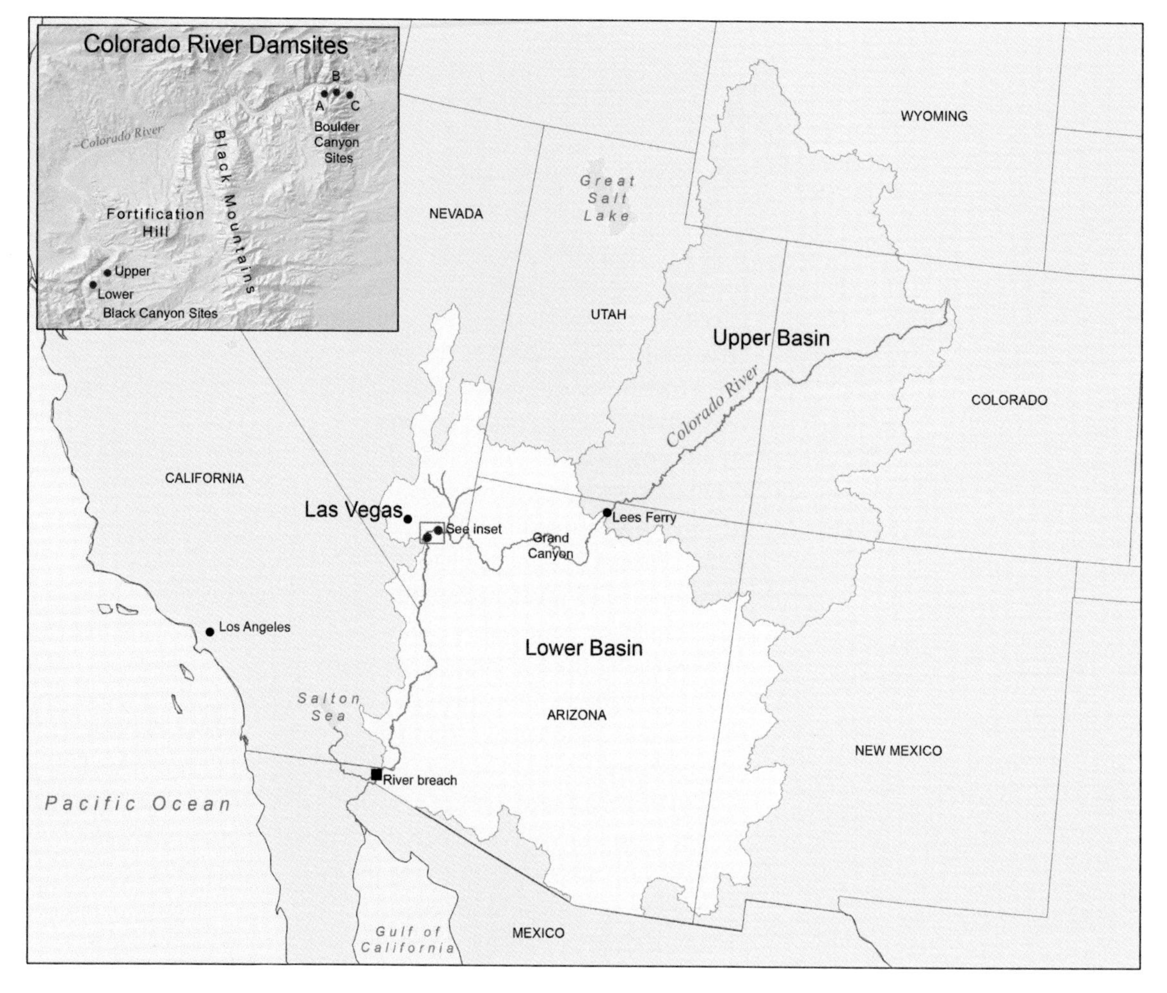

guaranteed a certain delivery of water based on how much is in Lake Mead; there is no guarantee there will be enough water for all three states.

The US Geological Survey (USGS) was given the responsibility to survey potential damsites. By 1920, possible sites had been narrowed to five locations in Boulder Canyon and two in Black Canyon. Survey camps in both canyons were established and extensive surveys were conducted, narrowing the candidate locations down to Site C in Boulder Canyon and Site D in Black Canyon. Both sites were considered sound, but in November 1928, the USGS chose Site D. In addition to being closer to railroad lines with easy access, the bedrock was shallower and would require less concrete to build, resulting in a larger reservoir with the same dam height. Although the selected damsite was in Black Canyon, the name Boulder remained associated with the project.

The Boulder Dam Project Act of 1928 authorized the construction of the dam at site D. Now it was time to build.

The basic plan for building Hoover Dam was to create new transport routes to the damsite; construct housing for the workers;

Map 5.6. Taming the Colorado. The Colorado River watershed was divided into upper and lower basins in the 1920s, and several damsites were investigated near Las Vegas. *Inset*: the five damsite locations surveyed by geologists in Boulder and Black Canyons. Based on watersheds obtained from the USGS NationalMap.gov website and a 1948 Bureau of Reclamation report.

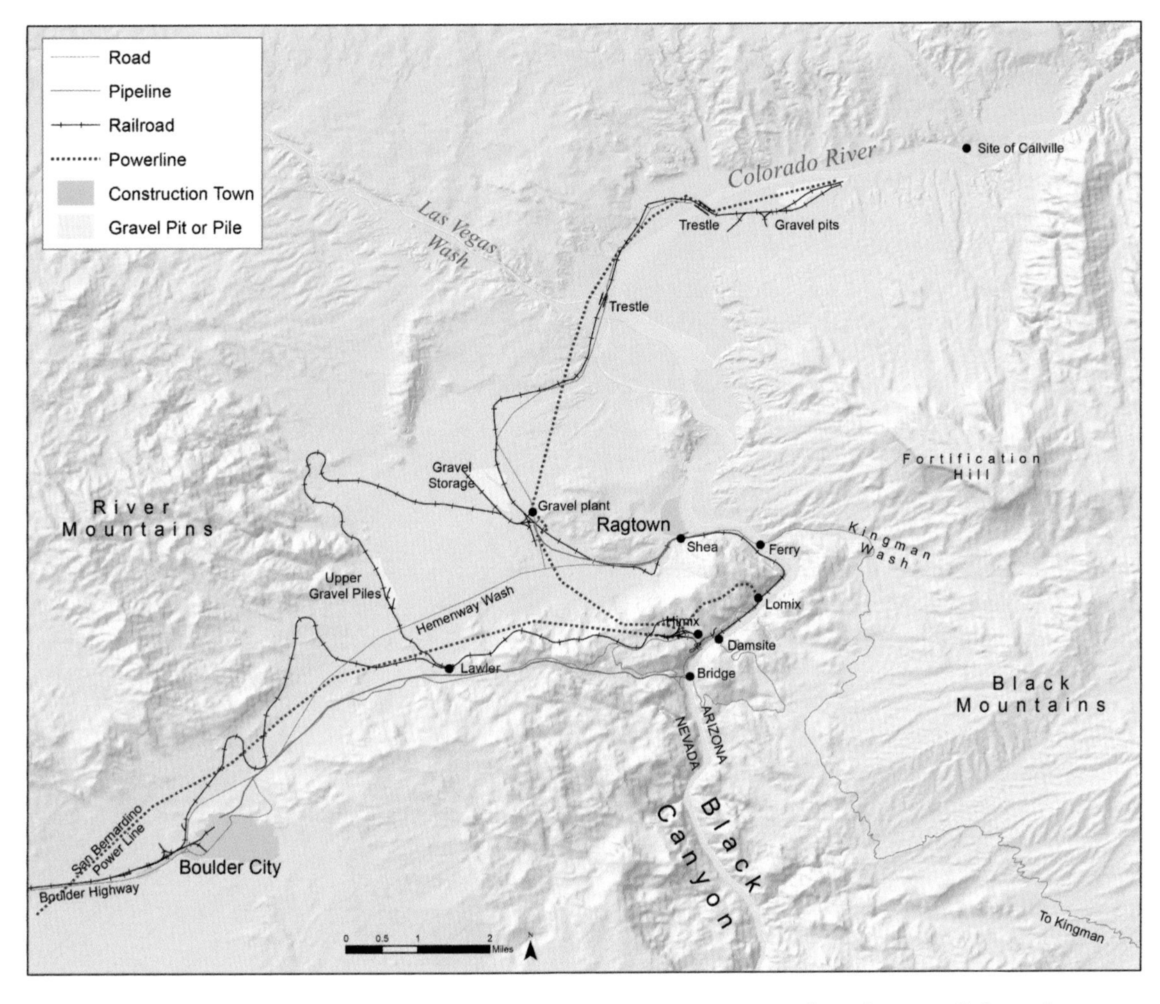

Map 5.7. Building Hoover Dam. The enormous infrastructure required to build the dam included roads, railroads, power lines, gravel pits, and towns for workers to live.

build tunnels through the canyon walls and use small dams above and below the damsite to divert the Colorado River around the construction area; excavate the now dry riverbed down to bedrock; pour a nearly endless procession of buckets of concrete to build the dam; and finally install the electrical power plant.

Construction on this project began September 17, 1930, when work commenced on a railroad line at a junction on the Union Pacific Railroad south of Las Vegas. The new line was built leaving the Las Vegas Valley through Railroad Pass and running 22.7 miles to a point just west of the future Boulder City. From there the Bureau of Reclamation built a 10.5-mile line down Hemenway Wash and over to the damsite, called the US Government Construction Railroad. The contractor selected for the dam's construction, Six Companies, Inc., built another 26 miles of rail line. New roads included Boulder Highway, built by the Nevada Highway Department from Las Vegas to Boulder City in 1931, a road down Hemenway Wash to the river, another through the mountains to the damsite, and one to the river below the damsite through a 1,974-foot tunnel. Other infrastructure required to support construction included a 222-mile power line from San Bernardino, completed in June 1931.

Housing for the workers was to be in the new model town of Boulder City, located outside the canyon because of its higher elevation and the abundance of flat land (map 5.8). But prospective workers desperate for employment began arriving long before housing was ready. By early 1931 a temporary settlement of tents sprang up along the river just above the entrance to Black Canyon, officially named Williamsville but more often known as Ragtown. Up to 1,500 people lived there, along with the nearby River Camp, a Bureau of Reclamation worker's camp built near the head of the canyon. Both were short lived; Ragtown was cleared out by authorities on August 12, 1931, following a strike, and River Camp was replaced by new housing in Boulder City in late 1931.

Two diversion tunnels were built for the river to bypass the construction site, with excavated rock and gravel from Hemenway Wash used to construct diversion dams. The upper and lower diversion dams and the two bypass tunnels were completed in November 1932, and the riverbed was drained and excavated to bedrock by the next June. The way was now clear to build the dam.

This required further infrastructure, including gravel pits to supply the gravel needed to make concrete, a gravel plant to wash and sort this gravel, and two cement plants to mix the enormous amounts of concrete out of which the dam would be made. These plants were called Lomix and Himix, with the former in Black Canyon and the latter on the rim of the canyon. The gravel plant was the center of the entire dam construction operation. This facility received regular rail shipments of gravel that were sorted and washed before being loaded onto more rail cars for shipment to the Lomix or Himix cement plants at the damsite.

Arizona Gravel Pit, 1932. Courtesy of the National Archives and Records Administration. Photograph by Bureau of Reclamation.

The major component of concrete is gravel, and so a large source for gravel meeting exacting standards was needed for the dam. Sixteen gravel deposits in the area were examined, with the chosen site upriver on a terrace above the river on the Arizona side, to be reached by a rail line. Wooden trestles were built to carry the line over Las Vegas Wash and the Colorado River; the former was washed out and rebuilt, the latter survived several annual floods without failure.

The gravel plant was connected by rail to the damsite via two lines. The Loline ran down to the Colorado River and then descended Black Canyon to the Lomix cement plant, which was used in the early stages of construction. This section was double-tracked and required a short tunnel. The Hiline climbed up Hemenway Wash and then crossed through several tunnels to the Himix cement plant, which saw increasing use as the dam rose. A junction on this line also connected the network to Las Vegas and other regional centers.

The first concrete for the dam was poured in November 1932 and continued nonstop until February 1935. Nine trains moved gravel and other supplies through this system around the clock. After the electrical power plant was installed, the dam began to generate electricity in September 1936. By this time more power lines had been built to send electricity to Los Angeles, Las Vegas, Kingman, and Needles (map 8.17). The original power line for the damsite was later used to transmit electricity to Los Angeles. The power lines were not the only construction infrastructure to be reused: the high road to the damsite was incorporated into US 93 and 466 (map 7.11).

Gravel Plant, the heart of the Hoover Dam railroad network, 1932. Courtesy of the National Archives and Records Administration. Photograph by Bureau of Reclamation.

Remaining construction facilities such as the cement plants, gravel plant, and gravel pits were stripped of anything salvageable and abandoned before the lake rose above their elevations in mid-1935. In case of any new demands or delays, piles of sorted and cleaned gravel were created above the future high-water mark on the line to the Himix plant. These gravel piles remain and can be seen to the west of Lakeshore Drive after entering Lake Mead National Recreation Area. Construction on the project did not end until 1961 when the last generators were installed. These were brought in using the rail line to the former Himix plant, after which the tracks were removed.

After basic transportation had been provided to the damsite, the next step in building the dam was to provide for worker housing. A new town was planned and desperately needed, as a tent settlement sprang up along the river just above Black Canyon once the dam project began (map 5.7). The name of the new town was based on the original damsite in Boulder Canyon; neither the town nor the project was renamed when it was decided to put the dam in Black Canyon.

The location of Boulder City was chosen because of its higher elevation, giving some relief from the oven-like temperatures along the Colorado River, and the abundance of flat land. The town plan was inspired by that of Radburn, New Jersey, which was built in 1929 as a model community using the idea of superblocks. This was a modular design with houses grouped around the outside of blocks and schools or parks on the interior. Pedestrian travel and cars would be kept apart, commercial and industrial land uses separated, and a civic center would be centrally located.

Boulder City became the first model town built by the federal government using these concepts. The townsite had a triangular plan, with government buildings and a park at the highest point and roads radiating away from it to the south. The main streets were named after the states in the Colorado River Basin,

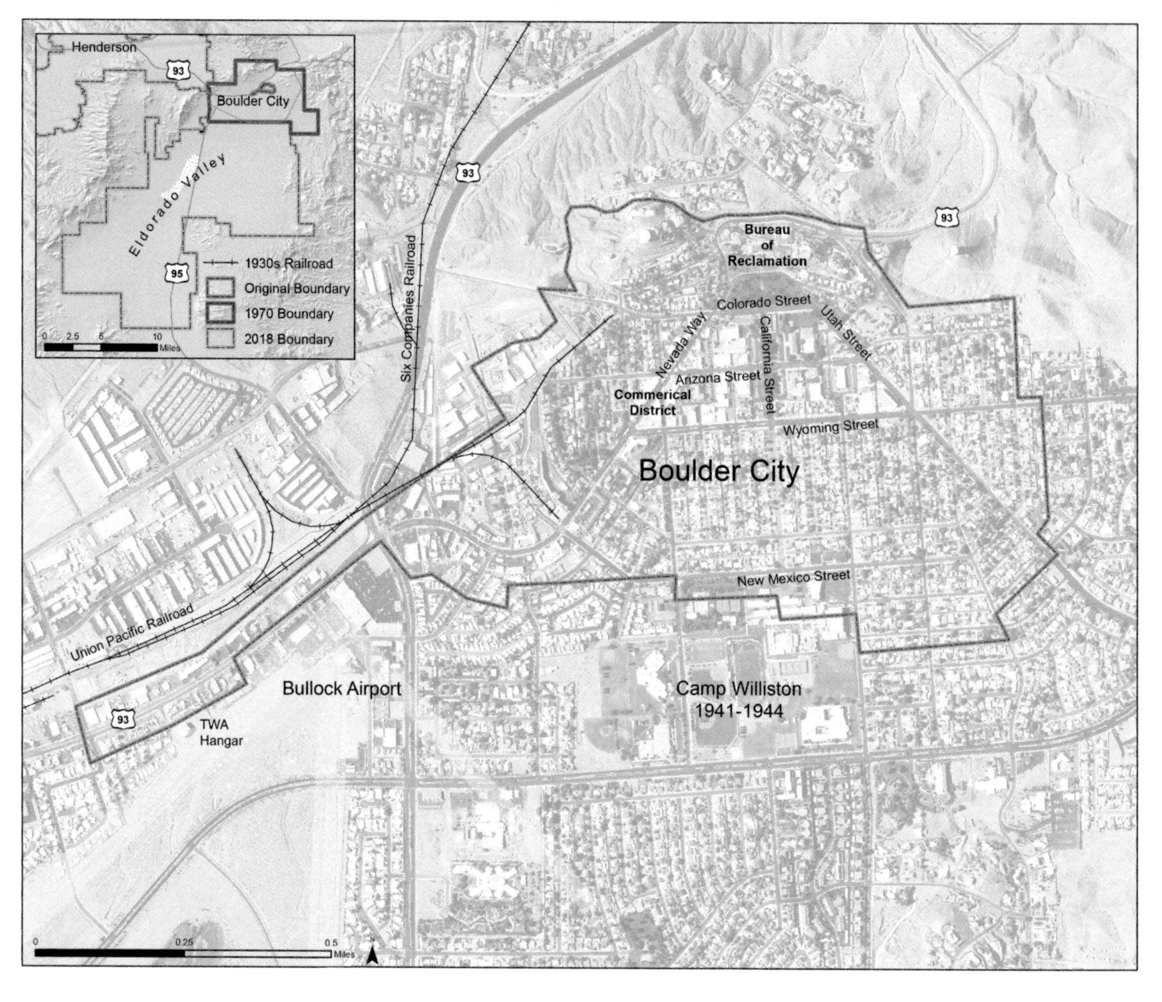

Map 5.8. Boulder City. The town's original boundaries, railroads, and airport runways are depicted with a modern aerial photo as a backdrop. Inset: The city boundaries have grown tremendously since the 1930s.

commemorating the reason for the town's existence. The government controlled all requests to operate businesses, and commercial buildings for retail, restaurants, and other needs were intended for several blocks on Boulder City Square. Most were built along Nevada Way and Wyoming and Arizona Streets. Some versions of the original city plan called for a forested park to be built around the city between New Mexico/Avenue L and Fifth Street/Avenue M, with two major curving streets south of this area to enclose more residential areas and a golf course built south of that; these developments did not come to pass.

The new Boulder Highway connected the town to Las Vegas in 1931, and other transportation options were created when Bullock airport opened on December 10, 1933. Grand Canyon Airlines became the first flightseeing airline in the region when it started service to the Grand Canyon and back from there in 1936. Trans World Airlines (TWA) became the second airline to serve the Las Vegas area when it began a Los Angeles-to-Newark, New Jersey, service with a stop in Boulder City in August 1937, a daily service it called the "Grand Canyon Route." Two of the runways were abandoned over time, but the east-west runway survived until the city's new airport opened in 1990. A hangar dating from the TWA period still stands at the former airport.

Boulder City in 1932. Courtesy of the National Archives and Records Administration. Photograph by Bureau of Reclamation.

The town reached a peak population of about five thousand during the dam's construction, but it fell to about half of that when primary construction ended. It slowly grew again in the postwar years, but its success posed a challenge. The Bureau of Reclamation, with users of Hoover Dam electricity paying for the city's administration, governed the city. The bureau had never intended to be in the permanent town administration business, and the dam's customers were not interested in paying for it. A new city government was established in 1960 to solve this dilemma. While the town had been a federal reservation, no gambling was allowed and alcohol sales were limited; these policies were popular and remained a part of the town's appeal for many. The Boulder City Historic District was created in 1983 and contains most of the original townsite and original buildings (map 11.8).

Another unique characteristic of Boulder City in Southern Nevada is its stringent growth controls, limiting home construction to a maximum of 120 each year, with consequent high home values. The city remains the smallest city by population in the Las Vegas area but has spread south and into Hemenway Wash; it is now more than two hundred square miles, making it the largest city in Southern Nevada by land area (map 8.2). Most of northern Eldorado Valley is within the city limits, but the city's restriction on growth effectively removes this area from urban development. It does, however, contain an enormous electrical infrastructure (map 8.17).

The completed Hoover Dam, 1942. Courtesy of the National Archives and Records Administration. Photograph by Ansel Adams.

While Hoover Dam took years of steady work to be created, Lake Mead came into existence remarkably quickly. The lake began filling on February 1, 1935, when the diversion gates were closed, and the rising waters reached the head of Boulder Canyon by the middle of the month. The arrival of the spring flood rapidly raised the water level to Pearce Basin by midsummer, where it remained until the floods of 1936. This raised it well into the Grand Canyon to Bridge Canyon rapids, the lake's greatest-ever extent. In this short span of about eighteen months, the previous century of exploration and settlement was buried under water. Settlements, farms, ferries, roads, trails, the old terraces, and local landmarks disappeared.

The geography of the new lake was similar to the earlier landscapes in that it had several broad basins—Boulder, Virgin, Temple, and Gregg—connected by narrow canyons. The Virgin River valley became the Overton Arm of Lake Mead while Las Vegas Wash became Las Vegas Bay. The Temple and Fortification

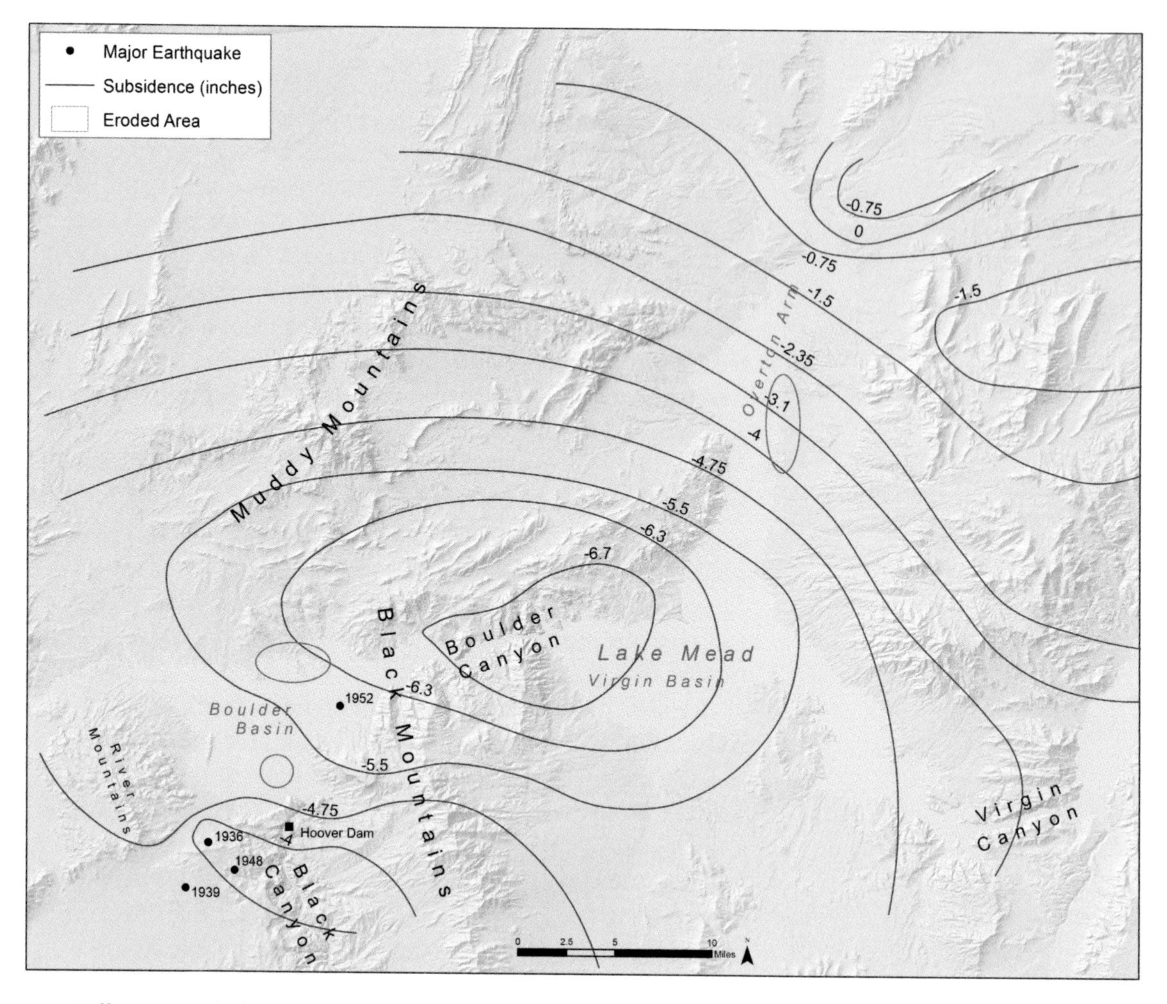

Map 5.9. Lake Mead's Early Years. The land surface sunk (subsidence) because of the new reservoir's weight. Several areas where erosion and slumping were a problem are identified in blue outlines.

Hill remained above water, while many small islands appeared throughout the lake.

Wave action and saturation of the soil soon changed the lake's shoreline. Landslides above water were noted in Virgin Canyon, near Hualapai Wash and on the Temple, and several loosely consolidated cliff areas have eroded because of wave action in Boulder Basin and in the Overton Arm. Those cliffs in the Overton Arm included salt deposits, especially near the Calico Mine. The saturation of soil by water caused underwater landslides or slumping along the north end of Boulder Basin and downstream from Sentinel Island, where sand dunes were once visible. Wave-cut terraces are clearly visible throughout the lake, much as they are around ancient dry lakes in the Great Basin. Some hills and islands have been cut down by thirty feet or more by wave action, especially in Las Vegas Bay and the east side of the Overton Arm.

The immense weight of water and sediment caused subsidence in the lake area by as much as 6.7 inches near Boulder Canyon, which is literally the center of gravity for the reservoir.

Lake Mead's subsidence was accompanied by a number of significant earthquakes (and hundreds of minor ones) occurring into the 1960s. These suggest that ancient faults have been reactivated by the subsidence—the Las Vegas Valley has experienced even greater subsidence, but that is because of groundwater pumping and not related to Lake Mead (map 6.5). Similar effects have been found with other reservoirs, though the earthquakes were not always minor. On May 12, 2008, a magnitude 8.0 earthquake hit Sichuan, China, following the filling of a nearby reservoir. That earthquake killed almost seventy thousand people.

Large lakes can have significant impacts on local climatic conditions (such as "lake effect snow" in the Midwest), and it was thought that Lake Mead would be large enough to produce a measurable increase in precipitation. This effect did not turn out to exist, nor did the predicted smoother flying weather expected to result from the cool lake waters reducing updrafts in the desert. Changes in vegetation, however, especially salt cedar's arrival, accompanied the untamed river's transition to a reservoir. This invasive species has proved extremely resilient to changes in lake elevation, even being able to survive partial immersion for several weeks.

The potential for recreation at the new reservoir was recognized early, but the Bureau of Reclamation was not interested in developing it. Instead, an agreement was created with the National Park Service (NPS) to develop the lake with new roads, campgrounds, marinas, and other recreational facilities. This new recreation area was initially known as Boulder Dam Recreation Area and in 1947 became Lake Mead National Recreation Area. This was a new kind of park unit in the nation's park system; up to the 1930s the system consisted of national parks showcasing outstanding natural beauty and national monuments preserving sites of scientific, historical, or archaeological significance. But in 1933 new kinds of places began to be added to the system, including battlefields, parkways, historic sites, and finally recreation areas. That the NPS would take over recreation for the new reservoir was not without irony and controversy: the agency had fought (and lost) political battles to prevent reservoirs from being created within national parks. Nonetheless, the NPS sought additional opportunities to subcontract recreation resources for the Bureau of Reclamation, including Glen Canyon National Recreation Area farther up the Colorado River.

The lake opened up a very large area to boat travel, including many areas never traversed by roads. The NPS put the Civilian Conservation Corps (CCC) to work developing recreational facilities in Hemenway Wash (map 4.7) beginning in 1936. Others followed in Callville, Las Vegas, and Echo Washes; the large washes

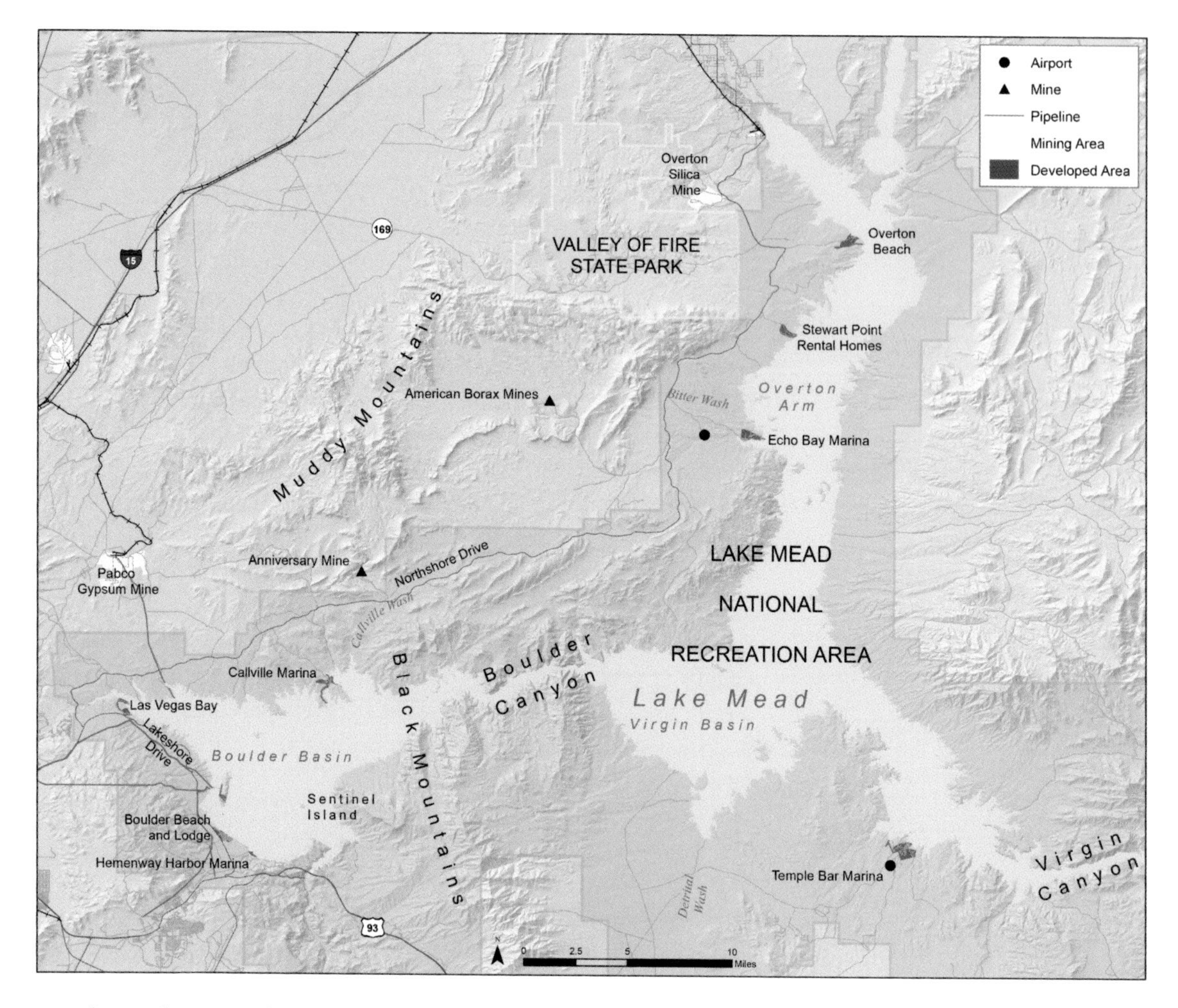

Map 5.10. Lake Mead Developed. The new lake was quickly developed for recreation, with new roads and marinas. A variety of users also built pipelines to obtain water from the new lake.

that earlier provided access to and from the river now provided sheltered bays for boaters. A lodge on the lakeshore in Hemenway Wash opened in 1941 but permanently closed in 2009 because of a receding lakeshore and competition from hotels in Las Vegas. The Bureau of Reclamation also leased homes to individuals, making Lake Mead one of the few national park units with a resident population. These homes remain at Stewarts Point and Meadview in Arizona. Airplane landing fields were also built at Echo Bay, Temple Bar, and near Pearce Ferry. Recreation development continued after World War II, including the completion of the Echo Bay and Las Vegas Bay facilities (map 9.4).

The Bureau of Public Roads proposed two new roads along the lakeshore, one to run along the west side from Overton to Boulder City and the other on the east from Overton to Pearce Ferry, but only Lakeshore Road was built from US 93 near Boulder City to Henderson. This road was finished in 1943. Northshore Drive was built in the postwar era along the west side of the lake from Callville and Overton. These roads were new and had no relation to pre-lake roads; they were built along the valley slopes well above the lake and old riverbed, with spur roads leading down to the lakeshore.

The names of mountains, washes, and other features above the water line survived the filling of the lake, and many new place-names also appeared on the map. The lower section of Las Vegas Wash became Las Vegas Bay, while Callville Bay memorializes the lower part of Callville Wash, and Echo Bay flooded Echo Wash. Sentinel Peak became Sentinel Island. The Big Cliff salt mine is memorialized with Salt Bay and Cove. The Temple, Campanile Rock, and Fortification Hill are among the few place-names along the lakeshore that have survived (though the rock formations themselves have undergone erosion as the lake filled). Many new place-names have appeared for formerly unnamed washes along the lakeshore as well as islands that were previously anonymous ridges, such as Beacon Rock, Boxcar Cove, or Middle Point. New marinas and boat ramps took their names from nearby features, such as the Las Vegas Bay, Callville Bay, and Echo Bay marinas. Separated by a few miles, Temple Bar Landing has no relation to the submerged Temple Bar. Several developed access points were left high and dry as water levels have fallen in recent years, prompting Overton Beach and Pearce Ferry to disappear from the official park map. Las Vegas Bay marina, now a mile from open water, may be the next to be removed from the map.

But while most of the place-names in this area were based on earlier ones, the two most prominent are new: Hoover and Mead. The dam had originally been named for Boulder Canyon, renamed for President Herbert Hoover in 1931, renamed Boulder Dam in March 1933, and finally for Hoover again on April 30, 1947. The lake was named after Elwood Mead, an engineer who supervised many large dams, on February 6, 1936, a month after his death.

All reservoirs are temporary. Once water is impounded, the reservoir will gradually fill with silt and lose its water storage and flood control ability. The muddy Colorado has been depositing its sediment load where the flowing river reached slack water since Hoover Dam was finished in 1935. The lake reached its farthest extent upriver near Bridge Canyon in a part of the lower Grand Canyon called Granite Gorge. As more sediment was deposited, the canyon floor filled in and the slack water moved down river. A similar process occurred along the Virgin River, though involving much less sedimentation.

A 1960 report concluded that Lake Mead would not fill up with sediment for at least four hundred years. In the early years after the lake filled, it was thought that clays deposited on the bottom of the lake might be mined, with the expectation that this would further prolong the life of the reservoir. This never

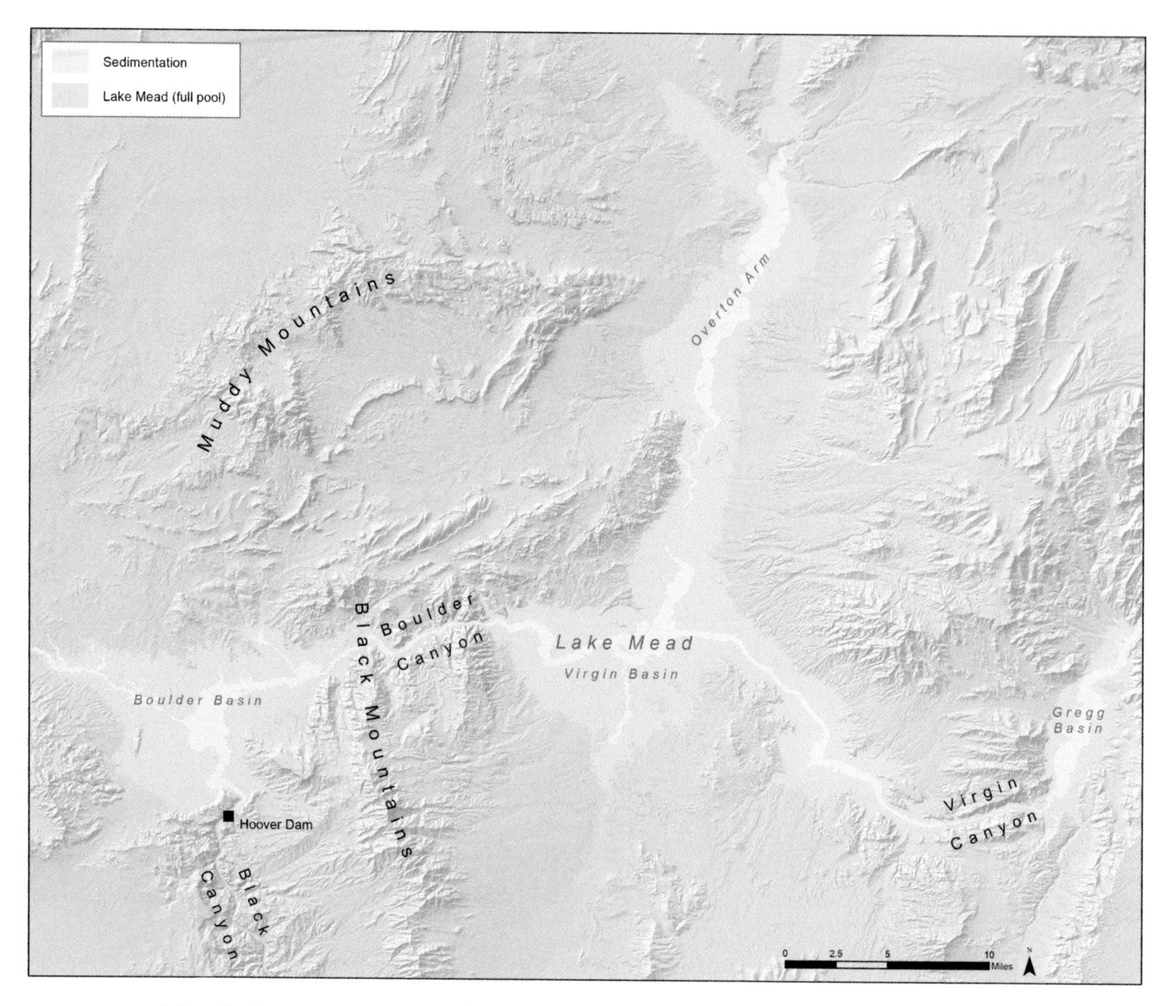

Map 5.11. Sedimentation. As the Colorado and Virgin Rivers flow into the still lake, the mud they carry slowly settles onto the floor of the lake. Several areas are buried under this sediment.

occurred, but the later construction of Glen Canyon would intercept much of the river's sediment load and prolong Lake Mead's life. A 2001 study estimated it would take several thousand years for sediment to fill Lake Mead. There is no need to worry about the lake filling with mud anytime soon.

What would the lake look like if it had filled up with sediment? A 1948 report provided a glimpse of what might happen. Sedimentation would not only fill the reservoir but rise above the current lake level upstream, as the new ground level created by sediment would slope downward at the 1.25 feet-to-the-mile slope of the Colorado River to the crest of the dam. Sediment would extend up to river mile 221.5, well within the Grand Canyon, extending the reservoir's effects 13.5 miles farther upriver than it did in 1937, and reaching an elevation of 1,365 feet. Sediment would create a natural dam across the Overton Arm, leaving a small lake filled by the Virgin River. This has already occurred on a smaller scale at Grand Wash Bay below Pearce Ferry because of drawdown (map 5.12).

A hypothetical map of the lake at this terminal stage showed

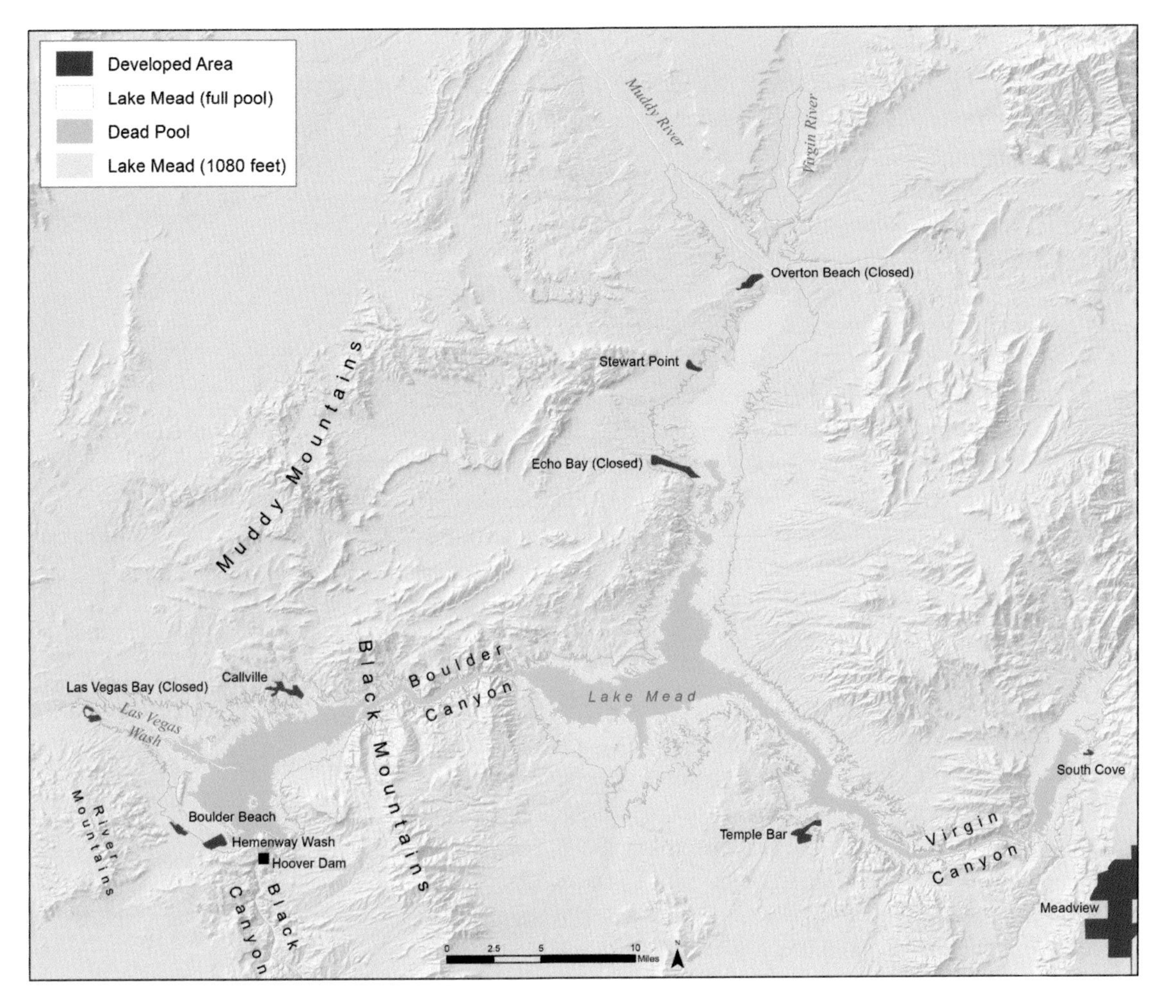

Map 5.12. The Dead Pool. Lake Mead has not been full since 1983 and has been steadily shrinking because of prolonged drought. Should this continue, the lake would eventually reach dead pool, the height of the lowest turbine intake, at which point power production will shut down.

Lake Mead as a small, shallow body of water extending from the dam to the upper end of Black Canyon. The dam would have very limited water storage capability and would be of very little use for flood control. But because most Colorado River water would flow over the spillways, the lake would not likely silt up past this point. This terminal stage of Lake Mead was expected to take place in about 520 years, or by the year 2455. However, this future is not likely to occur; a prolonged drought has dropped the lake to levels not seen since it started filling, and all forecasts are that it will continue to drop.

A drought since 2000 has substantially lowered the surface level of Lake Mead to record low levels; in July 2016 it dropped to a new record low of 1,071 feet, and in 2021 it dropped even lower. This drawdown in the lake's water levels has changed the geography of Lake Mead in many ways. Saddle Island is no longer an island; the tiny Boulder Islands have grown and merged into a larger island; new islands have appeared in the Virgin Basin; and the Colorado River now flows freely many more miles before reaching the lake's still waters. Grand Wash Bay, a former arm of the lake, has been cut off from the river by a sandbar and is now a dry lakebed. The Overton Arm has retreated several miles, resulting in the well-known reemergence of the St. Thomas townsite. The site of Pueblo Grande is also above water, as is at least one of the old salt mines along the Virgin River.

Another noticeable impact of drawdown has been that the lake's marinas and boat ramps have been left high and dry. Las Vegas Marina is now more than a mile from the lake. The marina was moved in 2002, and it was marked as a boat storage area on official park maps, while Overton Marina no longer shows up at all. Some boat ramps have been extended, and Hemenway Marina now has one more than a half-mile long. Temple Bar, Echo Bay, and Callville Marinas are still open, but the floating docks have moved farther and farther away from the onshore buildings and boat ramps. Backcountry roads will likewise be extended downhill to the current lakeshore.

If (or when) Lake Mead drops below 895 feet, no water will be able to flow through Hoover Dam, though the lake level could still drop because of evaporation and reduced river flow. This condition is known as dead pool. If that happens, the Boulder islands will become a peninsula, Overton Arm will be reduced to a bay, and Gregg Basin will be the upper end of the lake. Although more of the Colorado and Virgin Rivers will be free flowing again, the landscape will have been tremendously altered by decades of erosion, and the bluffs along the rivers will be reduced to rounded hills. A huge thicket of salt cedar will likely invade the river bottom all the way to the lakeshore.

Hoover Dam and the Mike O'Callaghan-Pat Tillman Memorial Bridge in 2010, showing drawdown of lake level and the characteristic "bathtub ring." Courtesy of the Bureau of Reclamation. Photograph by the Federal Highway Administration.

In the 1920s a geologist noted rock inscriptions near the mouth of Boulder Wash dating back as far as 1864, but he did not describe what they said or who wrote them. This location was flooded by Lake Mead and forgotten. Many underwater places have been exposed by falling water levels in recent decades. What is left under Lake Mead, and will this inscription ever be seen again?

Following the construction of many large reservoirs after World War II, archaeologists began to be concerned about the loss of archaeological and historical sites. Very little was known about what might happen to artifacts submerged or buried under sediment. Artifacts could be subject to wave action, varying pH levels and temperatures, and buried under sediment. Much depends on exactly where an artifact is within the reservoir. Locations within the reservoir can be divided into five different zones, corresponding to permanent storage or dead pool (Zone 1), the fluctuation or drawdown zone (Zone 2), and a higher zone of occasional flooding (Zone 3). Zone 4 includes land above the highest water, while Zone 5 is the river corridor below the dam.

The zone of occasional flooding includes areas above the

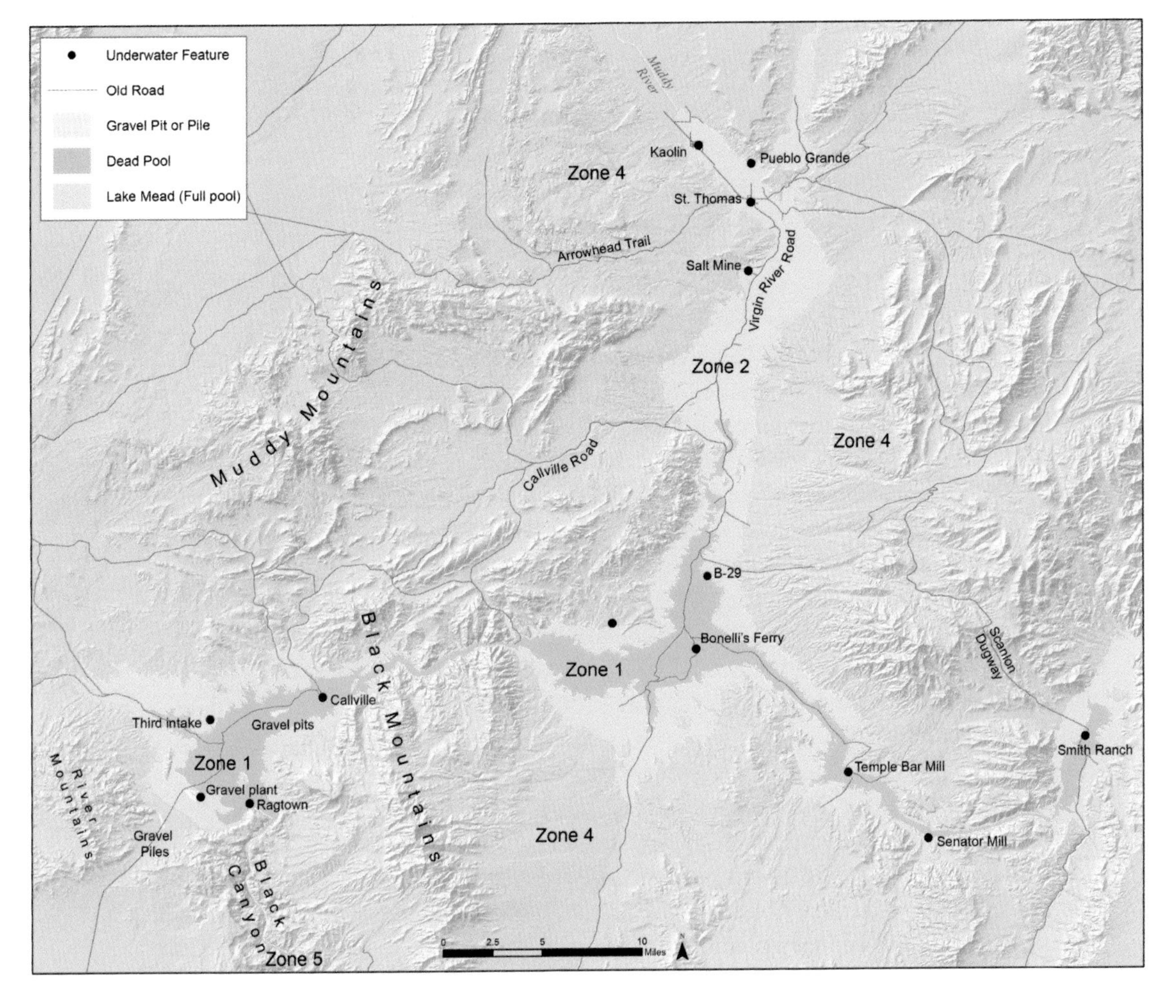

Map 5.13. The Return of the Lost City. Lake Mead can be divided into several archaeological zones. Zone 1 includes the lake bottom under dead pool, which will likely never be exposed to sunlight again. Zone 2 is the drawdown zone, where archaeological artifacts have been exposed by dropping water levels and waves. Zone 4 is land above the highest lake level, and Zone 5 is land below Hoover Dam. Zone 3, elevations above normal full pool elevation because of floods, only existed in 1983 and would not display at this scale.

normal pool elevation that may be underwater during flood conditions. While in some reservoirs this zone is inundated regularly, in the case of Lake Mead it has only happened twice, during the test of the floodgates in 1941 and high water during the summer of 1983. On July 24 of that year the lake reached its highest-ever level of 1,225.83 feet, with both floodgates open. This zone is not shown on the map because it is barely visible at this scale.

The drawdown zone lies between the full pool and dead pool elevations, or 1,219 to 895 feet above sea level at Lake Mead. In this zone water levels constantly fluctuate, turning hills into islands or islands into peninsulas, and places may appear or disappear depending on water levels. The most famous of these places in Lake Mead is the "Lost City" of St. Thomas, which reemerged from the waters in 1945, 1946, 1947, 1952, and 1964, and continuously since 2002. The abandoned town has become an attraction and the NPS has put up signs for visitors. It also published an informative guide about the town, allowing visitors to visualize what was there before. More places will emerge from the dropping waters of the lake. A water tank on a hill above the old gravel plant emerged from the waters in the early 2000s, and the site of the cement plant and railroad shops below it may emerge from the water in the future. This is also a zone in which wave erosion of hills and underwater slumping (an underwater mudslide because of unstable sediments) have been common. The shoreline terrain has been altered, and these processes have largely erased features such as the Mormon Temple Rocks and Salt Cave.

Building Ruin at St. Thomas exposed by drought, 2014. Photo by Prosfilaes.

The dead pool elevation in Lake Mead is 895 feet. Below this line, the water cannot be drained from the reservoir and so is not subject to drawdown. Bonelli's Ferry, Callville, Smith Ranch, Ragtown, the upper diversion dam, and the gravel pits are in this zone and so will remain under water. However, this also means they will not be subject to erosion by wave action as in the drawdown zone. Locations in this zone will eventually be buried under sediment, and some already have, including Callville, the Senator Mill and Temple Bar Mill placer gold mining operations, and the rock inscriptions noted near Boulder Wash. We will never know what was written there or who wrote them.

Downriver from Hoover Dam, relatively little has changed. The river immediately below the dam is closed to the public, and a number of construction features are preserved, including spoils piles and the old Boulder City water intake. Farther downriver can be visited by boat. Aside from a heavily regulated river flow and occasional salt cedar, the Colorado still looks much as it did when Jedediah Smith and Lieutenant Ives saw it.

6 Quenching Las Vegas's Thirst

The name Las Vegas derives from the well-watered meadows that once existed in the valley, and without this water there would be no city here. But as vital as water is to the existence of Las Vegas, it is strangely hard to see today. Most surface water in the area has long since disappeared, and modern water and sewer systems are underground and usually out of sight.

Where is Las Vegas's water, and what has happened to it over the years? This chapter covers the original water supplies of Las Vegas and the development of later water systems that supply millions of residents and visitors with water and remove their waste. The development of these systems has had huge impacts, drying up all the springs and creeks, eliminating some fish while introducing others, causing land elevations to change, and creating a new flowing river. And although it is not a topic most people consider, the development of the valley's flood control system is one of the city's biggest success stories and has made continued growth possible.

Before Hoover Dam, all of the Las Vegas Valley's water ultimately came from rain and snow in the Spring Mountains. These mountains receive much higher rainfall than the city because of their height, and the winter runoff flows down canyons and sinks into the desert soil of the Las Vegas and Pahrump Valleys (map 2.4). This groundwater was the source of Corn Creek, Tule, Kiel, Las Vegas Springs, and water sources along Duck Creek, which made settlement in Las Vegas possible.

The Las Vegas Springs were a group of mound or cauldron springs, with the water flowing out of the top of a low hill built up by sand and wind, and averaging ten feet high. This type of spring was common in Southern Nevada; Tule, Corn Creek, and several springs in Pahrump were also of this type. The springs were the source of Las Vegas Creek, which ran several miles through a mesquite and willow woodland. Pupfish once swam in the springs and creek, and a marsh or wetlands existed along the creek sometime before the mid-nineteenth century.

The Las Vegas Springs are thought to be as much as fifteen thousand years old, and until the 1960s they provided a reliable source of water for people and wildlife. Paiutes used them as camping sites, for agricultural irrigation, and for the plants that provided building materials and tools (map 3.2). Newcomers, who increasingly took over the valley's springs from 1855, also valued them. By the railroad's arrival in 1905, these springs had become the Las Vegas Rancho (map 3.8).

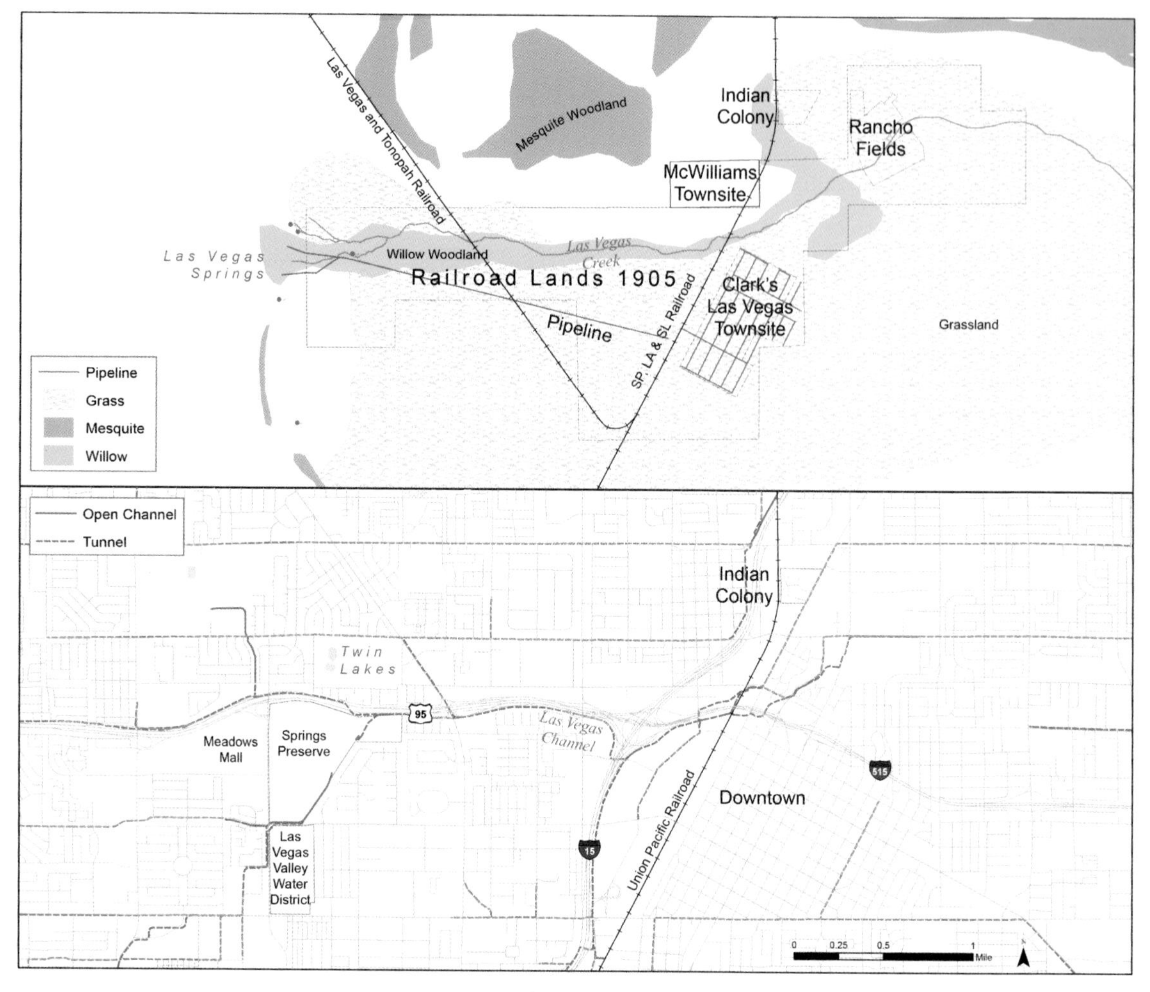

Map 6.1. The Meadows. *Top*: Las Vegas when it was founded in 1905; *bottom*: the same area in 2020. Las Vegas was founded next to a reliable water source and flowing stream, but all traces of that, along with the grass and trees that grew nearby, are gone today.

The San Pedro, Los Angeles and Salt Lake (later Union Pacific) Railroad created the Las Vegas Land and Water Company to take over ownership of all the land the railroad owned east of the tracks and provide water to the town. The first water supply was a ditch branching off the creek running to the railroad yard, with redwood pipes carrying the remaining water to the town. The springs were improved over time with fences, walls, and roofs installed to keep out people and animals. As the demand for water grew so did the water system. Pipelines were extended to the McWilliams Townsite (the Westside) in 1927, and the city's first reservoir was built in 1929. Some of the original redwood pipes remained in use for the city's water mains until 1940.

Cottonwood Springs was another of the valley's important early water sources, having been a stopping point on the Old Spanish Trail from 1829. This spring, now in the little town of Blue Diamond, was also used as a water source for the limestone mine at Sloan via a pipeline (map 4.2). Another pipeline ran from Willow Springs to a quarry in Red Rock Canyon. Water also moved by railroad in early Las Vegas; the Arden and

Jean rail stations initially received water from the Cottonwood Springs pipeline to Sloan, where it was pumped into rail tank cars for transfer to those stations.

Early water supplies were constantly coming up short because of the town's growth, the adoption of swamp coolers to fight the summer heat, and the planting of cottonwood trees on city streets for shade. To increase water supplies, wells were drilled in the valley beginning in 1907 (map 6.2). These increased the limited water supply and also provided water where it was not otherwise available and where pipelines were not economical—generally everywhere outside the railroad's land holdings and away from the valley's other springs.

The diversion of water and pumping with wells eventually took their toll on the creeks and springs. Las Vegas Creek first went dry in the summer of 1935 and ceased flowing altogether about 1950. The Las Vegas Springs went dry by 1962, ending the fifteen-thousand-year history of this life-giving oasis. What is left of these springs and wetlands is preserved today by the Springs Preserve; the nearby Meadows Mall also takes its name from the English translation of Las Vegas.

Las Vegas springs and creeks provided all the water needed by the early town and railroad, but this water source was entirely controlled by the railroad. The railroad refused to serve people outside of its lands, including the McWilliams Townsite, though it later relented. Other sources of water were needed, and a well was dug to provide it. This well's success prompted many more to be drilled, and the ease of finding water throughout the valley led to a short-lived agricultural boom in the city (map 10.5). These early wells were flowing or artesian, meaning that water rose to the surface on its own. Only many years later would pumps be required to raise the water to the surface.

Wells were found throughout much of the valley by the 1950s, but they were concentrated around Las Vegas Springs, Tule Springs, Duck Creek, along the Strip, and in North Las Vegas. These were areas where groundwater was known to be abundant and close to the surface as well as growing areas outside the city limits where city water was not available. The growing Strip increasingly relied on pumping for landscaping, pools, and other hotel-casino operations.

The water table's decline as more water was pumped out than naturally entered the ground, first noticed in the 1930s, has continued ever since. Wells had to be drilled deeper and more power was required to lift the water from greater depths, at ever increasing cost. The link between increased pumping and groundwater decline was clearly understood, but since no other source was available there was no choice but to continue pumping.

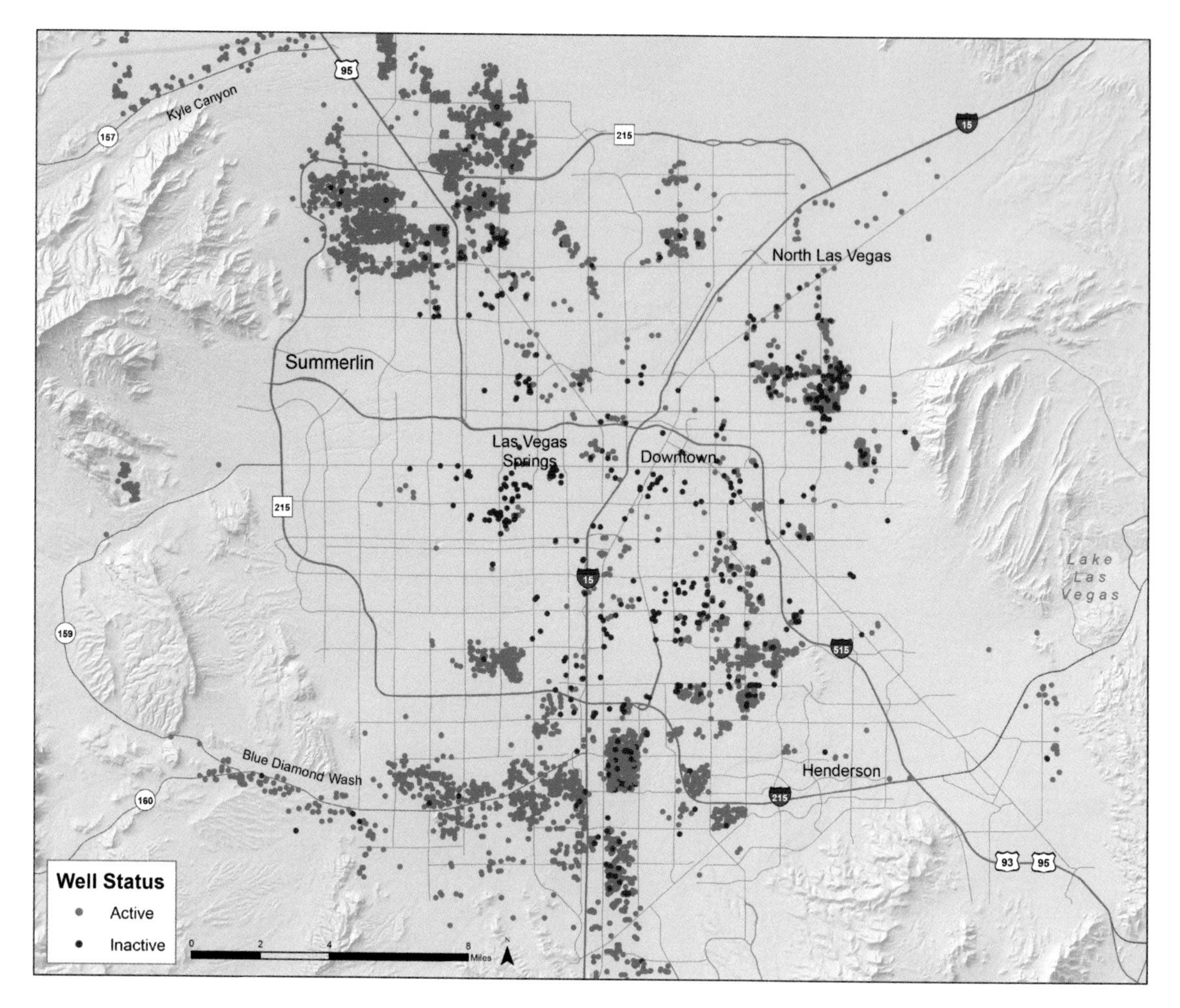

Map 6.2. Wells. Once the springs were fully used, more water was found by pumping. Many older wells are inactive, replaced by expanding water systems, but wells remain common in the northwest and far south ends of the valley.

These declines were not necessarily a bad thing. It was city policy in the 1950s to pump water at a rate that would lower the water table and dry up the remaining springs and creeks. This helped ensure no water would escape the city's efforts to tap it for economically productive uses; a flowing spring was considered a waste of water. This was part of a larger strategy to maximize opportunities for population growth. Groundwater resources would decline in the short term, but in the long run this would generate the population, wealth, and political power with which major water projects could be developed to sustain this growth. So far this strategy has paid off (maps 6.4 and 10.11).

The pattern of wells today shows several clusters where pumping remains important. Kyle Canyon, the northwest valley, western Henderson, and the area south of Nellis Air Force Base are important pumping regions. The master-planned communities of Summerlin in the west and Anthem in Henderson as well as older parts of Henderson are noteworthy for the absence of any current or former wells. Summerlin was vacant desert land when purchased by Howard Hughes in the 1950s, and it was not

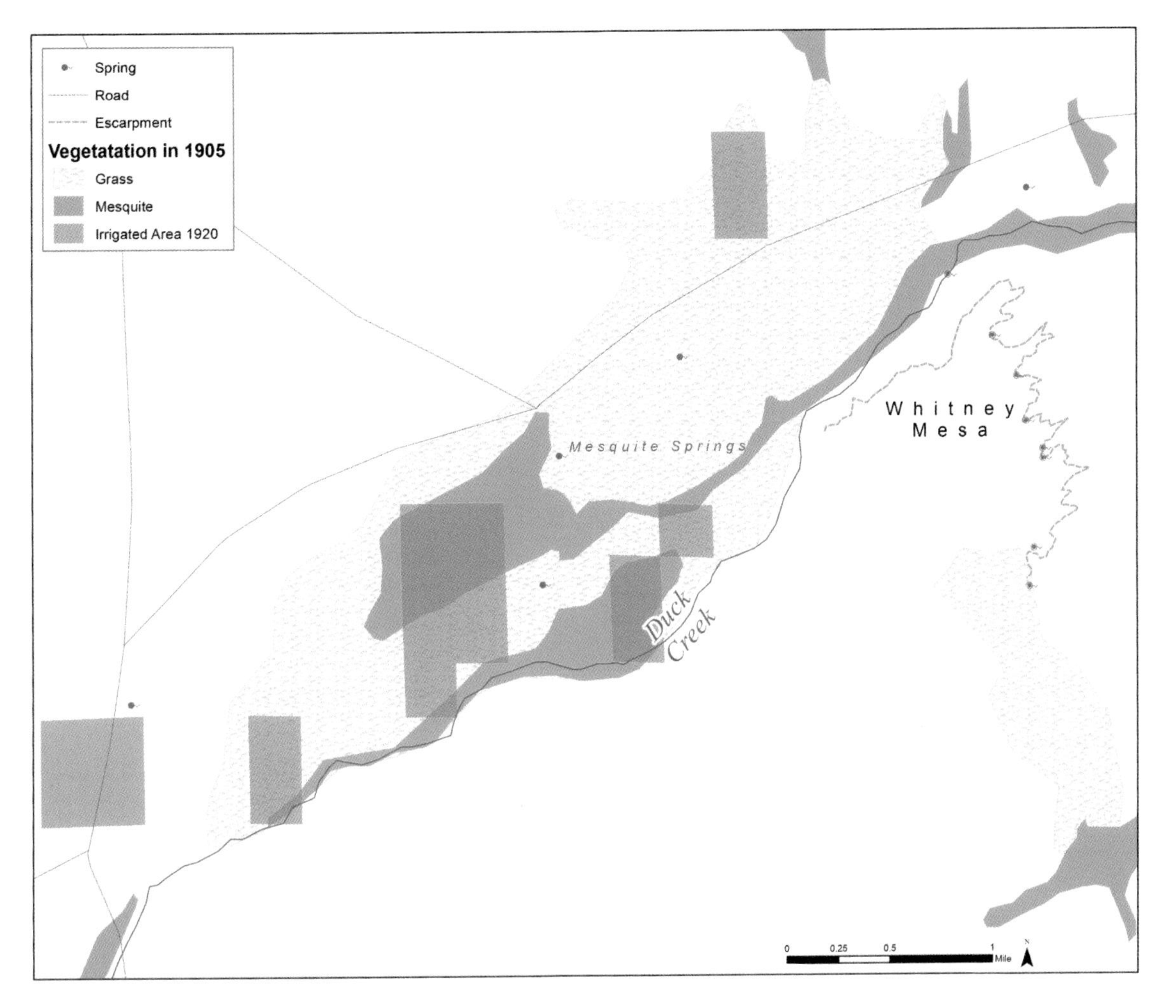

Map 6.3. The Original Las Vegas. Duck Creek was the first place named "Las Vegas" and once had springs, a flowing stream, and grass and mesquite woodlands. Early settlers attempted to irrigate a few parcels of land circa 1920.

developed until the 1980s (map 10.2). Water is, however, important to the design of these communities; while Summerlin and Anthem have no wells, they do have many bridges because of land-use regulations calling for the preservation of existing flood channels.

One of the few areas to see an increase in the height of the water table is the southeast part of the valley along Las Vegas Wash, because of both decreased pumping and infiltration of treated sewage (map 6.9). Groundwater levels also increased throughout most of the city during the 1990s when the Las Vegas Valley Water District began pumping Colorado River water into the ground to recharge the aquifer.

The name Las Vegas was first used to describe this place in 1830, but exactly where was it? Neither the town founded by the railroad in 1905 nor the competing Westside, the 1855 Mormon fort, nor even the springs that John C. Frémont mapped in 1844 is the original Las Vegas. That was several miles to the south, at what was once called Duck Creek and later known as Paradise Valley and today just northwest of the Henderson city limits.

Duck Creek was a flowing creek fed by several springs, surrounded by mesquite and willow woodlands. It was this area that the first Europeans to enter the valley, in Antonio Armijo's expedition in early 1830, named Las Vegas. Later travelers on the Old Spanish Trail crossed the valley farther north and transferred the name Las Vegas to another creek and meadows they encountered there. That was the location Frémont saw and mapped, and what would later become the town of that name (map 3.5). Although it escaped the attention of many explorers and travelers, Paradise Valley was an important location to Paiutes, who camped and farmed in the area (map 3.2). Little remains of their presence except evidence of irrigation ditches in an undeveloped section of Sunset Park.

After Las Vegas was founded in 1905, settlers were quick to occupy well-watered areas of the valley including Duck Creek. The creek had been entirely homesteaded and irrigated agriculture established in many properties by 1920, with water supplied by wells. The Union Pacific's Boulder City branch was built through the valley in 1931 but seems to have little effect on it, and neither did the construction of the BMI industrial plant in the 1940s (map 4.8). In the 1950s farms were still present, and much of the woodland and several flowing springs still existed, though mostly downstream of today's Sunset Park. Much of this vegetation survived until the early 1990s before being overwhelmed by urban growth. Duck Creek is now a concrete channel running past subdivisions and industrial areas.

It is often said that Paradise Road (and the town of Paradise) was named for the Pair-O-Dice Club, the first gambling establishment on Highway 91. However, United States Geological Survey (USGS) topographic maps show that Paradise Road used to be Paradise Valley Road and ran between the city and the Paradise Valley area. Warm Springs Ranch gave its name to a nearby street, though the springs are long gone. These street names provide clues to the valley's early history.

There is still water in Paradise Valley today. A lake in Sunset Park is one of the larger bodies of water in the valley and noteworthy for its Easter Island statues. These were originally displayed in front of the Aku Aku lounge at the Stardust on the Strip until 1980 (map 11.9). They are not from Easter Island but were carved out of stone quarried in northern Nevada. Today, they are relics of Las Vegas's past.

Sunset Park is a nice place to rest in the shade and enjoy the sight of water, just as Armijo's men did on January 7, 1830. The park includes some of the last mesquite woodland left in the valley, and the name "Las Vegas" must have seemed very appropriate to those men. But the tranquility they enjoyed is frequently

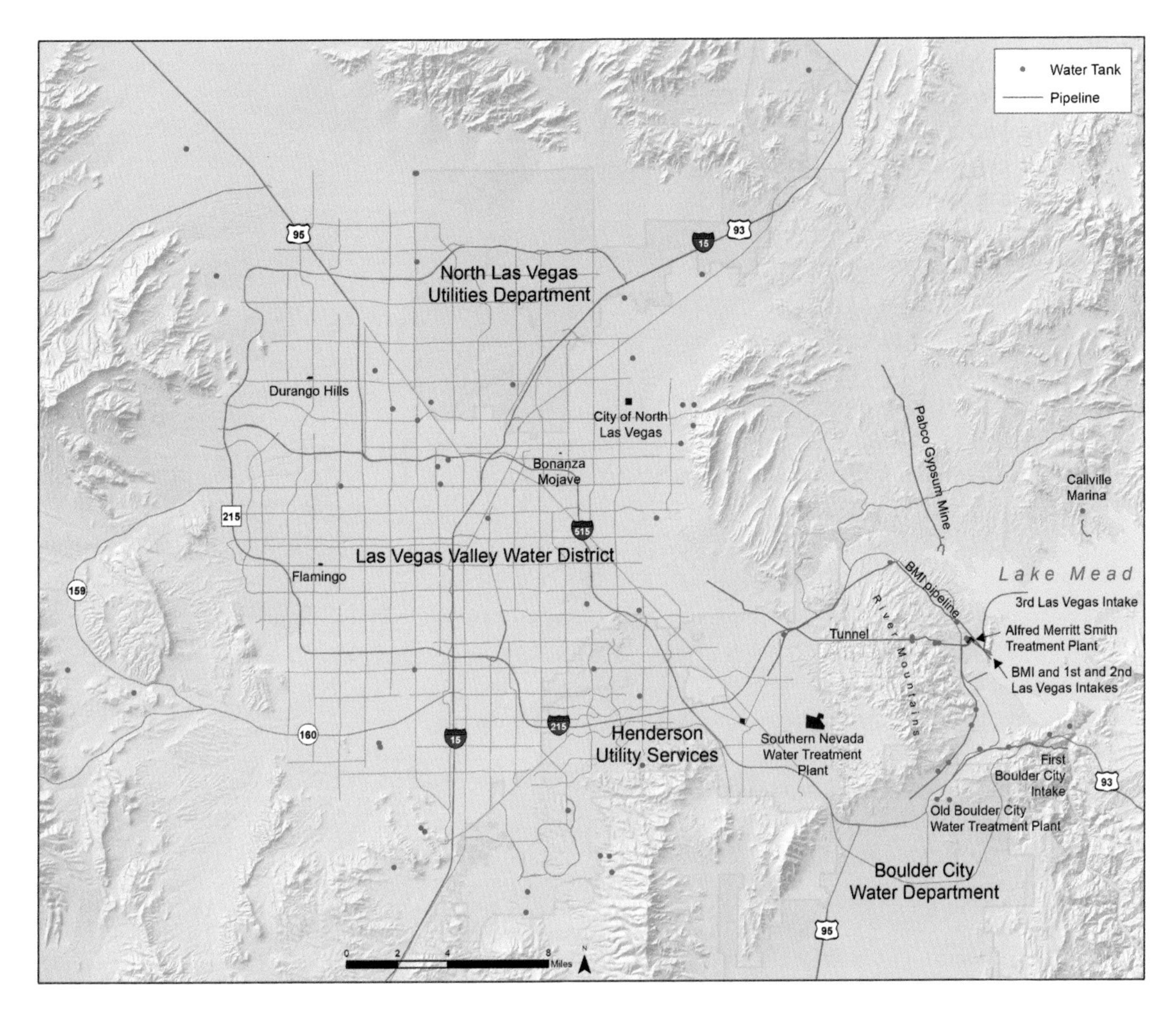

Map 6.4. Water Supply. Several water districts provide water to Clark County residents using water tanks and pipelines. Lake Mead is the source of much of the area's drinking water.

broken today by airplanes landing at Harry Reid (formerly McCarran) International Airport; imagine what Armijo might have thought of the sight of that never-ending stream of airliners bringing more than forty million visitors a year to a vast city of more than two million people.

Besides the now dried-up springs and falling water table, the only other major source of water in the Las Vegas region is the Colorado River. This 1,450-mile-long river begins in the snowy mountains of Wyoming and Colorado and makes its way southwest through a series of canyons, the biggest of which is Arizona's Grand Canyon. Near Las Vegas, it makes a large turn to the south and flows toward the Gulf of California, though very little water makes it to the ocean anymore. But the river was miles away and 1,300 feet lower than the small city and had an extremely variable flow, making it of no use to Las Vegas at a time when transporting water relied on gravity. This began to change when Hoover Dam was built.

The first local use of the Colorado River was for the new town of Boulder City (map 5.8) when the Hoover Dam project was initiated. A pipeline carried water up from the river to the city treatment plant at the north end of town. An intake was located on the river below the lower cofferdam until the dam was completed, when the intake was moved to a penstock on the Nevada side.

The next use of river water was for the BMI plant in Henderson, built to process magnesium for the defense industry during World War II (map 4.8). The valley's water supplies were already being fully utilized, so a pipeline from Lake Mead was built in 1942 to supply both the plant and new town of Henderson. A water intake was built at the south end of Saddle Island and a pipeline built up Las Vegas Wash. After the war emergency passed, the plant shut down temporarily but Henderson continued to use the lake's water.

The railroad remained the supplier of water to the town of Las Vegas until 1954 when the Las Vegas Valley Water District was created. This agency bought the Las Vegas Springs and nearby wells from the railroad. These, along with many additional wells, supplied the city with water until the following year when Las Vegas began receiving a modest amount of Lake Mead water from the BMI pipeline.

The likelihood of water shortages continued to threaten the city despite this extra water and the continuing growth of wells. The solution was the Southern Nevada Water Project, an ambitious plan to build a pipeline from Lake Mead through a tunnel underneath the River Mountains. New reservoirs and water treatment plants would be required, and the project would serve the entire Las Vegas area, including Boulder City.

The four-mile-long and ten-foot-wide River Mountains tunnel was completed in 1969, the Alfred Merritt Smith Water Treatment Center in 1971, and the water intake tunnel underneath Saddle Island the following year. At the west end of the River Mountains tunnel, the water line splits into the Henderson and Las Vegas laterals to supply different areas of the valley. Boulder City received a new water line as part of this project, and the old treatment plan was shut down. The Colorado River has been the region's major source of water ever since.

The first Lake Mead water intake was built at an elevation of 1,050 feet above sea level, and a second and deeper intake tunnel was built later at 1,000 feet to ensure a continued water supply if Lake Mead should shrink. This proved a wise decision, and the continuing drought in the twenty-first century prompted a third intake, this one a three-mile-long tunnel dug underneath the lake. This intake is at 860 feet above sea level, ensuring water can still be pumped from the lake even if it shrinks to the dead pool level (map 5.12).

The Southern Nevada Water Authority (SNWA) was created

in 1991 to better manage Clark County's water and ensure that future supplies can be found when needed. Various municipal departments, each receiving water from the SNWA, distribute drinking water to customers. The cities of Henderson, Boulder City, and North Las Vegas each have their own water departments and treatment plants; the remainder of the valley is served by the Las Vegas Valley Water District, which also provides water to Blue Diamond, Searchlight, Jean, and most of the remainder of unincorporated Clark County. Among the district's facilities are the Flamingo and Durango Hills treatment plants that treat sewage for use as irrigation water for golf courses, conserving the city's water resources.

A few smaller water systems survive. The Virgin Valley Water District provides water to Mesquite and Bunkerville, and the PABCO Gypsum mine and mill east of the city has its own water line from Lake Mead, with a pump on a barge in Government Wash. The Callville, Echo Bay, and Temple Bar Marinas on Lake Mead also have their own water lines from offshore barges.

The US Supreme Court set the amount of water that can be pumped from the Colorado River in Nevada in 1964. This was made at a time when the state had about 300,000 people and Clark County only about 130,000. The state was allowed to use just 4 percent of the water flowing through it, while California was granted 58.7 percent and Arizona the remaining 37.3 percent. Most of water you see in Lake Mead technically belongs to Southern Californians and residents of Phoenix. And they're not going to share it. Despite the tremendous growth of the state, Nevada has no interest in attempting to renegotiate its share of the Colorado; although the state has grown, so have Arizona and California, and there is every reason to believe Nevada could end up with a smaller share of the river.

Unfortunately, the state's water allocation was also made at a time when estimates of the river's annual flow were much higher than reality. The third water intake will guarantee the Southern Nevada will be physically able to take river water no matter how low Lake Mead goes, but the allotment may be legally reduced to preserve water for other users in Arizona and California if levels decrease. For these reasons the city has begun looking elsewhere for long-term water supplies.

One solution is to reduce use through conservation. Households in Las Vegas have very high water daily usage, with 70 percent going to landscaping, but through conservation the city has been able to reduce its use of Colorado River water despite a substantial population increase. The city may also return to groundwater, and water rights in several northern Nevada valleys were

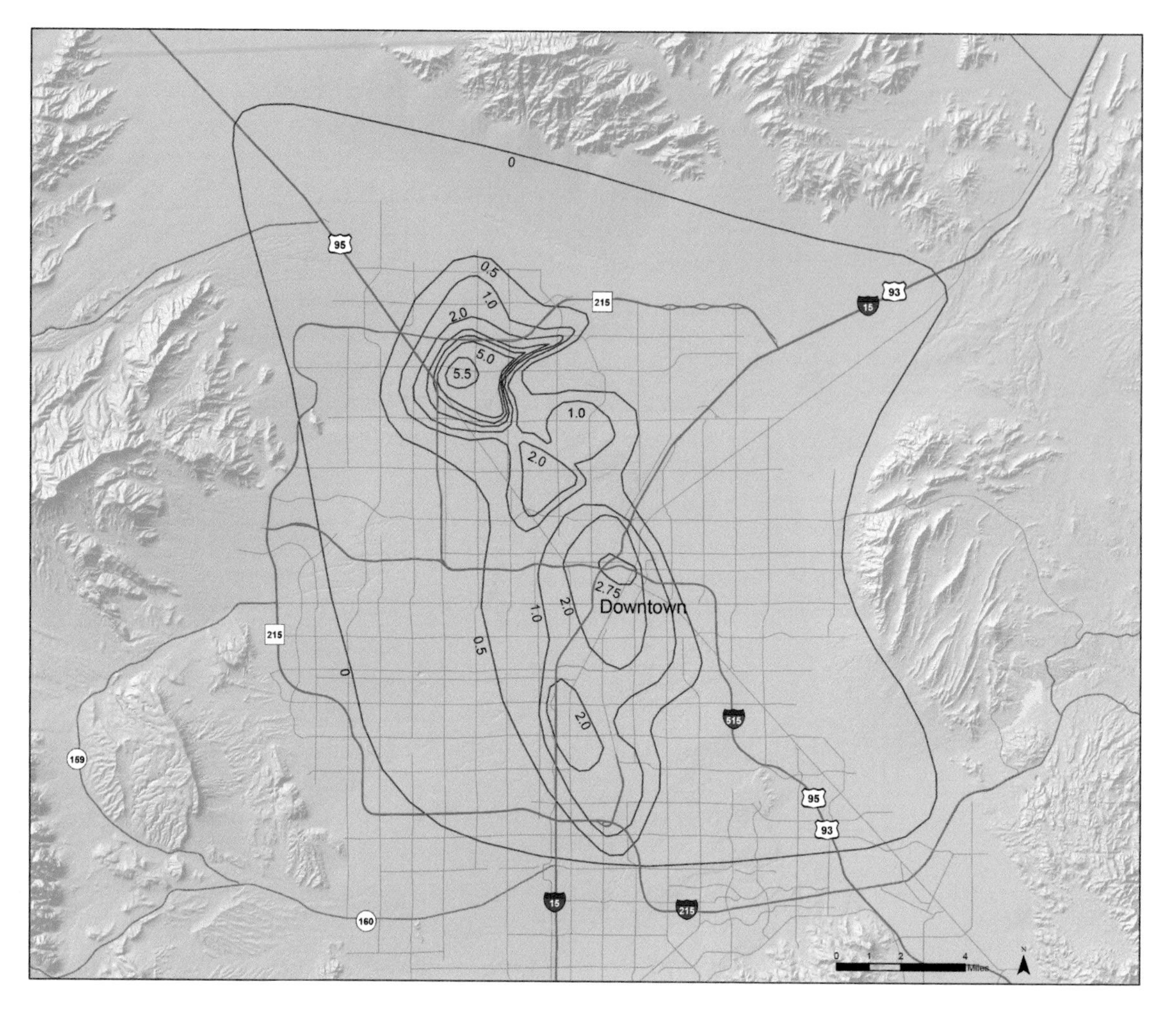

secured with the plan of building a long pipeline to carry this water to Las Vegas (map 10.11).

More than one hundred years of groundwater pumping has taken its toll on the water resources of the city as well as the ground itself. As water is allowed to drain from confined aquifers or pumped out of the ground at a faster rate than it is replenished, the remaining sediment becomes compacted. This results in subsidence, when the land surface sinks. This is a problem in Las Vegas because there is no bedrock under the city; the valley is simply sediment that has washed down from the mountains as they slowly erode.

Subsidence in Las Vegas was first noticed in 1948, using data from precise benchmarks surveyed in 1935 by the Coast and Geodetic Survey for the Hoover Dam project. The goal was to understand subsidence around the lake because of the weight of the water (map 5.9), but measurements revealed that Las Vegas had sunk as much as three inches since the city had been surveyed in 1915. The sinking continued at even greater rates as groundwater pumping increased.

Map 6.5. Overdrafting and Subsidence. Pumping water at a faster rate than it can be replenished has caused the land surface to sink, or subside. The contour lines, labeled in feet, show areas where subsidence has been greatest.

The area affected by subsidence also increased, doubling to four hundred square miles from 1963 to 1980. The first several decades of subsidence saw the greatest drops around downtown Las Vegas, but after the 1960s it began to occur along the Strip and along US 95 north. Pictures abound showing well foundations that now stand several feet above the ground.

Subsidence depends on geology as well as the amount of pumping. In areas with silt and clay, twenty feet of groundwater decline may result in one foot of subsidence; for dirt and gravel, it may take sixty feet of downdraft to lower the ground one foot. A related problem is the opening of fissures or cracks in the ground as areas subside at different rates, first noticed in the late 1950s. Damaged buildings and street pavement were common by the 1970s in several areas of North Las Vegas where groundwater pumping has been very heavy.

Radar measurements by orbiting satellites assess subsidence today. The rate of subsidence has decreased, and in a few areas may have even reversed slightly as the city has pumped water back into the ground (though by only several millimeters). Subsidence is greatest in the summer when pumping is greatest and aquifer recharge minimal; subsidence may decrease or even rebound slightly as the aquifers refill in the winter. Faults are also important in identifying where subsidence is occurring; faults may separate soils that are more compactable from those less so, which also affects groundwater flow and the amount available for pumping or recharge. The area with the greatest subsidence today is in the northwest part of the city, more than five and a half feet, near US 95 and the Interstate 215 beltway. Downtown has the second-greatest subsidence (2.75 feet), and a north-south corridor within the city has also substantially dropped.

Ground subsidence is not unique to Las Vegas but a common phenomenon in cities that have relied on groundwater pumping. It has been observed in central Arizona and numerous areas in Colorado. Subsidence because of oil wells has also occurred, and in Long Beach, California, even caused a thirty-seven-foot drop.

Nevada has many large lakes, and during the ice ages was home to the 8,500-square-mile Lake Lahontan, one of the largest lakes ever to exist in North America. In modern times, Lake Mead became the world's largest reservoir when it was created behind Hoover Dam in the 1930s. It remains the largest reservoir in the country by capacity when it's full, though its size of 247 square miles is tiny compared to Lake Lahontan. But Lake Mead is not the only lake in the Las Vegas area, and it was not even the first. This was at Lorenzi Park, developed in the early 1920s as a resort area north of the Las Vegas Springs. The lake survives today as Twin Lakes, a city park.

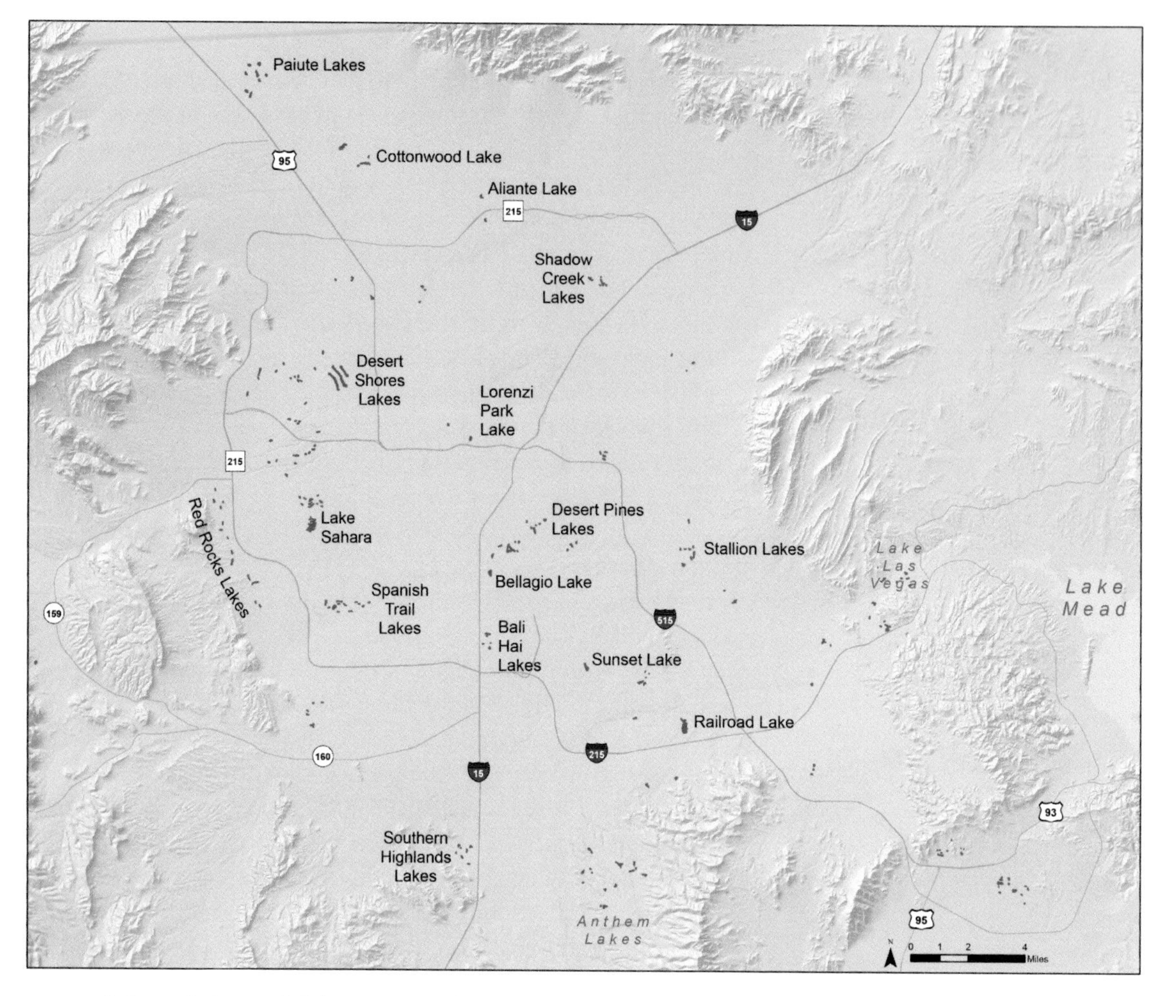

Map 6.6. Lakes of Las Vegas. In addition to Lake Mead, many small lakes dot the valley, often in golf courses or next to casinos.

Most of the city's lakes are much newer, appearing only in recent decades, and many are associated with golf courses. Lake Sahara is an exception and was built in the late 1980s when the Lakes community was developed on West Sahara Avenue. The lake is about thirty acres, the largest in the valley. Sunset Lake, southeast of Harry Reid International Airport, is 9.7 acres and a recreation oasis. The Strip has several bodies of water, of which the Bellagio Lake is the largest at eight acres.

Lake Las Vegas covers 320 acres, making it the second-largest body of water in Southern Nevada. It was built in Las Vegas Wash, the valley's drainage into Lake Mead. A large dirt dam was built across the wash in 1991 to hold the reservoir. It is not, however, filled with the storm drainage and treated sewage flowing down Las Vegas Wash (map 6.7) but water from Lake Mead. The storm water and sewage pass underneath Lake Las Vegas in two large tunnels.

Most of these lakes are stocked with fish and permit fishing, though Desert Shores Lakes has been afflicted with occasional fish die-offs because of natural processes. Elsewhere, fish have been

Lake Las Vegas, 2009. Courtesy of the Library of Congress. Photo by Carol Highsmith.

relocated to the Las Vegas area because they are in danger from human activities. A small concrete tank below Hoover Dam on the Nevada side was stocked with Devils Hole Pupfish in 1972, a species found only in an underwater cave in Amargosa Valley northwest of Pahrump. There are less than one hundred of these fish remaining in Devils Hole, and the Hoover Dam refuge was one of several established to preserve the species should a calamity occur in the cave. The Devils Hole Pupfish have been fortunate so far; those living at Hoover Dam were relocated downriver to Willow Beach in 2006, but all died that year.

One of the least glamorous aspects of Las Vegas is the need to dispose of the city's sewage. Fortunately, an efficient system has been developed for doing so, involving primary treatment (allowing solid material to settle out of wastewater) followed by secondary treatment (using anaerobic bacteria to digest the remaining biological matter). The result is sewage sludge (sent to landfills) and treated wastewater or effluent, which can be released back into rivers or used for irrigation.

This process is followed in Las Vegas, and as with water service, it is provided by several agencies. The cities of Las Vegas, North Las Vegas, Henderson, and Boulder City maintain their own systems and treatment plants. The remainder of the valley (and most of the county) is served by the Clark County Water Reclamation District, created in 1956. This agency has 2,200 miles of sewer lines with the main sewage treatment plant also at Las Vegas Wash, at the east end of Flamingo Road. This plant can treat up to 150 million gallons a day and treats sewage well enough to allow it to be used for lawn and golf course irrigation (though drinking it is not recommended). The Desert Breeze facility is smaller and located in the southwest corner of the valley. Its primary purpose is to provide treated sewage water for local golf courses. Sludge from all of these treatment plants is trucked to the Apex landfill (map 8.20).

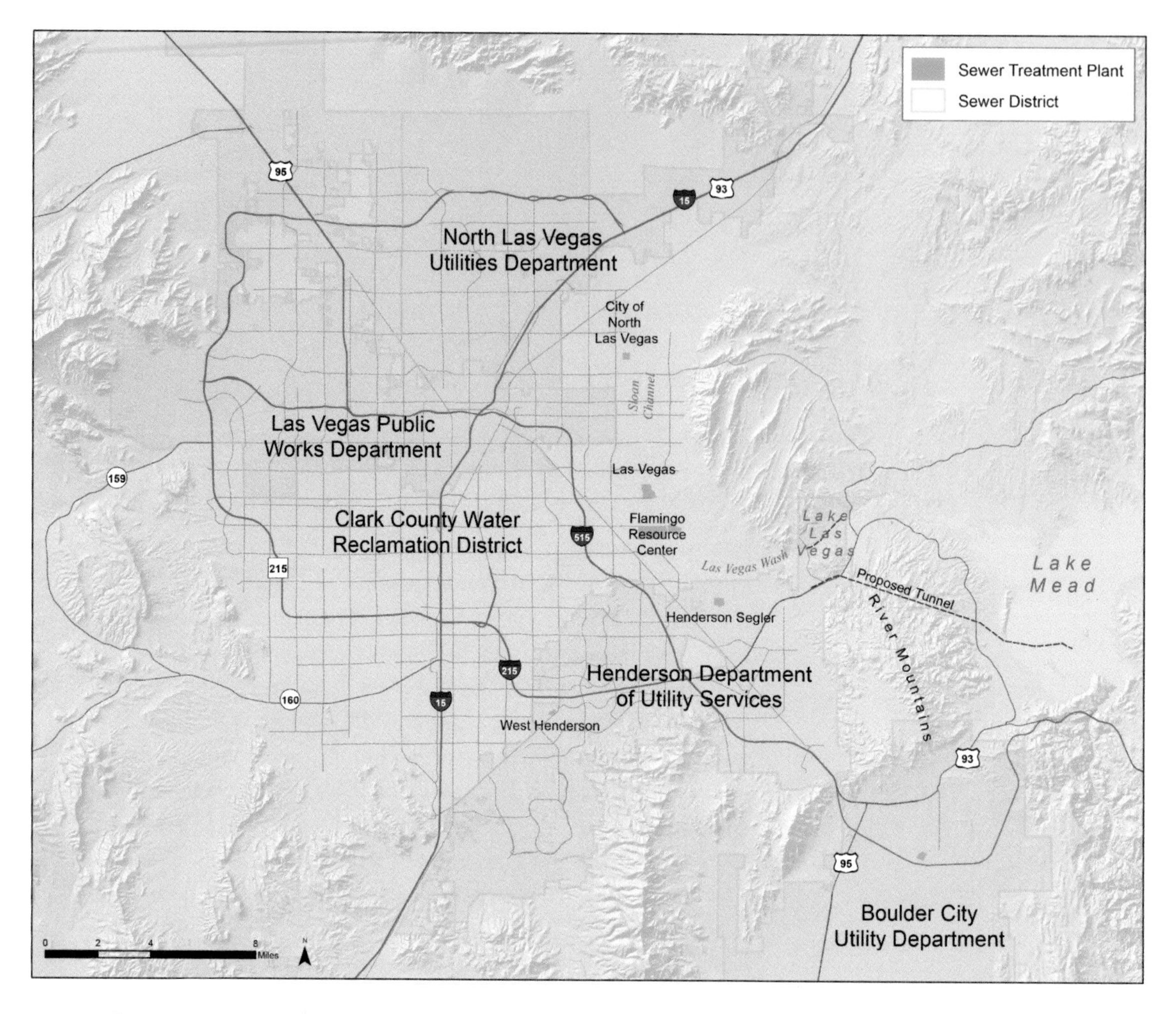

Map 6.7. Sewage. Several water treatment districts handle the city's sewage. Most treatment plants are located along Las Vegas Wash, which drains the treated sewage into Lake Mead.

Sewer systems rely on gravity flow to move sewage from homes and businesses to a treatment plant, so within Las Vegas Valley sewage treatment plants are clustered along the Las Vegas Wash, the valley's lowest point and natural outlet. The treated wastewater from each plant is allowed to flow into the wash and Lake Mead. The North Las Vegas plant is the farthest upstream and dumps its effluent into the Sloan Channel, which has been a point of concern for neighboring residents not pleased with the smell. Once in Las Vegas Wash, the sewage will eventually flow into Lake Mead; the lake is therefore both the source of the city's drinking water and its sewer. What happens in Las Vegas does indeed stay in Las Vegas!

Living in the Mojave Desert means coming to terms with two kinds of water problems: not enough and too much. The problem of not getting enough water has long been important and newsworthy; the problem of too much water is often ignored and quickly forgotten. But the city has had frequent and often devastating flash flooding; the desert soils and steep slopes around the city produce heavy runoff that flows across the valley to its outlet

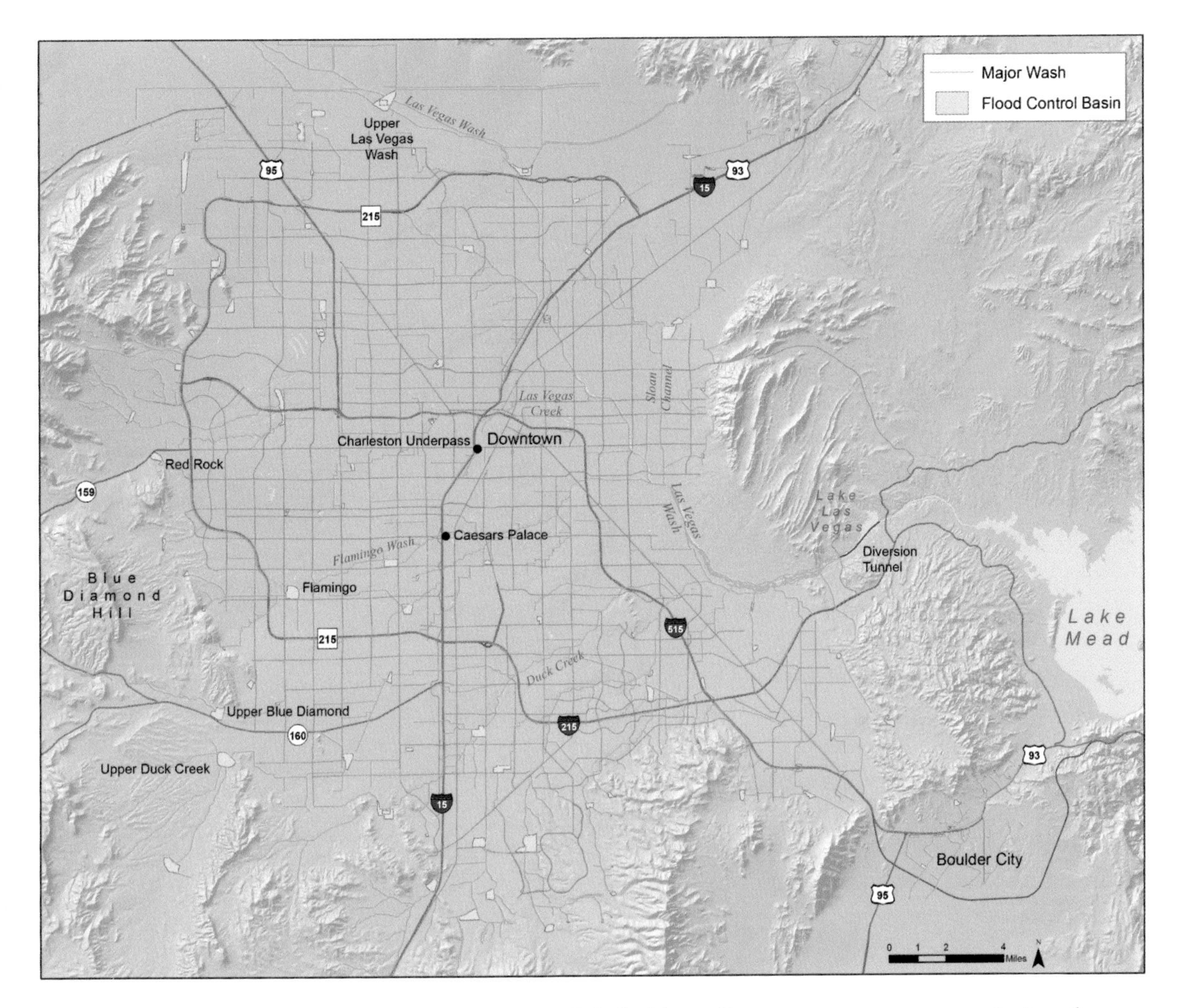

Map 6.8. Flood Control. The valley has an elaborate flood control system with basins around the edge of the valley to trap floodwaters. These are slowly released through a series of washes and tunnels to drain down Las Vegas Wash.

at Las Vegas Wash. These flows are most common during the summer thunderstorm season, with August the most common month, followed by July; December and May are the least likely months for floods to occur.

Early flash floods were notable for washing out railroads and highways outside the city. Flooding in the Meadow Valley Wash northeast of Las Vegas in 1910 destroyed or damaged one hundred miles of rail line. Another deluge several years later did more damage and prompted the railroad to rebuild the line at an even higher elevation, requiring fifteen tunnels and thirty-nine bridges. Early highways were regularly washed out. High water in 1916 and 1921 heavily damaged the truss bridge over the Virgin River at St. Thomas. California Wash on Highway 91 northeast of town was another regular flooding problem; new bridges were built in 1945 to eliminate this problem, and are still part of Interstate 15.

Railroads and highways were improved to protect them from flooding, but as Las Vegas grew it became more and more vulnerable to this problem. Many new neighborhoods and streets were built across dry washes and in low-lying areas, and residents frequently paid a higher price than they expected when summer storms arrived. Even where builders took into account flooding, reality often overwhelmed designs. The new and allegedly flood-proof Charleston Boulevard underpass proved this when it filled with water after a thunderstorm on June 13, 1955, the first of many times this would happen.

Perhaps the worst flood in the city's history was the Caesars Palace flood on July 3, 1975, when afternoon thunderstorms dumped up to three inches of rain west of the city. The raging waters from this deluge flowed eastward across the valley, filling the Flamingo and Tropicana Washes. Flamingo Wash crossed the Strip just north of Caesars Palace, where the casino had extended its parking lot across the wash bottom. The floodwaters submerged seven hundred cars and killed two people in North Las Vegas.

The solution to continuing flooding headaches was a comprehensive flood control project. The Army Corps of Engineers came up with a plan in 1959 that involved building a huge levee along the west side of the city to stop washes flowing down from the Spring Mountains (map 10.2). No construction would have been allowed west of the levee, severely restricting development and leading to the plan's rejection.

Nothing was done until a county flood control program was created in 1985. A regional master plan that called for forty years of construction projects was developed. The primary features were dams and detention basins around the periphery of residential areas to catch floodwaters and reduce the flow of washes through to a manageable level. Dozens of these were to be built on major washes. The plan also called for newer subdivisions to preserve washes as flood channels within them. Summerlin and Anthem Lakes are two areas that have done so, requiring a number of small bridges.

The construction of the flood control dams and basins has been a success, and even ended the regular inundation of the Charleston underpass, which last occurred in 1999. The program has also removed flooding and drainage from the minds of many residents. Many of the dry washes that once crossed the desert valley have been converted into concrete ditches or tunnels and are no longer even visible. No trace of Flamingo Wash remains at Caesars Palace, which has been expanded over the covered channel. Some of these tunnels have even become the habitation of the city's growing homeless population (map 8.8). The city's future growth is limited by an urban growth boundary (map 8.1), meaning that more flood control dams will not have to be built farther out as the city expands.

The city appears to have solved its flooding problem, but

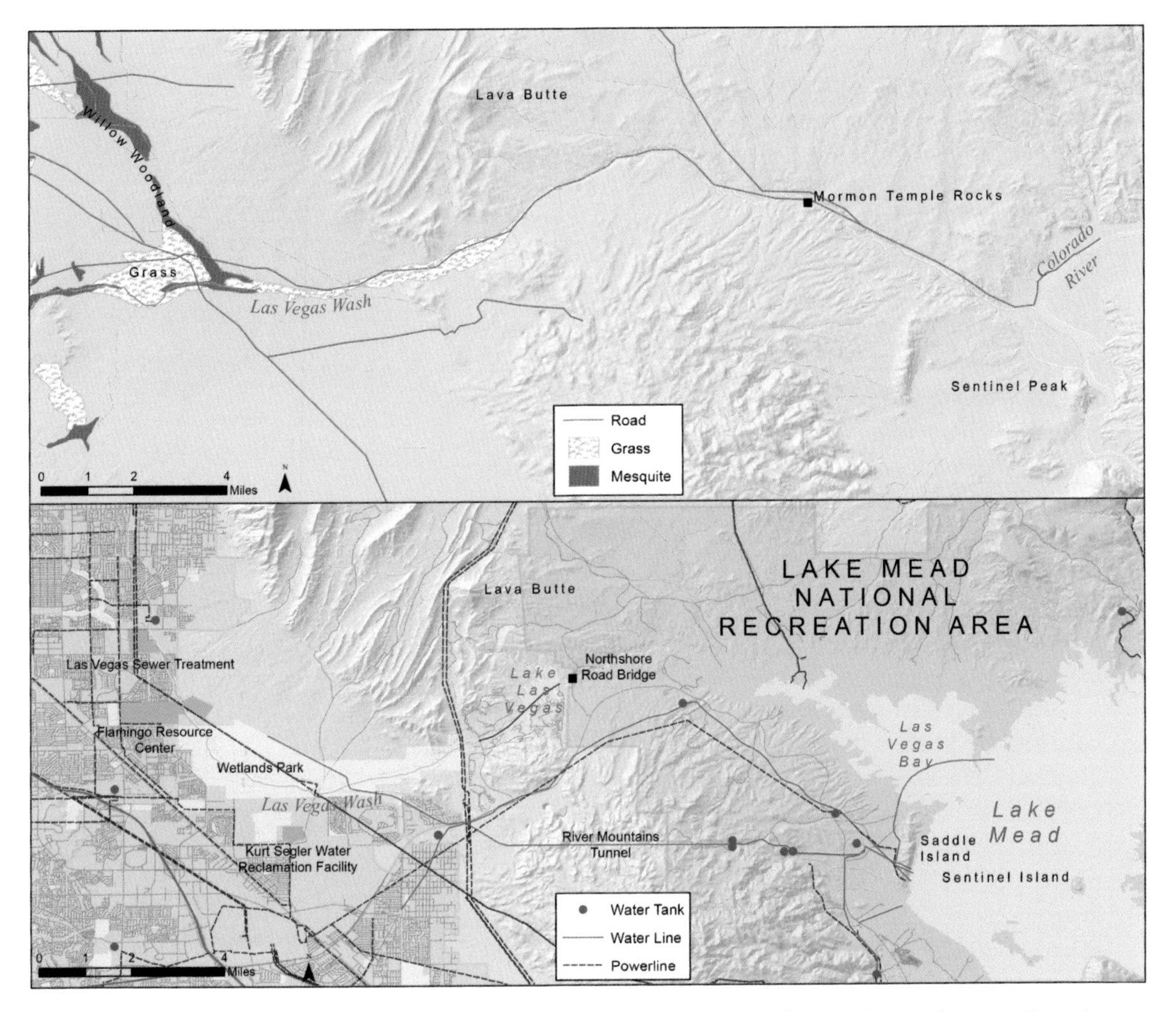

Map 6.9. Las Vegas Wash. The natural drainage for the Las Vegas Valley has been heavily developed. *Top*: Las Vegas Wash circa 1905; *bottom*: the same area in 2020.

every Las Vegan knows their luck can always change. There is the potential that severe storms may increase in the future as the climate warms (map 2.4), requiring improvements to the city's ability to stop the raging torrents sweeping down from the mountains. Future residents may rediscover the problem of too much water in the desert.

The growth of the city has created many new lakes, and it has also accidentally created a new river: Las Vegas Wash. Until recently this had been dry except when carrying floodwaters out of the valley. But by the 1970s, a decade or more after Las Vegas Creek had dried up, the Las Vegas Wash had become a flowing stream in the desert.

This was not because of changing groundwater conditions or flowing springs but because of the construction of several sewage treatment plants that dumped effluent, or treated sewage, into the wash (map 6.7). Storm drains also collected water from lawn irrigation and added that flow to the wash, and the discharge of cooling water by an electrical generation station also makes a contribution. Las Vegas Wash now has a flow equivalent to that of the Virgin River. It supports a variety of plants and animals along its length (a similar but much smaller stream flows from the Boulder City sewage treatment plant south into Eldorado Valley).

Like other flowing rivers in the desert, Las Vegas Wash was dammed, but this dam had a twist: it was specifically designed to *not* block the wash's flow, and instead to ensure the wash was kept separate from the new reservoir. When Lake Las Vegas was created behind a dirt dam in the 1990s, two tunnels were built to carry the wash underneath it. A second dam was built upstream from the reservoir to ensure the wash went through the tunnels. A sediment basin removes most of the sediment from the wash's flow before passing underneath the lake; this reduces the amount being deposited in Lake Mead but has also increased erosion downstream, threatening the Northshore Road bridge and even Lake Las Vegas itself should erosion reach the dam. Several structures were built to protect the bridge in 2002, but by 2006 they were in danger of failing, requiring still more structures to control erosion.

Las Vegas Wash and Lake Las Vegas dam, 2007. Photo by Stan Shebs.

Another problem is what happens after the wash empties into Lake Mead. Although the sewage has been treated, it is still rich in nitrates, affecting water quality and wildlife in the lake. The wash deposits wastewater upstream of the city's drinking water intakes, leading to the possibility of water contamination. To resolve this problem the Systems Conveyance and Operations Program (SCOP) was initiated to improve water quality in Lake Mead by building a 9.3-mile-long tunnel through the River Mountains and under Lake Mead to disperse treated sewage downstream deep in the lake and downstream of water intakes. Four hundred million gallons of treated sewage would be pumped into the lake each year; this will help improve the health of the lake and could reduce erosion problems in Lake Mead Wash. Unfortunately, it will also harm the wetlands environment that has developed in the wash by diverting the flow that supports it. The issue remains to be resolved.

7

Transporting Las Vegas

Las Vegas was founded in 1905 as a railroad town after almost a hundred years as a stopping point on wagon roads and horse trails. Through its first several decades, it remained a place on the way to somewhere else. The improvement of transportation technology in the twentieth century brought many new ways of getting to the growing community. New railroad lines, highways, and airways connected Las Vegas to the outside world, first pioneering rough paths across the desert and then being improved and taken for granted.

This chapter explores the development of the city's transportation connections with a series of maps detailing the placement of railroads, the evolution of highways, the creation of the freeway system, and the development of the modern aviation system that brings millions of visitors each year.

The city of Las Vegas began as part of a business scheme by a Montana senator. That man, William A. Clark (1839–1925), had become one of the richest men in American as one of Montana's copper barons. He diversified his interests to include power companies and railroads. One of these, the San Pedro, Los Angeles and Salt Lake Railroad, was built to connect Salt Lake City with the seaport of San Pedro, California (now part of the port of Los Angeles and Long Beach).

Construction began in 1901 with crews laying track south from Milford, Utah, and in 1903 north from Barstow, California. Those building south reached a site that became Caliente, Nevada, in August 1901, Moapa in May 1904, and founded the town of Las Vegas on May 15, 1905. Crews building north reached Nevada by late 1904, and the two crews finally met and completed their line on June 30, 1905, between the rail stations of Sloan and Jean. Many such stations were built along the line, most of them simply telegraph stations, and the majority have disappeared from the map. Arden was the only substantial one of those near Las Vegas, with gypsum mills and other facilities supporting the nearby mines.

The railroad between Barstow and Milford encountered some challenging terrain. From Daggett the line followed the sandy and usually dry Mojave River before entering Afton Canyon, where several bridges over the Mojave and a tunnel were needed. Farther north the line ascended Cima Hill, a long, steep grade that required helper engines on northbound trains. Although the tracks across Las Vegas Valley are nearly straight, low mountain passes had to be surmounted at the south and northeast ends

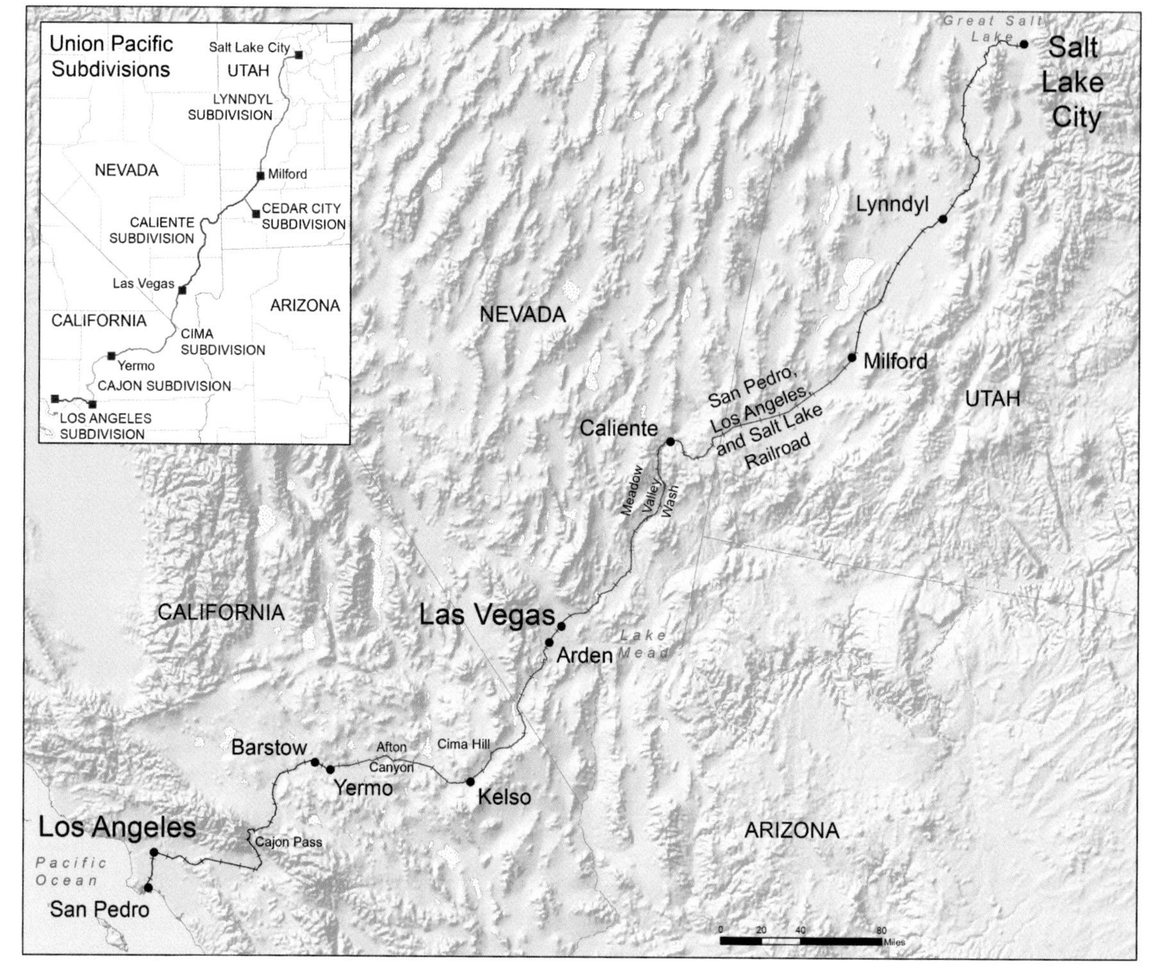

of the valley (near Sloan and Apex, respectively). A number of curves were built in both places to reduce the grade to reasonable levels, and a 304-foot-long tunnel was required through a ridge east of Sloan until it was bypassed in 1944 (map 7.3). North of Moapa, the track was laid in Meadow Spring Valley, a long narrow canyon that turned out to be a channel for raging floodwaters after storms. This line was washed out repeatedly until rebuilt at a higher elevation in the canyon, requiring fifteen tunnels and thirty-nine bridges over Meadow Spring Wash. It is still occasionally washed out, most recently in 2005. Another steep grade east of Caliente required helper engines for northbound trains.

The line became part of the Union Pacific Railroad system in 1921. This railroad was one of two that built the original transcontinental railroad completed in 1869 and is now the largest of four major railroads in the United States. The line through Las Vegas was now part of a much larger system and soon found itself with several impressive new passenger stations. The Milford and Caliente Depots were built in 1923 and served as a passenger stations, railroad offices, and hotels for railroad workers.

Map 7.1. The Salt Lake Route. The San Pedro, Los Angeles, and Salt Lake Railroad created Las Vegas as one of several stations and maintenance centers along the line. Portions of this route, shown in black lines, already existed. It is still a busy route, and the inset shows subdivisions into which the line is divided.

The second Las Vegas train station, 1940s. Tichnor Brothers postcard.

Milford's station was demolished in 1970, but the other was given to the town of Caliente in 1970 and is used by several city offices. A similar building opened in Kelso, California. This closed in 1985 but reopened as the visitor center for Mojave National Preserve in 2005. Las Vegas has had several passenger stations, but none as impressive as these multipurpose facilities. The original 1904 Spanish Mission–style station was replaced in 1940 with a streamline modern station that was in turn replaced by a new station attached to the Union Plaza hotel in 1971.

Once the railroad was completed, it became the principal means for traveling to or from Las Vegas. Northbound passengers could obtain connections to points east at Salt Lake City, while southbound passengers could connect to other lines at Los Angeles. The city hosted several named streamliner trains, beginning in 1936, when the *City of Los Angeles* began operating between Chicago and Los Angeles, with a stop in Las Vegas. Another train was the *City of Las Vegas*, which operated daily between that city and Los Angeles from 1956 to 1968. Passenger service to Las Vegas was shut down on May 1, 1971, when Amtrak took over that nation's passenger services but ended those through Las Vegas.

Amtrak did eventually serve the city, operating the *Las Vegas Limited* on weekends between Los Angeles and Las Vegas from May to August 1976. Daily service resumed on October 28, 1979, when the *Desert Wind* began running between Salt Lake City and Los Angeles. It was neither fast nor convenient. Northbound trains arrived about 7 p.m. while southbound trains departed at 7:46 a.m., and it took over seven hours for southbound passengers to reach Los Angeles. The service ended May 12, 1997, when the *Desert Wind* was canceled because of budget cuts, making Las Vegas the second-largest metro area to lose Amtrak service (Phoenix is the largest). Proposals exist to resume Amtrak service or by private companies interesting in developing a new high-speed service (map 10.7).

The railroad had an enormous presence in the new town of Las Vegas. It was the primary employer, ran the water system, and was essential to the construction of Hoover Dam and the creation of Henderson. However, many jobs were moved out of town in 1922 after a prolonged strike, and the railroad's presence has diminished over the years. This process accelerated during the 1990s when the downtown rail yard was shut down and sold to the city for redevelopment. Many new underpasses and overpasses (map 7.6) were built to eliminate the railroad as a barrier to crosstown traffic, but also made it invisible to many. Even the renaming of downtown's Union Plaza Hotel to merely the Plaza Hotel ended a connection to the railroad.

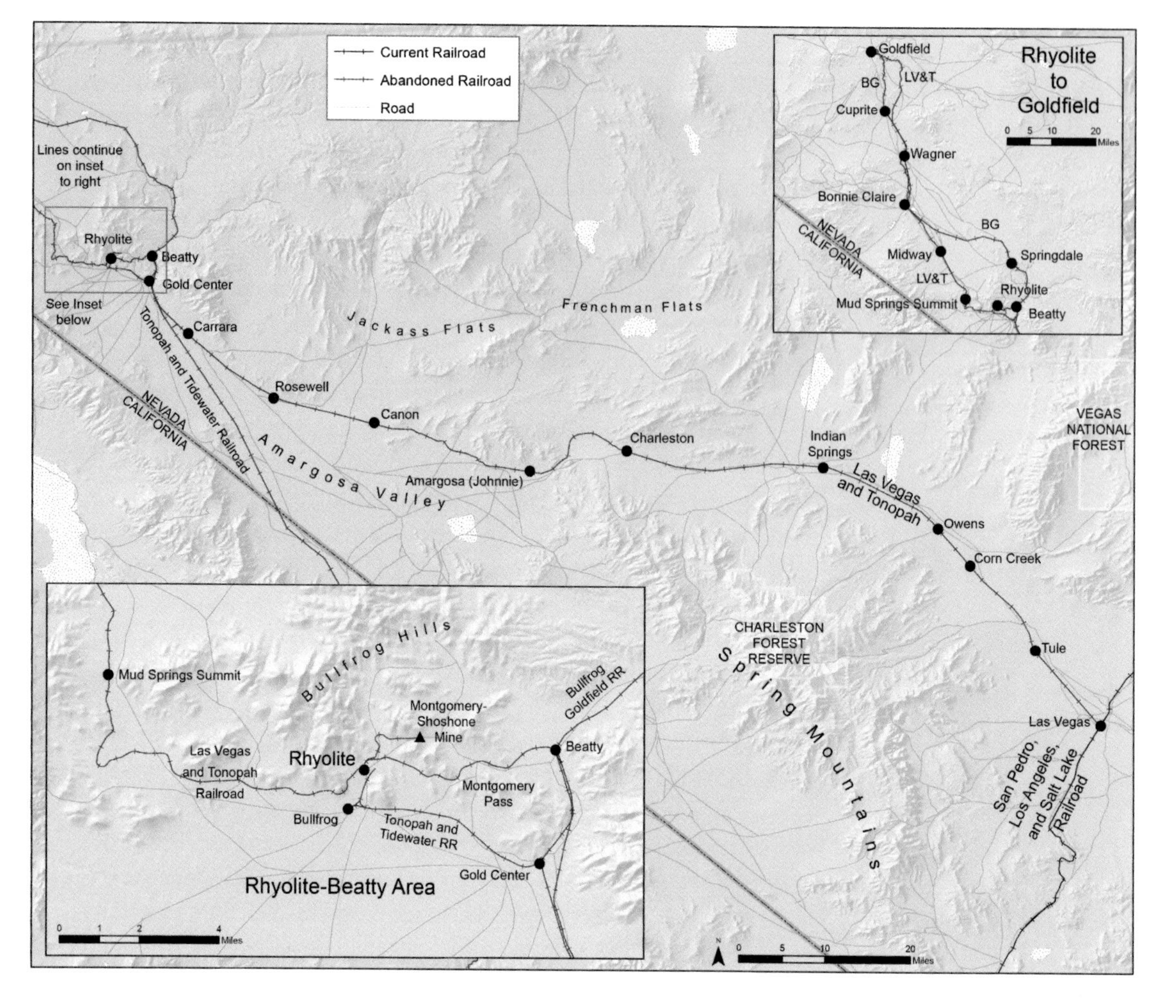

Map 7.2. The Goldfield Route. The Las Vegas and Tonopah Railroad was built to connect Las Vegas with the mines of Rhyolite and Goldfield. It created many new stations, but in little more than a decade it had been abandoned.

The railroad does still have a presence in the city. Many freight trains pass through Las Vegas each day, and a yard at Arden allows for local freight to be cut out of or added to through trains. An intermodal terminal in North Las Vegas allows shipping containers to be loaded or unloaded, and an automobile-unloading terminal also exists. Many industries rely on the railroad for shipping products and raw materials, and they are served by their own spurs. Although no trains stop in downtown anymore, Las Vegas remains a division point between the Cima Subdivision south of the city and the Caliente Subdivision to the north, each with its own train dispatcher and crew-change locations. Railroad workers north of downtown even operate on Mountain Time to keep the same schedule as those farther up the line in Utah.

The second railroad in Las Vegas was the Las Vegas and Tonopah (LV&T), built to connect it to the booming mining towns of Rhyolite and Goldfield. At 198 miles, it was by far the longest of the branch lines built from Las Vegas. The line connected to the San Pedro, Los Angeles and Salt Lake line at its yard in Las Vegas and shared its station. The LV&T ran northwest through upper Las Vegas Wash, past the stations of Tule, Corn Creek, and Indian Springs before turning west. The next stop, Charleston, was the highest on the line at 3,629 feet above sea level, followed by Point of Rocks and Amargosa. The line gradually turned north and passed Rosewell and Beatty before reaching Rhyolite in December 1906.

Owens station was added to the LV&T in May 1906 to serve a sawmill in the Spring Mountains (map 4.3) and in March 1908 Canon station was added to serve the Quartz Gold Mining Company, which had a mine in the Calico Hills to the north. Carrara was added in 1913 where a marble quarry was started. One mixed passenger and freight train ran each way every day between Las Vegas and Goldfield, taking about four hours to reach Beatty and eight hours to arrive in Goldfield.

During the height of the Rhyolite boom, the town had many amenities and was a major destination. The *Alkali Limited* train ran daily from Goldfield through Rhyolite to Las Vegas and Los Angeles. Lumber, food, and eager investors were carried in; ore and disappointed investors were hauled out. One of the more unusual commodities shipped out on the LV&T was ice; after Las Vegas's ice-making plant burned down in July 1907, the precious commodity was shipped in from an ice-making plant near Rhyolite.

But the mines in Rhyolite and elsewhere were short lived, and so was the railroad that depended on them. The LV&T ceased operations in 1918 and pulled up the rails. The engines and cars were sold as well; at least one engine later worked on the Hoover Dam construction railroad (map 5.7). Finally, the railroad sold the roadbed to the highway department for use as a road (map 7.10). Aside from the old depot at Rhyolite and a shack at Corn Creek made of salvaged railroad ties, little remains of this line today.

Several short mining railroads were built south of Las Vegas in the next several decades. The narrow-gauge Yellow Pine Railroad was finished in 1911 and connected gold mines near Goodsprings to the mill in that town and with the San Pedro, Los Angeles and Salt Lake line (later Union Pacific) at Jean. The line was not long but had steep grades that made operations difficult. The railroad shut down in 1934, although the mines continued in operation for a few more years.

Most of the mining around Las Vegas was not gold but less exciting minerals such as gypsum, salt, borax, and even sand. Las Vegas lies amid several valuable deposits of gypsum, used by the construction industry to make drywall (map 8.19), and these deposits have attracted a number of railroads over the years. The first of these was a narrow gauge line operated by the Arden

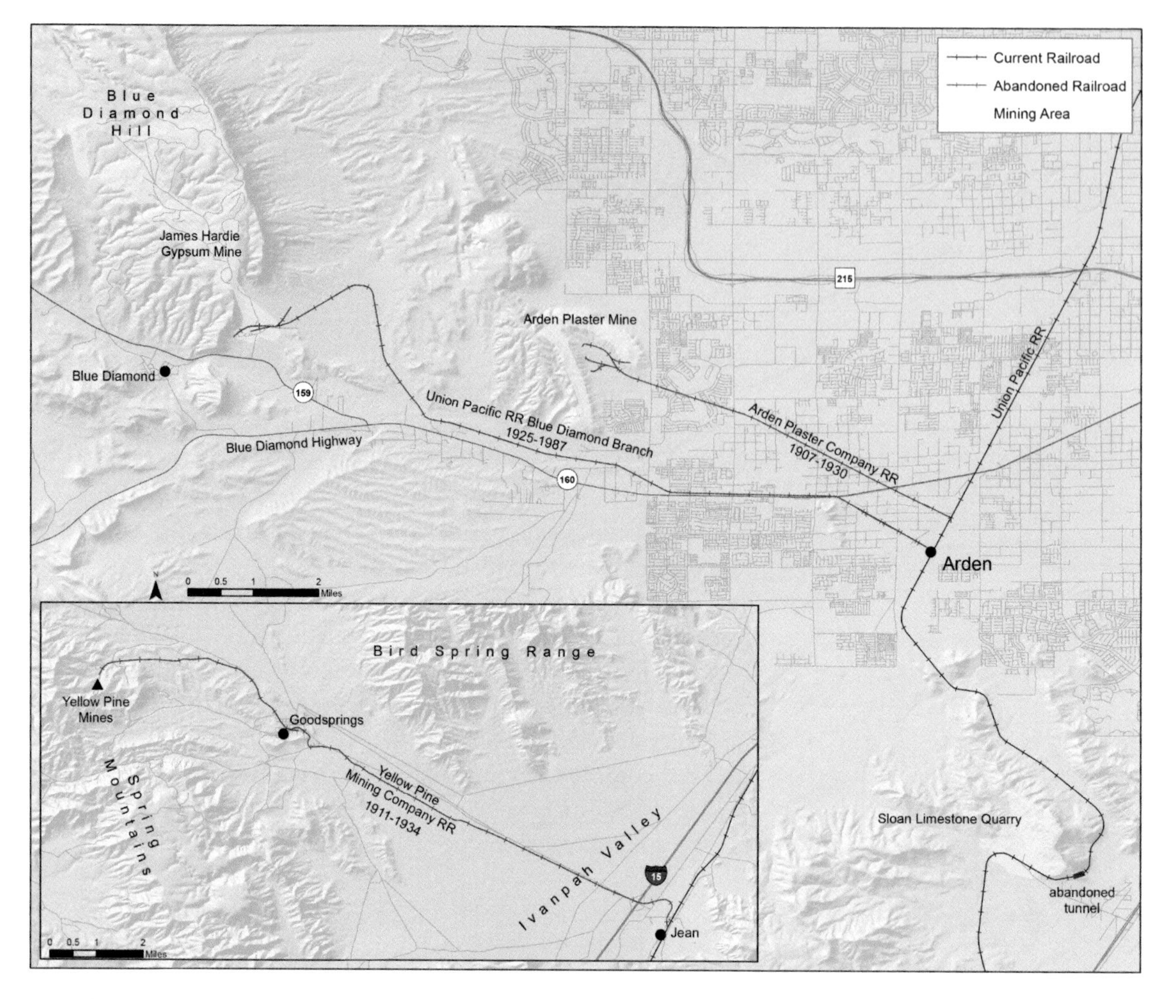

Map 7.3. Mining Branches. Three branch lines were built from the Union Pacific Railroad to reach mines southwest of Las Vegas, serving several new towns. Blue Diamond and Goodsprings remain, but the railroads have disappeared.

Plaster Company from Arden west to mines in the Blue Diamond Hills. This started operations in 1907 and lasted until 1930. It was joined in 1925 by the standard-gauge Blue Diamond branch of the Union Pacific Railroad, which also ran from Arden west to mines in Blue Diamond Hills. This line was removed in April 1987, although the mine remained open for two more decades. Farther to the south on the Union Pacific mainline is the Sloan limestone mine, which began operating in 1914 and is still active.

The Mead Lake Subdivision was completed by the San Pedro, Los Angeles, and Salt Lake Railroad in 1912 to serve the farming communities in the Muddy River Valley. It runs through the Narrows of the Muddy River, crossing the river three times, and ran as far south as St. Thomas. This town was founded in 1865 and was soon followed by Logandale, Moapa, and Overton (map 5.3). All of these were part of Mormon colonization efforts, with an emphasis on cotton production, but this did not succeed. Farmers eventually switched to food crops, and in the early 1900s about five thousand acres were being farmed in the Muddy Valley. Cantaloupes were the first major crop, but they were

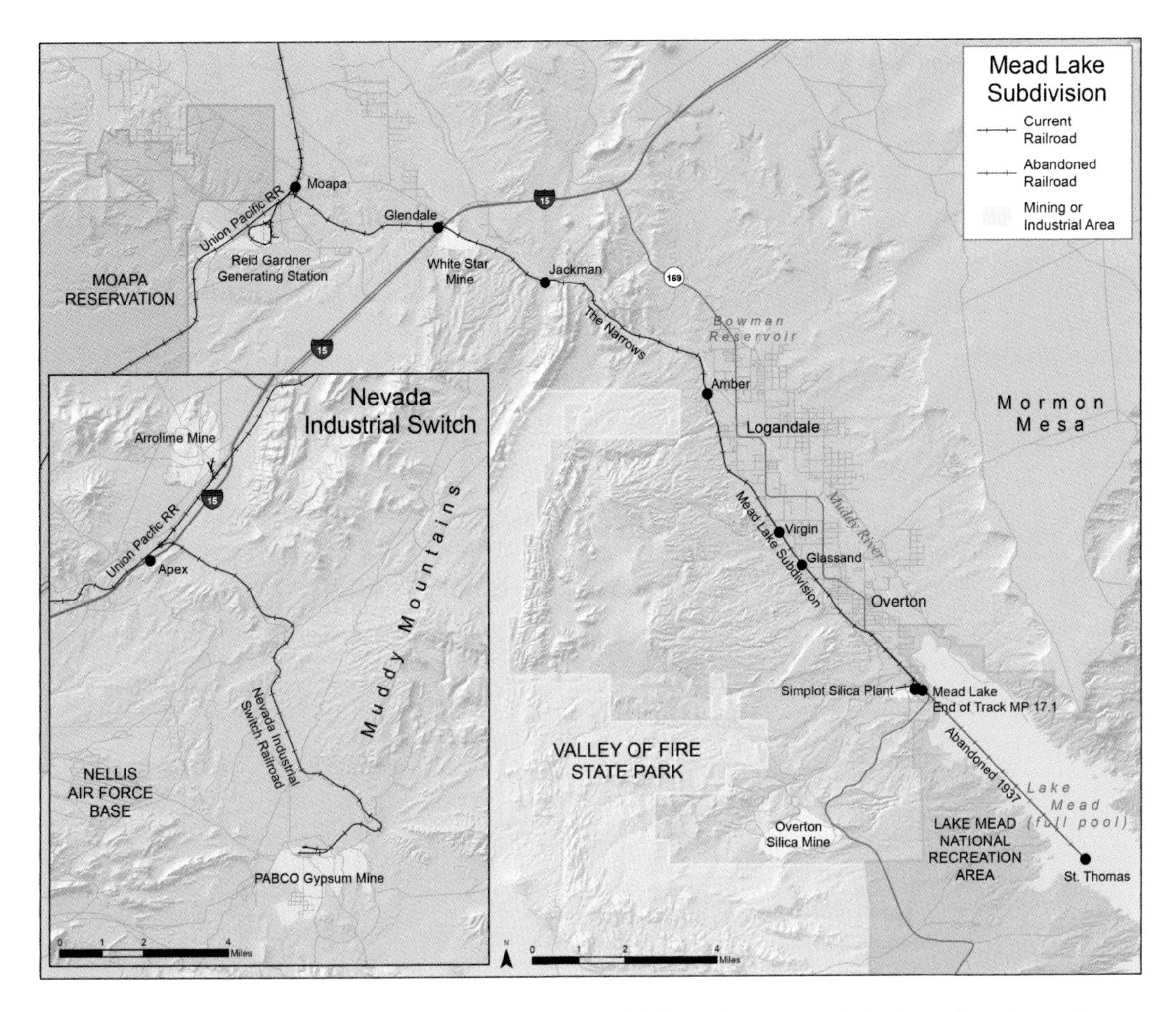

Map 7.4. Northern Branches. The Union Pacific Railroad built the Mead Lake subdivision to serve farm towns along the Muddy River. In 1937, the last few miles were abandoned to Lake Mead as it filled behind Hoover Dam. *Inset*: a line was built to the PABCO gypsum mine east of Las Vegas in later years.

replaced later by alfalfa and grains. While the railroad greatly reduced transportation costs to ship produce and grains to Las Vegas and other hungry desert towns, it also allowed even cheaper competition from California's Central Valley to be imported, undercutting the Muddy farmers. The last six miles of the line were abandoned and removed in 1939 before the waters of Lake Mead submerged it. The remainder of the line continues to serve the farms in this area as well a gypsum mine near Interstate 15 and a silica sand mine at the end of the line.

Several newer rail customers can be found to the northeast of Las Vegas. The Reid Gardner Generating Station opened in 1965 near Moapa. It was Southern Nevada's only coal-burning power plant and required regular coal trains from mines in Utah, Colorado, and Wyoming. A loop around the plant allowed trains to unload coal and then return to the main line headed north. It was shut down in 2017.

The newest branch line in the Las Vegas area serves the PABCO Gypsum mine and mill east of Las Vegas. This line branches off the main line at Apex and runs south 11.3 miles to the mine and

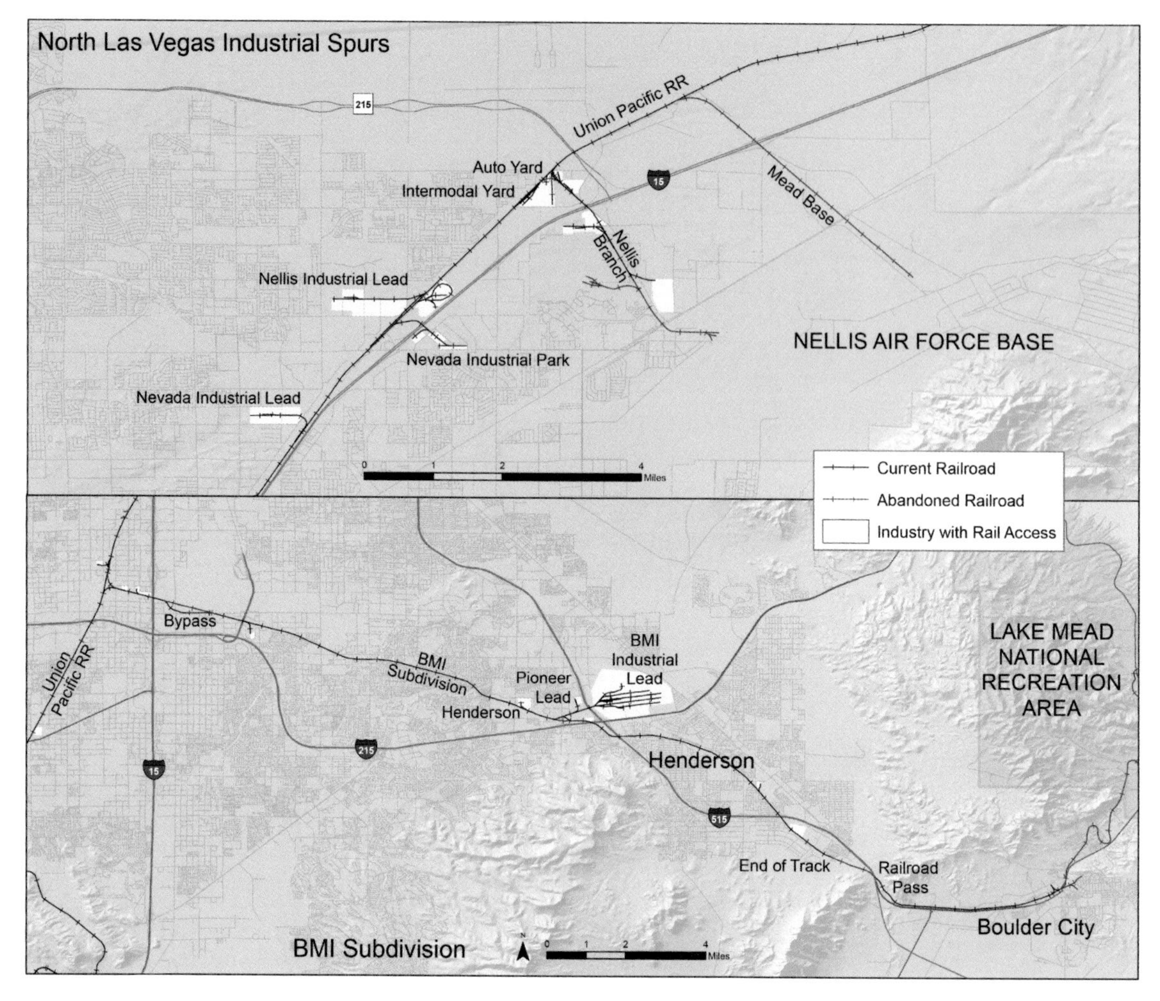

Map 7.5. Industrial Spurs. Many short railroad spurs have been built to serve industries, military bases, or railroad yards.

drywall factory. Originally operated by Union Pacific, it is now owned and operated by the Nevada Industrial Switch company. A small spur serves the Arrolime gypsum mine just north of Apex.

The Union Pacific built a branch to serve the construction of Hoover Dam in the 1930s, now known as the BMI Subdivision. This branched off the mainline at a point now called Boulder Junction and headed southeast to cross through Railroad Pass and on to the new town of Boulder City. From there it connected to the Six Companies Railroad, built and operated by the main contractor for the dam's construction, which in turn connected to the US Government Construction Railroad (map 5.7). The completion of Hoover Dam was not the end of the line for this railroad; a spur was built in 1941 on the Union Pacific branch to serve the massive new BMI plant in what became Henderson (map 4.8). The installation of the last generator in Hoover Dam in 1961 removed the need to keep the line to Boulder City open, and the rails across US 95 at Railroad Pass were paved over, only to be restored when Interstate 11 was built through the pass. West of Railroad Pass, the line remains in use, now known as the BMI Subdivision. In addition to the industrial park at the former BMI plant, a number of other industries have short spurs.

Spurs allow a number of industries along the Union Pacific mainline through Las Vegas to receive or ship by rail. Most of these are toward the north end of town, in part because of the presence of Nellis Air Force Base. The Nellis branch was built in 1941 and served the air base as well as several industrial customers. The southern end of this line into the base has been removed, but the industrial customers remain, among them a scrapyard and the Calnev Pipeline terminal (map 8.18). An intermodal yard has been added where this branch meets the Union Pacific mainline. To the northeast was the Lake Mead Base branch, now entirely removed. This was built in the 1950s to serve a navy nuclear weapons storage site. The storage site remains as Area 2 of Nellis Air Force Base, but the rails were eventually removed. An overpass allowed Interstate 15 to cross over the tracks, and once the tracks were gone Speedway Boulevard was built in their place. The Nellis Industrial Lead farther south serves several warehouses and lumberyards, while the Nevada Industrial Lead has an automobile-loading facility as well as several industrial customers. This branch was built in 1988 as a replacement for the downtown yard. Just north of downtown, another spur runs west under Interstate 15 to reach the *Las Vegas Review-Journal* newspaper printing plant.

Las Vegas was divided by the railroad into east and west from its outset. While Clark's townsite on the east contained the majority of the population, the McWilliams Townsite to the west attracted many settlers in its earliest years (map 4.1). The busy tracks were a barrier to those who lived in what would be known as the Westside. They felt their neighborhood to be isolated from the rest of the city. This barrier was finally overcome in 1937 when an underpass was built on Clark (later Bonanza) Street, allowing people to cross the tracks safely without delays. This was a two-lane undercrossing, later expanded to four lanes, and became part of US 95 when that route was established.

The next safe and reliable crossing was an underpass on Charleston Boulevard, completed in 1950. It was not provided with pumps since it was thought to have good drainage; June 1955 marked the first of many times it was filled with water after a thunderstorm (map 6.8).

As the city grew so did the frustration of traveling between east and west in the city. Crossing the tracks often meant a long delay for trains slowing to a stop at the downtown yard. There were only nine places to cross the tracks between Cheyenne Avenue and Blue Diamond Road in 1963, and the Bonanza and

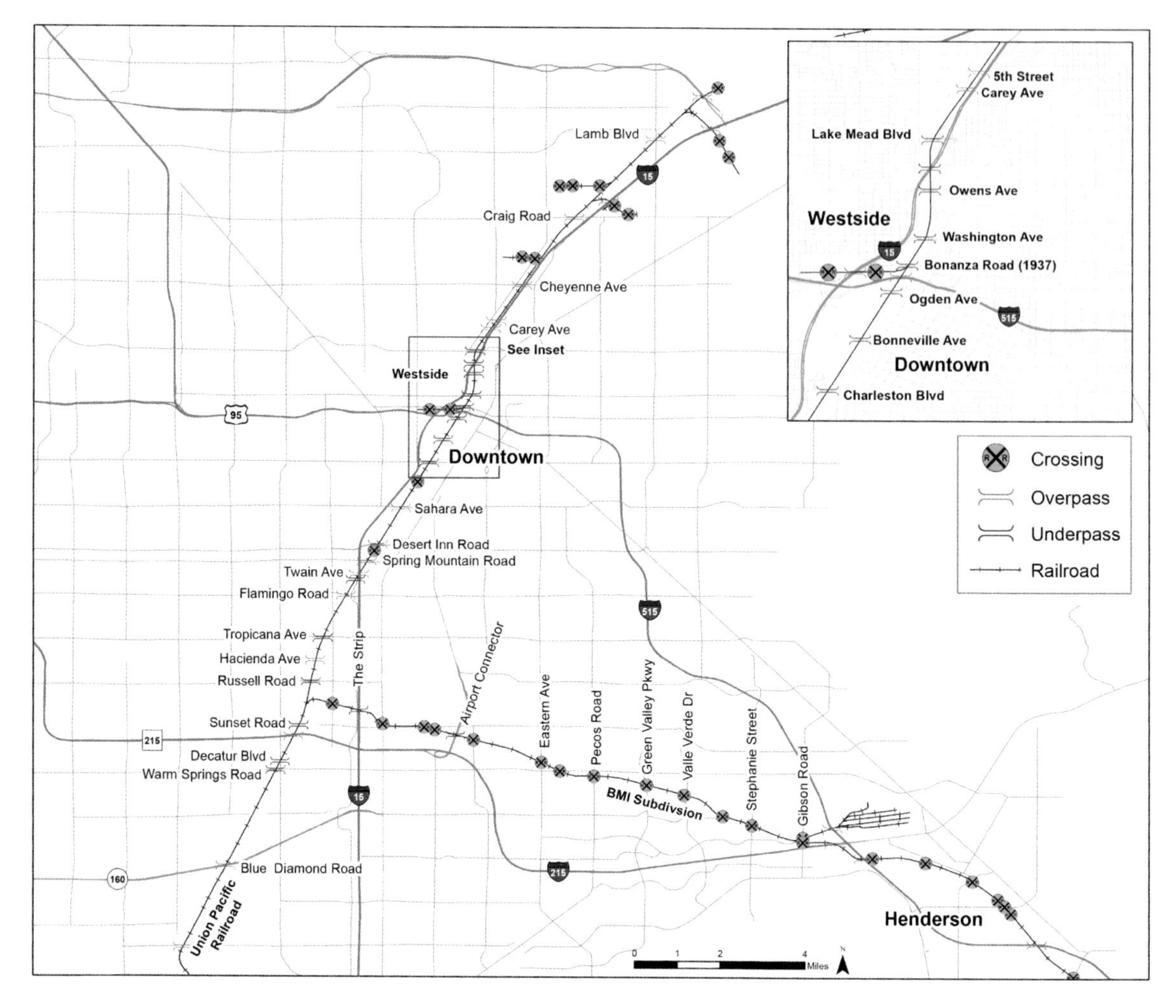

Charleston underpasses were used by half of all vehicles. A few more overpasses and underpasses were built later, beginning with the Sahara overpass in 1966. But most grade-separated crossings have been built since the late 1990s during the city's greatest building boom. Almost all of these were at new rail crossings needed in rapidly growing areas, and only a few replaced grade crossings such as at Spring Mountain Road. Several new underpasses were built downtown after the rail yard was closed and vacated. Only two grade crossings remain along the Union Pacific line in Las Vegas, at Wyoming Avenue and Desert Inn Road.

While this building program has been relatively modest compared to rail line relocations in Los Angeles or Reno, where tracks have been placed in trenches to separate them from streets, the railroad no longer divides the city as it used to. Safety has also been improved, especially compared to nearby Phoenix, which has five of the thirteen most dangerous railroad crossings in the nation. These efforts were limited to the Union Pacific mainline; the BMI Subdivision line to Henderson has many grade crossings

Map 7.6. Crossing the Tracks. Many highway underpasses or overpasses have been built along the Union Pacific mainline to make it easier to cross the tracks.

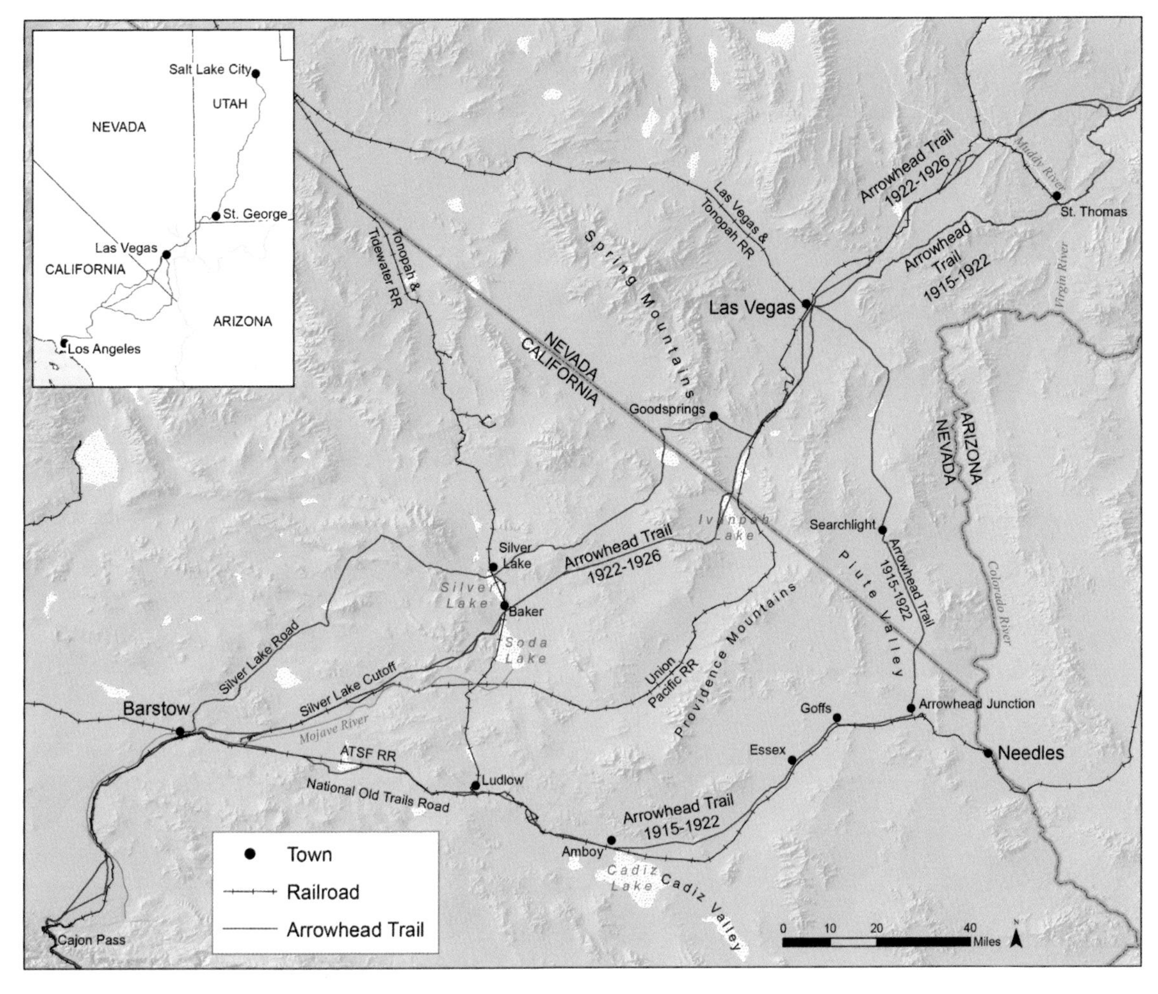

Map 7.7. The Arrowhead Trail. The first long-distance highways to pass through Southern Nevada are displayed, based on USGS topographic maps and the articles by Lyman.

and no under/overpasses except at freeways. Fortunately, this line sees only infrequent trains and is restricted to 10 mph.

Eliminating rural railroad crossings was a priority of the Nevada Highway Department in the 1920s and 1930s, as it was in many states. While much of this involved roads in northern Nevada, several overpasses or underpasses were built near Las Vegas. The Apex and Mud Lake underpasses north of Las Vegas on US 91 were both built in 1924 and the Jean overpass on Highway 91 to the south circa 1928. The Jean overpass was replaced by an underpass in 1937, which is still in use. A new Apex underpass was built in 1939, complete with decorative cactus gardens (which unfortunately no longer exist). An overpass was built at Glendale over the Mead Lake branch circa 1940 on US 91, at which time it was possible to cross the state on US 91 without the risk of having to stop for a train.

The earliest efforts in creating long-distance automobile routes in the United States stemmed from the efforts of individuals and organizations. The first of these was the famous Lincoln Highway, founded in 1912 as a transcontinental road between New York

City and San Francisco (passing through northern Nevada), and many others soon followed. Las Vegas's first highway connection was one of these, known as the Arrowhead Trail. Promotion of the Arrowhead Trail began in 1916, when Las Vegas residents responded to other auto trails being created across central and northern Nevada. The road was announced as connecting Los Angeles with Salt Lake City via Las Vegas and was promoted as being free of bad grades, high altitudes, or sand.

From San Bernardino the road followed the National Old Trails Road, an earlier named auto trail, through Cajon Pass to Victorville and Barstow. Although the most direct route to Las Vegas was the Silver Lake Road to the northeast, the Arrowhead Trail continued to follow the route of the National Old Trails Road east along the Santa Fe railroad, where towns, water, and communications could be found. Motorists headed for Las Vegas or Utah would eventually turn north at Arrowhead Junction, where the Arrowhead Trail ran north to Searchlight and into Las Vegas via Railroad Pass.

From Las Vegas the road headed northeast and crossed through the Muddy Mountains into a valley of red sandstone, originally known as the Red Road region and later named the Valley of Fire by highway promoters (figure 7.7A). From there it ran down to the town of St. Thomas, a small farm town along the Muddy River that had been founded in 1865 by Mormon settlers (map 5.3). By 1918 the town had several grocery stores, restaurants, a garage, hotel, and other businesses available to motorists. The Arrowhead Trail passed through town on the main street and over the Muddy River on a small bridge before climbing a ridge on the south end of Mormon Mesa and then across the Virgin River on a substantial truss bridge. The road then ran northeast to a second Virgin River crossing at the town of Mesquite and then on to Utah.

Surviving section of Arrowhead Trail in Valley of Fire State Park. Photo by author.

In 1921 the Hoover Dam project called for a dam with a lake at an elevation of 1,300 feet (higher than actually built), which would put St Thomas, Overton, and part of the Arrowhead Trail underwater. For that reason, the road through St. Thomas was bypassed in 1924 by a new Arrowhead Trail that followed the railroad from Las Vegas to the Muddy River and then up and over Mormon Mesa to Mesquite. Another change took place to the south when the Silver Lake Cutoff opened in 1922, providing a more direct route between Barstow and Las Vegas. Built by San Bernardino County, it bypassed Silver Lake in favor of the nearby town of Baker, which has depended on this highway traffic ever since. The Arrowhead Trail was rerouted along this road and became US 91 and still later rebuilt as Interstate 15.

The Arrowhead Trail did not attain the fame or prominence

of other named roads. It was not even considered the best road between Los Angeles and Salt Lake City. The Midland Trail held this title, and ran north from Los Angeles via Palmdale and Mojave into the Owens Valley before turning east at Big Pine and climbing Westgard Pass between the White and Inyo Mountains. From there it crossed several other desert basins and ranges before reaching Tonopah and Ely. The Arrowhead Trail was used mainly in winter when snow closed Westgard and other passes. Likewise, travelers heading east preferred the National Old Trails Road.

The National Park-to-Park Highway also shared the route of the Arrowhead Trail between Las Vegas and St. George, Utah. An association created in 1920 to link the twelve major western national parks backed this highway. A large loop connected parks such as Yellowstone, Glacier, Mount Rainier, Yosemite, the Grand Canyon, and Rocky Mountains National Parks, but also included a spur that followed the Arrowhead Trail from Arrowhead Junction, California, to Utah's Zion National Park. The Arrowhead Trail also became part of the Pikes Peak Ocean to Ocean Highway in 1924, a transcontinental highway created in 1914 between New York City and San Francisco, but which in 1924 was rerouted south at Salt Lake City to run to Los Angeles instead. Local interests favored the southern route along the Arrowhead Trail because it provided more opportunities for Utah businesses to make money from travelers.

In 1926 the route of the Arrowhead Trail became US 91, part of the new national numbered highway system. The Arrowhead Trail and the National Park-to-Park Highway Association ceased activities at that time, but the Pikes Peak Ocean to Ocean Highway association lasted into the 1930s. Parts of the original road and the town of St. Thomas finally slipped underwater in 1938 when Lake Mead filled up behind Hoover Dam.

Except for the name Arrowhead Junction, which remains on some maps, little of the old highway remains. The section of road through the new Valley of Fire State Park was bypassed and abandoned except for a solitary historical marker. The only stretch near Las Vegas still in use is part of New Gold Butte Road south of Mesquite. The original Arrowhead Trail has truly become a lost highway.

Nevada's state highway system was created in 1917. It consisted of only four routes connecting Reno and Carson City to communities in northern and central Nevada, and it included what had been the Lincoln Highway and Midland Trail. Lida was the southernmost town on the system; there was little automobile travel farther south after the decline of Rhyolite and other central Nevada mining towns. The challenges of building roads

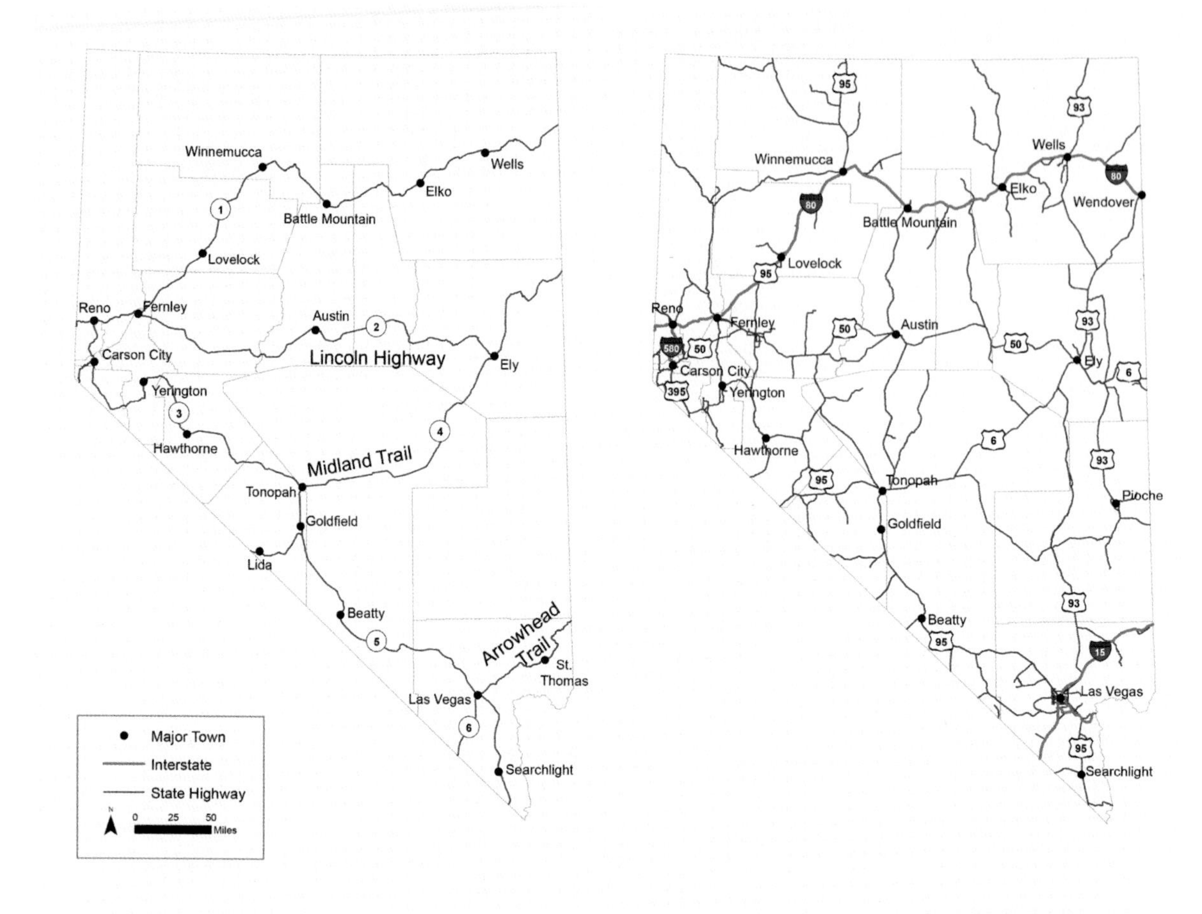

through hundreds of miles of empty desert to the tiny town of Las Vegas seemed insurmountable. However, an opportunity to build a road all the way to Las Vegas soon presented itself in the form of the Las Vegas and Tonopah Railroad. This line had shut down in 1918, and the highway department bought the 120-mile route in 1919. It was converted to a highway by removing the railroad ties, regrading it, and rebuilding several bridges. This road was added to the map in 1919 as Nevada Route 5, which also continued south to Searchlight and the state line.

A new state highway crossing Southern Nevada from California to Utah was also added in 1919, becoming Nevada Route 6. Route 6 included the Arrowhead Trail, the first automobile road to connect Las Vegas to the region (map 7.7). The road's new status as a state highway brought some improvements, such as a replacement for a wooden bridge over the Muddy River at St. Thomas and two wooden trestles across the Virgin River near Bunkerville and Mesquite. These were 884 and 964 feet long, respectively, the longest bridges in Nevada when they were completed in 1921.

Other Nevada highway routes soon appeared. In the Las Vegas

Map 7.8. Nevada's State Highways. *Left*: Nevada's highway system in 1919; *right*: the 2020 system. Maps based on the Biennial Reports of the Nevada State Highway Commission and Nevada Highway Department road maps.

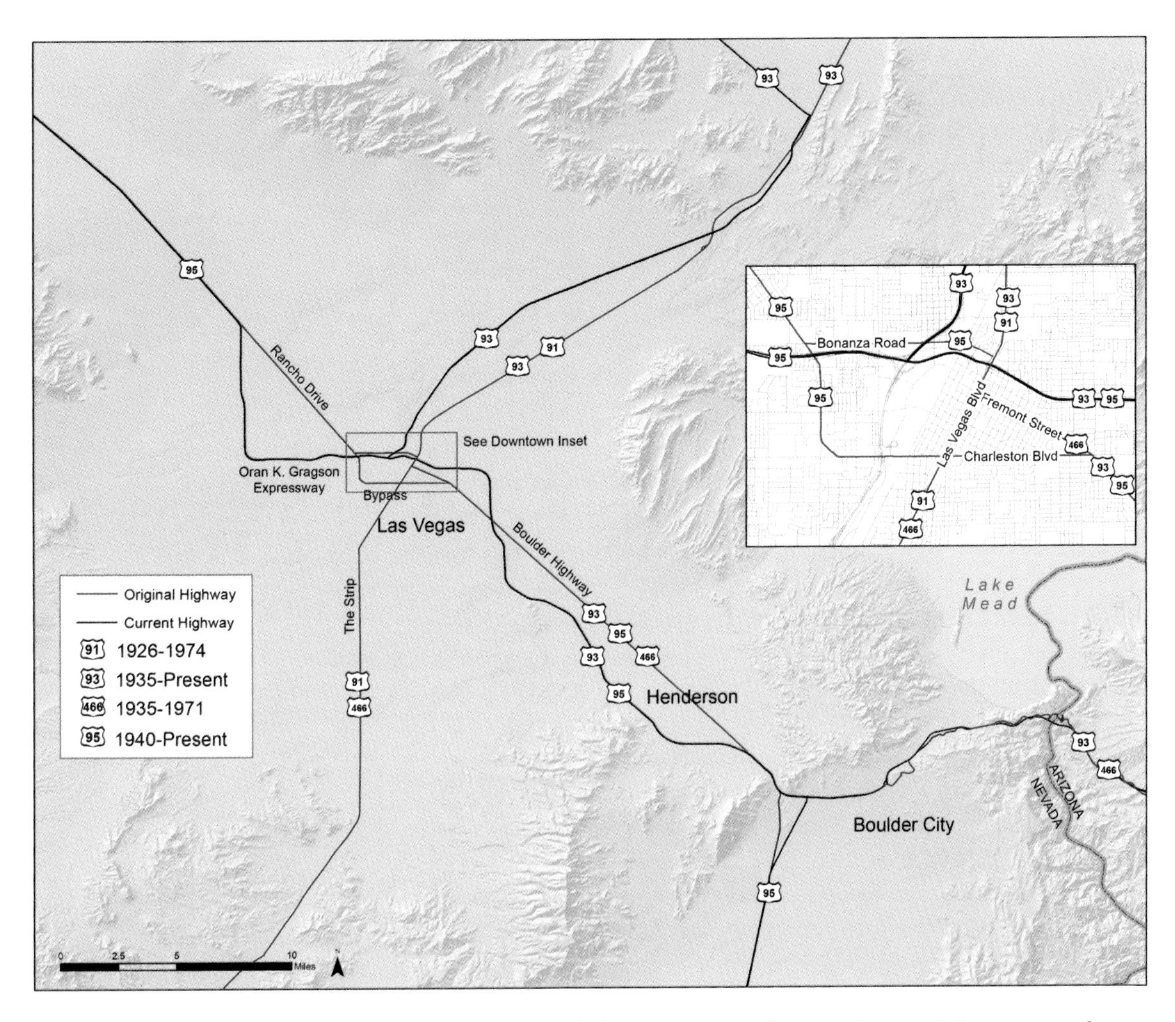

Map 7.9. Numbering Las Vegas Highways. Highways through the city used a variety of numbers from 1926 to 1956. Based on Nevada Highway Department road maps.

area, these included many secondary roads to outlying areas such as Nevada Highway 12 from Glendale to St. Thomas and Nevada Highway 16 through the Pahrump Valley to the California line. The Hoover Dam project prompted the construction of Boulder Highway, which in 1932 became Nevada Highway 26. Several state routes were also built into the Spring Mountains. The United States Numbered Highway system appeared in 1926 and replaced many of the early route numbers, and a statewide renumbering in 1976 also changed many of the lesser roads.

Although it was left off the early state highway system, Southern Nevada has become the most populous part of the state and a very important part of the state's highway system. Clark County contained 7,282 miles of roads in 2019, making up 18.5 percent of the total mileage in Nevada, as well as 68.2 percent of all motor vehicle miles traveled in the state.

The Arrowhead Trail was one of at least 250 named routes popularized by associations across the country by the 1920s. Many of these overlapped, and some roads carried the signage of as many as eleven different routes, while in other cases multiple routes used the same name. Towns along a potential route were lobbied to provide funding, and those withholding it might be bypassed. As spending on roads by states increased during the 1920s, the existence of named routes was increasingly seen as a problem to be resolved.

A solution was found when the American Association of State Highway Officials (AASHO) was founded in 1914. A plan to number 75,884 miles of roads was approved by 1925. Numbering for the system was specified in a systematic fashion. Even numbers were assigned to routes running east-west. US 2 ran near the Canadian border, and US 90 along the Gulf Coast. Odd numbers were given to north-south highways, beginning with US 1 along the Atlantic Coast and extending to US 101 on the Pacific Coast. The distinctive shields used on signs were also designed at this time.

Three of the four US numbered routes that passed through Las Vegas were odd-numbered north-south routes. The Arrowhead Trail and Nevada Route 6 became US 91, now part of a much-longer route running from Long Beach to the Canadian border. Soon after the World War II ended, casinos began appearing south of town along this highway, as casino owners hoped to pick up traffic from Los Angeles before they reached Las Vegas. These early casinos eventually became the Strip.

Other numbers soon appeared along Southern Nevada roads. US 93 was another 1926 route, but only from Wells, Nevada, north. It was extended south to Glendale in 1932, and to Kingman, Arizona, over the newly completed Hoover Dam in 1935. It was routed along Boulder Highway to downtown Las Vegas, where it continued north along US 91. US 466 appeared in 1935, running from Kingman to Las Vegas along US 93 and then south to Barstow, California, along US 91. The final route was US 95, extended in 1940 south from Idaho to Needles, California, where it intersected the legendary US 66. Several Nevada highways were re-signed as US 95, among them Nevada Route 5 from Goldfield south. This route ran through Las Vegas and used the Bonanza underpass before sharing the Boulder Highway with US 93 and 466.

The mileage of the numbered system has continued to change, and AASHTO (the successor to ASSHO, now the American Association of State Highway and Transportation Officials) continues to make decisions about new numbers, deletions, or route changes. US 466 was eliminated as a route in Nevada in 1971, several years after Arizona and California had taken their signs down. US 91 no longer appears on the map either, removed in 1974 when Interstate 15 neared completion (though it does still

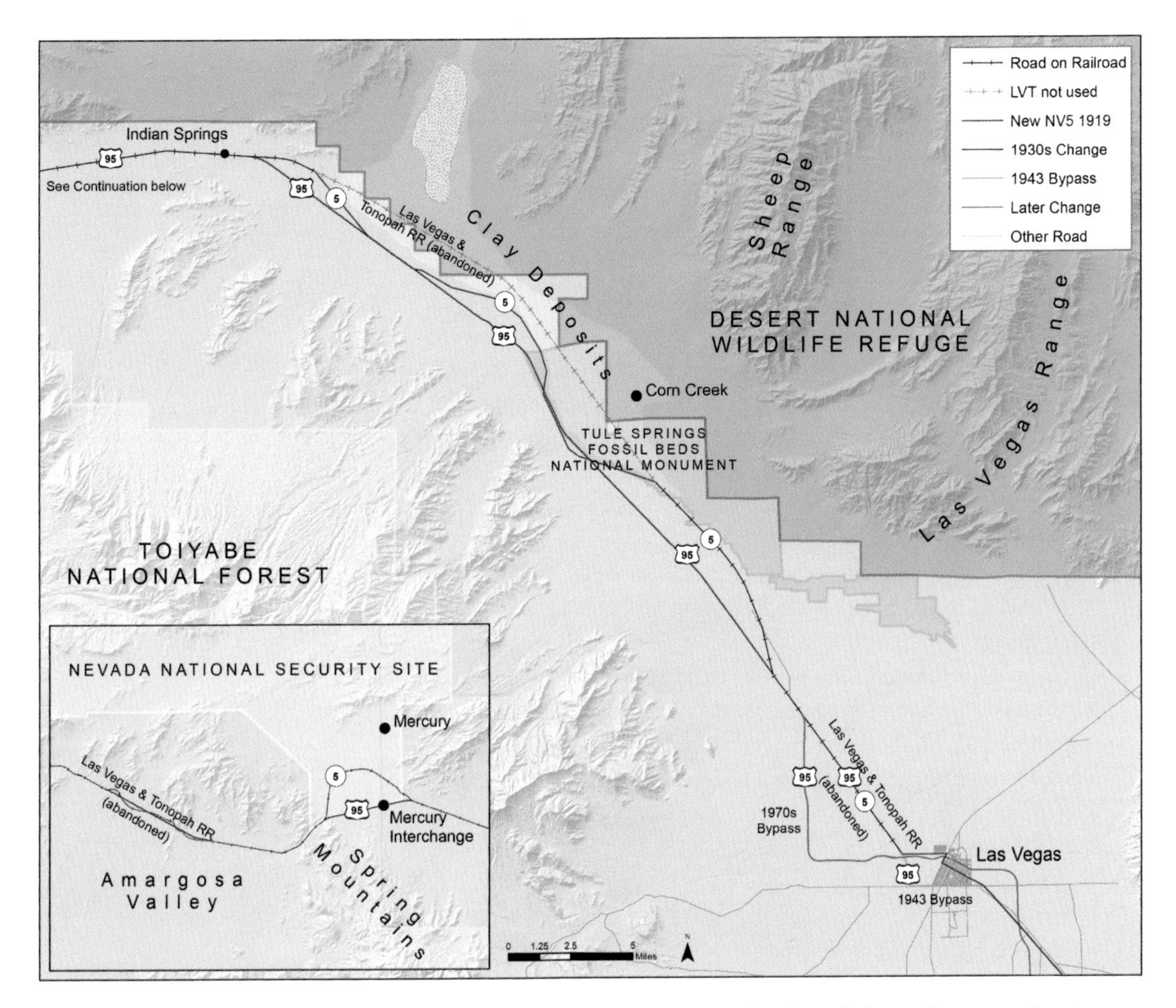

Map 7.10. US 95. The road between Las Vegas and Indian Springs has gone through an amazing transformation, from a dirt road built along an abandoned railroad line to a freeway. *Inset*: the road near Mercury. Based on the first to twenty-fifth biennial reports of the Nevada State Highway Commission, published from 1919 to 1967, and Nevada Highway Department road maps.

survive in Utah and Idaho). It has largely been forgotten, but its legacy as the Las Vegas Strip remains famous around the world.

While a new freeway replaced US 91, US 93 and 95 became freeways. When the Oran K. Gragson Freeway was completed west from downtown and then north to Rancho Drive, it became the new route of Highway 95. When the freeway was built southeast from downtown, US 93 and 95 were shifted from Boulder Highway onto this new road. A new Highway 93 was built in 1967 to the northeast of Las Vegas, and the old highway from Glendale became Nevada Highway 168. The most recent change was the shift of US 93 to the new Hoover Dam Bypass bridge (the Mike O'Callaghan–Pat Tillman Memorial Bridge). As Interstate 11, eventually connecting Mexico to Canada, is developed, more changes will come. New shields denoting the route have already been placed along the highway around Boulder City.

All of Nevada's highway connections to the rest of the state, Utah, Arizona, and California have changed drastically since they were pioneered by private associations or the state highway department. One highway was built on a railroad, portions of two highways were flooded by Lake Mead, and one was even built by NASA as part of a nuclear testing project.

Nevada Route 5, now US 95, the main road connecting Las Vegas to the rest of the state, was added to the state highway system in 1919 because of the abandonment of the Las Vegas and Tonopah Railroad (LV&T) in 1918 (map 7.2). The state highway department bought the 120-mile route between Las Vegas and Beatty in 1919 for use as a highway, though separate projects were still necessary to connect this road to Goldfield and the rest of the state highway system. The route required some work for highway use since all of the railroad ties had been left in place. During the winter of 1919-20 these were removed, the existing railroad trestles were widened to fourteen feet and refloored for auto use, several new bridges were built, and the road graded to a width of sixteen feet (which reduced the height of the railroad bed as well). Many rural highways in Nevada were one-lane wide at this time, and this was considered sufficient for the light traffic of the time. The former LV&T produced an excellent highway, with an easy grade and wide curves. Only a few problem areas existed, the main one between Indian Springs and Las Vegas, where the line ran across the valley bottom through an area of mud and silt. The old LV&T route was abandoned for a new road south of Indian Springs to Tule.

This road was rebuilt and paved in the early 1930s. Much of the former road can still be seen in the Skeleton Hills on either side of the road in Amargosa Valley. The road became much straighter, but also now had some steeper grades. Nevada Route 5 became US 95, part of the new national numbered highway system, in 1940.

The road became much busier after World War II when the Nevada Test Site began operations. Many employees had long commutes from Las Vegas on this road, leading to a high accident rate and the road being known as the "widowmaker." To improve this situation, the road from Las Vegas to Mercury became a four-lane divided highway in 1963, called the Mercury Expressway. NASA and the Atomic Energy Commission paid for 90 percent of this widening project's cost, using money from the Nuclear Engine for Rocket Vehicle Application (NERVA) underway in the Jackass Flats area of the Nevada Test Site (map 9.14). This highway likely will become part of Interstate 11 when this freeway is constructed (map 10.8). Originally built to help humanity reach Mars, it will ultimately help make the drive to Reno a little quicker.

The road between Boulder City and Kingman, Arizona, has been transformed more than any other road leading to Las Vegas. The first version of this automobile road appeared during the

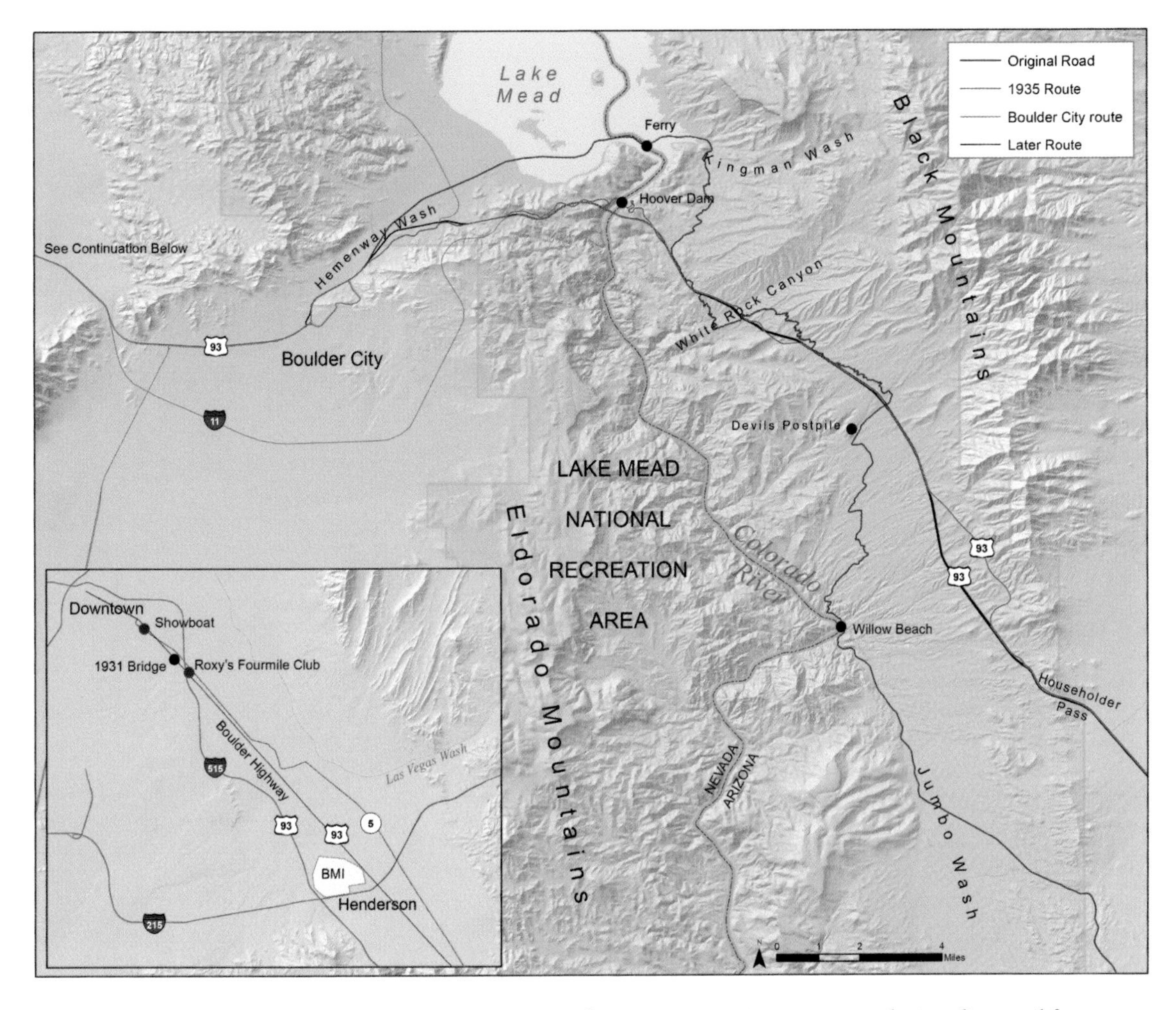

Map 7.11. Boulder Highway and US 93. A road from Las Vegas to Kingman, Arizona, first appeared in the 1920s and required a ferry to cross the Colorado River. It has since been transformed into a modern highway. *Inset*: roads between downtown Las Vegas and Henderson.

construction of Hoover Dam. It was a meandering dirt road from Kingman down Jumbo Wash to the Colorado River at Willow Beach and then north up another wash, passing a rock formation known as Devils Postpile. It crossed several ridges and washes on the slopes of the Black Mountains before descending Kingman Wash to a Colorado River ferry just above the damsite. From there drivers enjoyed a paved road leading up Hemenway Wash to Boulder City. The completion of the dam inundated the river crossing, and the paved Hemenway Wash road was later demolished, though the Kingman Wash road is still used to access a trailhead. A new paved road, US 93 and 466, crossing over the Hoover Dam's crest, opened in 1935. The Nevada portion of the road already existed, originally having been built for the dam's construction. The Arizona portion of this road was substantially improved and rerouted in the 1950s. This has been gradually turned into a four-lane divided highway; the Hoover Dam Bypass section opened in 2010 and greatly eased the long delays over Hoover Dam. Sections of the old road are often still visible from the modern highway.

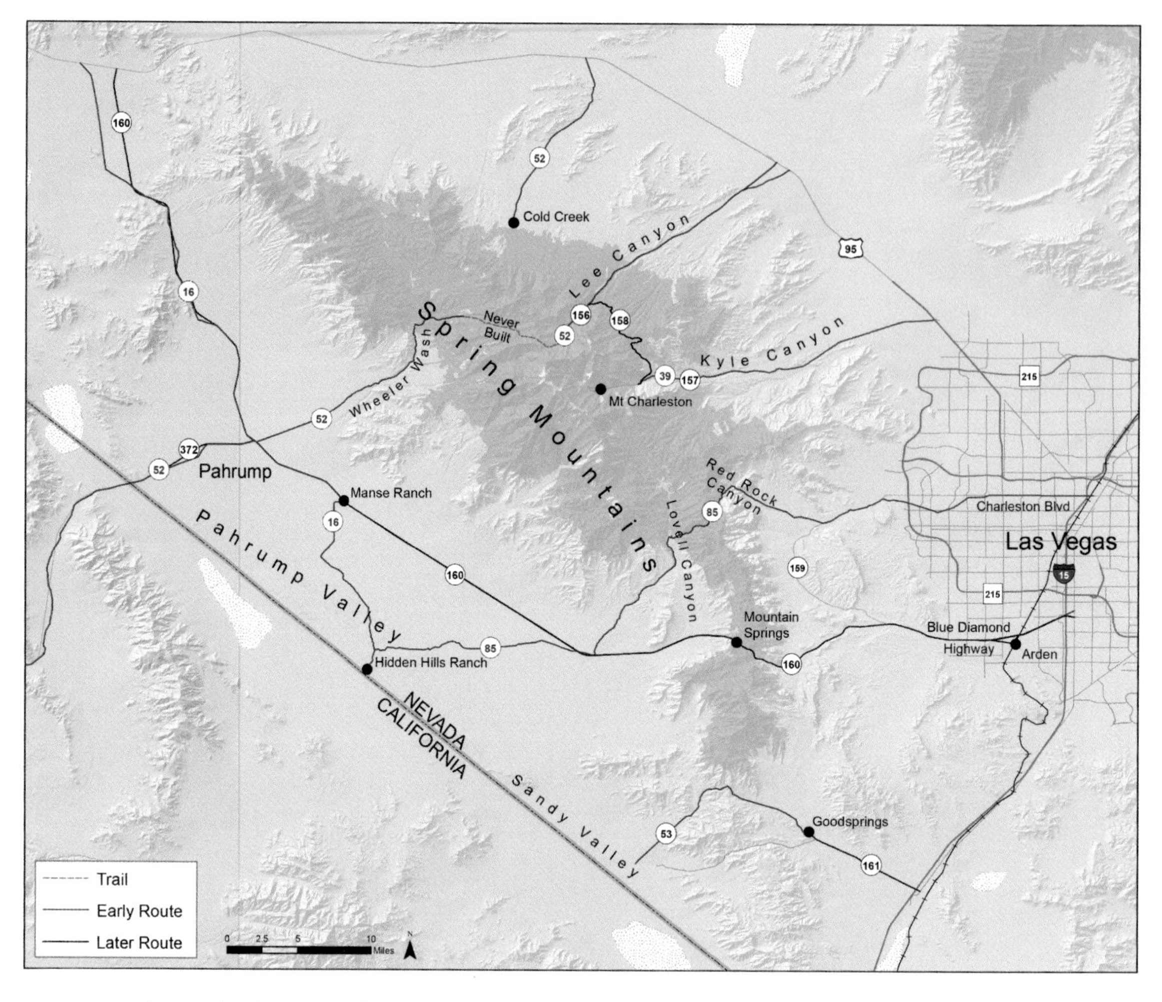

Map 7.12. Spring Mountain Roads. Highways into and through the Spring Mountains have changed substantially over the years. Several early state highways have disappeared.

Highways leading west from Las Vegas and into the Spring Mountains have also changed considerably over the years. Although the Old Spanish Trail had used Mountain Springs pass since the 1830s, the route through the pass fell out of use in the early twentieth century. One of Southern Nevada's forgotten state highways provided the first automobile route to Pahrump. This was Nevada Route 85, which first appeared on state road maps in 1947. This dirt road branched off West Charleston Boulevard, went up through Red Rock and Wilson Canyons, over the mountains into Pahrump Valley, then down to a junction with Nevada Route 16 at Hidden Hills, which ran north to Pahrump. These roads remained part of the state highway system until 1964.

The current road to Pahrump is named for its original destination at the Blue Diamond mine west of Las Vegas. It was extended through Mountain Springs to Nevada Route 85 in 1953 and completed to Pahrump in 1955. It has been slowly rebuilt since the 1990s as a four-lane highway to handle the huge increases in traffic between Las Vegas and the even faster-growing Pahrump Valley.

The east side of the Spring Mountains also had some early

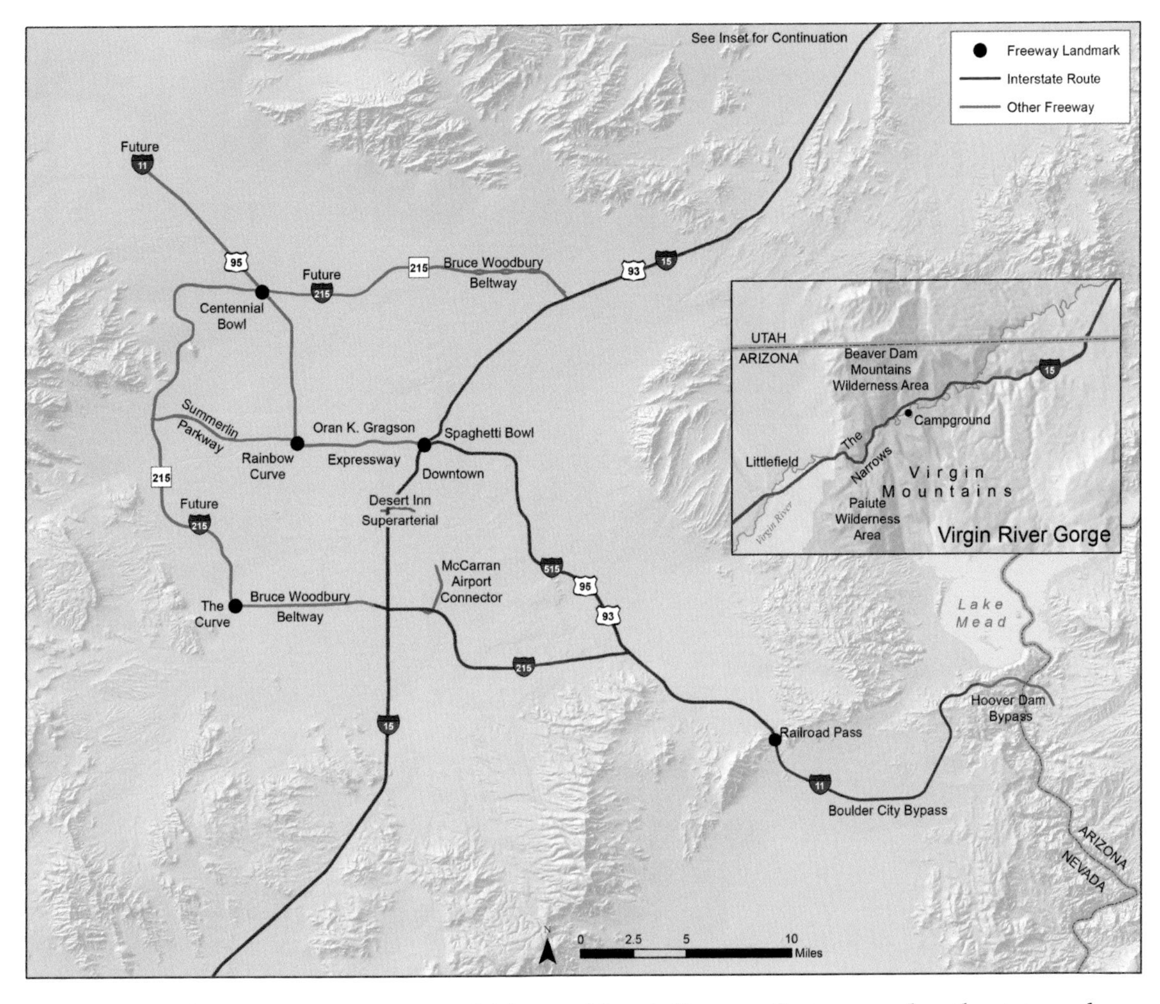

Map 7.13. Las Vegas Freeways. Freeways built as part of the Interstate Highway System are in red, and others are in blue. The Bruce Woodbury Beltway is being built as a county road but will become an interstate when it is completed. Based on Nevada Highway Department road maps and online news stories about the progress of highway construction.

state highways. Nevada Route 52 first appeared on the state road map in 1935 running from Indian Springs to Cold Creek, but that number was transferred to Lee Canyon Road in 1937. Later maps showed Nevada Route 85 crossing the Spring Mountains and descending to Pahrump and the California state line, but the middle section was never built and disappeared from maps in 1976. Nevada also renumbered many of its state highways that year, and Nevada Route 85 was replaced by Nevada Route 372 west of Pahrump, by Nevada Route 156 in Lee Canyon, and Kyle Canyon Road was changed to Nevada Route 157.

Although freeways began to appear in cities such as Los Angeles in the late 1930s, they would not appear in Las Vegas until the 1960s as part of a new government highway program, the Interstate Highway System. This dated back to the 1930s, not long after the United States Numbered Highway System was set up, when planners began thinking of a new national highway system designed for long-distance travel. This would connect major cities but have bypasses around them so drivers would not be stuck in city traffic. This became official when President

Franklin D. Roosevelt signed legislation to create the National System of Interstate Highways in 1944, though little in the way of construction happened for several years. It was not until 1956 when the Highway Trust Fund was enacted under President Dwight D. Eisenhower to pay for this new system (and changed the name to the National System of Interstate and Defense Highways) that construction began in earnest. Like the US numbered roads, interstates are all state highways but built to a common standard. The system was to be completed by 1972, but this date was not met and interstate routes continue to be built.

The routes for this system had already been chosen in 1947, though numbers were not assigned until 1957. The basic numbering system first devised in 1925 for the US numbered system was applied to the interstate System, with the exception that the numbering of north-south routes begins on the West Coast and increases eastward, and that for east-west routes begins in the South and increases northward. Interstate 80 was to cross Nevada along the route of US 30, while Interstate 15 would replace US 91 between Los Angeles and Salt Lake City. This highway would run from San Diego to Canada. The small town of Las Vegas at that time did not warrant a bypass or beltway planned for many larger cities.

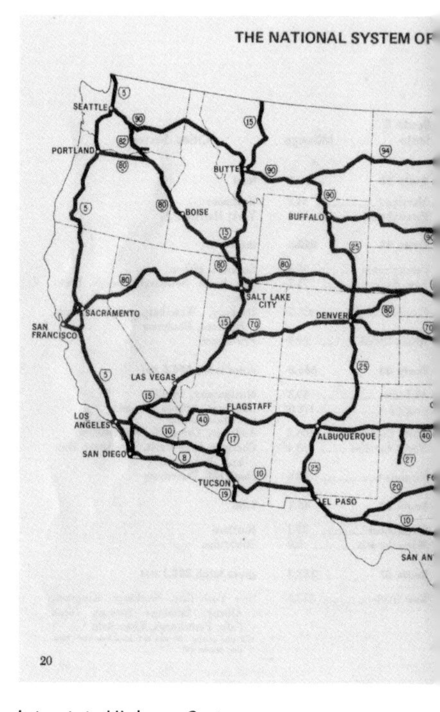

Interstate Highway System, 1970. Courtesy of the Federal Highway Administration.

Rural sections of Interstate 15 were completed relatively quickly. Interstate 15 was finished from the California stateline to the south end of the Strip by 1963, and from the north end of town toward Overton, but was only slowly finished through Las Vegas. In 1967 it was open as far north as Sahara Avenue and from Lamb Boulevard north, leaving a gap through downtown. The section between Sahara Avenue and Charleston Boulevard was finished by 1971, and the gap had closed to the stretch between Bonanza Road and Cheyenne Avenue in 1972. This section of the highway was built with a broad curve around the railroad yards on the west side of downtown; these yards are now gone, but the curve remains.

Interstate 15 was completely open in Las Vegas (and throughout Nevada) in 1974, but has never been truly finished: maintenance and widening has been never-ending. The Interstate 15/US 95 interchange, or Spaghetti Bowl, is one example. It was completed in 1968 and built to handle 60,000 vehicles a day but overloaded by the 1990s. It was completely rebuilt by 2000 and now carries more than 300,000 vehicles a day. The almost $1 billion Project Neon, completed in 2019, which supplanted Hoover Dam as the largest public works project in state history, widened 3.7 miles of Interstate 15 between Sahara Avenue and the Spaghetti Bowl. It is the busiest interchange in the state, and although the name doesn't appear on road maps, the Spaghetti Bowl is labeled on aviation maps as a landmark for pilots (map 7.19)!

The completion of Interstate 15 and rapid growth of the city stimulated plans for other freeways. The first of these was originally called the Fremont Expressway and was built from downtown west. It later became US 95 and known as the Oran K. Gragson Freeway. The highway was complete to Rainbow Boulevard by 1978, but the stretch north to Rancho Drive was not completed to freeway standards until 1990. New interchanges have been built to the north, extending the freeway farther. This freeway was extended east from downtown as well, and by 1986 had been completed to Boulder Highway and now also known as Interstate 515. Many other routes sprang up during the building boom of the 1990s, among them the city's beltway. The first section opened in 1994 in conjunction with the Airport Connector tunnel under McCarran International Airport runways. The entire route was open by 2003, but will not be built out to freeway standards until about 2025. Other freeways are the Desert Inn Super Arterial, opened in 1995 under the Strip and over the railroad tracks; Summerlin Parkway, completed in 2004; and the Hoover Dam Bypass bridge and freeway in 2010. The Boulder City Bypass connected this to Interstate 515 in 2018.

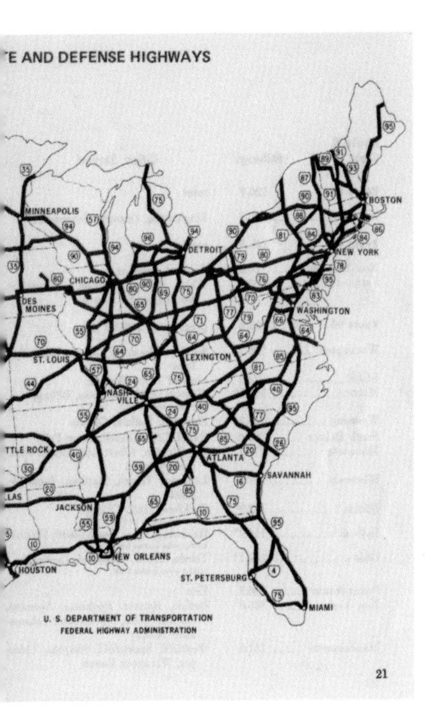

Las Vegas was a latecomer to beltway construction, and the city differs from others in how these freeways were built. The beltway is being built by Clark County rather than the state highway department, making it one of only a handful of county-built freeways in the nation. When completed, it will be signed as Interstate 215 and taken over by the state highway department. Summerlin Parkway has an even more unusual origin: a homeowners association and the master-planned community developer built the initial portions. Toll roads have become popular in Southern California and elsewhere to finance new construction, but this has not yet been tried in Nevada.

Interstate 15 offers a spectacular view of the Strip, but northeast of Las Vegas it passes through the spectacular Virgin River Gorge in Arizona, cut by the Virgin River through the Beaver Dam and Virgin Mountains. Interstate 15 was the first highway built through this canyon, squeezing in along the Virgin River and crossing it six times. The surrounding mountains are wilderness, with a campground located next to the freeway. In addition to its beauty, this stretch of Interstate 15 is also unusual for its history. It is located in the northwest corner of Arizona, and the freeway was not a priority for that state as it was for Nevada and Utah, where it was considered a vital route. Nevada and Utah had to help Arizona pay for construction of this section of road

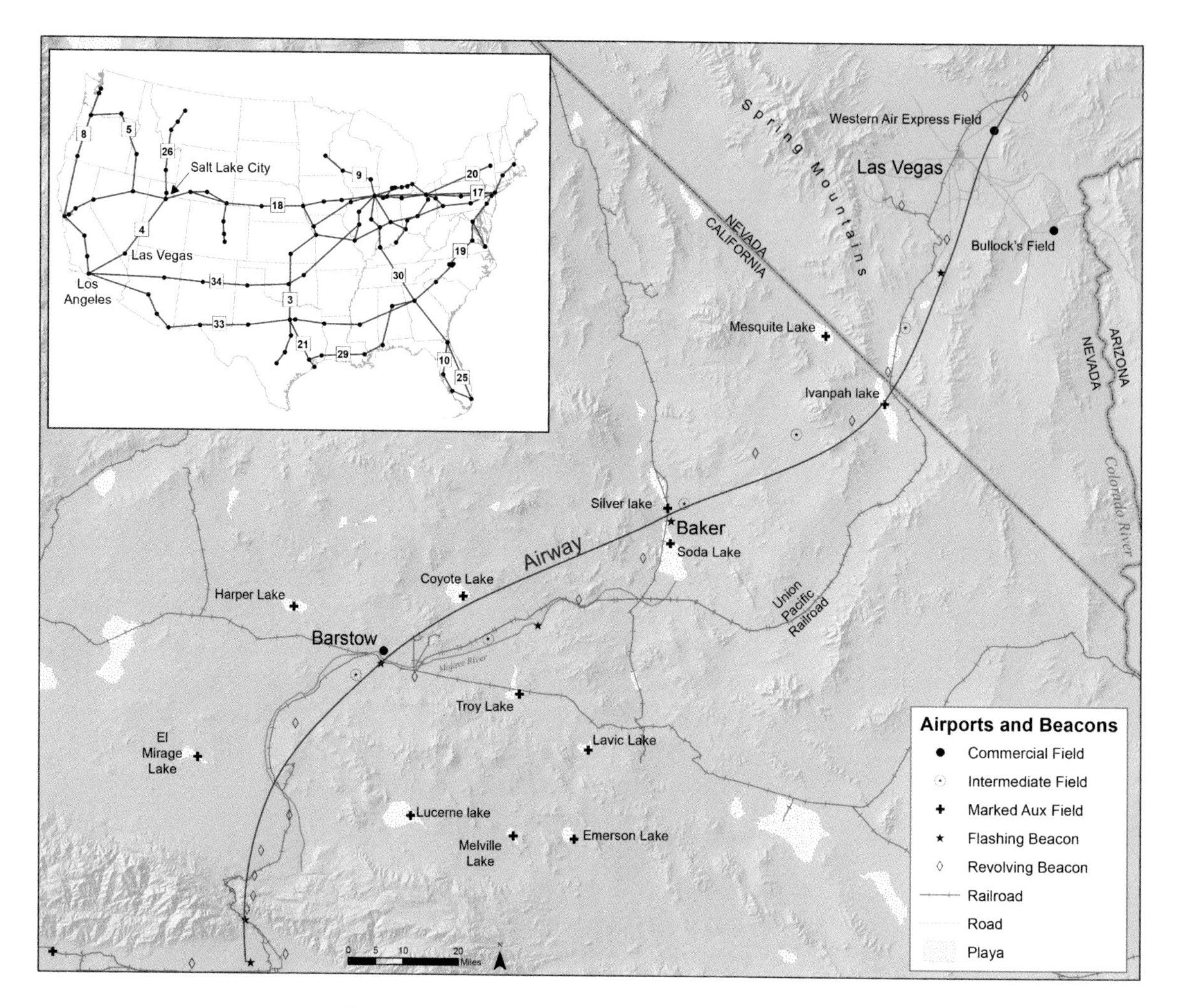

Map 7.14. CAM-4. The first airline route through Las Vegas required many facilities (the symbols are based on those in use in early aviation charts). Based on US Air Navigation Maps 132 and 133, produced by the Aeronautics Branch of the Department of Commerce in 1929.

to ensure it opened quickly. This was also a particularly expensive section of road when it was completed in 1973, and the Federal Highway Administration designated it as a nationally and exceptionally significant feature on the Interstate Highway System. It is one of only three canyon freeways to receive this designation; the others are Interstate 80 through the Truckee Canyon east of Reno and Interstate 70 through Colorado's Glenwood Canyon. Southern Nevada benefited from the latter project because extending Interstate 70 west from Denver to Utah provided better access to and from Southern California via Interstate 15. This route was completed in 1992 without any help from Nevada.

The commercial airline system that brings visitors to Las Vegas from around the world was created in the 1920s as part of efforts to create transcontinental airmail routes. These routes required the construction of numerous emergency landing fields, lighted airway beacons spaced about ten miles apart for nighttime flying, and giant concrete arrows built on the ground to point to the next airport. The first such route was established in July 1924 between New York City and San Francisco, with intermediate stops every few hundred miles. The line in the West largely followed the first transcontinental railroad finished in 1869, passing through northern Nevada with stops at Elko and Reno. A company that later became United Airlines operated the service.

Other airmail routes soon followed, including a Salt Lake City-to-Los Angeles route with a stop in Las Vegas, specified as Contract Airmail Route 4, or CAM-4. This generally followed the Union Pacific Railroad and US 91, heading north from Vail Field in Los Angeles, which opened in 1926 near Huntington Park (the airport shut down in the early 1950s and was converted to an industrial area). The airmail route headed east to San Bernardino, then north through Cajon Pass and on to Victorville, down the Mojave River to Barstow, and then northeast along US 91 to Las Vegas. Northeast of Las Vegas, the airway followed the railroad and highway to Glendale and then the highway across Mormon Mesa and past Mesquite. Although the highway detoured around the Beaver Dam Mountains to reach St. George, the airmail route went straight across the mountains, requiring pilots to climb to more than ten thousand feet above sea level and contend with turbulent air and bad weather. Flying elevations were lower from Cedar City north to Salt Lake City.

Several intermediate landing fields were constructed for emergency use. The route passed many dry lakes between Los Angeles and Las Vegas, and several were marked on maps as auxiliary landing fields. At first the route was flown only in the daytime, but rotating light beacons were placed near the few towns and on desert mountaintops for nighttime use.

A Douglas M-2 airmail plane in Western Air Express markings at the Smithsonian National Air and Space Museum in Washington, DC. This type of plane provided the first passenger service to Las Vegas. In the upper left background a supersonic Concorde airliner can be seen; this type of plane made visits to Las Vegas in later years. Courtesy of the Smithsonian National Air and Space Museum.

A new company, Western Air Express, began airmail service over this route on April 17, 1926, with one flight in each direction each day using Douglas M-2 biplanes. The original schedule was for eight hours and twenty minutes, with Las Vegas the only stop. The auxiliary landing fields were well used, with thirty-eight emergency landings recorded the first year.

It wasn't long before passengers wanted to ride along, and the airline accommodated them by allowing for one or two passengers to sit on top of the mail bags. On May 23, Ben Redman became the first airline passenger to land in Las Vegas. Wearing heavy clothing, a parachute, and ear plugs, holding a tin can the airline issued him for lavatory facilities, and looking around the hot, dusty, desert outpost, he might have wondered if he would be the last. But they did keep coming. It wasn't long before Elliott Roosevelt, son of the president, became the one hundredth passenger recorded. Hundreds of millions have followed him at an ever-increasing rate, exceeding 45 million a year in 2015.

Thirty-four airmail routes were designated by 1934. The

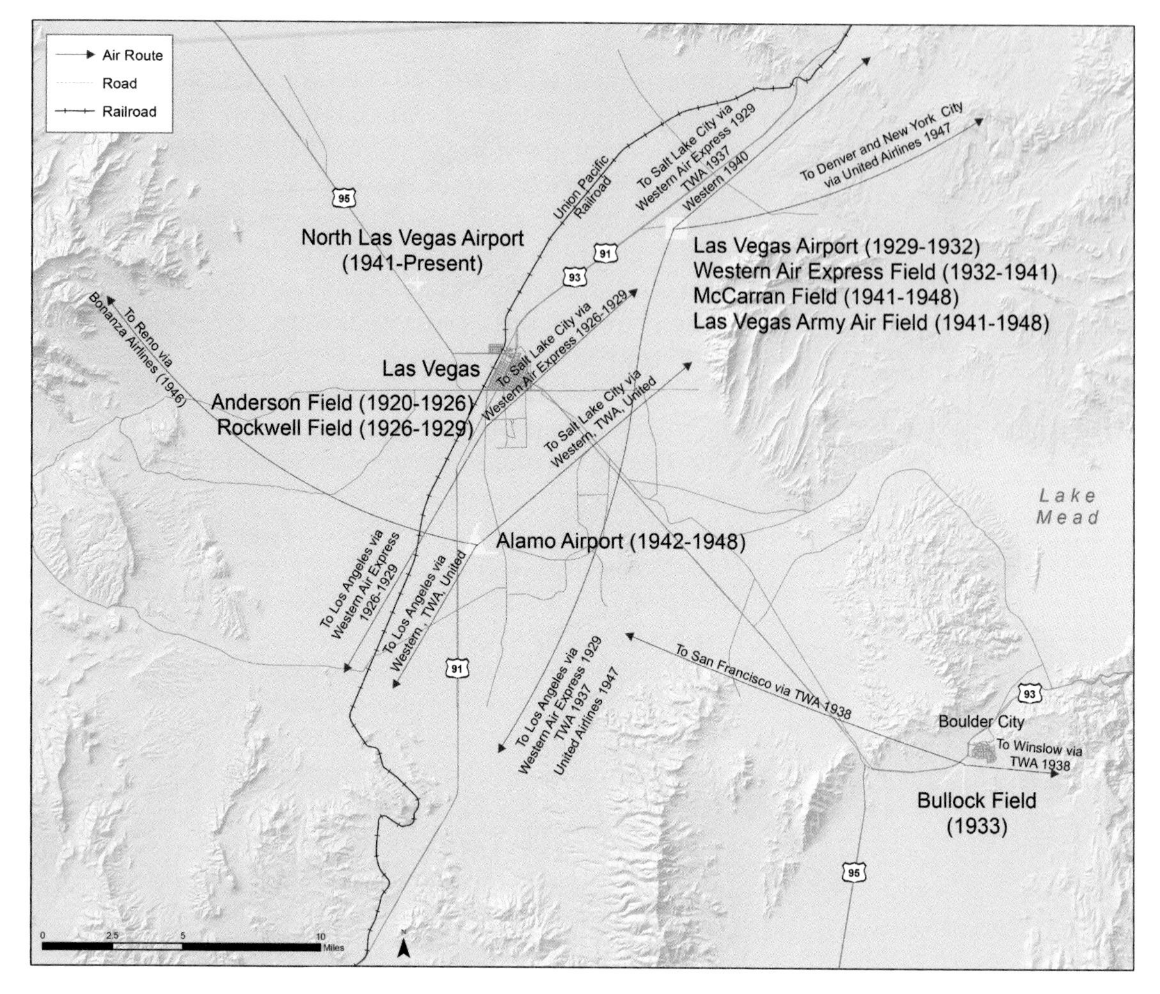

Map 7.15. Early Aviation. In the 1920s and 1930s, Las Vegas had a number of competing airports with flights to a number of different cities.

contracts held by various airmail companies were canceled that year because of a political scandal, and the US Army temporarily performed the service. The private companies continued their passenger business, and passenger airlines that today are vital to Las Vegas's economy emerged. The beacons and concrete arrows of the early airways have mostly disappeared, but remains of at least five survive in the Las Vegas region.

The aviation era began in Las Vegas on May 7, 1920, when the first airplane arrived and landed at the new Anderson Field. This was at the southwest corner of what would later be Sahara Avenue and Paradise Road. The forty-acre field did not have runways; rather it was a cleared stretch of desert allowing landings and takeoffs in any direction that winds dictated. It was known as Rockwell Field by 1928 and had been the site of many flying exhibitions. It had a small building and two towers for a beacon light and windsock.

Scheduled airline service to the city began with the creation of airmail routes by the US postal service (map 7.14). The New York–San Francisco transcontinental route crossed far north of

Las Vegas but a Los Angeles–Salt Lake City route was also specified, with a fuel stop in Las Vegas. Western Air Express began service over this route on April 17, 1926, with one daily flight each way. Two more seats were later added to allow for passengers.

Rockwell Field closed in 1929, but a new airport opened eight miles north of town along Highway 91. This became known as Western Air Express Field after that company shifted its service there, and it became the town's airport for many years. A 1930 timetable shows a two-hour, twenty-five-minute flight from Los Angeles to Las Vegas with a ticket price of $20 (equivalent to $327 in 2021).

Western Air Express merged with Transcontinental Air Transport in 1930 to form Transcontinental and Western Air (which in turn later became Trans World Airlines and better known as TWA), but in 1934 Western Air Express once again became an independent airline. It changed its name to Western Airlines in 1941, and it continued to serve Las Vegas, which advertised its Los Angeles–Salt Lake City route as the "Boulder Dam Route."

More airline services came to the Las Vegas area after Bullock Field in Boulder City opened on December 10, 1933. Grand Canyon Airlines became the first flightseeing airline when it started service from there in 1936, and TWA became the second airline to serve Las Vegas when it began a Los Angeles-to-Newark, New Jersey, service with a stop in Boulder City in August 1937, a service it called the "Grand Canyon Route." Las Vegas had two commercial airports during this time, and TWA flew connecting passengers between Bullock Field and Western Air Express Field. If traffic continues to increase, Las Vegas may someday again have two competing airports (map 10.9).

The city of Las Vegas purchased Western Air Express Field in 1941, making it a municipal airport, and it was renamed McCarran Field after longtime senator Pat McCarran. The army was given permission the same year to operate a gunnery school there, known as Las Vegas Army Air Field, alongside the regular airport operations. This airport was soon joined by two more in the town: North Las Vegas, which opened on the inauspicious date of December 7, 1941, and Alamo Airport, opening in 1942 as a pilot training operation south of town along Highway 91.

After World War II, the army shut down its operations at McCarran Field, but the airline flights continued to increase. Bonanza Airlines, Las Vegas's first hometown airline, began flights between Reno and Las Vegas from Alamo Airport in August 1946 (Bonanza later merged with two other airlines in 1968 to form what would become Hughes Airwest).

The newly created US Air Force decided to reopen its base at McCarran Field in 1948. The air force wanted total control of the

Map 7.16. The Jet Age. Nonstop service to and from Las Vegas in 1978. Based on information from Daniel Bubb (1996) and the Airline Timetable Images website.

airfield this time, so the city went looking for a new airport. The solution was to purchase Alamo Airport south of town and build it into a modern airport, to be renamed McCarran Field. On December 19, 1948, the new airport was dedicated, and the old McCarran Field then became Nellis Air Force Base.

McCarran Airport started with twelve scheduled daily flights. Many new carriers and destinations appeared over the years. Bonanza began flights to Phoenix in 1949, and United Airlines arrived in Las Vegas with a Los Angeles–Denver flight that passed through the city. When the present airport terminal opened in 1963, there were 128 daily flights, increasing to 236 in 1970. Western Airlines, the first to serve Las Vegas, continued as one of the most important carriers. Las Vegas was also very important to Western; one year during the 1930s, five slot machines in Western's ramshackle terminal provided the only profits for the airline. Western began its popular "It's the only way to fly" advertising program in 1956, and it remained a staple of Las Vegas TV commercials into the 1980s. When airlines began operating jets, it was considered crazy to think a small isolated town such as Las Vegas

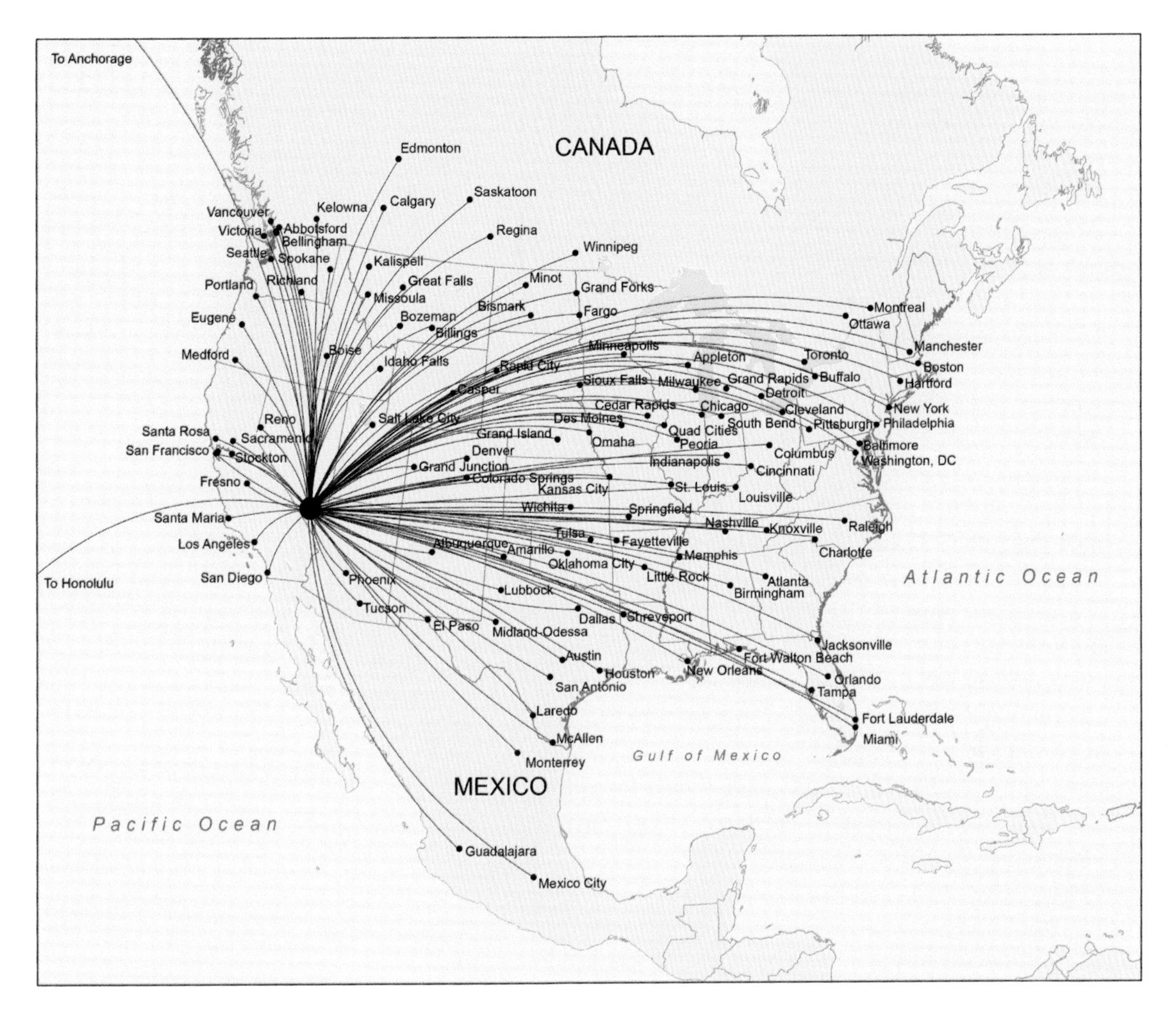

Map 7.17. Nonstop Service. Nonstop service to and from Las Vegas in 2018. Based on information from the McCarran International Airport website. Airport statistics are from the Clark County Department of Aviation website.

would ever see them, and the city was indeed bypassed when Western began operating its first 707 jets in the 1960s. But eventually the jets did arrive, and Las Vegas entered the jet age. Hughes Airwest, with its bright yellow planes, was another important airline for Southern Nevada. The name referred to Howard Hughes, who lived for a time atop the Desert Inn (map 11.3).

The Civil Aeronautics Board (CAB), a government agency known for resistance to change, tightly regulated airline routes, schedules, and fares from 1938 to 1978. After services were deregulated, airlines were free to fly where they pleased and charge what they wanted. This led to enormous and rapid changes in the airline industry as well as enormous growth in the number of airlines and passengers flying. By 2018, Las Vegas had service from twenty-eight airlines to 117 cities in the United States, Canada, Denmark, Mexico, Norway, South Korea, Sweden, and the United Kingdom (not shown on the map). It was ranked as the ninth-busiest airport in the country and twenty-sixth-busiest in the world.

Among major airports, Las Vegas is very unusual in remaining an origin-destination airport. While 64 percent of passengers at the Atlanta airport are connecting between flights, only about 20 percent of passengers at McCarran International Airport (which was in the process of being renamed Harry Reid International Airport in 2021) are there to change planes. Las Vegas is only a hub for Allegiant Air, the corporate headquarters for which is in Summerlin. Of those cities that have more passengers than Las Vegas, all are hubs except Los Angeles.

Service to eastern cities greatly expanded after 1978, but Los Angeles and San Francisco are the two most popular nonstop destinations. A few nearby cities that once saw many flights to and from Las Vegas, Burbank, and Palm Springs, dropped off the network. A ticket to fly between Las Vegas and Los Angeles will likely be much less than that in 1926 (adjusted for inflation). The travel time and level of comfort have greatly improved; parachutes are no longer required, and lavatory facilities have improved a bit from the days each passenger was issued an empty tin can.

One of the more unusual airlines operating out of Harry Reid International Airport officially has no name, but is commonly referred to by its call sign of "Janet." This is operated by defense contractor AECOM under contract to the air force to fly workers to Area 51 and the Tonopah Test Range airport (map 4.12) daily. This service started in 1972 and continues with six Boeing 737s painted white with a red stripe, as well as five smaller Beechcraft propeller planes. The planes fly out of a separate terminal on the west side of the airport accessed off Las Vegas Boulevard, and are easily spotted from the Strip. But don't even think of trying to get a ticket on this airline; no amount of money or frequent flyer miles will get you on one.

Like most other big-city airports, Harry Reid International has been transformed over the course of its seventy-five-year history. It opened as Alamo Airport before becoming McCarran Airport in 1948. The original airport terminal was on the southwest corner of the field, off of Highway 91. Traffic growth at the airport necessitated improvements, and a new jet age terminal opened on the east side in March 1963, served by a new access road off Paradise Road. This terminal was expanded in the early 1970s to include the large round rooms housing the A and B gates. Paradise Road was cut when the airport was expanded in the 1960s, and growth since the 1970s has caused the closing of many more streets and demolition of several neighborhoods.

A massive expansion plan initiated in 1978 continued until 2012, adding C gates in 1987 and D gates in several stages from 1998 to 2011, terminal 2 for international flights in 1991 (and closed in 2012), and terminal 3 for domestic and international flights in 2012. International flights are handled at the E gates

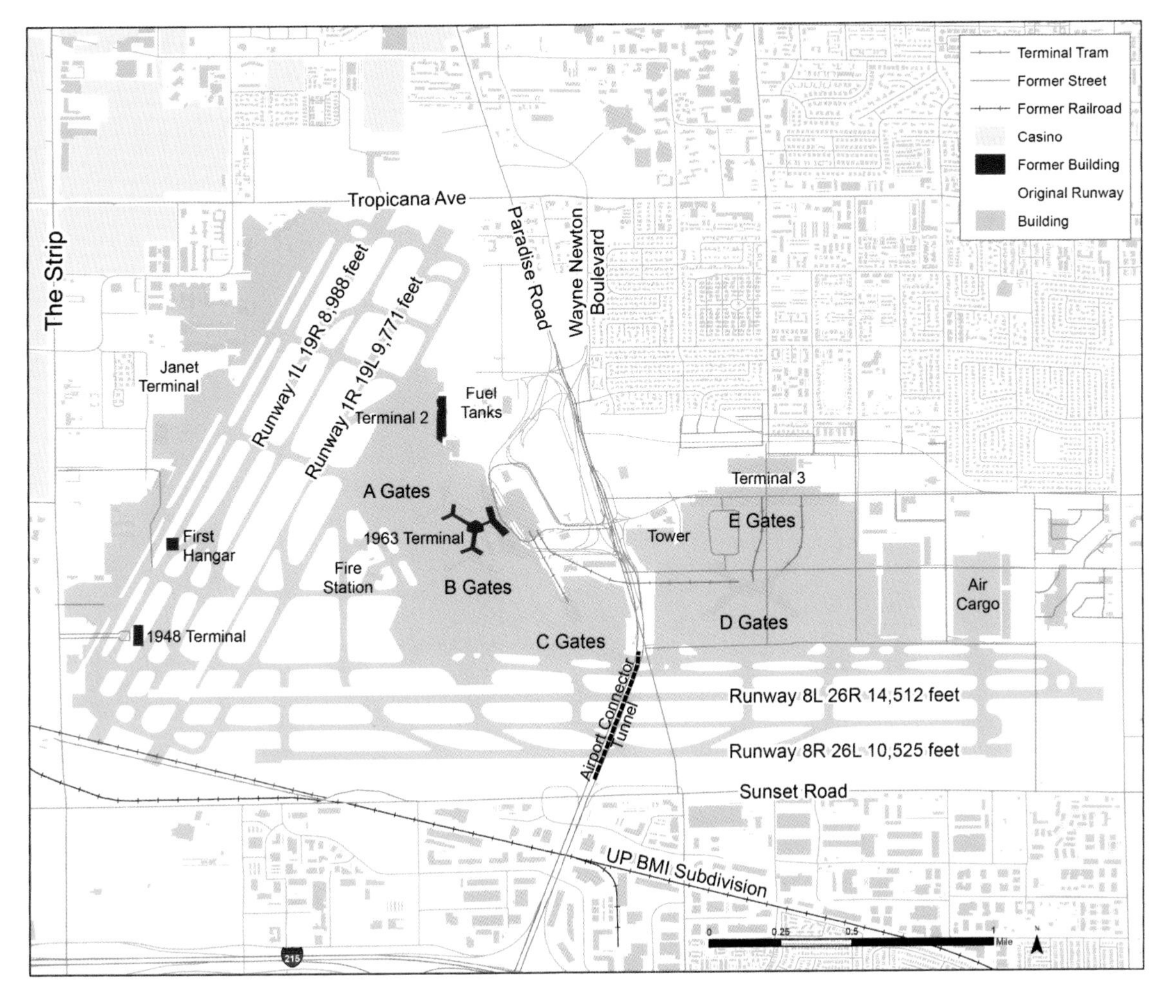

Map 7.18. Harry Reid International Airport. Based on current and historic imagery in Google Earth, USGS topographic maps from the 1950s, the McCarran Airport website, the FAA airport diagram for McCarran, and the Red Flag Nellis 2013 In-Flight Guide.

but will be expanded to include several of the northeast D gates. Three people-mover lines opened, the first in 1985. This expansion required growth in other airport facilities, including a new east-west runway in 1994 and the Airport Connector road running underneath it.

The original east-west runway, 8L/26R, is now 14,512 feet long (or almost 2.75 miles), the fourth-longest civilian runway in the United States. This length is required because of the airport's altitude (2,181 feet above sea level) and hot summer temperatures; the combination decreases the performance of planes. The main east-west runway was moved to the south to allow for greater clearance from taxiways. A short stretch of the Union Pacific Railroad's BMI Subdivision line was rerouted to avoid the ends of the lengthened diagonal runways. Beware of military aircraft using this runway; it is the preferred runway for those carrying live ordnance!

New air traffic control towers in 1983 and 2016, and a rental car terminal opened south of the airport in 2007. A new air cargo facility opened on the east side of the airport in 2010, shared

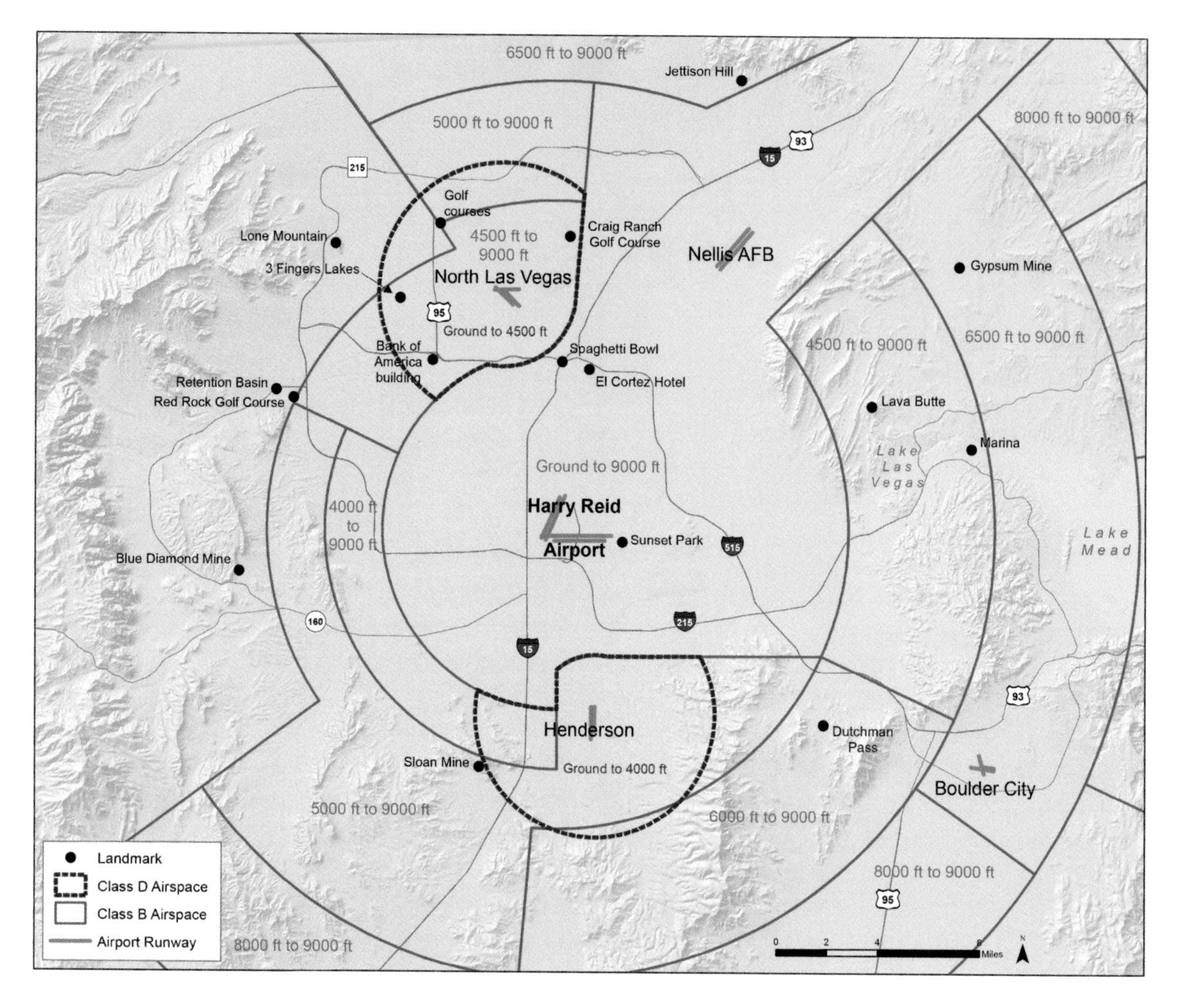

Map 7.19. Airspace. The air above the city is closely monitored and regulated by air traffic controllers. Map based on the 2016 Las Vegas Terminal Area Chart, obtained from the Federal Aviation Administration.

by UPS, FedEx, American, Allegiant, and Southwest Airlines. Although it is one of the busiest passenger airports, Las Vegas is not an important air cargo destination, ranking forty-first in the United States. This is mainly because of geography; major cargo airports are either at airports near the center of the country's population, such as Memphis, Tennessee, or Louisville, Kentucky, or are major international gateways on the periphery of the country, such as Los Angeles and Miami.

Along the west side of the airport are several flightseeing companies, the Janet airline terminal, charter companies, and two fixed-base operators (FBOs) providing fuel and other services to the public. There is no vacant land surrounding the airport, so expansion would require an entirely new facility. This has already been planned for a location far south of the city (map 10.9).

The sky may be the limit inside a casino, but the skies above Las Vegas are very limited by air traffic rules. Some of the most tightly restricted airspace in the nation can be found here. The airspace immediately above Las Vegas is Class B airspace: air traffic controllers monitor all aircraft movements from the ground or a specified altitude up to nine thousand feet above sea level. It is centered on Harry Reid International Airport, includes most of the Las Vegas Valley, and extends thirty-six nautical miles to the northwest. Planes arriving and departing from the major airports follow regular routes through this airspace. About 75 percent of traffic departing Harry Reid is to the west, climbing over largely vacant, county-owned land, but many of these flights are ultimately headed east. In the past these would make a left turn to the south before turning east, but to increase the airport's capacity a third of these turns were to the north beginning in 2007. This prompted vocal complaints from Summerlin residents who suddenly found themselves under these flight paths.

In addition to Harry Reid International, several other airports operate in the area. North Las Vegas Airport is the main general aviation airport in the city. Henderson Sky Harbor was a newcomer when it opened in 1969 and remains in use as Henderson Executive Airport. Like Harry Reid International and North Las Vegas, it is owned by Clark County. Both of these airports are in Class D airspace, which is controlled by a tower from the ground up to 4,000 or 4,500 feet above sea level (above these elevations, it may be part of Class B airspace). Military planes departing Nellis Air Force Base usually take off to the southwest and then make a sharp turn to the northwest, leaving the city between Shadow Creek and Craig Ranch Golf Courses. Those carrying live ordnance must take off to the northwest to avoid overflying the city.

The R-4808 and R-4806 military airspace areas, which cover the Nellis Air Force range and adjacent areas (map 9.13), dominate skies to the north. Civilian planes are prohibited in these areas, home to military flight training. In the center of that area is the R-4808A airspace, restricted even to military pilots. Air force pilots call this area Dreamland, perhaps the most tightly restricted airspace in the country. It includes Groom Lake, better known as Area 51 (map 4.12). To the west and south of Death Valley and much of the Mojave Desert lies in the R-2505, 2502, 2524, and 2515 airspace zones. Within national parks, civilian pilots must fly at least 500 feet above the ground, while military pilots are required to maintain 3,000 feet above the ground (though this is routinely ignored).

To the east, air traffic must contend with the Grand Canyon, one of the most complicated and restricted nonmilitary rural airspace zones in the country, and the only rural area with its own air navigation charts. All air traffic below 18,000 feet must follow special regulations, and in several zones the airspace below either 8,000 or 14,500 feet above mean sea level (MSL) is restricted. These restrictions exist for several reasons. The large number of aircraft flying in the area must be controlled to prevent midair collisions, the deadliest of which occurred between a United Airlines and TWA airliner over the eastern Grand Canyon on June 30, 1956. All 111 aboard both flights died. The other reason for restricting flights is noise. Noise surveys at Cape Royal in Grand Canyon National Park reveal that jet aircraft can be heard 30.3 percent of the time, propeller planes 9.7 percent, and helicopters 1.1 percent of the time. These totals are much higher along the North and South Entrance roads. The National Park Service is committed to preserving the natural sounds as well as sights of the Grand Canyon, and this has meant closely regulating scenic overflights. Only to the south is the airspace relatively open, but this may change if a new Las Vegas airport is built (map 10.9).

Pilots must be aware of many hazards at lower elevations near Las Vegas. The city lies in a valley surrounded by mountains, with elevations up to 11,916 feet above sea level at Charleston Peak, 9,912 feet in the Sheep Range to the north of the city, 8,517 feet on Potosi Mountain to the southwest, 7,946 on the Virgin Mountains east of Lake Mead, and 5,456 feet in the Black Mountains east of Hoover Dam. These mountains have been the scene of many airplane crashes (map 7.20). Many high buildings in the city must be avoided, including the Stratosphere tower 1,149 feet above the ground. While this is the tallest building east of the Mississippi River, it is not the tallest structure in the Las Vegas area. In the vicinity of Glendale are two television antennas, one standing 1,400 feet above the ground and the other 1,315 feet. And until its demolition in 2012, the tallest structure in Southern Nevada was the 1,527-foot-tall BREN (Bare Reactor Experiment, Nevada) Tower in the Jackass Flats section of the Nevada Test Site (map 9.14).

In addition to avoiding these obstacles, pilots at low altitudes also look for landmarks. Those shown on the navigation charts include the Blue Diamond mine, the Spaghetti Bowl interchange, the Bank of America building at the Rainbow Curve, several golf courses, and what is labeled as "3 fingers lake," which in reality is four lakes. Residents around those lakes, in the Desert Shores community, may not realize they are landmarks for pilots.

The airspace above Las Vegas and surrounding towns is also of interest for those on the ground looking up. In addition to the city's natural and manmade skyline (map 8.14) are countless airliners and military aircraft. What the observer won't see are stars at night. The bright lights of the Strip have long since overpowered the natural night sky, though this can still be enjoyed a few hours away at Death Valley National Park.

McCarran International Airport served about 45 million

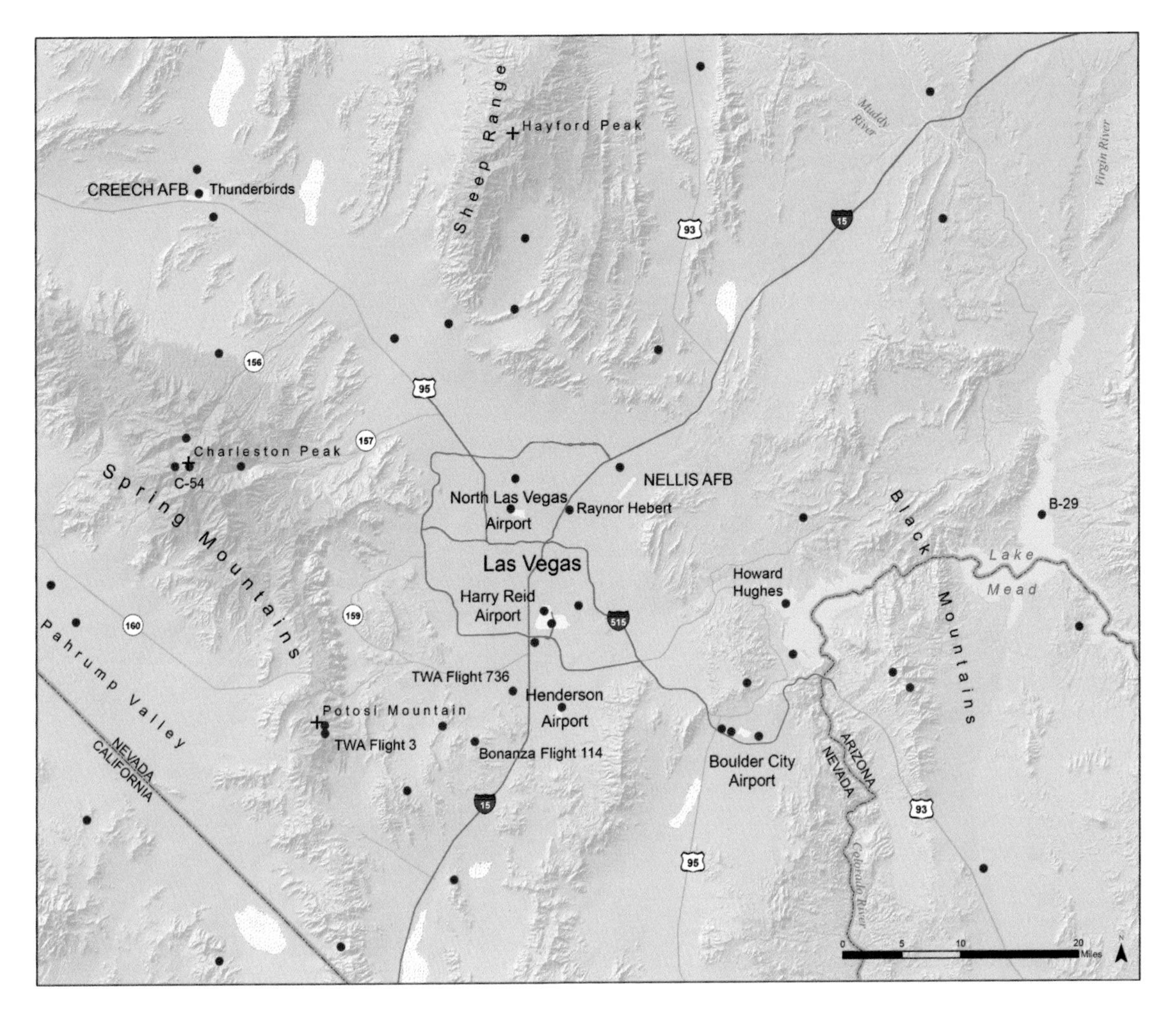

Map 7.20. Airplane Crashes. Each red dot depicts an airplane crash. There is no single database of airplane crash sites. Based on the Plane Crash Sites in Nevada website, supplemented by a search of the National Transportation Safety Board records, and news accounts.

passengers in 2015, but unfortunately not all flights to or from Las Vegas arrive safely. The aircraft involved vary tremendously. Many are light planes that rarely make headlines, while countless military planes have crashed to the north of the city on and near the Nellis Air Force Range.

Airplanes are safest midflight at high altitude of course. Many airplanes became victims of the region's geography when they flew into mountain ranges surrounding Las Vegas (map 2.1). This was particularly a problem in the early decades of flight when unpressurized airplanes flew below the height of these mountaintops. Others occurred during takeoff or landing where there was no margin for error or mechanical problems. Such crashes are found near airports. Some, such as midair collisions, fall into neither pattern.

There have been several airline crashes in Las Vegas. The first was TWA Flight 3 in 1942. It was a westbound transcontinental flight between New York City and Burbank, California, with several stops along the way. The flight normally stopped at Boulder City in the daytime but if it arrived after dark would land at the original McCarran Field (now Nellis Air Force Base), the only lighted field in the area. This was the case on January 16, and after a short stop in Las Vegas the plane proceeded southwest toward Burbank, but on the standard compass heading for flights departing Boulder City. This was not the correct route for flights leaving McCarran, and the plane flew into Potosi Mountain, about 730 feet below the 8,500-foot-high summit. All twenty-two aboard were killed, including actress Carole Lombard. The fact that most airway beacons had been shut off because of wartime restrictions may have contributed to the crash.

The next airliner incident was on November 15, 1956, when TWA Flight 163 crashed on runway 7/25 at McCarran while making an emergency landing after an engine failure, without injuries to passengers or crew. United Airlines Flight 736 was not as lucky on April 21, 1958, when a DC-7, a large pressurized four-engine propeller plane flying between Los Angeles and Denver, had a midair collision with an Air Force F-100 fighter jet on a routine training flight from Nellis. The two planes collided at twenty-one thousand feet above sea level, went out of control and plummeted to the ground. All forty-two passengers and five crew aboard the DC-7 and the two air force crewmen were killed. The DC-7 hit the ground near the present intersection of Decatur Boulevard and Cactus Avenue, where there is a small memorial on an empty lot.

The next was Bonanza Air Lines Flight 114 on November 15, 1964. This Fairchild F-27 turboprop was arriving in Las Vegas from Phoenix and was landing at night from the southwest when it crashed into a hill southwest of Sloan, killing all twenty-nine aboard. The plane was on course but had descended too low. More recent airline crashes have involved smaller planes. On August 30, 1978, a small Las Vegas Airlines plane crashed on takeoff from the North Las Vegas airport, killing ten. The "Janet" airline (map 7.17) also suffered a fatal crash on March 14, 2004, when a plane was landing at the Tonopah Test Range airport. It was caused when the pilot suffered a heart attack, and the five other people onboard were killed.

Among the more well-known crashes related to military activities in the area were the July 21, 1948, crash of a B-29 bomber, which hit Lake Mead's surface while flying low. The crew survived, but the plane quickly sank. The wreck was only rediscovered in 2001. On November 15, 1955, less-fortunate crew members were on an air force C-54 transport carrying military personnel and civilian workers between Burbank and Groom Lake (Area 51) (map 4.12). The plane was off course and flying in conditions with poor visibility when it crashed into Charleston Peak just below the ridgeline. All fourteen aboard were killed.

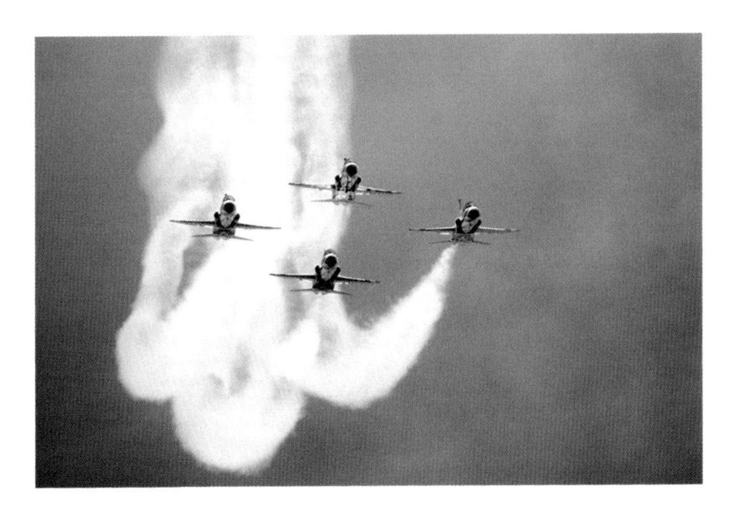

Thunderbird T-38s in diamond formation, 1980. Photo by Bill Stephenson.

A memorial to those on board was erected in 2015 in the community of Mount Charleston.

On January 18, 1982, the Thunderbirds Air Force demonstration team suffered a four-plane crash during low-altitude acrobat training in their T-38s at Indian Springs (now Creech) Air Force Base. The planes were to pull out of a loop at one hundred feet off the ground but a mechanical problem prevented the lead pilot from doing so. The other pilots were watching him rather than the ground, and all four pilots died when they flew into the ground at 400 mph.

Many foreign military planes come here to participate in Red Flag exercises or were brought here secretly for flight testing. Two former Soviet MIG-17s crashed on the Nellis range in 1979 and 1982, along with a MIG-23 in 1982, part of a small fleet being used for secret flight tests and training missions in the 1970s and 1980s.

An F-105 crashed in a North Las Vegas residential area in 1964, killing the pilot, Raynor Hebert. The single-engine jet had been taking off to the south when it lost power; the pilot could have ejected but stayed with the plane to steer it away from Lincoln Elementary School. Had he not taken this action, the crash might be well known today while Lincoln Elementary would be nothing but a sad memory.

Another noteworthy fatal crash was one by future Las Vegas resident Howard Hughes, on May 17, 1943. While test-flying a Sikorsky S-43 amphibian out of the Boulder City airport, he crashed the plane during a landing attempt on Lake Mead in the Las Vegas Bay area. A crewmember and a government inspector were killed and the plane sank, but it was soon raised and restored to flying condition. It still exists and is being restored by a private owner in Florida. The body of one of the fatalities was never recovered and remains on the bottom of Lake Mead.

Contemporary Geographies of Las Vegas

8

This chapter examines a range of topics about Las Vegas today, including boundaries, movie theaters, street grids and street names, population, mining industry, neighborhoods, skyline, visitors, and electricity. One of the most important topics is the question of land ownership and the Las Vegas Valley Disposal Boundary. While early settlers could claim land from the public domain under the Homestead Act and similar laws, that era ended in 1976. Since that time, the amount of private land around Las Vegas has not increased much. It will not change much because Congress has drawn a boundary around the city within which the Bureau of Land Management BLM is only allowed to sell land at periodic auctions. The boundary may be revised to allow for growth, but any expansion will soon reach the national forests, recreation areas, conservation areas, and military bases around the city. Other important boundaries, those of cities and towns, will also be addressed in this chapter. It is often surprising to realize that the Strip, Harry Reid International Airport (the new name for McCarran International Airport), and UNLV aren't actually in Las Vegas, or within any city limits. Even many Las Vegans would find it difficult to identify exactly where they are.

Because this is a book about Las Vegas, casinos must inevitably be mentioned. The spread of the casinos, the growth of the Strip and the development of neighborhood casinos are depicted. Where do visitors to Las Vegas come from? The Las Vegas Convention and Visitors Authority (LVCVA) tracks these numbers, and they are mapped out here. Las Vegas's competition for tourists around the country and world is also featured. Not all of the topics are that fun. Several maps look at the disposal of waste, fires and explosions. Many Las Vegans enjoy the city's tremendous number of airline connections (map 7.17), but the noise those airplanes make is another matter. Certain parts of the city are much more affected by these than others. Older parts of Henderson and the northeast near Nellis Air Force Base have the most air pollution, the most noise, and the city's worst explosions.

Who owns Las Vegas? The question depends on timing; Las Vegas was once the territory of the Paiutes, though they did not think in terms of owning the land. The territory that became Nevada was claimed by Spain and then Mexico until the end of the Mexican-American War and Treaty of Guadalupe Hidalgo

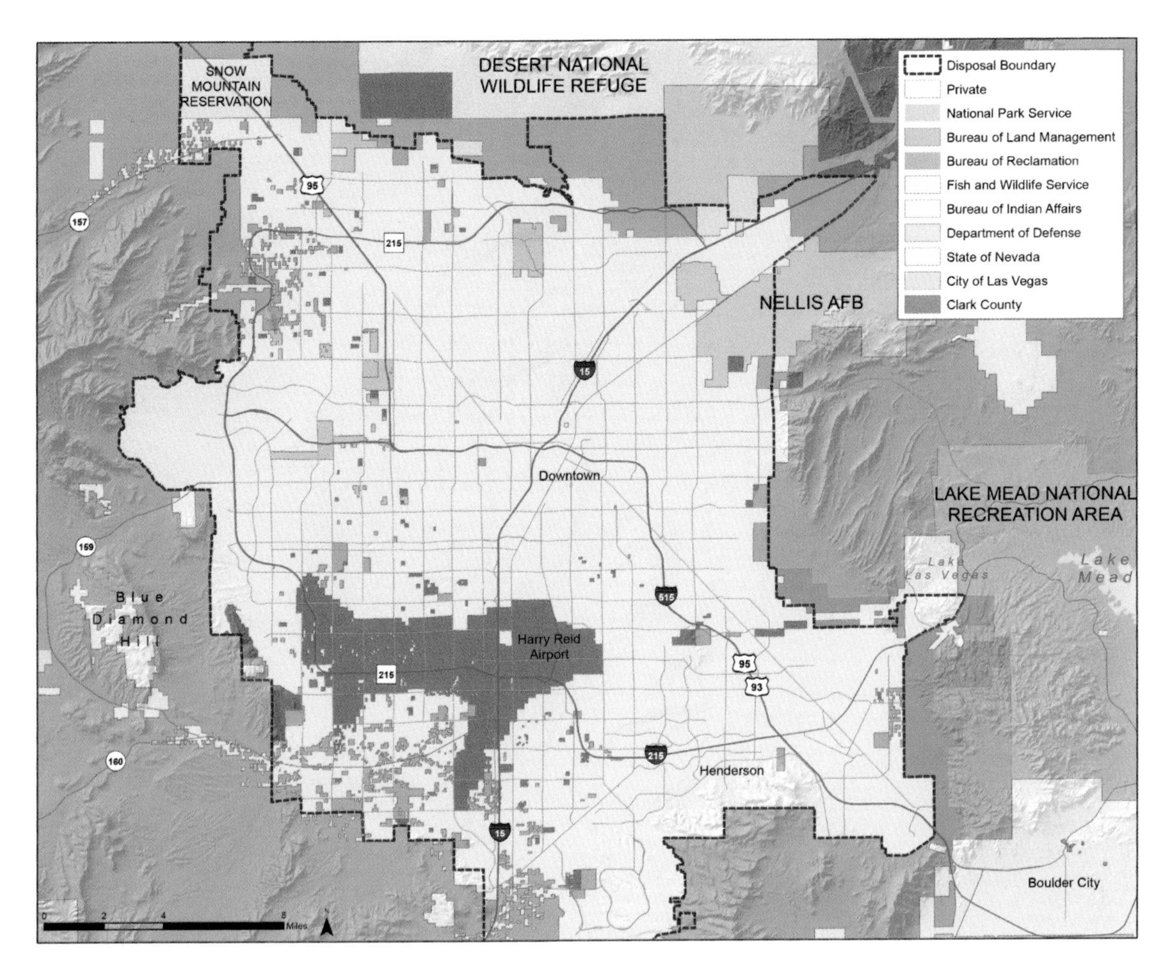

Map 8.1. Land Ownership. Land is owned privately and by a variety of local and federal agencies. The red-dashed line is the Las Vegas Valley Disposal Boundary, a line drawn by Congress within which the Bureau of Land Management may periodically sell land in the future. The city may only grow within this boundary.

in 1848 when it became part of the United States (map 3.4). At this point, all lands not already settled became part of the public domain, under the control of the federal government. This meant nothing until 1855 when a group of Mormon pioneers arrived and began farming an area along Las Vegas Creek. They effectively took ownership of it, though Mormon settlers typically did not bother with formal homesteading or property ownership.

The arrival of the railroad in 1905 brought new concepts of private property to the valley. The railroad purchased the land along the creek, and lots in two rival townsites were sold off to the highest bidder. A variety of owners soon claimed much of the rest of the valley. The Public Land Survey System PLSS (map 3.12) made it very easy to claim land, usually in the form of squares or rectangles in multiples of forty acres. Much of the valley had been claimed, especially near springs, by 1910.

Like most western cities, Las Vegas today is an island of privately owned land surrounded by federal lands. The US Department of Defense manages Nellis Air Force Base and the enormous Nellis range to the north of the city, both of which were created in World War II (map 4.8). The US Fish and Wildlife Service manages the Desert National Wildlife Refuge, the biggest outside of Alaska, and several smaller refuges. The US Bureau of Reclamation became involved in Las Vegas in the 1930s when the Boulder Dam Reservation was created, though most of this reservation was subsequently transferred to the National Park Service, which also manages the new Tule Springs Fossil Beds National Monument to the north of the city (not shown on the map because of how recently it was created). The US Forest Service manages the Spring Mountains and originally the Sheep Range as well. The biggest federal landowner is the BLM, which manages all federal lands not controlled by another agency. On the map, its land merely rings the city. But on a map of Nevada, it would be shown as the largest landholder in the state.

Clark County controls several major pieces of land in the valley, most notably Harry Reid International Airport and the lands underneath the flight paths to the west and south. Planes arriving and departing from follow regular routes; about 75 percent of traffic departing the airport is to the west, climbing over largely vacant county-owned land. Living under an airport flight path is not pleasant, and this county-owned land helps eliminate the problems of residential development near airports that plagues many cities (map 8.21).

Why hasn't more land entered private ownership around Las Vegas? Congress passed a law in 1976 that ended the Homestead Act and prevented the BLM from selling off the remaining public lands except under specific conditions. Among these are land swaps when parcels of value to the government, such as scenic land adjacent to a national park, can be traded for lower-value BLM land elsewhere. This occurred northeast of Las Vegas when vacant BLM-controlled land planned for the Coyote Springs development was obtained by swapping it for a smaller parcel of private land in the Florida Everglades. To obtain land for a potential reliever airport in Ivanpah Valley, Clark County used a more direct approach, convincing Congress to pass a law directing the BLM to sell it the land.

Another law has authorized the BLM to sell land around Las Vegas to the public at periodic auctions. All of this land is within a congressionally designated border, called the Las Vegas Valley Disposal Boundary. Within this boundary the BLM will sell (or dispose of) government land, but outside the boundary it cannot. Congress may enlarge the boundary to enable future growth, and proposals to do that were initiated in 2018. The county wants to add 38,636 acres, with the largest addition to the south along Interstate 15.

Growth will be concentrated inside this area. The Disposal

Boundary extends northwest along US 95 past the Snow Mountain Paiute reservation, allowing for growth along the Interstate 11 corridor. Some land outside the boundary is already in private hands, along Blue Diamond Highway and in Kyle Canyon. Boulder City is outside the boundary, though it has enacted strict growth controls that will limit development (map 5.8).

Even when the Disposal Boundary is expanded, it will still soon reach nearby protected lands, such as Toiyabe National Forest, Red Rock and Sloan Canyon National Conservation Areas, Tule Springs Fossil Beds National Monument, and the Nellis Air Force Range (map 9.13). These areas will provide the ultimate growth boundary. Similar federal lands have had the same effect on other western cities. Some parks may even be created for this reason; the establishment of Mojave National Preserve along Interstate 15 southwest of Las Vegas was justified in part by the desire to prevent Las Vegas and Los Angeles from ever merging.

As the valley fills up, its low-density growth will be coming to an end. The population will surely continue to grow, but fitting more people in the same area means higher densities. This in turn will require multistory buildings and even less distance between adjacent homes. The same has already happened to Los Angeles, which now has the highest population density of any major American urban area.

Another possibility would be for growth to jump to adjacent valleys where land is available. This has been taking place for the last several decades in Pahrump, which has grown from a few thousand people to almost forty thousand (map 4.18). However, private land in Pahrump is also limited, restricting the future growth of this community. Boulder City contains an enormous amount of vacant private land in Eldorado Valley, but the strict growth controls make any development here unlikely.

Many visitors to Las Vegas lose it all, but they didn't lose Las Vegas itself. The Paiutes did, and today they have only two small reservations in Las Vegas Valley and the Moapa Reservation north of Overton. The odds of getting any more of it back are slim.

There are four cities in the Las Vegas area: Las Vegas, North Las Vegas, Henderson, and Boulder City. Las Vegas became an incorporated city in 1911 and had 662,368 people in the 2020 census. The city limits have grown to 141.8 square miles, but extend only as far south as Sahara Avenue; the city includes only downtown, parts of Summerlin, the Meadows Mall area, and the northwest valley.

North Las Vegas was incorporated in 1946 and now has 260,098 people. It covers just less than one hundred square miles

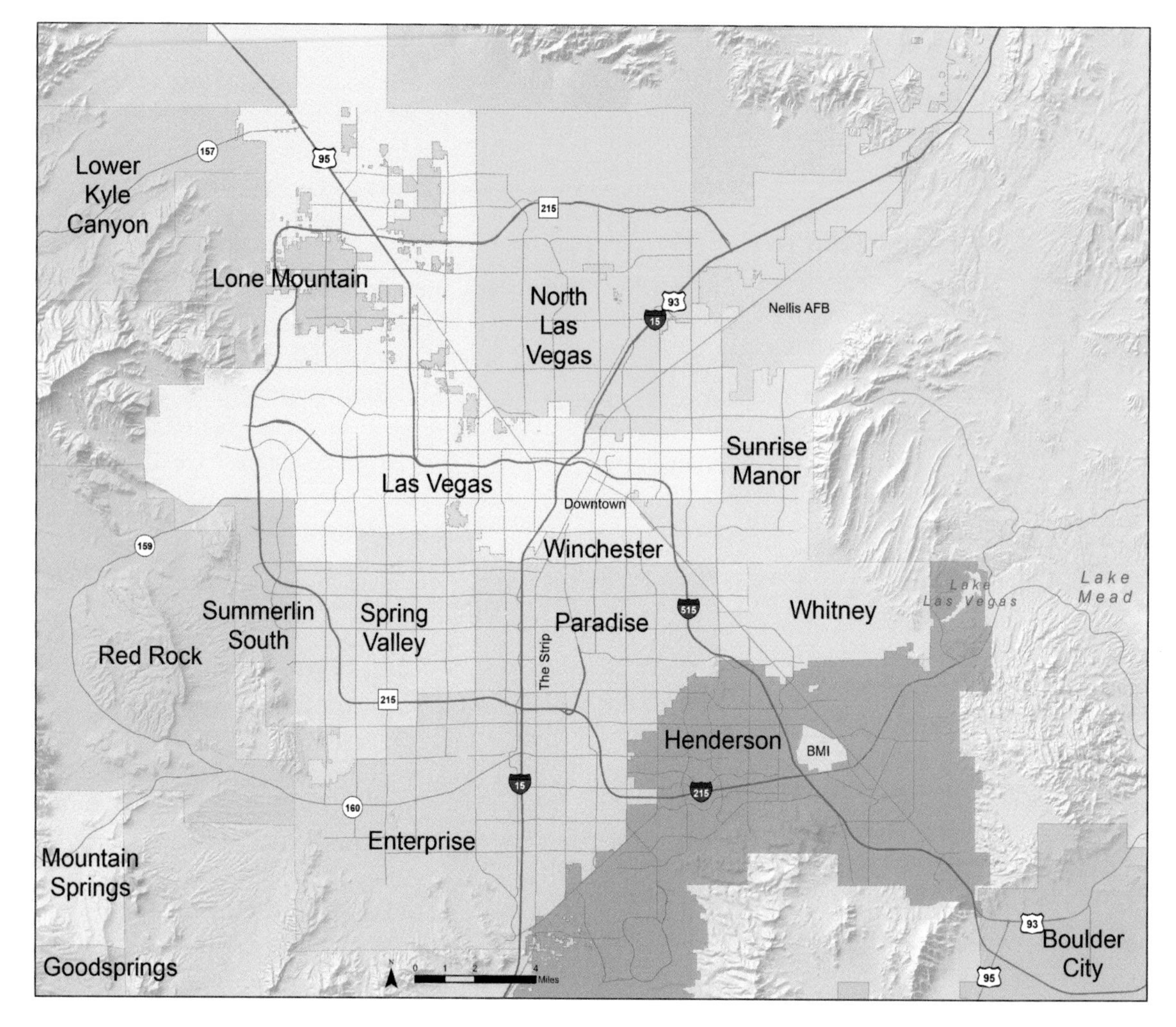

Map 8.2. City and Town Boundaries. There are four incorporated cities in the Las Vegas area and many unincorporated towns that contain the Strip and much of the valley's population.

of the north central part of the valley but also extends northeast along Interstate 15. It borders but does not include Nellis Air Force Base, which like the Strip also remains outside any city boundary.

The coming of World War II brought the BMI plant to the Las Vegas Valley, along with a town founded to house its workers (map 4.8). The town was originally called Basic before being renamed Henderson, after former senator Charles Henderson. The mill shut down in 1944, but the town managed to survive the loss of employment and slowly grew before exploding in population in the 1990s. It became an incorporated city in 1953 and today has 329,172 people. For most of its existence it was a small municipality along Boulder Highway but in the 1990s began an expansion to 106.6 square miles that extends as far west as Interstate 15.

The Bureau of Reclamation created Boulder City to house workers for Hoover Dam (map 5.8). After the dam was completed, the population plummeted as workers moved away but then slowly grew again. It became an incorporated city in 1959, and in 2020 had 16,410 residents. Despite this modest population, the boundaries cover 208.3 square miles, by far the largest of any city in Southern Nevada. Most of this is empty desert land in Eldorado Valley, which because of the city's growth controls is effectively off limits for urban growth.

Casino owners have always resisted the expansion of the city of Las Vegas to encompass the Strip, and it remains outside any city. But that doesn't mean that there are no local governments; Nevada law also specifies that unincorporated towns may be created. These are an outgrowth of the state's mining past and based on the reality that mining-town governments would only be needed for a short time: when the mine played out and the town's residents moved on, the town government would be abandoned. State law makes it easier for residents to create a town than an incorporated city, though they have less power and fewer responsibilities.

Events worked out differently in the Las Vegas area, where there was little need for a mining town government, but most of the valley outside of city limits is now within one of several towns. The first of these was Paradise, established in 1950. This includes most of the Strip and had a population of 201,810 in 2020. Winchester was created in 1951, Spring Valley in 1981, and Enterprise in 1996. Although Summerlin, Anthem, and Green Valley are well-known names in the valley and have distinct identities, they are neither cities nor towns but master-planned communities. To the west of Las Vegas, the community of Pahrump is also a town, established in 1962 (map 4.18).

Other boundaries no longer appear on the map. All 186 Nevada school districts were consolidated into seventeen countywide districts in 1956. The Clark County School District replaced fourteen districts in the county and today has 318,040 students (the fifth-largest school district in the country). A 1971 study recommended merging several districts, including those of Lincoln and Clark County, to create eleven even larger school districts statewide, but this did not occur (neither did several plans through the decades to break up the Clark County School District into smaller districts). A merger that did take place was when the Las Vegas Metropolitan Police Department was created in 1973 by merging the Clark County Sheriff's Department with the Las Vegas Police. North Las Vegas, Henderson, and Boulder City still have separate police departments.

There were 2,315,963 people living in Clark County in 2020, and they are a diverse crowd. The Census Bureau classifies people as white, Black or African American, Asian, Native American or Alaska Native, Native Hawaiian or Pacific Islander, and multiracial. Hispanics are an ethnicity and may be of any race, though

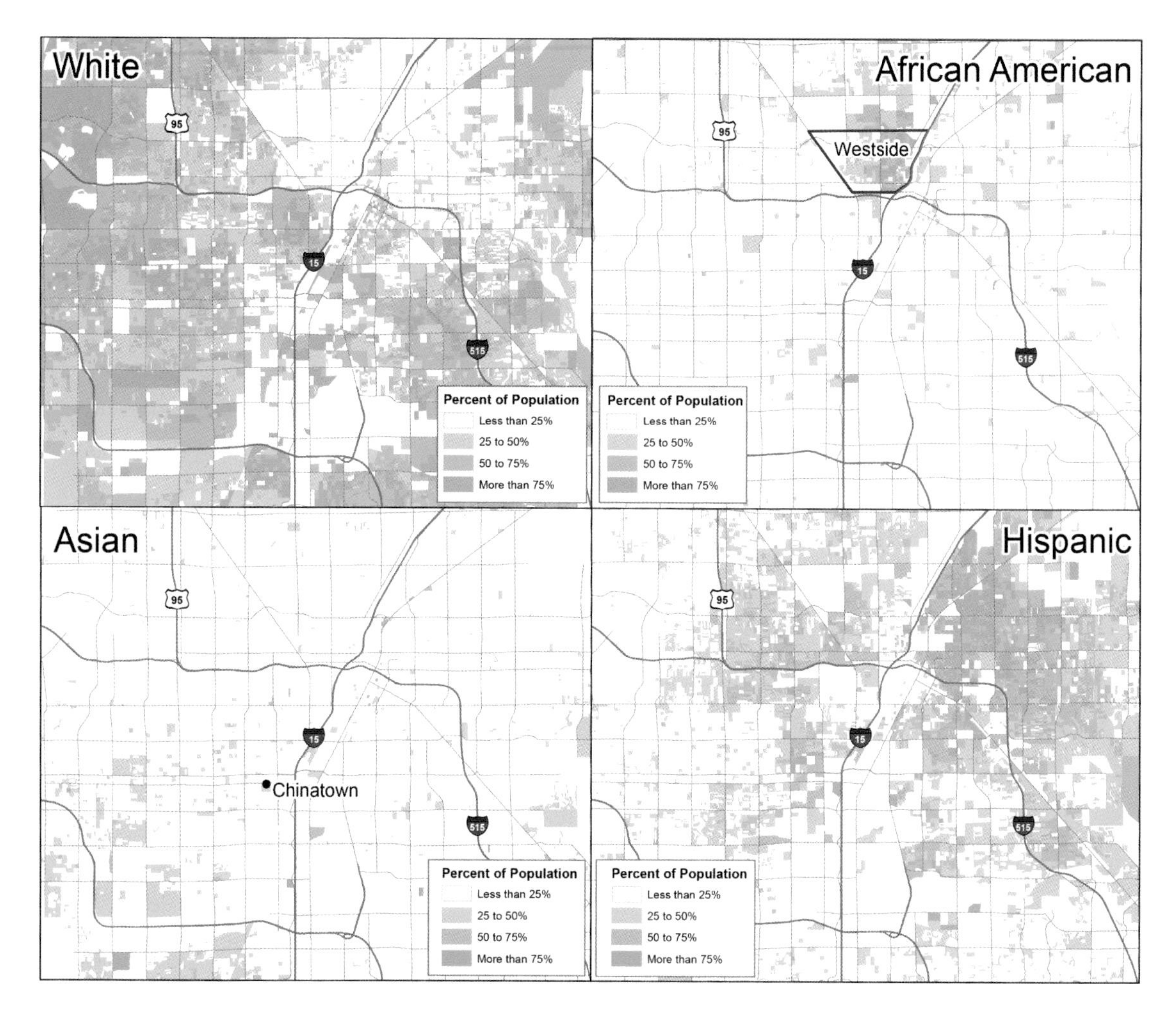

Map 8.3. People. The percent of the overall population that is made up by each of these races in the central part of the urban area in 2010 is shown. The darker the shade of green, the higher the percentage of that race within a neighborhood.

most are white. The county's population as measured by the Census Bureau in 2010 was predominantly (60.9 percent) white, non-Hispanic. Not surprisingly, they are found throughout the urban area, though many fewer live in the Westside community.

African Americans were a part of Las Vegas from the earliest years, first arriving as railroad laborers. Their population remained small for decades until many arrived from the Deep South to work at the new BMI plant at Henderson. A small segregated community, Carver Village, was built for these workers, and the east side of Henderson still has a substantial Black population. Blacks were largely forced into the Westside neighborhood on the other side of the tracks in the 1940s. Although the city desegregated in the 1960s, the Westside is still visible on the map as a mostly African American neighborhood. Many also live in adjacent parts of North Las Vegas; Martin Luther King Boulevard connects these areas while honoring the civil rights leader who visited Las Vegas in 1964. In 2010, 10.5 percent of the county population was classified as African American.

Although the area was once part of Mexico, the Hispanic

population has emerged since the 1970s and especially the 1990s. The county is 29.1 percent Hispanic today, with the majority of that population living in northeastern part of the city between downtown and Nellis Air Force Base. Largely Hispanic neighborhoods are also found west of the Strip and in the central part of the valley along Maryland Parkway.

Las Vegas had Chinese and Japanese residents in 1920 and likely earlier, but most of the county's Asian population growth is more recent. In 2010, Asians made up 8.7 percent of the county population and are common in the west and southwest parts of the valley, though many also live on the east side. A Chinatown emerged along Spring Mountain Road in 1995, mostly between Decatur and Valley View Boulevards. It is centered in a shopping center, Chinatown Plaza, but few Asians live in the area.

Native Americans made up a significant share of the early population, though they lived in their own villages on the town's outskirts. A small reservation to the north was set aside in 1905, and a new reservation was created north of the city along US 95 in the 1980s. Only 0.7 percent of the Clark County population was Native American in 2010; their small numbers would make them invisible on the map, and so they are not shown. In 2010, another 0.7 percent of the county was made up of Native Hawaiians and Pacific Islanders, also too little to show on the map. Although they're a small group, they have nonetheless built a strong link to Las Vegas. The California Hotel downtown pursued Hawaiians as a clientele, and many came to stay permanently. Las Vegas is sometimes known as the "Ninth Island of Hawaii" in recognition of these ties.

Visitors to the city still may be shocked that people live here. But they do; there are schools, hospitals, churches, libraries, parks, cemeteries, movie theaters, malls, Walmarts, professional sports, and other necessities and amenities of life. Take a look at Map 8.4: here is a view of Las Vegas you probably haven't seen, showing all of these places but no casinos.

Of course the casino industry is important to life here; many locals work in the casinos, and they also have their own casinos tourists rarely visit. The Bingo Palace (later Palace Station) was one of the first, and from the late 1980s many new ones were built, including all of the Station properties. These also contain many of the city's movie theaters (map 8.13) and other entertainment.

One advantage of living here is Harry Reid International Airport; not only is it conveniently located, it has an enormous number of nonstop connections to destinations on four continents (map 7.17). Locals who move to another city of similar size will be surprised by how few air connections are available in their new home. The city's excellent air service is also an asset when it

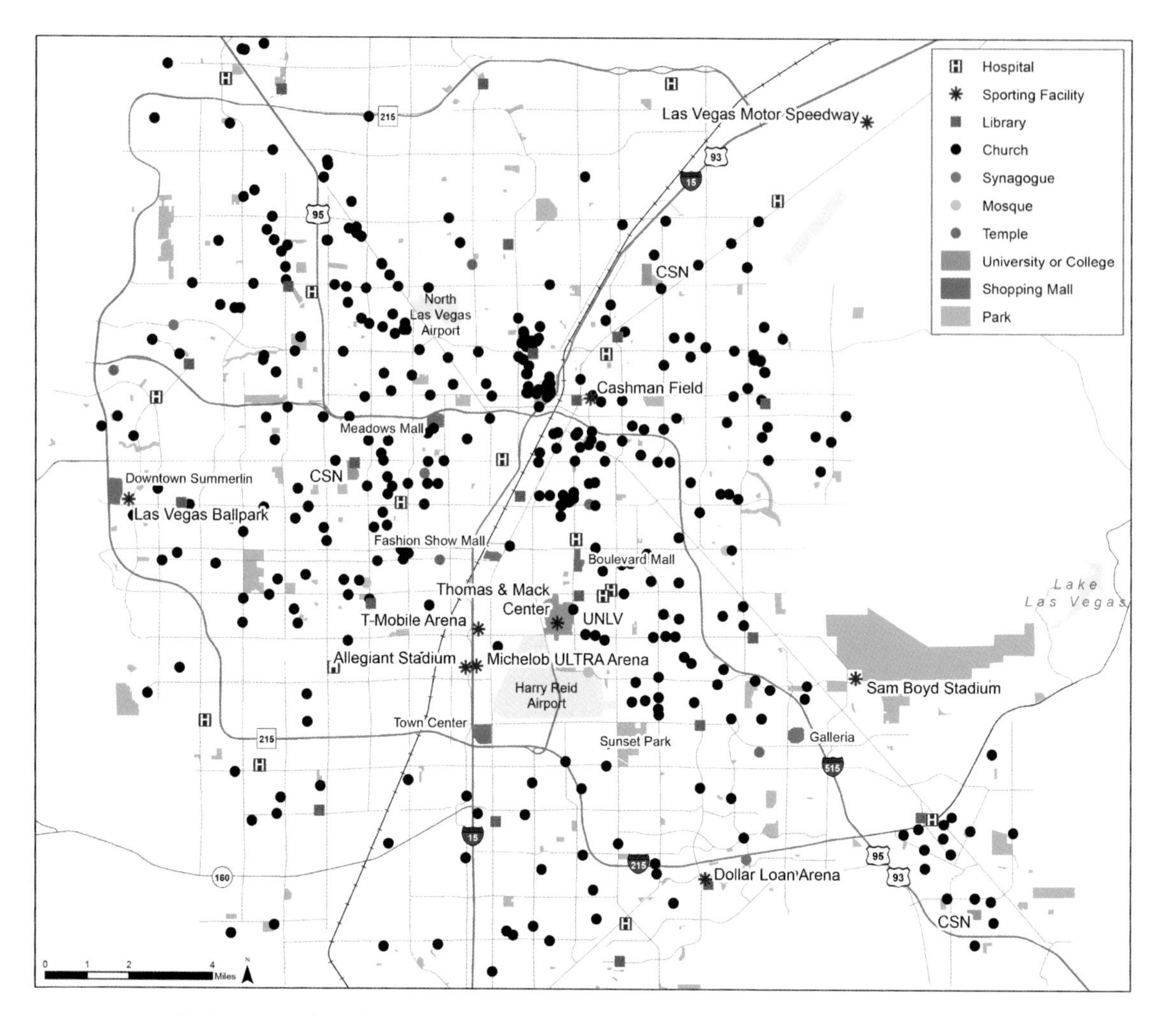

Map 8.4. Living in Las Vegas. Have you seen a map of Las Vegas that doesn't show a single casino, and instead shows places that people living here care about? Here it is, showing malls, hospitals, houses of worship, and places to cheer on the local teams.

comes to finding good healthcare: Las Vegas has long been underserved, has a shortage of doctors for every specialty but plastic surgery, and is overwhelmingly reliant on for-profit hospitals that see patients much the way casinos view gamblers. A quick trip to another city may be the best way to visit a good doctor.

The city does not rank as one of the nation's healthiest, and even groceries are hard to find in parts of the city. These "food deserts" include the Strip, parts of North Las Vegas near the Westside, and several neighborhoods near Nellis Air Force Base. According to the 2021 Robert Wood Johnson Foundation's health rankings, Clark County is only the sixth-healthiest of Nevada's seventeen counties.

Las Vegans love to read. Although the city has always had a shortage of bookstores, the Las Vegas–Clark County Library District ranks eleventh in the nation by circulation and fourteenth by visits in 2016. The city of Las Vegas also has many parks, ranking forty-second out of the one hundred largest cities in access of residents to parks in 2020 (Henderson was ranked twenty-seventh). This is based on the size of parks, parks within a ten-minute drive of people's homes, and the acreage per person. These include not just city parks but also Lake Mead, Lee Canyon, and Red Rock Canyon, whose attractions are especially evident on weekends when looking for a parking spot. By comparison, nearby cities such as Los Angeles (ranked forty-ninth), Phoenix (fifty-sixth), and Reno (sixty-eighth) have much lower access to parks.

Locals also enjoy several college and professional sports teams in addition to a long history of hosting high-profile boxing and Ultimate Fighting Championship (UFC) cards. Las Vegas has had a minor league baseball franchise (the Stars, the 51s and now the Aviators) since 1983. The Vegas Golden Knights of the NHL became the first major league team in 2017, the Las Vegas Aces of the WNBA arrived in 2018, and the Las Vegas Raiders began play at the new Allegiant Stadium in 2020. Older Las Vegas residents fondly remember the glory days of the UNLV Runnin' Rebels basketball team under coach Jerry Tarkanian, including the 1990 national championship.

The Las Vegas Valley has many churches, as well as a smaller number of synagogues, mosques, and Buddhist and Hindu temples. There are no complete and reliable statistics about attendance or religious institutions, but the number of religious denominations has clearly grown since the 1920s (map 4.4). A wide variety of Protestant denominations are represented, Roman Catholic churches are common, and the Church of Jesus Christ of Latter-day Saints has made a strong comeback after abandoning Las Vegas in 1855. Many consider Las Vegas the most religious city in America: after all, at any given time of the day there are tens of thousands of people praying. Does the fact that they are in casinos rather than houses of worship lessen their sincerity?

Like most large metropolitan areas, it has become polycentric: economic activity or shopping is not dominated by downtown but instead there are many competing centers around the valley. The Strip is clearly one, while Boulder Highway, Maryland Parkway, and the Downtown Summerlin area are also important. Most of these are well-served by the city's freeway system, and the beltway is emerging as the city's new main way to get around (though again, one that few tourists ever see).

Whatever amenities other big cities have, Las Vegas probably has too. But with brighter lights and slot machines.

One provision of the Public Land Survey System (PLSS) (map 3.12) was that section lines could be used as public roads. In rural areas, this often meant a grid of north-south and east-west roads at one-mile intervals cutting through farmland, most visible when flying over the Great Plains. These section-line roads in the West became the arterial streets as cities expanded.

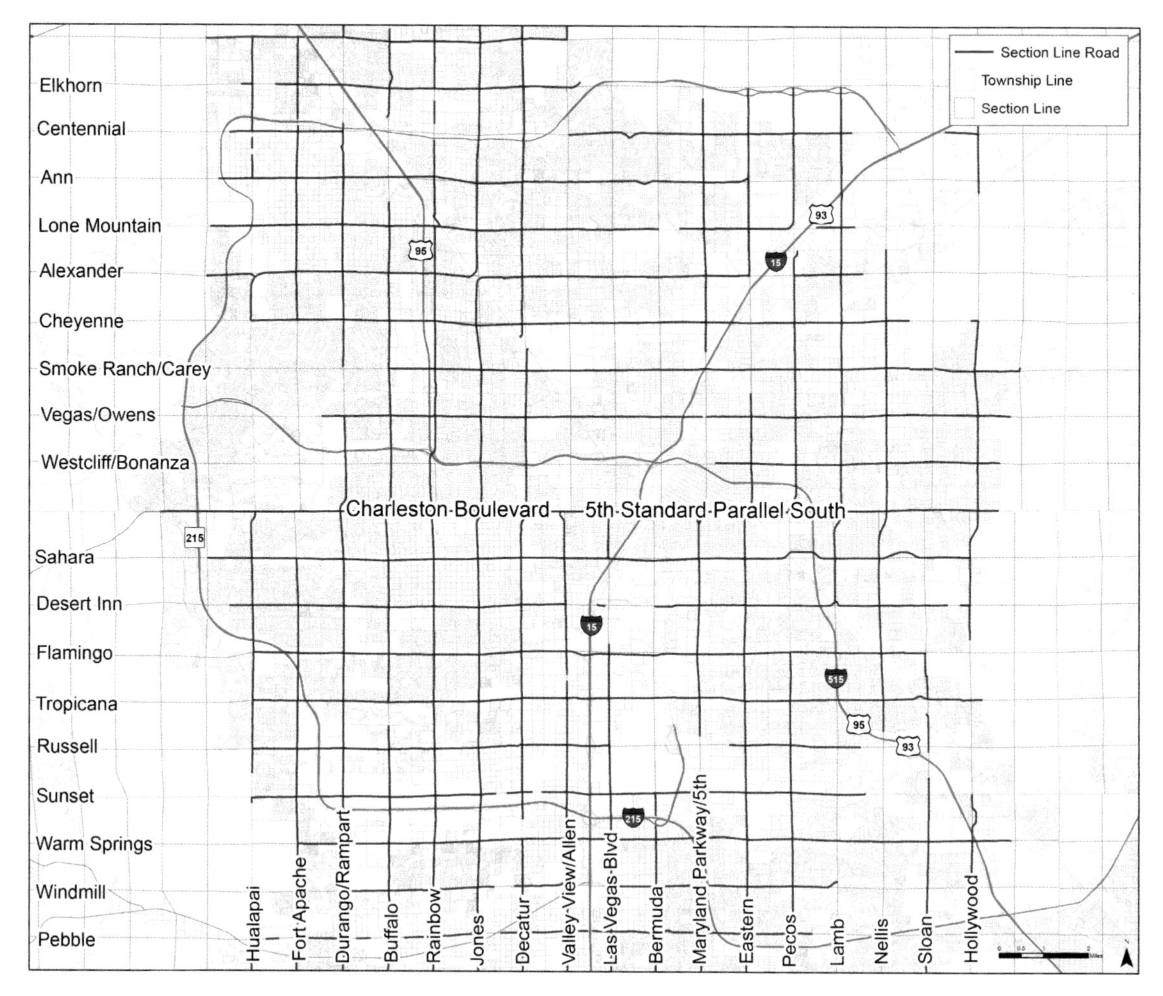

Major streets in many cities are one mile apart for this reason; Las Vegas is no exception.

Map 8.5. Section Line Roads. Most main roads in Las Vegas are spaced one mile apart on a grid that was first laid out by nineteenth-century surveyors. Many north-south roads crossing Charleston Boulevard have curves because the curvature of the earth required the surveyors to modify their grid north and south of this line.

The original Las Vegas townsite was not based on section lines, and streets were parallel or perpendicular to the railroad. Boulder Highway, US 95, and Las Vegas Boulevard were built before the city had expanded out of downtown and were not built along section lines. But Charleston Boulevard, a section line, emerged as a major street in the 1950s. Many (but not all) of the major streets built in Las Vegas since then run also along section lines with one-mile spacing. Charleston Boulevard, Sahara Avenue, Tropicana Avenue, and Flamingo Road are well-known east-west examples.

Another big influence on roads in rural areas were railroads. Early highways often followed them closely, and again Las Vegas was no exception. Rancho Drive follows the route of the former Las Vegas and Tonopah Railroad. US 91, later known as Las Vegas Boulevard and simply the Strip, paralleled the Union Pacific Railroad. But a few miles south of town, the surveyors curved the road to head straight south on section lines. From that curve, the distant trees and buildings of Las Vegas must have seemed a mirage; it has been home to the Mirage hotel-casino, occupying the prime spot on the Strip, since 1989.

Among these section line roads, Charleston Boulevard is unique. It was the first of these arterial streets to appear and became an early suburban commercial strip. But more importantly, the line it follows is the fifth standard parallel south of the Mount Diablo benchmark, meaning that it is a correction line in the PLSS. Because of the earth's curvature, north-south lines will converge as you move toward the North Pole; corrections are necessary at certain intervals to keep the townships and sections square. But if the sections and townships are kept square, the north-south section lines will not line up in adjacent townships. For that reason, streets running north of Charleston Boulevard do not directly meet those from the south. They have an offset or dogleg of about six hundred feet where they cross Charleston, requiring two turns for drivers.

These doglegs were gradually replaced with curves to allow continuous travel. Nellis Boulevard was the first to be curved in 1953, with a short section of the old street remaining in use under a different name. Lamb and Decatur Boulevards were curved across Charleston Boulevard in 1962, Eastern Avenue by 1969, Torrey Pines Drive and Buffalo Drive in 1975, Pecos Road and Valley View Boulevard in the late 1970s, and Hollywood Boulevard in the late 1990s. Rainbow Boulevard is unusual: it has a curve, but this is located where Rainbow crosses the US 95 rather than at Charleston Boulevard. The Strip has no dogleg because it is not aligned with the PLSS grid where it crosses Charleston.

Another area where the PLSS grid remains visible is in Muddy River Valley near Overton, where Nevada Highway 169 takes a stair-stepping route through the valley, though the curves have been rounded to make driving easier. This is common in agricultural areas of the West where highways cross diagonally through an area of farmland but remain on section lines.

Surveyors laid out the PLSS lines in the Las Vegas Valley January 18–24, 1882. The valley had a tiny population at that time and was empty except for a few springs, the Las Vegas Rancho, and a few wagon roads. The surveyors must have wondered what the point of their effort was, and their notes indicate they thought the desert valley offered little potential for farmers wanting to homestead land. Little could they have known that in those seven days they were laying out what would become the main streets for a city of several million people known throughout the world.

A grid of streets requires names. The surveyors who laid out the Las Vegas townsite in 1905 provided them by numbering streets after explorers, with one for John Frémont (map 3.5)

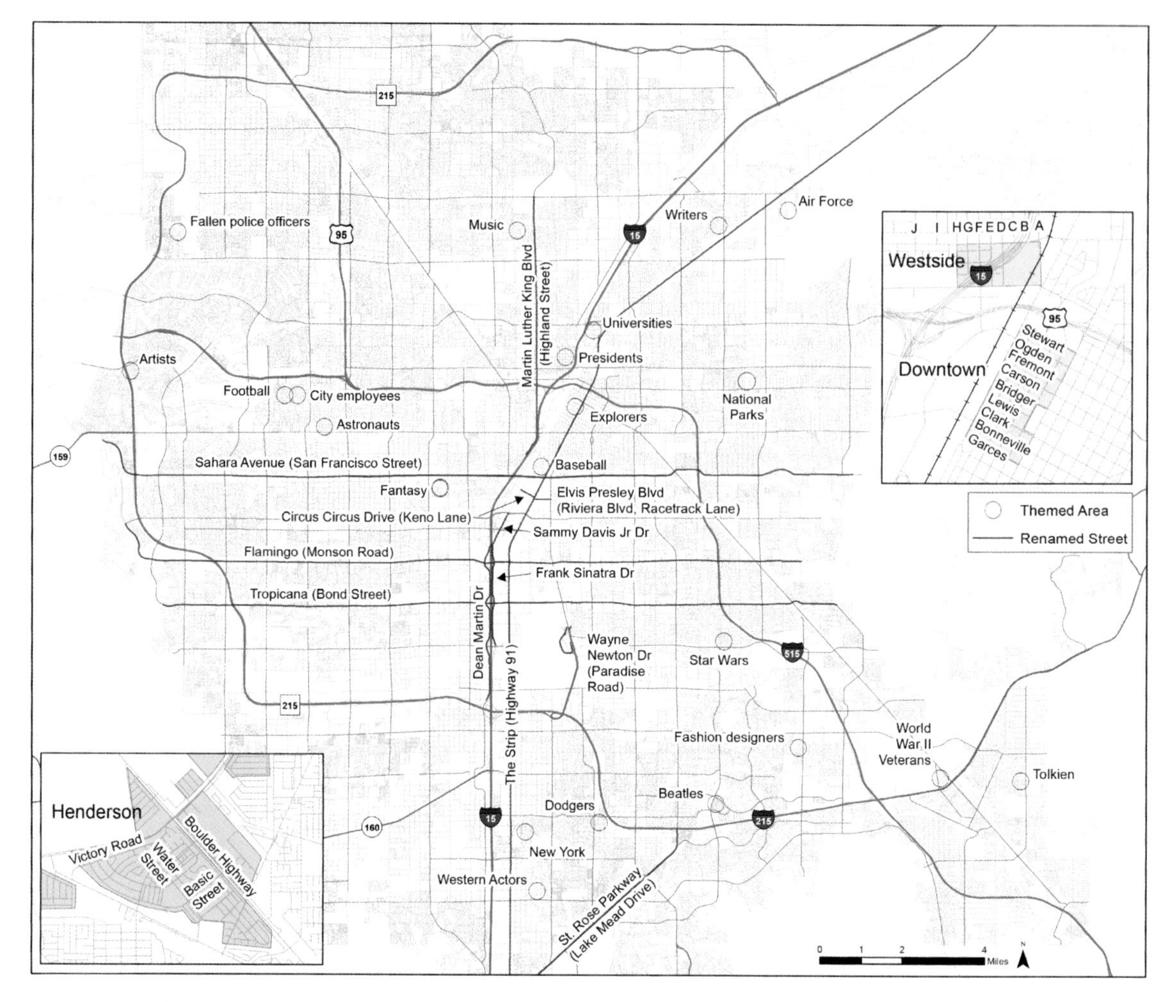

Map 8.6. **Street Names.** Some neighborhoods have street names based on a common theme. *Insets*: street names in downtown Las Vegas and Henderson.

assigned to the main commercial street (the street closest to the tracks is Main Street, but in Las Vegas this was never of any importance). Other streets are named for explorers Meriwether Lewis and William Clark, Jim Bridger, Benjamin Bonneville, Peter Ogden, Kit Carson, and Francisco Garcés. The exception is Stewart Avenue, which is likely named for Helen Stewart, longtime owner of the Las Vegas Rancho.

Cross streets running parallel to the railroad tracks were numbered from First through Fifth Streets. This is very common; the use of numbers for street names is found in almost half of all cities and towns in the United States, though it has gone out of style. Numbers are most common in western states, peaking in Utah, where more than 92 percent of cities use them. The use of numbers for streets in one direction and names for the other, as in downtown Las Vegas, is often referred to as the Philadelphia plan, after the first US city to use this style.

Boulder City and the Westside also had numbered streets, although those in the Westside were changed to letters in the 1930s to avoid confusion with the numbered streets in downtown

Las Vegas. Today in the Westside, A through N Streets remain—an area sometimes referred to as the Alphabets. Boulder City also has lettered avenues but only made it as far as M Avenue. Less than 10 percent of cities and towns nationwide use lettered streets, but it is more common in Nevada than any other state, with 52.6 percent of the state's cities having them.

The main streets in Boulder City's carefully planned townsite were named after states in the Colorado River watershed; Nevada Way became the town's commercial street. Streets in the original Henderson townsite include those named after metals and states, with the town's main street named after the most precious commodity in Southern Nevada, water. Basic Road honored the Basic Magnesium Inc. plant that was the reason for the town's existence, while Victory Road referred to the ultimate war cause for which the plant workers toiled.

As Las Vegas grew so did the need for more street names. Numbered streets are easy to add, but subsequent additions only went up to 30th Street. Most of these were parallel to the railroad tracks, and the use of numbering seems to have fallen out of favor once a new grid aligned to the compass was adopted.

Coming up with names requires more effort. Explorers or pioneers can be added, as with Gass Avenue, just south of the original townsite, which commemorates Octavius Gass, who took over the abandoned Mormon Fort in 1865 and made it into a prosperous ranch and farm (Gass Peak north of the city is also named after him). But eventually some creativity is required to keep coming up with a supply of new names. One solution is to have themed street names in a subdivision, named after presidents, writers, artists, astronauts, World War II veterans, police officers killed in the line of duty, city employees, Los Angeles Dodgers players, western movie actors, fashion designers, Air Force bases and pilots, universities, music, football, baseball teams, J. R. R. Tolkien characters, New York themes, fairy tales, Beatles songs, and *Star Wars*. All of these can be found in the valley.

Many streets have been renamed. In 1958 Highway 91 was renamed Las Vegas Boulevard, its official name, but was by then already known as the Strip. This name was invented by an early casino manager who used it as a mocking comparison to the glamorous Sunset Strip in Los Angeles. Little did he suspect it would come to be one the most famous streets in the world.

Many street names were changed to help promote new casinos. San Francisco Street was renamed Sahara Avenue, and Bond Road became Tropicana Avenue in 1960. These changes started a trend that continues to the present. Racetrack Road was renamed Riviera Boulevard after the new casino in 1961, while Keno Lane and Monson Road disappeared from the maps in 1974, replaced by Circus Circus Drive and Flamingo Road, respectively. The south end of Paradise Road became Wayne Newton Boulevard in 1985. Parts of Industrial Road became Dean Martin Drive and Sammy Davis Jr. Drive, intersecting with Frank Sinatra Drive. As the casino landscape changes, so do the streets named after them. The Riviera closed in 2015 and was then demolished; Riviera Boulevard Became Elvis Presley Way in 2016. The Sahara became SLS in 2014 but the street name was not updated (and thankfully won't need to be since the property was again rebranded to be Sahara Las Vegas in 2019). But not all name changes were for casinos or performers: Highland Avenue became Martin Luther King Boulevard through the Westside.

Street names can provide clues to a city's history. Sandhill Road commemorates an area of sand dunes that once existed in Paradise Valley. It is often said that Paradise Road (and the town of Paradise) was named for the Pair-O-Dice Club along the road. However, a 1952 US Geological Survey topographic map shows that Paradise Road used to be Paradise Valley Road, running south to the Paradise Valley area.

Compared to other large cities, Las Vegas has few historic neighborhoods. The city simply hasn't been around as long as many others, and most original residential areas have long since been torn down. This is especially true of the original townsite (map 4.1), though a few of the oldest homes have been relocated to the Springs Preserve and the Clark County Museum in Henderson. Much of the McWilliams Townsite or the Westside was demolished to make way for Interstate 15, and little original survives there.

Although the city does not have any intact early twentieth-century neighborhoods left, it does have many that are mid-century modern (or mid mod), a term that covers many related housing styles with single floors, open plans, often described as ranch houses. Several such neighborhoods have retained their identity and something of their character over the years. They include the John S. Park and Huntridge neighborhoods south of Charleston Boulevard, Scotch 80s, McNeil Estates, Las Verdes west of Interstate 15, and Paradise Crest on the east side. Today they are desirable (and often quite expensive) neighborhoods. Berkley Square is not only a classic mid mod neighborhood but the first to be built for the city's growing African American population; it and the Park subdivision have even been designated as historic neighborhoods by the federal government (map 11.8).

Several neighborhoods have been home to the more well-to-do residents of the city over the years. The Rancho Circle neighborhood was the first of these. It was constructed in the

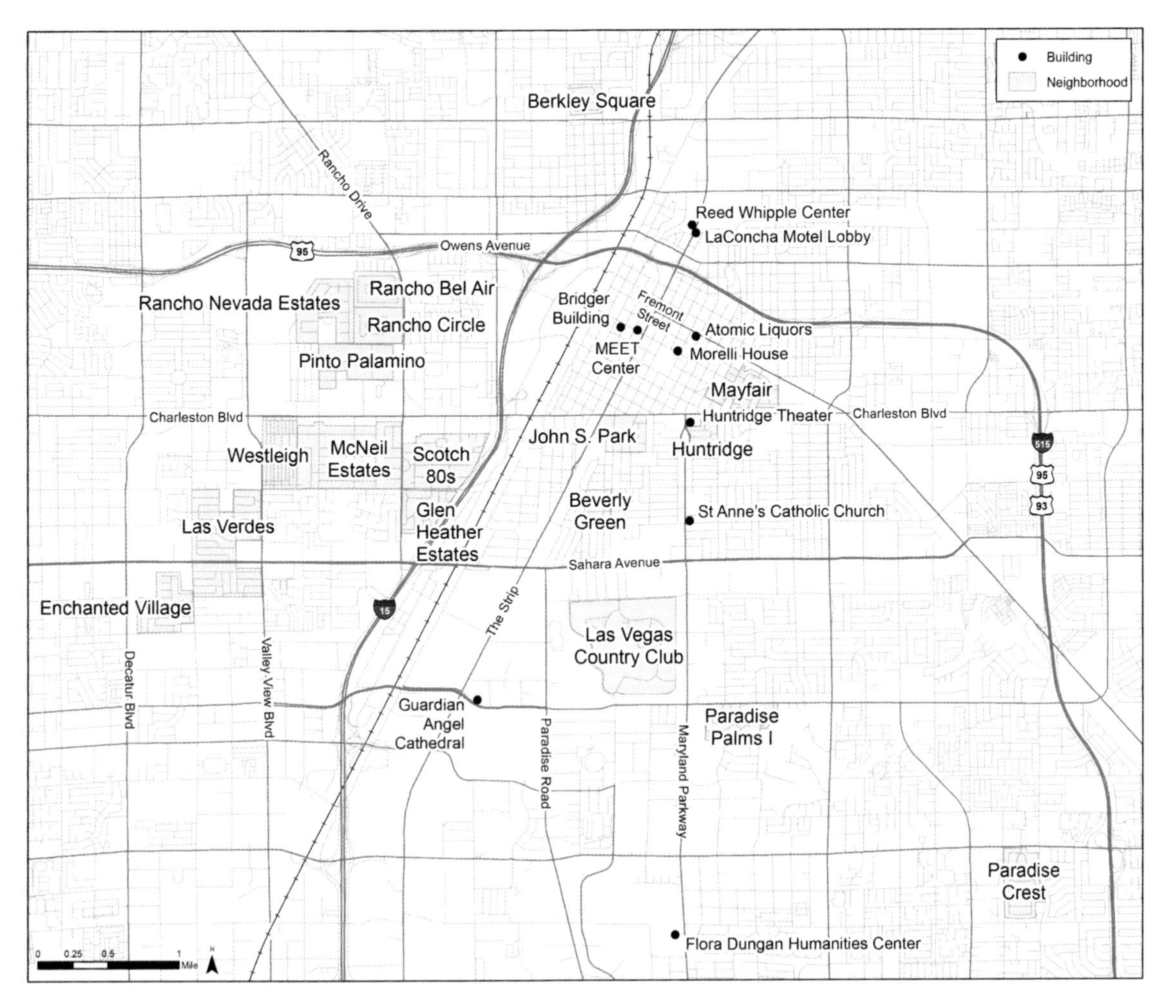

Map 8.7. Mid Mod Las Vegas. Las Vegas has a large collection of mid-twentieth century modern architecture, both commercial and residential.

1950s on the western edge of the city off of Rancho Drive and for many decades was known as the wealthiest neighborhood in town. The Scotch 80s has also been home to many well-known Las Vegans over the years: Oscar Goodman and his wife, Jerry Lewis, Sammy Davis Jr., Howard Hughes, Steve Wynn, and Nicolas Cage. The Spanish Lakes area has more recently become the trendiest place for the wealthy. This is bounded by Tropicana Avenue, Rainbow Boulevard, Hacienda Avenue, and Durango Drive. Whether it will achieve the renown of the mid mod neighborhoods remains to be seen.

Many commercial buildings from this era survive. The lobby of the La Concha Motel is one of the most outstanding. Designed by prominent African American architect Paul Revere Williams, this was at 2955 Las Vegas Boulevard South but moved to the Neon Museum when the rest of the motel was demolished in 2004. Many others can be found downtown and along Maryland Parkway, the city's commercial core when modernist styles were most popular. The Nevada Preservation Foundation, a nonprofit group, is dedicated to preserving Las Vegas's history. Among their

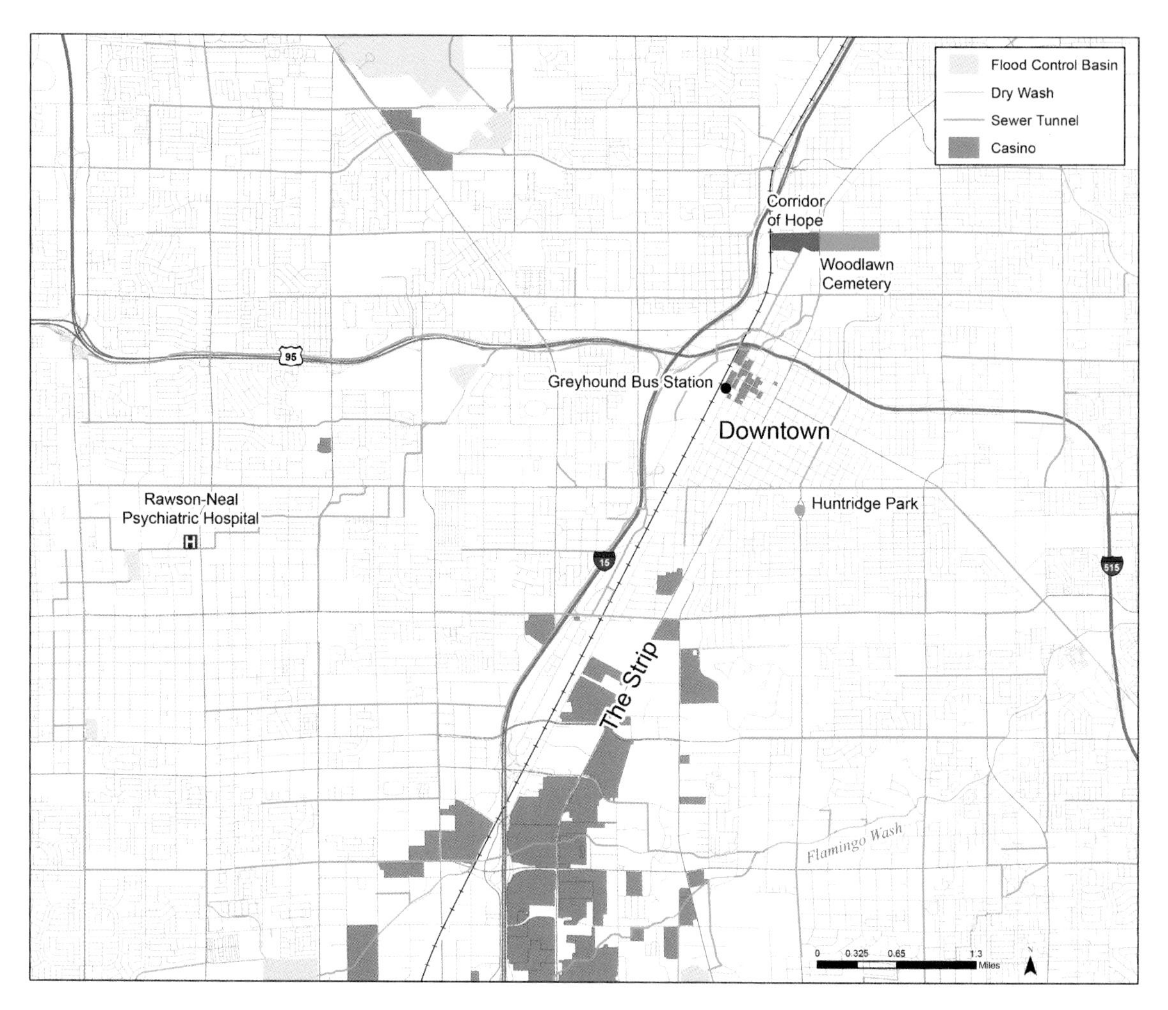

Map 8.8. Homeless in Las Vegas. Those without a place to live have often congregated near several parks or near casinos where they can look for money. Many occupy drainage tunnels under the city. The Corridor of Hope is an attempt to provide for the city's homeless population, while the Rawson-Neal Hospital ships them off to other cities.

efforts has been a list of outstanding examples of mid mod commercial architecture in the city (shown on the map).

Some win big in Las Vegas, and some lose everything. If you're reading this book you probably aren't in the second group, but you may have encountered those who are, living on the streets and under bridges. Homelessness is not new in Las Vegas. Thousands made their way to Southern Nevada during the Great Depression hoping for work on Hoover Dam; they founded the community of Ragtown along the Colorado River in 1931. Others created a squatter camp north of Las Vegas between Woodlawn Cemetery and Tonopah Avenue. But the odds are stacked against today's homeless.

There are more homeless than you think in Las Vegas; estimates range from 6,490 in a January 2017 count to more than 35,000, or 1.8 percent of the county's population. Regardless of the numbers, Las Vegas is home to one of the largest homeless populations of any American city. Thousands of these are children living on their own. Where do they come from? Many were already living in Las Vegas but lost their homes, sometimes from gambling addictions but also because of drug addictions, mental health issues, or because medical bills or job loss pushed them into foreclosure. Others came looking to get rich quick in casinos or other pursuits, but failed miserably.

Many congregate in out-of-the-way locations where tourists, and most locals, don't visit. This is often downtown and in nearby areas, sleeping where they can and heading to Fremont Street or the Strip in search of money. Panhandling is one way to get money, but many go "silver mining" in and around casinos, looking for chips or vouchers that gamblers may have lost or overlooked. Some have regular jobs where few customers would suspect their living conditions. However they get the money, gambling is a favorite use of it, though others concentrate on using their money to get a night or short stay in one of several low-rent motels and apartments around downtown. A homeless encampment also grew along Foremaster Lane, sometimes spilling over into Woodlawn Cemetery (where the homeless had camped in the 1930s). Another favorite campsite was Huntridge Park before it was closed to keep them out. The homeless also camped around and sometimes inside the nearby Huntridge Theater, complicating efforts to reopen it (map 8.13).

Some even live underground. The city has many washes carrying runoff from summer storms. As the city has grown, an extensive network of flood control structures built over the last forty years has prevented heavy flooding (map 6.8). Many of the dry washes have been covered over, creating a network of concrete storm sewer tunnels running underneath the city. There are more than 434 miles of these in the city, and some are quite large, up to eleven feet tall and more than thirty feet wide. Frank Sinatra spent time in a sewer tunnel in the movie *Guys and Dolls*, but Lady Luck's underworld in Las Vegas is not as appealing. The tunnels are dark, dirty, and still subject to floods that may kill their occupants, but they offer shelter to those who are without it. Nobody knows how many live beneath Las Vegas, but some people have lived in the tunnels for years and regularly commute to jobs they have been able to find in the world above. The person behind the nearest cash register might be one.

The city began opening what it called the Corridor of Hope along Owens Avenue in 2017. This groups together a number of services for the homeless, including housing for individuals, families, and seniors; food services; immigration and refugee counseling; and substance abuse treatment. These are provided by the city, The Salvation Army, Catholic Charities of Southern Nevada, National Alliance to End Homelessness, Lutheran Social Services, and other organizations. This location is far removed from the bright lights of the Strip or Fremont Street; few locals and fewer

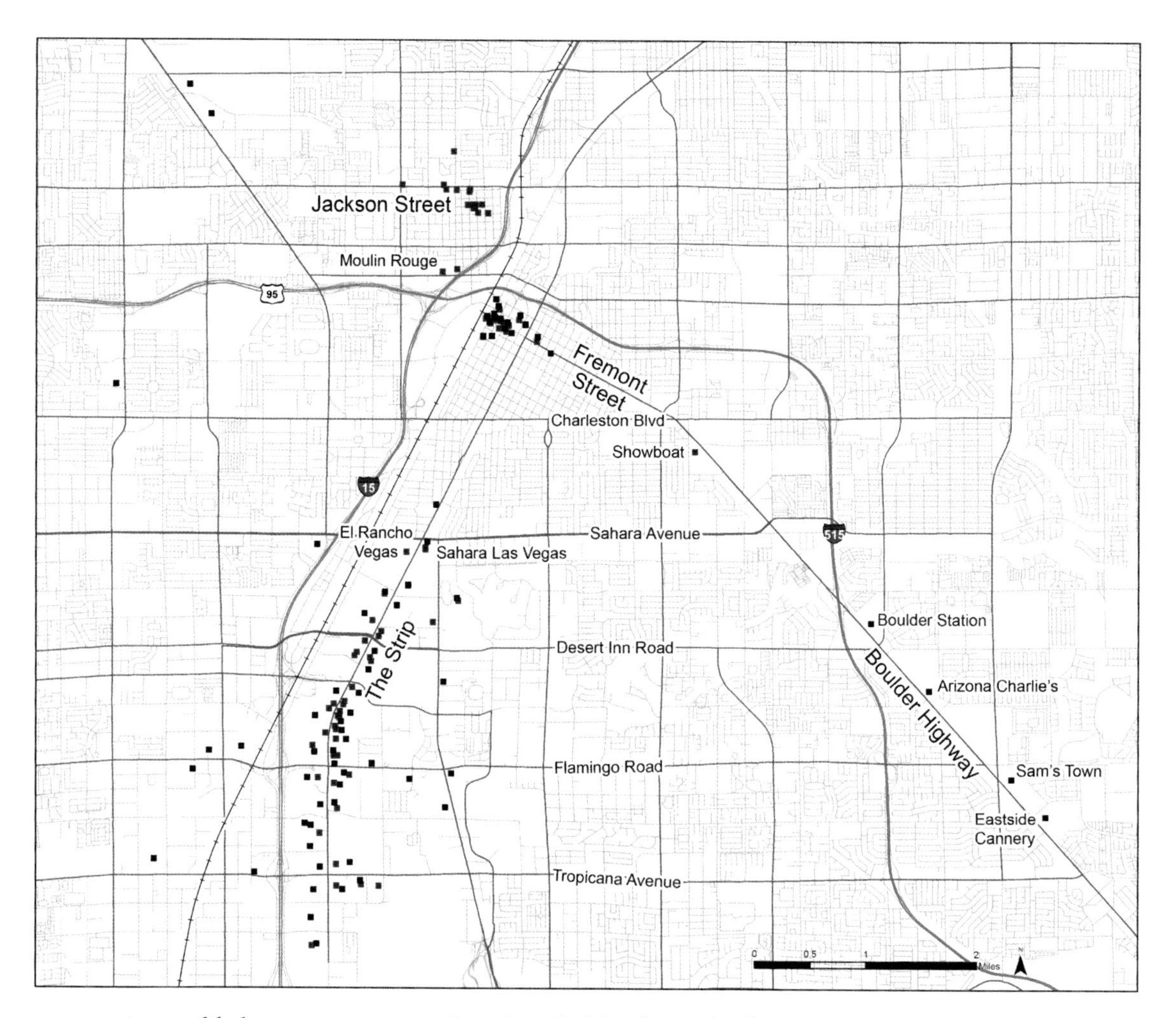

Map 8.9. Casinos. Las Vegas has had many casinos, though not all survive. Black squares represent casinos open today, and red squares denote those that have shut their doors. In addition to downtown and the Strip, there are a number along Boulder Highway and a former cluster along Jackson Street in the Westside.

tourists are likely to come across it. It is about half-mile north of the ruins of the Old Mormon Fort, where Las Vegas celebrates its past. For many, it will be a welcome step toward the future.

But for others, their future has been hijacked by doctors and bureaucrats with another solution: drug them and ship them out of the city. This was routinely done by doctors at the Rawson-Neal Psychiatric Hospital, which drugged up patients, gave them a bag with a few days' worth of medication and snacks, and put them on a Greyhound bus to destinations throughout the US. This practice, called "Greyhound therapy" or "patient dumping," is quite common across the country. Rawson-Neal dumped at least 1,500 patients in destinations throughout the forty-eight contiguous states, a third of which went to California. All were mentally ill and were hungry and homeless upon their arrival. Homelessness is one thing that happens in Las Vegas that does not stay here.

Gambling has a long history in the Las Vegas area, going back hundreds, if not thousands, of years with Paiute games of chance. It was common in Las Vegas from its founding in 1905 until it was outlawed statewide in 1910. When it was legalized again in 1931, it initially had little impact; massive government projects such as Hoover Dam, Basic Magnesium Inc. in Henderson, and Nellis Air Force Base brought the jobs and construction government leaders desperately wanted. It was not until after World War II that gambling became a major industry in the city.

The city originally limited gambling establishments to Fremont Street between First and Third streets, but it soon expanded to Fifth Street. The Northern Club was the first legal casino when it opened on March 20, 1931. Others soon followed: the Las Vegas Club later that year, El Cortez in 1941, Golden Nugget in 1946, Binion's Horseshoe in 1951, Fremont in 1956, The Mint in 1957, Four Queens in 1966, Union Plaza (now just the Plaza) in 1971, California in 1975, and Main Street Station in 1978. The Sundance (later Fitzgeralds and now the D) opened in 1980, Circa in 2020, and Resorts World Las Vegas in 2021.

The city's first neon signs were erected along this street, including Vegas Vic in 1951, leading to the nickname Glitter Gulch (a female companion named Vegas Vickie was located across the street from 1980 to 2017 and is now a centerpiece in Circa). The first high-rise buildings in the city were along Fremont Street; the Fremont hotel became the tallest building in Nevada when it opened in 1956, and Fitzgeralds was the tallest building in the city until the Stratosphere Tower was completed. To attract more tourists, the Fremont Street Experience was created in 1995 by closing the street between Main and Fourth Streets to vehicle traffic and putting a barrel vault canopy over it. The underside (Viva Vision) was covered in light bulbs (now more than forty-nine million LEDs) for nightly shows.

Not all downtown casinos were on Fremont Street. Lady Luck opened in 1964 and is now the Downtown Grand. Casinos also began to appear outside the city limits to be free of city regulations. The first of these was the Meadows in 1931, located where Fremont Street becomes Boulder Highway. A small casino even briefly operated at Lorenzi Park.

As all Las Vegans know, US 91 south of the city became the principal location for new casinos in the postwar era. The Red Rooster was one of the first, a small casino that opened in 1931 where the Mirage is now located, only to burn down later that year. However, El Rancho Vegas is generally considered the first hotel-casino to open along Highway 91. It was at what is now the southwest corner of the Strip and Sahara Avenue, where the Las Vegas Festival Grounds is now located. The hotel opened in April 1941 with 110 rooms, making it the largest in the city. The sprawling complex, mostly one story, centered on a casino with a swimming pool in front and motel wings in an arc behind the parking lot. A tower with a windmill displayed the name of the hotel. At the time, it was surrounded by empty desert and visible from a considerable distance. It burned to the ground in 1960 and was never rebuilt.

The Last Frontier followed in 1942 and was later known as the New Frontier and finally just the Frontier. The Flamingo opened in 1946 and was considered a significant step in marketing the city as more than just a dusty desert town. It was the first to involve theming that went beyond Old West styles. The Flamingo is now the oldest hotel on the Strip, although nothing remains of the original building.

New casinos followed quickly in the 1950s. The Desert Inn opened in 1950, the Sahara and Sands opened their doors in 1952, followed by the Dunes and Riviera in 1955. The Riviera included a high-rise tower, introducing a new kind of building to the Strip. Other major additions were the Stardust in 1958, Caesars Palace in 1966, Circus Circus in 1968, the International (later Hilton and Westgate) in 1969, and the first MGM Grand (now Bally's) in 1973.

A new era of themed high-rise megaresorts began when the Mirage opened in 1989. This launched a massive building boom of high-rise hotels that reshaped the Strip. The Rio and Excalibur appeared in 1990, the second MGM Grand, Luxor and Treasure Island in 1993, Monte Carlo (now Park MGM) in 1996, New York–New York in 1997, Bellagio in 1998, and Mandalay Bay, Venetian, and Paris Las Vegas in 1999, each introducing a new theme to the Strip. The Wynn followed in 2005, Palazzo in 2007, and Encore in 2008. The Great Recession largely ended this building boom, but several projects underway were eventually completed; the three hotels within the City Center project opened in 2009 and the Cosmopolitan in 2010. Resorts World on the former Stardust site opened in 2021, but several other proposed additions to the Strip have stalled.

Boulder Highway was an early casino location, with the Meadows, where Fremont Street becomes Boulder Highway, and the Railroad Pass Casino in Henderson opening in 1931. The Castaways Hotel (which opened 1954) was a longtime presence at the north end of Boulder Highway, while Nevada Palace (1977) and Sam's Town (1979) were more recent. Since the 1990s, several more have opened along this road as neighborhood casinos (see below). Henderson incorporated as a city in 1953 with no gambling allowed, but this changed in 1956. Several casinos soon appeared along downtown's Water Street, including the Lucky Star Club, Royal, Eldorado Club (now The Pass), Rainbow Club, and Emerald Island Casino (built on the site of the Lucky Star Club).

The Sahara Hotel, 2007. Photo by mrak75.

In the 1950s and 1960s, Jackson Street in the Westside emerged as the Black strip. A number of small gaming halls, a few hotels, and rooming houses catered to the city's Black population and visiting entertainers who were denied entrance to Strip casinos (map 4.11). The first racially integrated hotel-casino in Las Vegas was the Moulin Rouge, which opened in the Westside in 1955. This attracted a biracial crowd and tremendous attention from the media. Unfortunately, it only remained in business six months before shutting down. While famous, it simply did not earn enough to pay the bills. It reopened on several occasions in later years but never achieved the prominence of its brief 1955 incarnation.

A new geography of casinos emerged in the 1990s with neighborhood casinos catering to locals and scattered throughout the valley. These were designed not to appeal to visitors from distant places but to the city's residents. The first of these was the Showboat at the north end of Boulder Highway, and the Moulin Rouge was another early example. Another early example was the Bingo Palace (later Palace Station). Many new ones were built from the late 1980s, including all of the Station properties. One big difference between neighborhood and Strip casinos is in the names of the streets around them; while the Strip casinos were the source of many street names (map 8.6), the local casinos more typically derived their names from local streets or areas.

About 43 million visit every year and spend $35.5 billion annually to make all this possible. Where do they all come from? Most are from the United States. The map shows visitation in 2016 as collected by the Las Vegas Convention and Visitors Authority (LVCVA). The largest number of visitors come to Las Vegas from two groups of states: those with the largest populations (California, Texas, Illinois, Florida, New York, and others), and those closer to the city (from within Nevada, Arizona, Utah, for example). This is a common geographic pattern that has been used to predict attendance at many kinds of attractions, even national parks.

Las Vegas casinos are not without competition. Gambling in Atlantic City, New Jersey, was legalized in 1976, and a US Supreme Court decision opened the way for gambling on Indian reservations. Casino gambling has since been legalized in forty-one states, with 486 tribal casinos and 524 nontribal casinos nationwide in 2016. All but two states now have some form of gambling. Of course Las Vegas remains the top gambling center in the country, followed distantly by Atlantic City, Chicago, New York City, and the Washington/Baltimore metropolitan areas. These cities have no Strip, but there are casinos near each of them, such as the tribal Foxwoods Casino in Connecticut, the

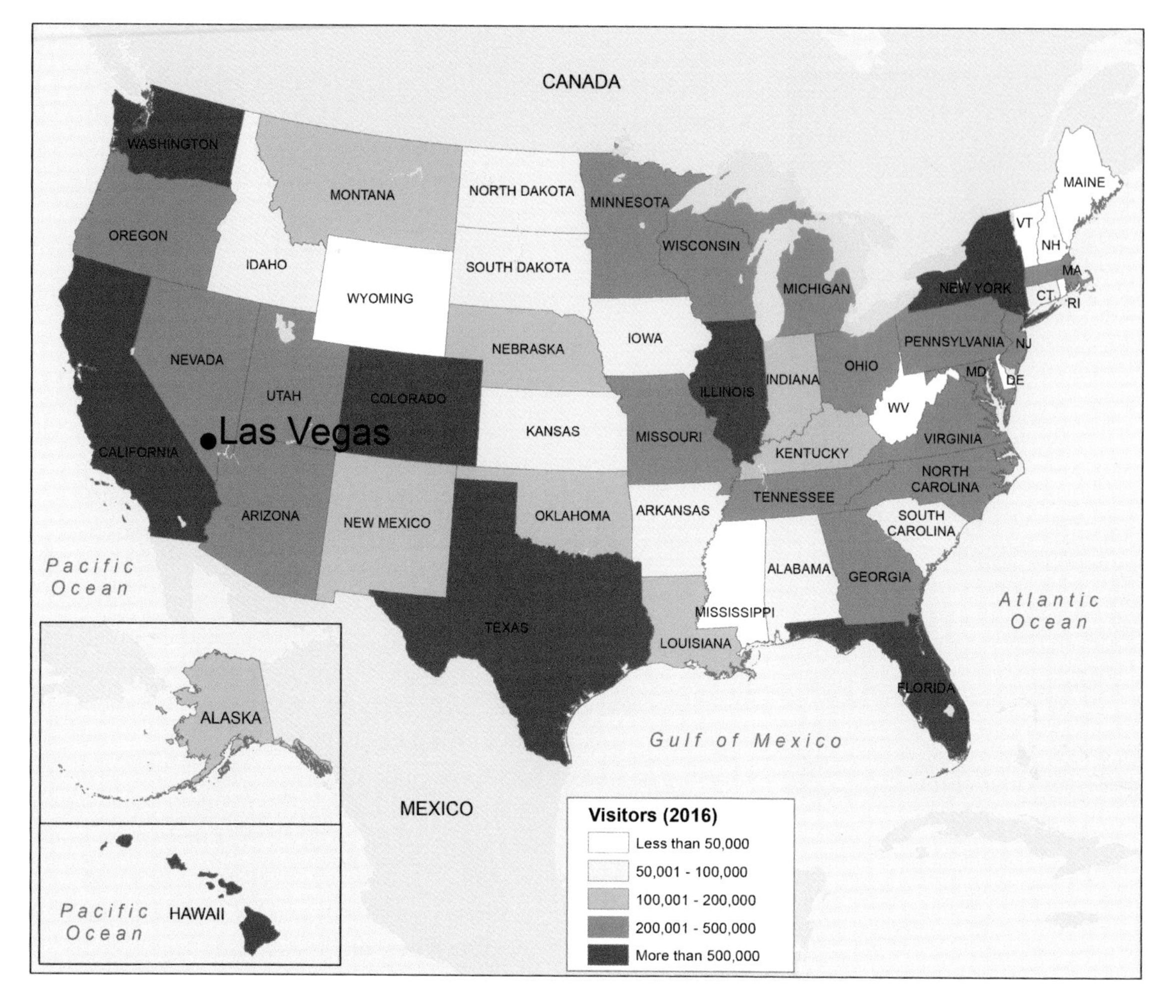

Map 8.10. Visiting Las Vegas. Each state's number of visitors to Las Vegas in 2016.

largest in the country by square footage. These regional casinos attract large numbers of visitors who do not need to fly to Las Vegas to gamble.

Among actual gambling centers are several in Mississippi, which allowed gambling since 1990. Tiny and dirt-poor Tunica County is home to six casino-resorts in 2021, including a Sam's Town, while Biloxi has a Golden Nugget, Hard Rock Hotel, and Silver Slipper. While Las Vegas was once the "Mississippi of the West" for its segregation (map 4.11), it now seems Mississippi is becoming the Las Vegas of the East because of its casinos. Needless to say, none of America's other gambling towns and cities can compare to the one and only Las Vegas.

Visitors come to Las Vegas from all over the world. The map shows the number of annual visitors from the top twenty-five countries; Canada, Mexico, Australia, China, Japan, and the United Kingdom lead the list. Several of these countries have nonstop air connections to Las Vegas; the others can be reached with only one extra flight.

Las Vegas has some tough competition, and is now the third-

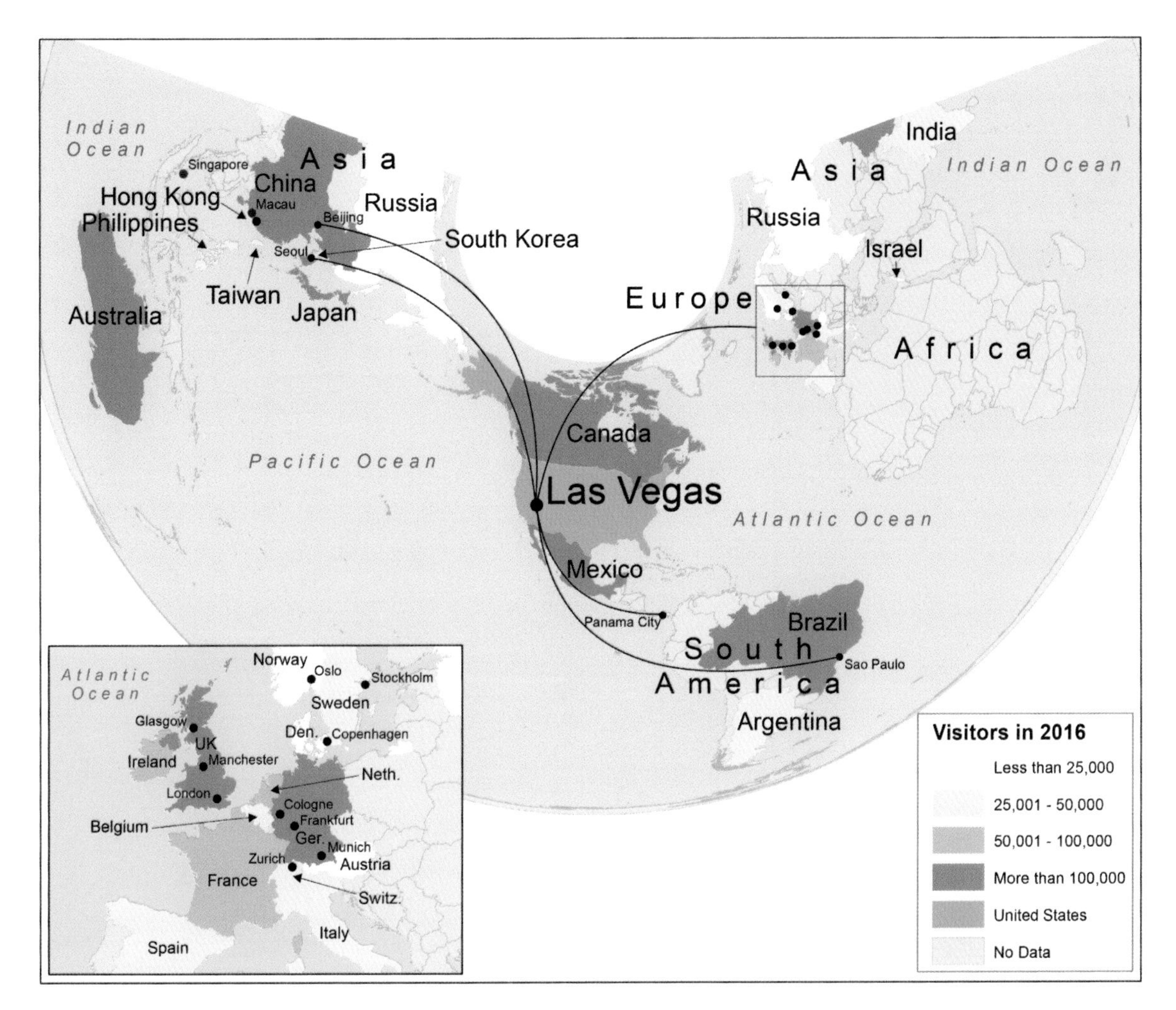

Map 8.11. Entertainment Capital of the World. Relatively few countries provide most of the international visitors to the city. The black lines show international nonstop air routes to Las Vegas, and the two leading competitors, Singapore and Macau, are depicted with red dots.

largest gambling center in the world. The city's closest peer, and number-one gambling city, is Macau, a special administrative region of China, where gambling has been permitted only since 2002. It is a much smaller city, with only 650,000 people, but has several advantages over Las Vegas. It is the only city in China where gambling is legal, and it is a short flight for more than one billion people. However, the city has less than one quarter of the number of hotel rooms because its casinos do most of their business with daytrip visitors; the average stay of a Macau visitor is 1.2 days, compared to 4.6 days in Las Vegas. Many Las Vegas–style casinos have been built in Macau, including a Venetian modeled after the one in Las Vegas, but the city's gambling landscape will likely shift in different directions. Table games are overwhelmingly popular (while slot machines make up two-thirds of casino revenue in Las Vegas), the casinos are filled with high-rollers playing for high stakes, and gambling makes up 96 percent of casino revenue (compared to only 42 percent in Las Vegas). The Macau airport is also limited in capacity with a single runway and cannot handle the largest planes. Ninety percent of

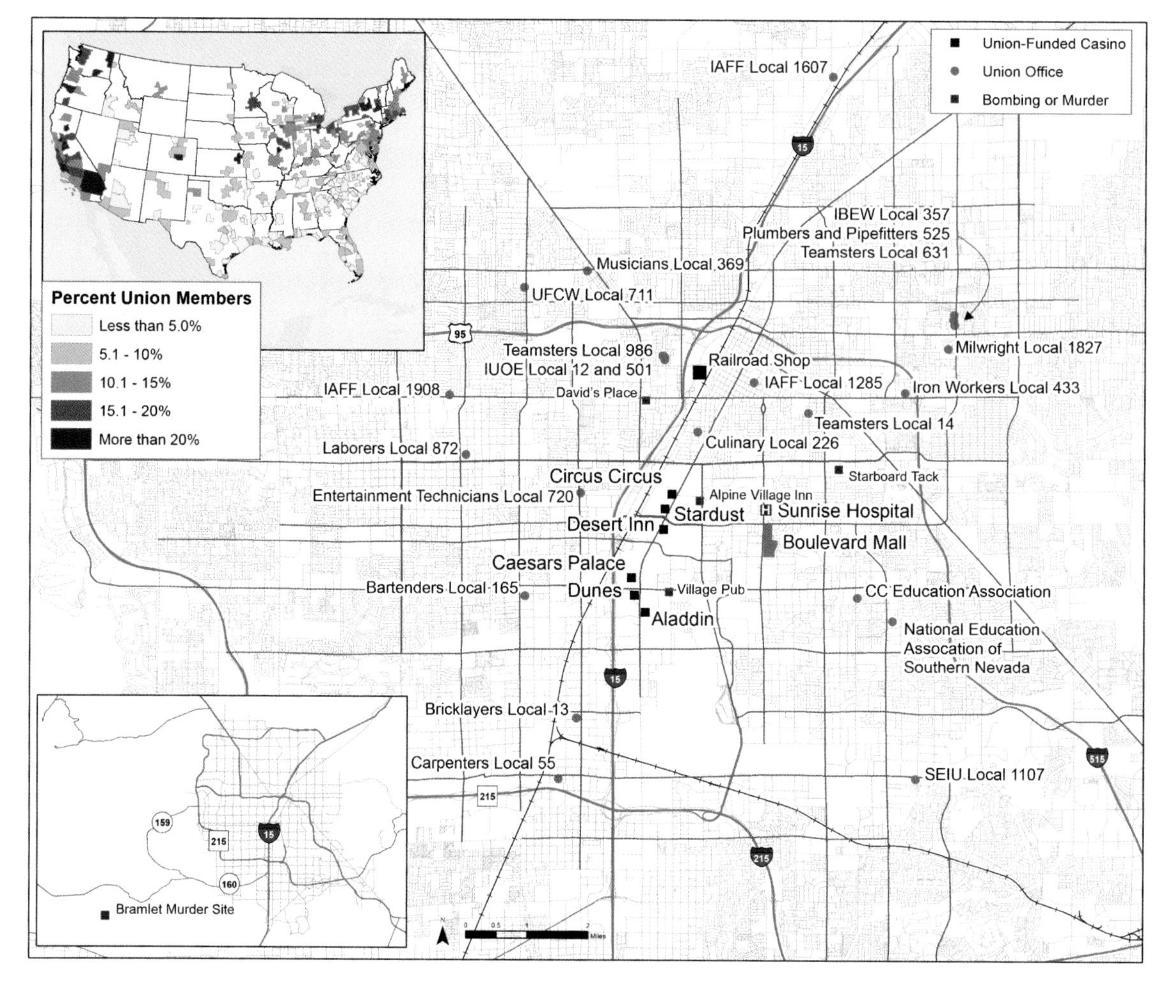

Map 8.12. Union Las Vegas. Major unions have played big roles in the history of Las Vegas. *Top inset*: the percentage of workers who are members in US metropolitan areas. *Bottom inset*: the location where union leader Al Bramlet was murdered.

visitors to Macau come from China or Taiwan, making the city very dependent on changing political conditions.

Another competitor is Singapore. The city has only two casinos, both of which opened in 2010. The Marina Bay Sands is surely one of the most spectacular casino buildings in the world, with three separate towers topped by a 1,100-foot-long park and pool. It was the most expensive casino ever built, and is also the world's most profitable casino, pushing it into the number-two ranking of world gambling cities.

Where did the expertise to run such large casinos come from? Las Vegas companies built most of them using money and management techniques learned on the Strip.

The history of Las Vegas is entwined with organized labor from the very beginning. Unionized railroad workers with the San Pedro, Los Angeles, and Salt Lake Railroad helped create the city in 1905. These workers occasionally went on strike for better conditions in the hot, dusty, remote desert town. The Great Railroad Strike of 1922 was a national work stoppage by sixteen unions representing railroad maintenance and repair workers in response to wage cuts at a time when the national economy was improving after the post–World War I recession. About 400,000 railroad workers across the country went on strike on July 1, including some in Las Vegas. The Union Pacific Railroad reacted by moving many maintenance jobs up the line to tiny Caliente, Nevada, and hiring scabs to replace the union workers there.

This loss of railroad jobs was a major blow to the small town of Las Vegas, and the good times did not return until the 1930s when the Hoover Dam project began. Unions attempted to organize jobs at the dam but had little success at a time when workers were desperate to take any job available. A 1931 effort by the International Workers of the World (IWW, or "Wobblies") did, however, succeed in improving working conditions somewhat.

Unions returned to the forefront in the postwar era. As casinos grew on the Strip, the workers joined a number of unions, reflecting the various trades involved in running a casino-hotel. For many years there were few labor problems; casinos were locally owned, and management was in close contact with workers. This began to change in the late 1960s; distant corporations began taking over casinos and building new ones, with managers imposing new standardized practices. Tensions increased, leading to the first major casino strike in Las Vegas in 1967 at twelve downtown casinos. Other strikes followed. Workers at the Dunes went on strike in 1969, followed by those at the Desert Inn, International, and Caesars Palace in 1970. On March 11, 1976, the biggest strike yet began, when workers walked out of sixteen casinos, including Circus Circus, Dunes, and Caesars Palace, and picketing for several weeks. This was topped in 1984 when workers at thirty-two hotels walked out. Strikes, and threats of strikes, remain part of life in the city.

Despite sometimes contentious battles between unions and casinos, the unions helped build many of them. The Teamsters Central States Pension Fund, guided by Jimmy Hoffa, provided loans to help build the Stardust, Dunes, Desert Inn, Circus Circus, Caesars Palace, and the Aladdin, as well as Sunrise Hospital and Boulevard Mall. Before the modern corporate era, banks were very reluctant to loan money to developers involved in gambling; the Teamsters were one of the few organizations willing to do so. Hoffa is thought to have been murdered by the Mafia when he disappeared in 1975; he had allegedly displeased the mob after a large payment was skimmed from Las Vegas casinos and made to President Richard Nixon. This was to have Hoffa's prison sentence commuted but did not lead to his reinstatement as the Teamsters' boss.

These sorts of rough tactics sometimes happened in Las Vegas. Al Bramlet ran the Culinary Union in the 1970s and emerged as one of the strongest labor leaders Las Vegas has had. He used his power to pressure casinos and restaurants to offer better deals, and he was responsible for the 1970 and 1976 strikes. He also targeted other businesses in an attempt to force unionization. One was the Alpine Village Inn restaurant on Paradise Road, in front of which he put picketers for twenty years. He was willing to take more drastic action when threats failed, and he hired the father and son team of Tom and Gramby Hanley to place bombs at a number of restaurants. David's Place, the Village Pub, and the Starboard Tack were all targets. Not all of these bombs exploded, and disagreements over paying for useless bombs apparently led to the Hanleys kidnapping Bramlet from McCarran airport, driving him out to Potosi Mountain, and killing him in February 1977. Hikers found his body several weeks later.

The Stardust paid off the last of its Teamster loans in 1986, the last casino to do so. Las Vegas has a fairly unionized work force today, much of it involved in casinos. There were 164,000 union members in Nevada in 2017, making up 12.7 percent of the workforce. These are concentrated in Las Vegas where unions have been able to raise wages for Las Vegas casino workers compared to those of nonunion Reno. Foremost among these is the Culinary and Bartenders Union with about 57,000 members in Nevada. About 24,000 of these work at MGM properties in Las Vegas and 12,000 at Caesars properties. The Service Employees International Union (SEIU) local 1907 represents 18,000 workers, including county employees and nurses. The map shows the offices of these and many other unions found in Las Vegas.

The number and percentage of unionized workers has been falling in recent years in Las Vegas, reflecting nationwide trends. Attacks by elected politicians on government unions have not yet spread to the state, but unions continue to represent police, firefighters, teachers, and city and county workers. Many Clark County teachers voted in 2018 to leave the Clark County Education Association and found a new union, the National Education Association of Southern Nevada. The process has so far been very contentious.

Las Vegas has appeared in countless movies over the decades, but Las Vegans have been going to the movies even longer, since 1907 when the Majestic Theatre opened on Fremont Street where the Golden Nugget is now. At least three generations of movie theaters are visible on the map. The first generation was downtown on Fremont Street, in between casinos, department stores, and other staples of small-town main streets. The Rainbow Theatre in downtown Henderson was similar and for many years the only theater outside downtown Las Vegas. The 1960 world premiere of *Ocean's Eleven* was held at the Fremont Theatre, but

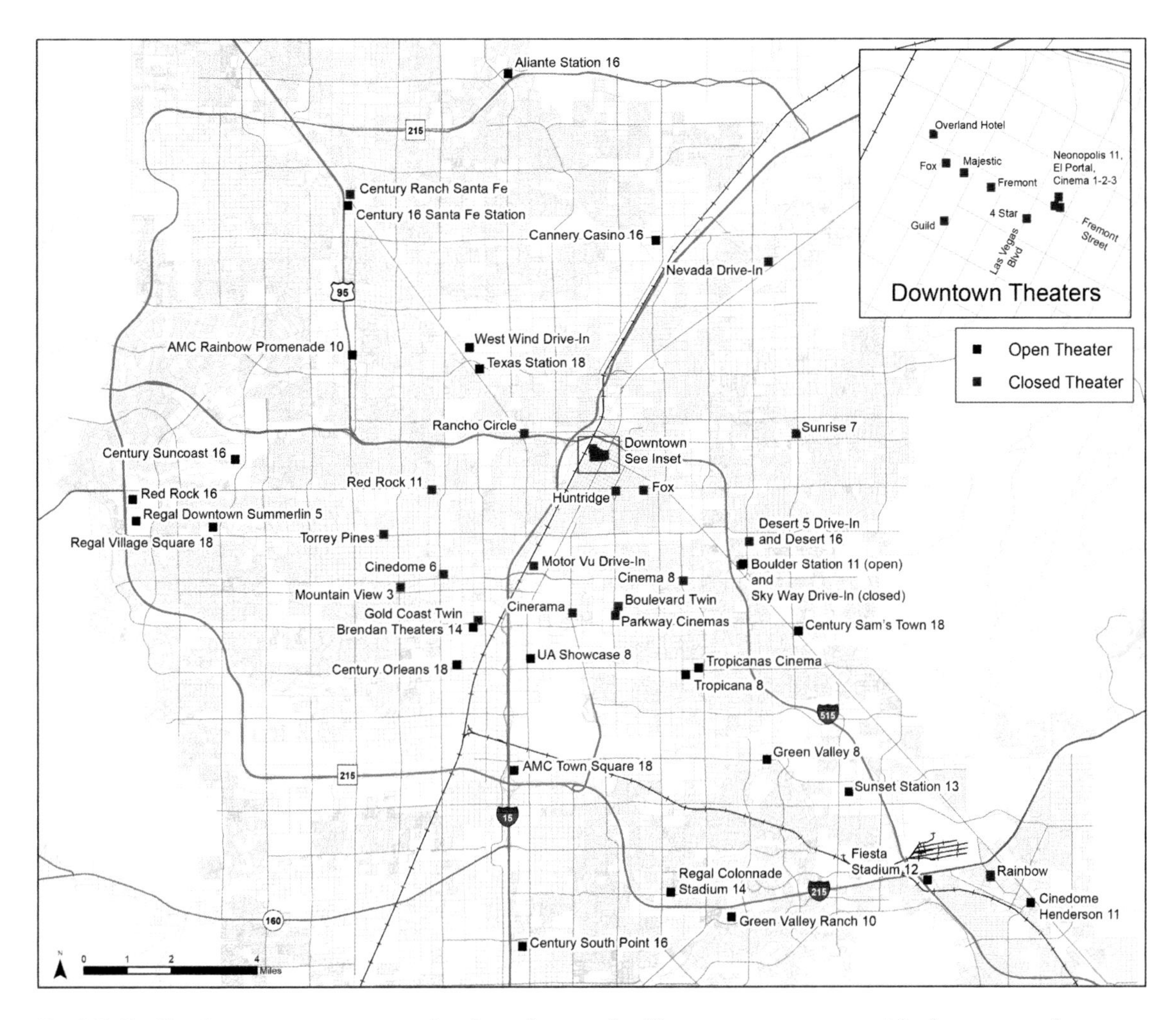

Map 8.13. The Silver Screen. Las Vegas has had many movie theaters over the years, though most are now gone. *Inset*: downtown theaters. Map based on the Cinematreasures.org website.

by then theaters had begun to appear outside downtown along new commercial strips on Charleston Boulevard and Maryland Parkway. Downtown theaters began to disappear, and the Fremont closed in 1980, one of the last of the old downtown theaters to survive.

The second generation began in 1944 when the first movie theater outside of downtown opened on Charleston Boulevard. This was the Huntridge, which remained in business until 2004. The building remains and may someday be restored to operation. During the city's postwar growth, many movie theaters opened up along Maryland Parkway, Boulder Highway, and in other outlying locations. These theaters provided several generations with their movies, but almost all are now gone. Several of these were drive-in theaters along Highways 91, 93, and 95. Drive-ins peaked in the late 1950s and early 1960s when about four thousand were open nationwide. Most have disappeared, and only the Las Vegas Drive-In survives off of Rancho Drive. It first opened in 1966 and is the oldest operating movie theater in the city.

Perhaps the most notable of Las Vegas's lost movie theaters was Red Rock 11, on West Charleston Boulevard. This started out as a single-screen theater in 1966 before growing to four, then six and finally eleven screens by 1973. It was noteworthy for its size at a time when having even six screens was considered large. But within one addition was a lobby built as a replica of an early-twentieth-century town square. A park occupied the middle, with benches and a popcorn cart and artificial grass. The entrances to six theaters were contained within building facades lining the "streets" that surrounded the park. Movie tickets were purchased from a window in another building facade labeled as the mayor's office. Those who were lucky to have gone to the movies there will never forget it. Sadly, it closed in 1999 after several years as a discount theater and was demolished. A shopping center now occupies the site. The sign in front of the theater can be briefly seen in the background of Clint Eastwood's *The Gauntlet* (map 11.6), which also provides a view of the old single-screen Cinerama Theater.

A new generation of movie theaters has opened since 1990, farther out in the suburbs. Twelve of these are in neighborhood casinos (map 8.9). One such casino theater replaced the Rancho Santa Fe 16 on North Rancho Drive after only five years in use, perhaps the shortest-lived theater in Las Vegas. It is evident on the map that not all areas of the city are equally well served. The Boulevard and Meadows Malls no longer have any theaters around them, and only one theater exists in the northeast part of town. Seven theaters are adjacent to the beltway, perhaps signaling that this road is becoming the new main thoroughfare of the city.

The Syufy theater chain (now part of Century Theaters) purchased all the first-run theaters in the city during the 1980s. The US Department of Justice filed suit against the chain alleging that it had created an illegal monopoly on theaters. Syufy won the case, by which time another chain, Roberts Theaters (later purchased by United Artists Theaters), had already purchased or constructed many theaters in the valley. The case was of interest only because the judge's twenty-page ruling incorporated the titles of two hundred movies into its text. Movie buffs may enjoy reading it and trying to spot those titles.

Las Vegas is surrounded by a spectacular natural skyline (map 2.1), with 11,916-foot-tall Charleston Peak the most prominent feature. It is visible throughout Southern Nevada and as far away as California's Mount Whitney 146 miles away. However, the city's man-made skyline is much better known.

It is also quite recent. For the first half century of the city's existence the tallest buildings were downtown and only a few

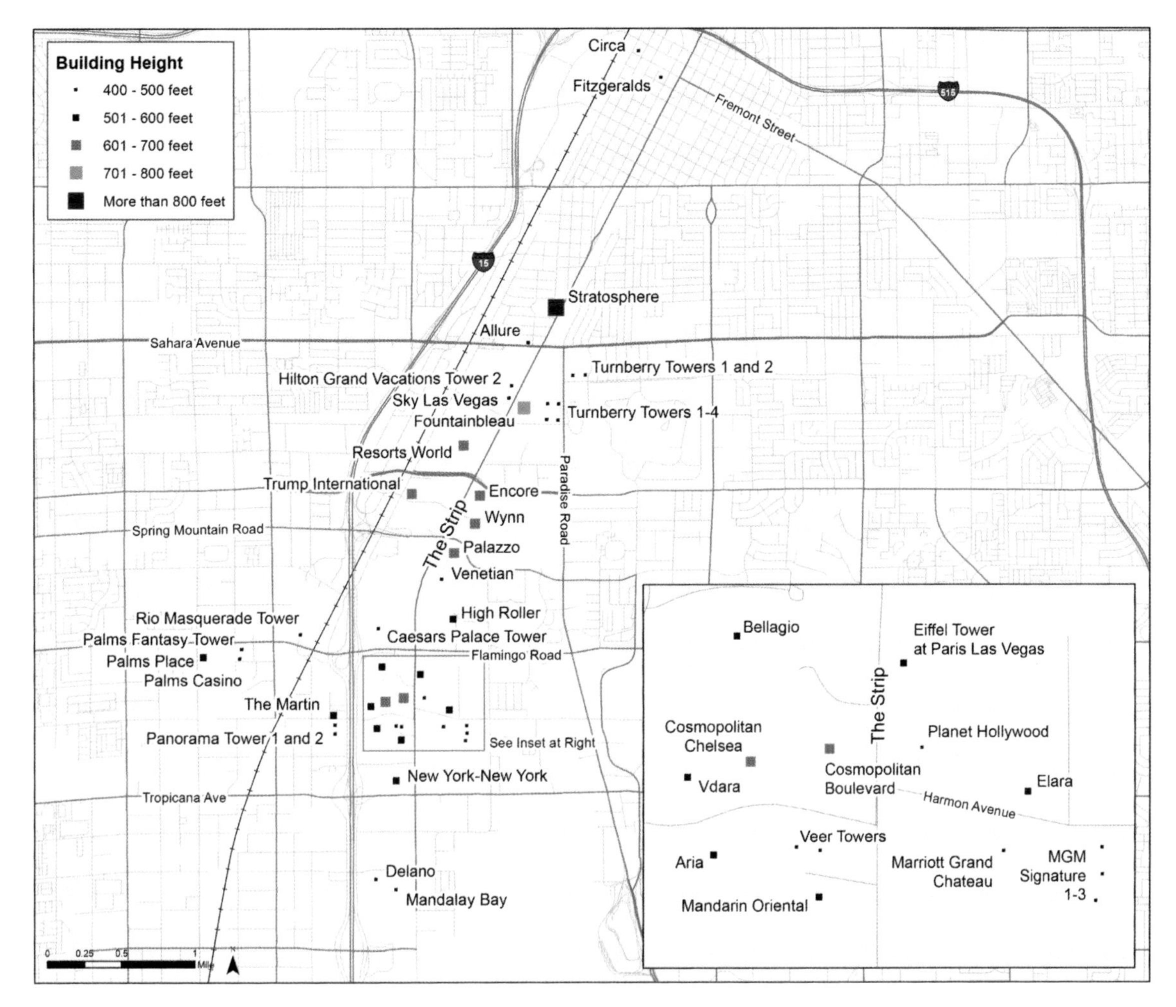

Map 8.14. The Skyline. Las Vegas has many buildings at least four-hundred-feet tall. Inset: a portion of the Strip near City Center with a high concentration of high-rises.

stories tall, though the Fremont hotel became the tallest building in Nevada when it opened in 1956. The Strip began as small one- or two-story motel buildings until the Riviera built a high-rise tower in 1955. Even then tall buildings were slow in coming. The Dunes built a twenty-four-story hotel tower in 1961, which was eclipsed by the International Hotel in 1969, which in turn was surpassed by Fitzgeralds downtown in 1979. This remained the tallest structure in the city until the Stratosphere tower was completed in 1996.

Many of Las Vegas's tall buildings were demolished to make way for new hotel-casinos during the building boom that began in the 1990s. Many of these towers were still relatively new when they were taken down. The most extreme example was the Harmon hotel-condo tower, which was demolished in 2014–2015 before it even opened at City Center after it was found to have been inadequately designed for its intended height. The plans were revised to make it shorter, then it was demolished because of the threat of collapse during an earthquake.

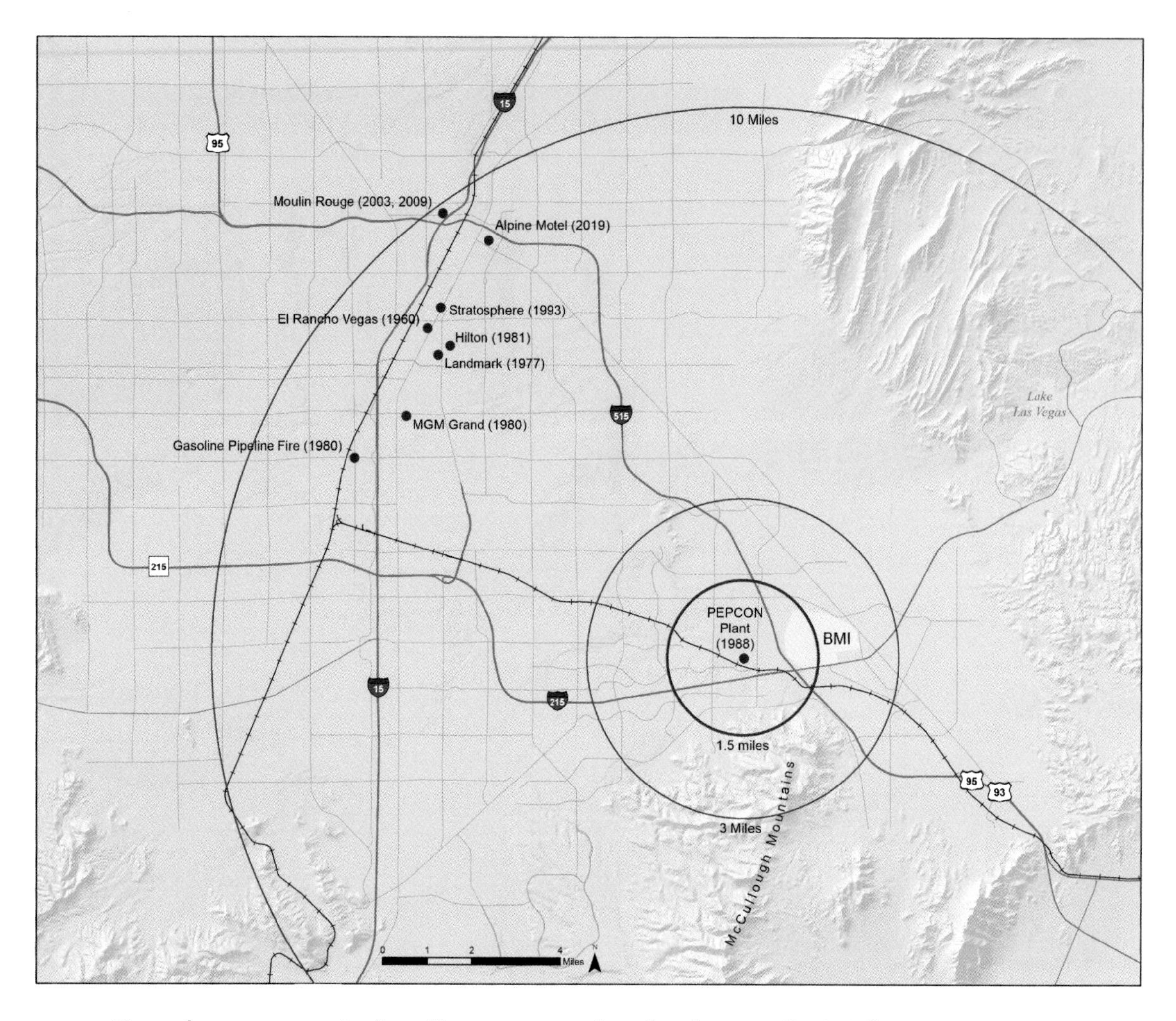

Map 8.15. Fire. There have been several hotel fires in Las Vegas's history, along with the PEPCON explosion in 1988. The red rings show distances from the rocket fuel plant.

Forty-four structures in the valley were more than four hundred feet tall in 2017, counting the Stratosphere, Eiffel Tower, and the Linq High Roller observation wheel.

The Las Vegas skyline is tall enough to present hazards to aviation. The Eiffel Tower at Paris Las Vegas was originally planned to be full size and stand 984 feet tall, but the Federal Aviation Administration (FAA) wouldn't allow it so near McCarran International Airport. It was instead built at half scale and tops out at 540 feet in height. The Luxor shines a bright spotlight straight up and also required FAA approval. But the most dangerous beam in Las Vegas shot downward: the Vdara "death ray." When the Vdara hotel opened in City Center on the Strip in 2009, the curving southern side of its 570-foot tower acted as a mirror and focused the sun's reflection on a small patch of ground. This patch moved with the sun during the day, crossing the swimming pool where many guests reported severe heat and even burns. The hotel added a film to the glass to reduce the reflection and placed umbrellas to shield pool guests.

Fire was a serious threat in western boomtowns because of their wooden construction, dry air, and inadequate water systems, and Las Vegas was no exception. The Las Vegas townsite burned to the ground (except for one building) when it was only a few months old. It was quickly rebuilt, and the likelihood of a catastrophic fire was reduced when water lines were laid and buildings began to be built of concrete or stone.

Other fires were more limited but still had an effect on Las Vegas. Many of the most spectacular were at hotel-casinos. The El Rancho Vegas, the first hotel on what would become the Strip, burned down in 1960. The hotel was not rebuilt, and its site at the southwest corner of Sahara Avenue and the Strip remained vacant until it became the Las Vegas Festival Grounds in 2015. Other fires have broken out from time to time, including one at the Sahara in 1964, Aladdin in 2003, Monte Carlo in 2008, Excalibur in 2014, on the twenty-sixth floor of the Bellagio in 2008, and in 2015 near the pool of the Cosmopolitan, but with no deaths because of any of these. The semi-abandoned Moulin Rouge hotel also suffered several fires in 2003 and 2009. The Alpine Motel downtown was occupied when a fire broke out in 2019, killing six. The Stratosphere tower looming over Las Vegas was still under construction and only 500 feet tall when it suffered a fire in 1993. The hotel had previously been Vegas World and had suffered a fire in 1981. Before opening this casino, owner Bob Stupak had operated one nearby before it burned down in 1974. The first two fires were considered suspicious, but no definite conclusions were ever reached.

The deadliest fire in Las Vegas's history was at the MGM Grand hotel (now Bally's) on November 21, 1980. A small electrical fire broke out about seven o'clock in the morning in the Deli restaurant on the ground floor in the casino. Because only a few areas of the resort had sprinklers, the fire spread through the casino quickly, incinerating everything at 2,800°F, and a fireball shot out the front entrance of the hotel. Smoke was sucked up the elevator shafts by a natural draft and onto each floor of the high-rise tower. As many as eight thousand people were in the hotel and casino at the time. The building's fire alarm never went off, and there were no smoke detectors. Guests in some rooms woke up to smoke and shouts in the hallways, while others learned of the fire while watching TV news. Others never woke up at all and died in their sleep from asphyxiation and toxic fumes.

Many guests were trapped in their rooms by smoke and fumes. The longest ladders available to the fire department could only reach the lowest nine stories. Some on higher floors tried climbing down from their windows using sheets knotted together. Others jumped. Hundreds of others fled their rooms through the hallways to find their way out of the building any way they could. Because of the thick smoke coming up the stairs, many went up until they reached the roof of the hotel where they were trapped.

Nine military helicopters from Nellis Air Force Base and ten police and civilian helicopters arrived to rescue those trapped on the rooftop. The majority were lifted off by air force CH-3E cargo helicopters flown by the 1st Special Operations Wing, visiting Las Vegas for a Red Flag exercise (map 9.13). These also used their rescue winches to pluck people from their hotel room windows or off balconies too high to reach by ladder. Helicopters flew in a loop from the rooftop to a parking lot south of the hotel to let off their passengers, then back up to the roof. Up to a thousand people may have been rescued from the burning hotel this way. Some helicopters flew this loop continually for three hours, eventually bringing down bodies as the hotel was searched room by room.

Eighty-five people died, and 650 people were injured. The vast majority of the deaths were caused by smoke inhalation and carbon monoxide poisoning. The fire was fortunately stopped shortly before it entered the high-rise tower, which would have vastly increased the death toll. Only eighteen died on the casino floor; lady luck played favorites, and some floors filled with toxic smoke more rapidly than others.

The hotel had numerous fire code violations, including a ventilation system that had been altered and allowed smoke to spread throughout the building. The casino was rebuilt and the hotel reopened, with great attention given to improved fire safety. It was bought by Bally's in 1986 and renamed, and a new MGM Grand was built elsewhere on the Strip; the hotel tower where sixty-seven died is now Bally's Jubilee Tower.

A mere ninety days after the MGM, fire locals were stunned by a fire that broke out at the Las Vegas Hilton (now the Westgate Resort). This fire, on February 10, 1981, was on the twenty-second floor. Eight people died, and two hundred were injured; air force helicopters again rescued people off the roof. Unlike the MGM Grand, this fire was intentionally set by a hotel employee. The arsonist was caught and is serving eight life sentences. The state legislature was spurred into action to enact strict fire safety requirements for buildings more than fifty-five feet high.

Another horrific Las Vegas fire was one near the BMI plant in Henderson. This World War II plant had been built to process magnesium, a variety of industries occupied the site and surrounding land after the war. One of these was the Pacific Engineering and Production Company of Nevada (PEPCON) chemical plant, which manufactured ammonium perchlorate oxidizer for the space shuttle's solid rocket boosters. After the loss of the space shuttle *Challenger* in January 1986 and the temporary end of launches, rocket fuel oxidizer was stored at the plants where these were manufactured, including Henderson. For undetermined reasons, a fire broke out at the PEPCON plant on May 4, 1988, and could not be contained. The employees evacuated, and when the fire reached the stored oxidizer, a series of explosions took place. The largest explosion was estimated to be the equivalent of a magnitude 3.5 earthquake or a one-kiloton nuclear bomb.

Two employees were killed in the blast, 372 were injured, and most buildings and vehicles within a 1.5-mile radius were destroyed or severely damaged. Some buildings as many as ten miles from the plant were damaged, a radius that covers places as far away as downtown. The ammonium perchlorate plant never reopened; a replacement plant was built near Cedar City, Utah, but it too suffered a fatal fire and explosion in 1997.

Many in Las Vegas are still nostalgic for the days of atomic testing when a mushroom cloud could be seen on the horizon; having such a blast within the city is not nearly so appealing. The city was extremely fortunate because much of the valley had not yet been developed and few people lived near the plant. The population density near the BMI plant is now much higher, and a one-kiloton blast there would have far worse consequences today.

Consolidated Power and Telephone, using a small generator, started commercial electricity service in Las Vegas in 1906. More generators were added, and service was expanded until 1914 when the company shut them down and bought all its electricity from the railroad's power plant. Early service was limited, and electricity was not available twenty-four hours a day until 1915. The town's power company became Southern Nevada Power Company in 1929 and is known today as NV Energy. Many more power plants were built to service the city's growing population and casinos.

Where does Las Vegas's electricity come from today? Most of its electricity is generated at nearby plants, mainly along Las Vegas Wash or to the northeast in Apex Valley. The city's power supply is similar to that of many other western cities, with coal being phased out, natural gas dominant, and solar becoming important. The oldest plant is the Edward Clark Generating Station in 1954, located on Stephanie Street in unincorporated Clark County just north of Henderson. The Sun Peak and Clark stations are on the east side of the valley, near the wash. The Harry Allen, Silverhawk, and Chuck Lenzie plants are all near Apex. An innovative power plant opened in 2012 near the Apex Landfill, burning methane released from the decomposition of garbage. Two natural gas power plants are to the southwest of the city. One of these, the Goodsprings Waste Heat Recovery Plant,

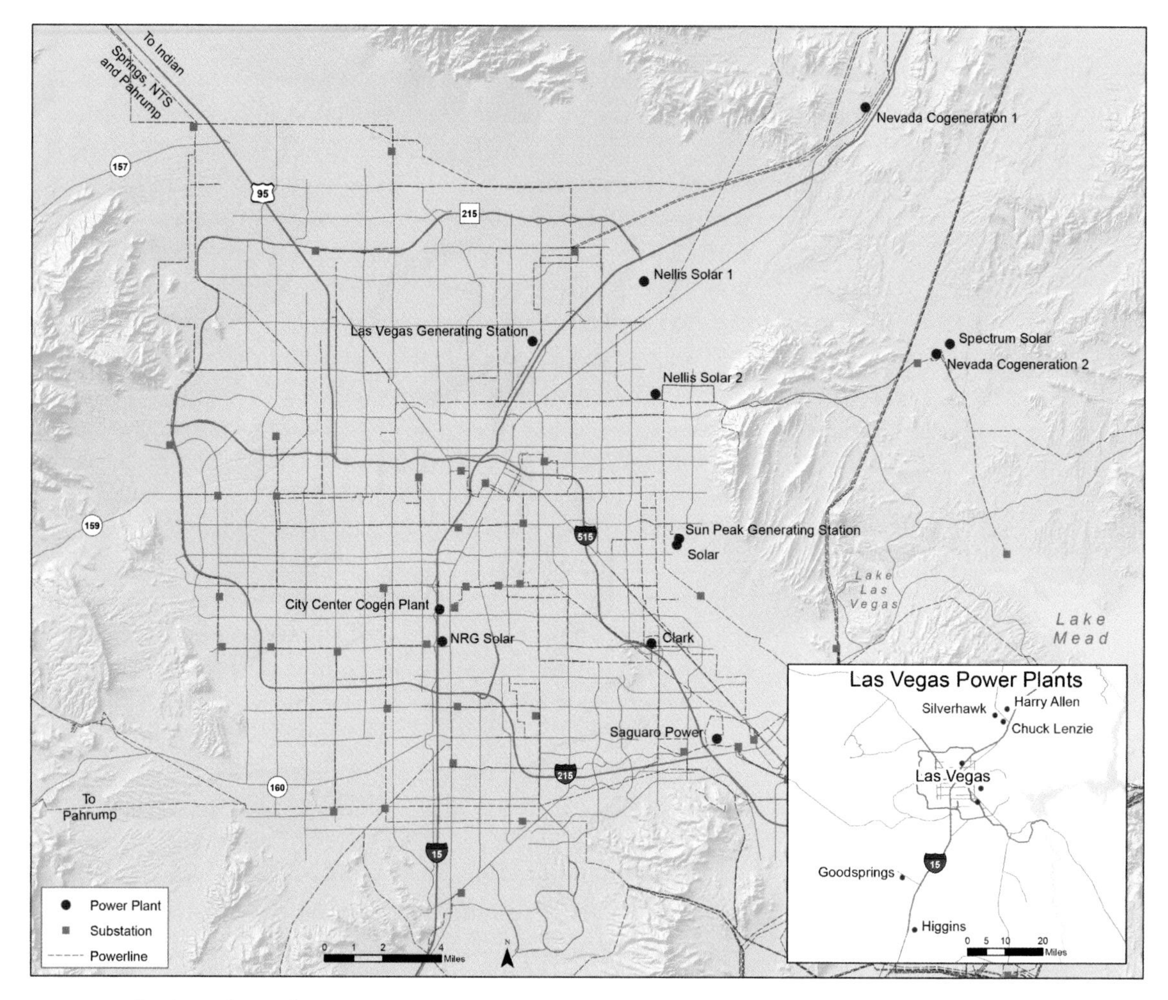

Map 8.16. Electricity. Power plants, major powerlines, and substations keep Las Vegas running. *Inset*: many of the city's power plants are outside the valley.

uses the heat from the Kern River pipeline pump compressors to generate electricity (map 8.18), while the Walter Higgins plant in Primm burns gas for the same result.

A few casinos operate their own power plants onsite to provide some or all of their power needs. These are cogeneration plants that generate heat, hot water, and electricity for the property. One of these is located at City Center, in the southwest corner of the parking deck; another is at the Rio to the north of the hotel tower. Solar power is also becoming more common in the city. Mandalay Bay has a rooftop solar array to provide for the hotel-casino's needs while Nellis Air Force Base has two solar power arrays that went online in 2007 and 2016. Not part of a commercial power system, they supply much of the base's electricity needs. Electrical self-sufficiency is an important Pentagon goal to ensure that the military's ability to function is not compromised by problems with the nation's commercial electricity system.

NV Energy has operated two coal-burning power plants outside the valley. The Reid Gardner plant opened in 1965 near

Moapa and was Southern Nevada's only coal-burning power plant, requiring regular coal trains from mines in Utah, Colorado, and Wyoming. It was shut down in March 2017. NV Energy also owned an 11 percent share of the Navajo Generating Station near Page, Arizona. This power plant was built after plans to build two electricity-generating dams in the Grand Canyon as part of the Pacific Southwest Water Project were canceled (map 10.6). The Navajo plant burned coal mined on the Navajo Indian Reservation and delivered by a 75-mile long, remote-controlled railroad. The power plant began operating in 1974 with electricity sent to Las Vegas via a power line built through southern Utah. Electricity was also sent to Phoenix, Tucson, and formerly to Los Angeles from this plant. Coal plants such as Navajo or the Reid Gardner plant are more expensive to operate than natural gas plants and are also much more polluting. The Navajo plant was shut down in 2019.

No mention has been made of Hoover Dam. Despite frequent statements to the contrary, Las Vegas gets very little electricity from Hoover Dam, and the dam does not supply any electricity to the Las Vegas Strip. The simple reason? The Strip did not exist and Las Vegas was only a small town when the dam was planned and its power output allocated.

Hoover Dam is the largest single source of electricity in Southern Nevada, but very little of it remains within the state. The Western Area Power Administration allocates the dam's electricity according to a law passed by Congress in 1934, updated in 1984, and again by the Hoover Power Allocation Act of 2011. This act governs power distribution until 2067. According to current allocations, 28.5 percent of the power generated by the dam is sold to the Metropolitan Water District of Southern California, while the cities of Anaheim, Azusa, Banning, Burbank, Colton, Glendale, Los Angeles, Pasadena, Riverside, and Vernon together receive another 21.8 percent. Southern California Edison, the largest power company in the area, also receives 5.5 percent. About 19 percent goes to Arizona, where the Arizona Power Authority divides it among thirty customers, most of them irrigation districts that use the power to pump water. The remaining 23.4 percent is assigned to Nevada and allocated by the Colorado River Commission of Nevada. Most of this power is sold to the Las Vegas Valley Water District, Clark County Water Reclamation District, Las Vegas, North Las Vegas and Henderson to be used for pumping water, with the remainder sold to Boulder City and the Black Mountain Industrial Center (formerly BMI). Needless to say, none goes to the Strip.

A network of long-distance power lines was created to distribute this power. The first of these was completed in 1931 to

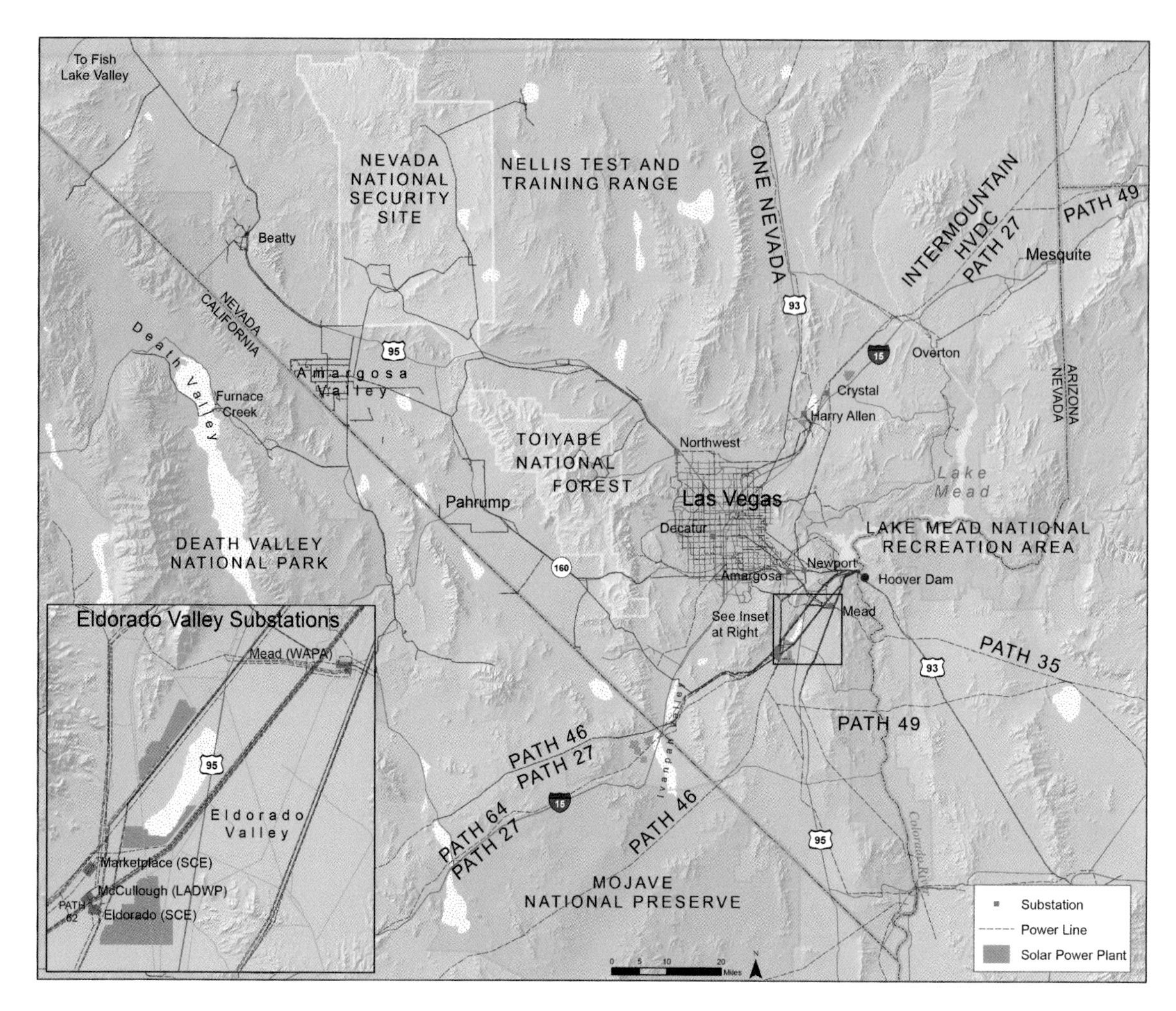

Map 8.17. Hoover Dam and Valley Electric. Hoover Dam is part of a regional electricity-distribution system, connecting it and many other major power plants to large cities throughout the West. *Inset*: major substations associated with Hoover Dam, along with newer solar power plants, in the Eldorado Valley south of Boulder City.

bring power to Nevada when a 222-mile power line was completed from San Bernardino, California, to the dam construction site and Boulder City. Once the dam's first generator started in 1936, the flow of power was reversed. Many more power lines have been built since to send ever more electricity to Southern California and later to central Arizona. Beginning in the late 1960s, this power infrastructure was rebuilt and several massive substations were constructed in the northern Eldorado Valley. Eight transmission lines connect the dam's output to the Mead substation south of Boulder City. From here several lines deliver power to local customers for water pumping, to the Black Mountain Industrial Center in Henderson, central Arizona, and to three nearby substations. The Marketplace, McCullough, and Eldorado substations are owned by Southern California Edison and the Los Angeles Department of Water and Power, the primary recipient of Hoover Dam electricity. Power flows from these substations to various points in Southern California.

The electric transmission system of the western United States is operated as a single system by the Western Electricity

Power lines in Eldorado Valley, 2003. Courtesy of the Historical American Engineering Record. Photo by Douglas Edwards.

Coordinating Council or WECC. Major electricity transmission lines within this grid are grouped into routes known as paths. Path 46 includes several parallel power lines between the Mead substation and Victorville, while other components of Path 46 run from Davis Dam to Los Angeles and from destinations in central Arizona to Southern California. Path 35 connects Mead substation with Phoenix. Path 62 provides a connection between the Eldorado and McCullough substations while Path 58 connects Mead and Eldorado. Path 49 included several lines. One brought power to Las Vegas from the Navajo Generating Station in Arizona to the Crystal substation. Another ran from that same power plant to the Eldorado substation, where power was sent on to Los Angeles.

The One Nevada powerline was completed in 2014 and runs 235 miles from Ely to the Harry Allen substation (adjacent to the Harry Allen power plant) north of Las Vegas. Several major powerlines connect the Crystal and Harry Allen substations in Dry Lake Valley with the substations in Eldorado Valley. These lines run east of Frenchman Mountain before skirting the east side of Henderson and Dutchman Pass. Not all power lines have local connections. Path 27 is a high-voltage DC power line that delivers power from a coal-burning power plant in Delta, Utah, to Southern California with no local connections. This was built in the 1980s and runs alongside the earlier power lines.

Solar power is a growing industry in Southern Nevada, with several large photovoltaic facilities around the city. The Nevada Solar One facility in Eldorado Valley began operating in 2007. The plant is owned by a Spanish company, which sells all of the electricity produced to NV Energy. It is adjacent to the Marketplace, McCullough, and Eldorado substations and connected to each. The Ivanpah Solar Power Generating Facility was completed in 2014 across the state line from Primm and adjacent to Interstate 15 and the lines of Path 27. All of the electricity from this plant goes to Southern California. Several plants are also found in Dry Lake Valley, connected to Crystal and Harry Allen substations.

The original 1931 power line that first connected Las Vegas to the outside world was once one of the longest power lines in existence. Building the 222-mile line in 225 days across the Mojave Desert was an enormous undertaking. The route did not follow US 91 or the Union Pacific Railroad and required that new roads be built for construction. Mules had to be used to haul equipment and supplies to the tops of some mountains. This transmission line still exists but has been rebuilt with new towers and was rerouted to pass through Mead and McCullough substations. Remnants of a telephone line that once paralleled the

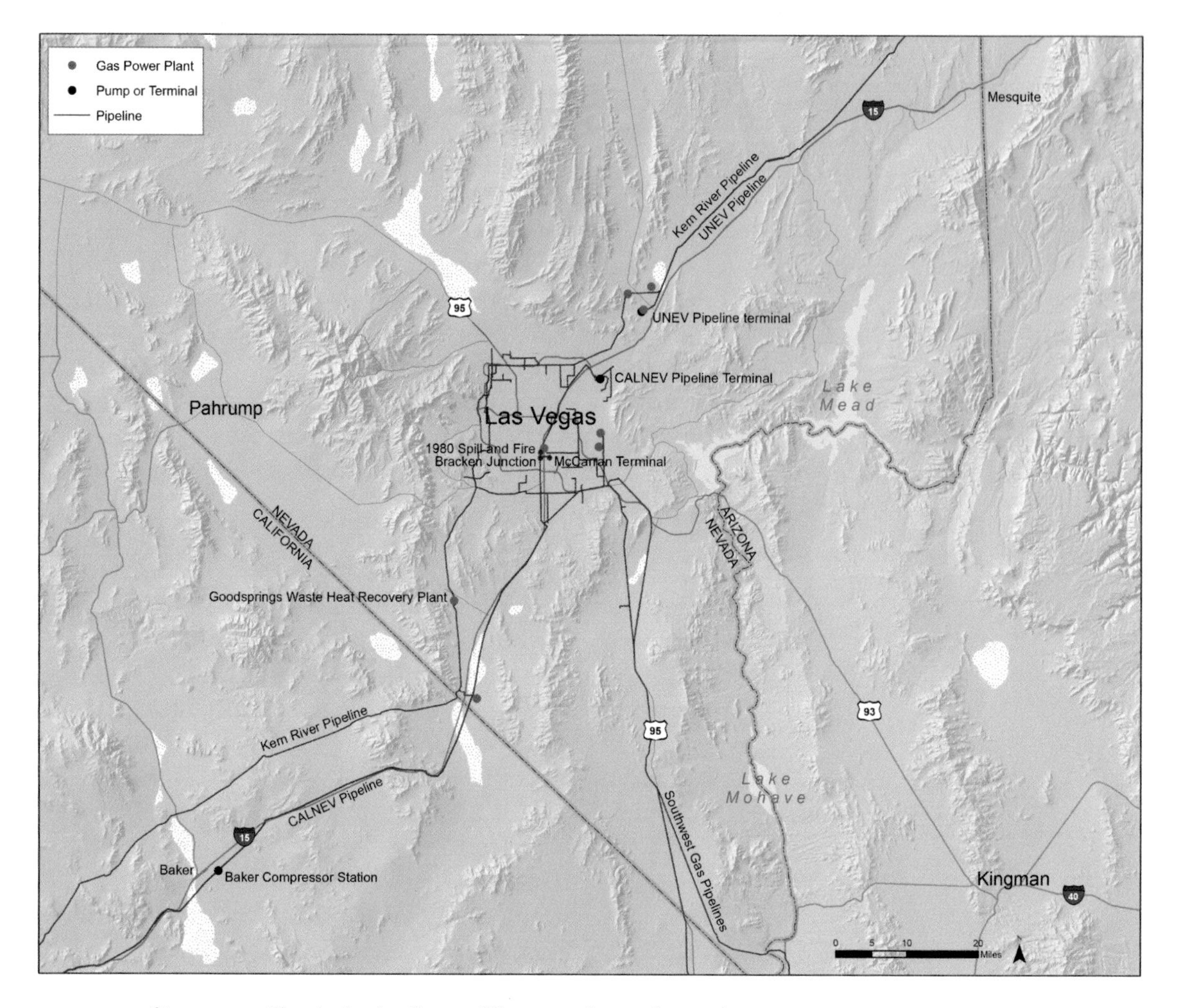

power line may still exist in the form of five wooden poles in the McCullough Mountains, all that remains of the original 1931 line.

Other power lines were added to serve growing rural areas outside the city. The Valley Electric Association (VEA) was created in 1965 to provide power to Pahrump, Amargosa Valley, Fish Lake Valley, and later (via Southern California Edison) to Shoshone, Tecopa, and Death Valley in California, with the electricity coming from Hoover Dam. It also provides electricity to the Nevada Test Site and Groom Lake (Area 51). The growth of Pahrump (map 4.18) required a second powerline in 1996 from Mead substation and a third line in 2013 connecting to NV Energy at the Northwest substation.

Drivers in Clark County used three million gallons of gasoline each day in 2006, with McCarran International Airport using another 1.27 million gallons of jet fuel every day. Vast quantities of natural gas are burned in the city's power plants (map 8.16) and in homes and businesses. These fuels must come from somewhere. Gasoline was brought to Las Vegas by railroad or truck

Map 8.18. Pipelines. Hidden out of sight underground are a number of gasoline and natural gas pipelines supplying the fuel to keep the city running.

in the early days, and today two pipelines carry gasoline, diesel, and jet fuel to the city, while others bring natural gas.

The first of these was the Calnev gasoline line in 1970, running from Los Angeles up Interstate 15 to North Las Vegas, where a terminal is located near Nellis Air Force Base. A branch continues past the terminal to supply Nellis with jet fuel, and another branch line runs from a point called Bracken Junction to McCarran. The line is now made up of two pipes, one fourteen inches in diameter and the second eight inches. The lines carry gasoline, diesel, and jet fuel at different times, with any mixing separated at the terminal. Fuel trucks fill up here and then distribute gasoline to local filling stations. To ensure a reliable supply, a second gasoline pipeline was completed from Utah in 2012. This UNEV pipeline is a 399-mile long, twelve-inch diameter pipe running from a refinery at Woods Cross, Utah, to a terminal at Apex.

Several pipelines bring natural gas to Las Vegas for heating and electricity production. Southwest Gas built the first line from Needles, California, in 1954, where it connects to a major gas pipeline connecting Texas and the West Coast. The Kern River pipeline carries natural gas from Wyoming to Las Vegas and Southern California. There are two parallel pipes, the first finished in 1992 and the second in 2003. Most of the power plants in the Las Vegas Valley run on natural gas. Several are located on the east side of the valley, others near Apex, and two near Primm, all near major gas lines. The Goodsprings Energy Recovery Station uses the heat from the Kern River pipeline pump compressors to generate electricity.

Pipelines are a safe way to transport liquids, but accidents happen. On December 22, 1980, work on the Calnev gasoline pipeline caused jet fuel to spill onto Tropicana Avenue and ignite into a massive blaze. On November 21, 2004, a leak on this same line along Interstate 15 between Baker and Barstow shut the freeway down until the spill could be contained. The worst event was on May 25, 1989, when construction work following a train derailment in San Bernardino caused the Calnev pipeline to explode and burn for seven hours, killing two people and destroying eleven homes.

The Kern River natural gas pipeline passes through the west side of the valley and underneath city streets in Summerlin. It has so far had no troubles, but a similar urban pipeline in San Bruno, California, exploded in 2010, killing eight people, injuring fifty-eight, and destroying thirty-five houses. It took two hundred firefighters about eighteen hours to put out the resulting fire, and the blast left a forty-foot-deep crater.

Aside from agriculture, mining is the oldest legal profession in the Las Vegas area, having been practiced for more than a

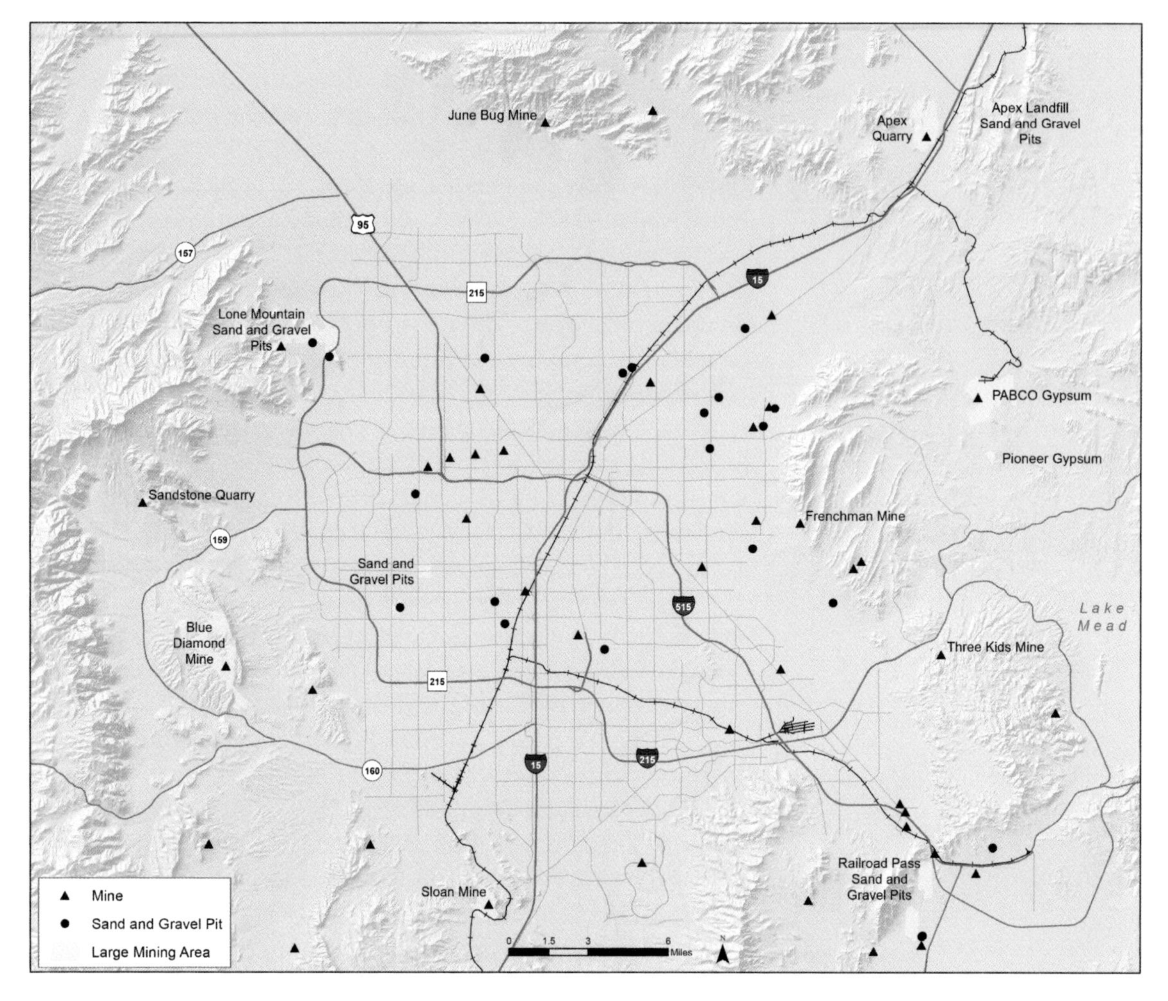

Map 8.19. Mining. This is one of the oldest activities in Las Vegas, though most mines have long since shut down. The Blue Diamond, Sloan, and PABCO mines continue to operate, as do several gravel pits used to supply the city with its building materials.

thousand years at a salt mine along the Virgin River (map 5.1). Jedediah Smith called this the Big Cliff salt mine in 1826, while others called it Salt Mountain. The salt extended along the bluffs for several hundred feet with tunnels dug into the cliff. It was intermittently mined commercially in the twentieth century before being flooded by Lake Mead.

A new mining era began in August 1856 when Mormon pioneers from the Las Vegas Fort (map 3.7) founded a mining camp near Potosi Mountain. They extracted lead ore and smelted it to make bullets. Despite difficult working conditions, nine thousand pounds of lead were shipped out before the camp was abandoned in 1857. Others worked the mines in later years, most prominently from 1913 to 1917 when the mine was an important zinc producer.

The arrival of the railroad lowered the transportation cost of ore and opened up many possibilities for mining. This was especially important in the Las Vegas area because it is rich in heavy materials such as gypsum, stone, and similar materials. A sandstone quarry was opened in Red Rock Canyon after the railroad arrived in 1905. A steam traction engine pulled wagons loaded with cut stone to Las Vegas for shipping to San Francisco by train. Even with these machines, the quarry only operated for about a year because of high shipping costs. It reopened in 1910 but again lasted only a few years. The road built became a county road, and part of it eventually evolved into Flamingo Road. The quarry site can still be seen along the Scenic Drive in Red Rock Canyon (map 9.6).

Borax prospectors located two extensive depots to the west of the Virgin Valley in 1920. Francis Marion Smith, the "Borax King" who had created profitable borax mines in Calico, Death Valley, and Boron before losing control of his business empire, quickly bought the depots. He used these new mines to generate enough money to found a second borax empire at Trona, California. No borax mining, however, has taken place in the Las Vegas area since 1928.

The Three Kids Mine has been a source of manganese metal east of Henderson. This metal is used to produce stainless steel and some aluminum alloys. The mine was first worked in 1917, again during World War II when it supplied manganese ore for the war effort to be processed at the BMI plant in Henderson, and again on and off up to 1961.

The main products now being mined in the area are gypsum and limestone, at five major mines, along with several smaller sand and gravel deposits. The Blue Diamond Mine near the town of Blue Diamond has been active since 1902. The present mine started in 1924 when the standard-gauge railroad line was built (map 7.3), along with an aerial tramway to connect the mill at the end of the tracks with the mine on top of Blue Diamond Hill. The mine expanded greatly in the early 1940s, along with a large gypsum mill and drywall plant built in 1941, leading to the need for more housing. The company town of Blue Diamond was created to supply this at Cottonwood Springs.

A change in company ownership brought a different perspective on the need for a company town, and it was privatized beginning in 1959 when residents were allowed to purchase their homes. The mine closed in 2004, but the mill and drywall factory are still active. Although there is considerably more gypsum left in the mountain, the current owner is attempting to build a large residential area on former mine property on top of Blue Diamond Hill (map 10.2).

The Sloan quarry started in 1914 with limestone, dolomite, and calcite mined. Much of the output is used to make asphalt or concrete for the city's growth. Although the mine is adjacent to the Union Pacific Railroad, the entire product is shipped out by truck. Residential growth has spread to within 1.25 miles of the mine, but it is not visible from the north side of the hills. A new but similar open pit mine was proposed several miles to the southeast but was rejected by the federal government in 2013.

The Apex quarry, operated by Lhoist North America, has 165 employees and is the biggest in the area. It produces limestone and dolomite. The PABCO Gypsum mine started in 1964 and includes one of the biggest drywall factories in the country. Drywall from the mill is used in Las Vegas and transported to California, Arizona, Utah, and Washington by train. The mine and mill obtains water by pipeline from Lake Mead, and the mill has its own electrical generating plant, built in 1994. To the south of the PABCO mine is the much smaller Pioneer Gypsum mine.

The Simplot silica sand mine in Overton has operated since at least 1969, with the sand used for a variety of industrial applications in the production of glass, electronics, steel production, and oil fracking. The sand is shipped out by rail on the Mead Lake branch of the Union Pacific Railroad, one of the few remaining customers on that line.

Many other sand and gravel pits are on the periphery of the city. Several are near Lone Mountain, including a pit about 270 feet deep. This operation is very apparent on a road map of Las Vegas; the beltway makes a conspicuous bend around it north of Lone Mountain. Other sand and gravel operations include several deep pits near Buffalo Drive (one 170 feet deep and another 225, both pits bounded by busy streets and surrounded by residential neighborhoods). It is interesting to see what future land uses will be devised once these pits have been dug as deeply as they can be. Sand and gravel operations are also farther out at Railroad Pass and at the Apex landfill. At the latter location, the city's garbage (map 8.20) is dumped into the pits excavated for sand and gravel. This sand and gravel is used in local concrete production, and much of Las Vegas has been built out of it. Much of the gypsum mined in the area is also used locally in the form of drywall for houses.

Where does the city's trash go? All garbage is delivered to the 2,200-acre Apex Landfill off Interstate 15 northeast of the city, where it is buried in deep pits. The landfill, opened in 1993, is the biggest in the country. The volume of daily trash dumped peaked in 2006 and has declined since then, partly because of recycling. At current rates, it will not be filled until 2263. Sludge from the city's sewage treatment plants (map 6.7) is also deposited here, reducing the amount of sewage flowing into Lake Mead. This trash and sludge are put to use: methane released from the decomposition of garbage is used to run a power plant, with the electricity going into NV Energy's distribution system.

View of Las Vegas from Sunrise Mountain landfill, 1998. Courtesy of the Environmental Protection Agency.

Where did Las Vegans dump their trash before 1993? The

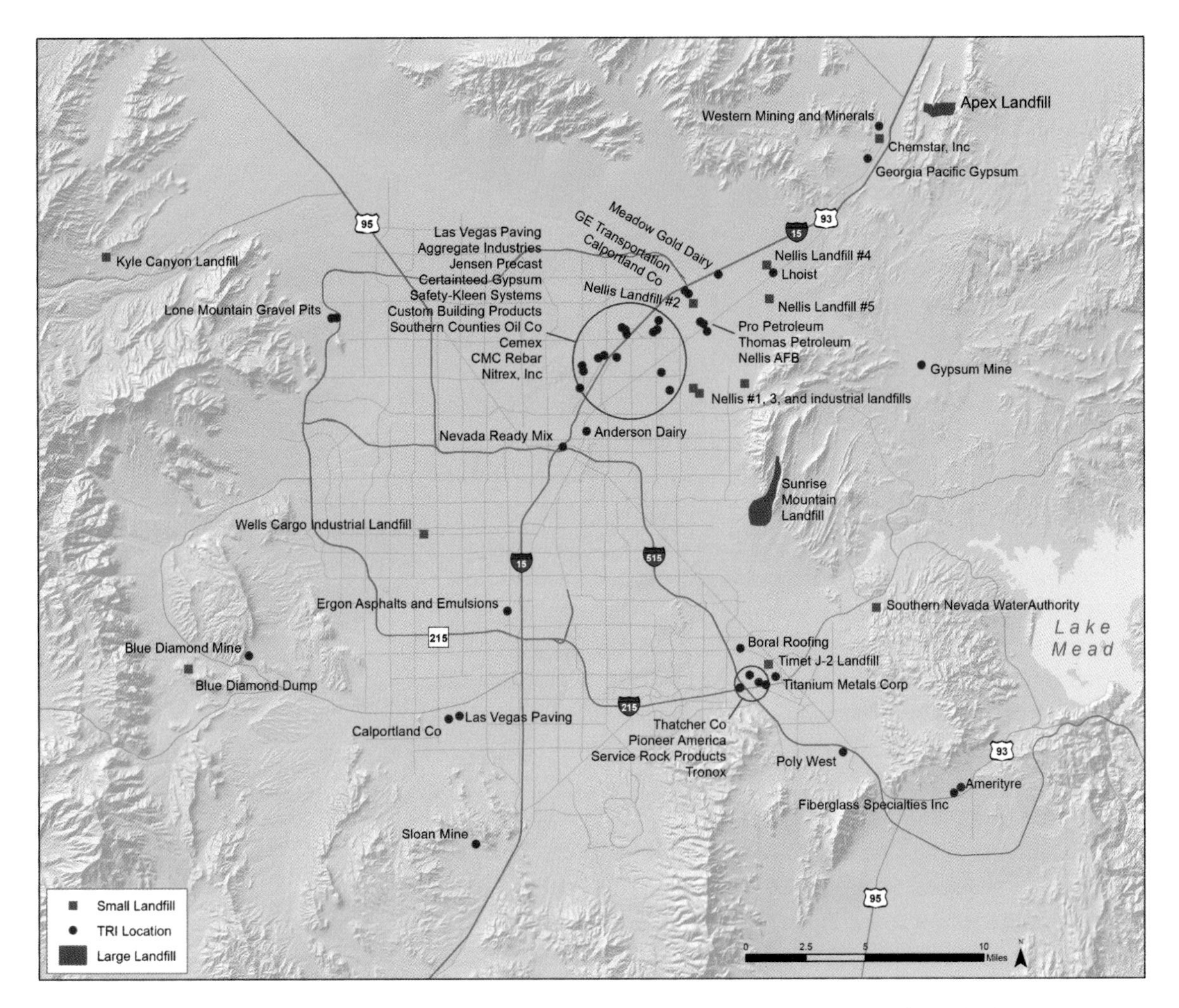

Map 8.20. Garbage and Pollution. The waste generated in Las Vegas has been taken to large landfills. Some of the waste is considered toxic and listed in the EPA's Toxic Release Inventory. Sources of toxic waste are shown in red. Locations of heavy concentrations of toxic waste are denoted with red circles west of Nellis Air Force Base and at the BMI plant in Henderson.

Sunrise Mountain Landfill opened in 1953 and reached 440 acres before it was shut down. The lower end of this was in plain view of much of the city and passengers arriving at Harry Reid International Airport; it has since been covered and landscaped and is no longer so visible. Unfortunately even after it was closed it was still vulnerable to flooding, and a 1998 flood washed towns of trash into Las Vegas Wash. The Environmental Protection Agency (EPA) oversaw a program to clean up this mess and prevent future occurrences. The towns of Blue Diamond and Mount Charleston have their own landfills, and other landfills are scattered, most of them handling construction debris or other specialized materials.

Although Las Vegas is not an industrial city, it does have many sources of pollution. The EPA keeps track of these sources through its Toxic Release Inventory (TRI), a program began in 1986 that requires any business or agency that generates any of 770 chemicals to report how much is generated. These could be released into the air, groundwater or waste water, or dumped in a landfill. These chemicals are those known to cause cancer or other harm to humans and include many metals (such as arsenic, lead, copper, mercury, selenium) as well as hydrocarbons and many other substances.

There are forty-seven TRI sites in Clark County, including the region's active mines, asphalt plants, industrial facilities, and several dairies (where nitric acid is often used for cleaning, resulting in nitrate compounds dumped into wastewater). The greatest concentration is at the Black Mountain Industrial Center in Henderson and in North Las Vegas along the railroad tracks and Interstate 15. Reports for each can be found at the EPA website.

Several tenants have polluted the Black Mountain Industrial Center through the decades. The soil and groundwater around it, as well as groundwater in Las Vegas Wash, has been contaminated with many hazardous chemicals. Extensive and expensive efforts to clean up the site, the evaporation ponds, and groundwater around the site, have been underway. Pollution from the site is sometimes very obvious. Titanium Metals Corp (TIMET), released chlorine gas in 1996 and 2012, smelled by residents up to seven miles away and sickening nearby schoolchildren. TIMET was also fined almost $14 million in 2014 for illegally producing and disposing of toxic polychlorinated biphenyl compounds (PCBs).

A form of pollution that few even notice is light pollution, the result of city lights creating so much illumination that only the brightest stars and planets can be seen at night. Needless to say, the bright lights of the Strip are one of the planet's biggest source of light pollution. It is doubtful that many Las Vegans consider this a bad thing, but other cities have enacted ordinances to protect their night skies. Starry skies and views of stars, the Milky Way, and other phenomena largely unknown to city people can still be enjoyed a few hours away at Death Valley National Park.

Harry Reid International Airport has connections to many cities around the world and is conveniently located, but living near the airport has disadvantages, one of them being noise. Planes arriving and departing from the airport subject residents underneath them to regular noise exposure that can have long-term health impacts.

The city's first airports were miles outside town and had few flights by relatively quiet piston planes. The arrival of jet airplane service in the 1960s raised the noise levels tremendously, but the city had not yet surrounded the airport. Improved turbofan engines replaced the earlier turbojets to provide the greater power needed for widebody jets and had desirable side effects of reduced fuel consumption, pollution, and noise. The Federal Aviation Administration (FAA) began banning the remaining older and noisier planes from US skies in the 1990s, spelling the

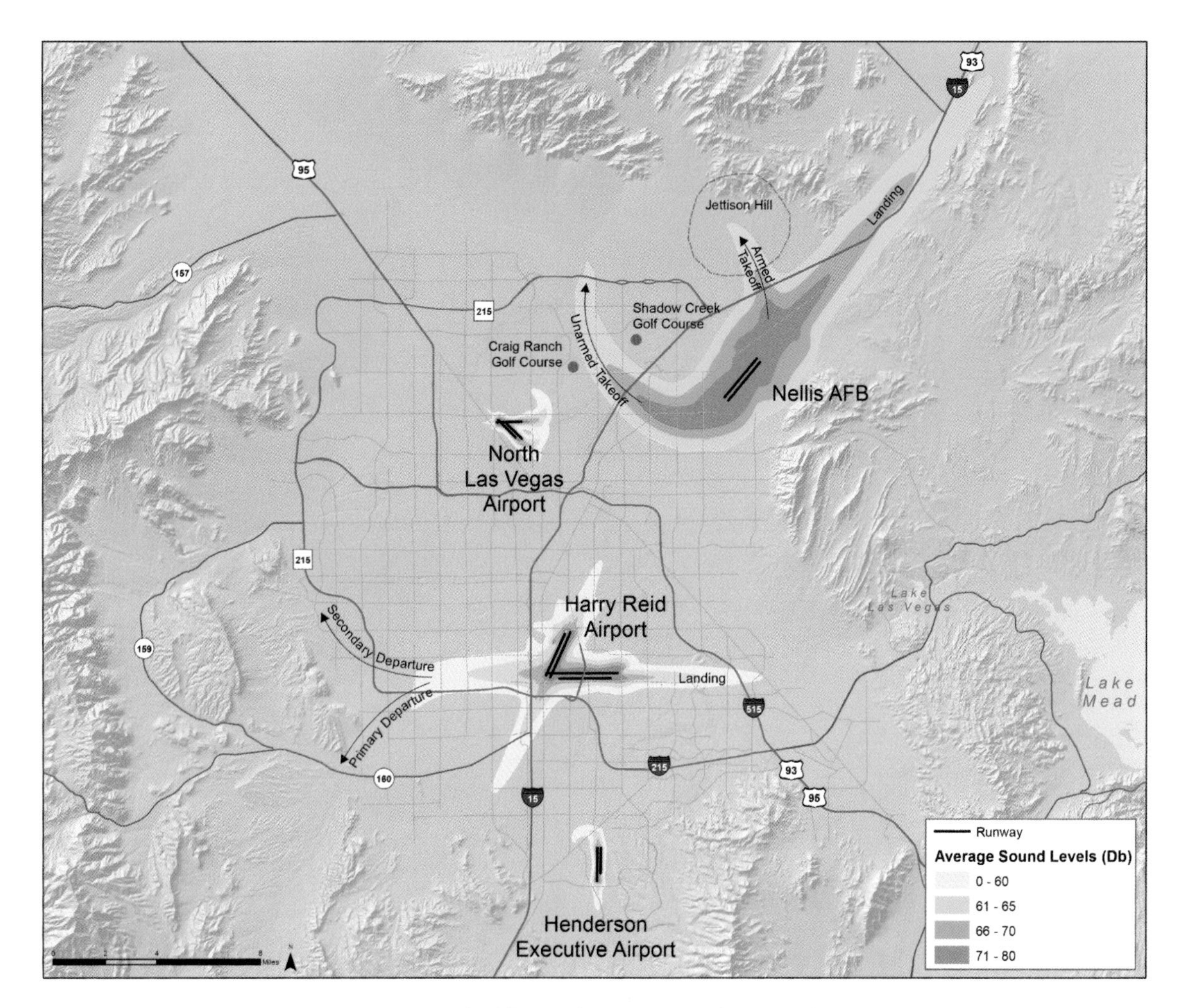

Map 8.21. Airport Noise. A busy airport means a lot of noise. Military planes carrying live weapons take off to the northeast; those unarmed take off toward Las Vegas but must immediately take a sharp right turn and fly between two golf courses on their way out of Las Vegas.

end of formerly common planes such as 727s. American airports have become far quieter than in the 1960s.

Yet complaints about noise are increasing in Las Vegas and many other cities. The explanation in most cases is urban growth, as more and more people are living closer and closer to airports. This is certainly the case in Las Vegas, where residential development now surrounds Harry Reid airport. Different solutions have been found for airport noise across the country, including soundproofing homes, restricting airport operation hours or flight directions, and even demolishing entire neighborhoods near airports. Although several neighborhoods were demolished decades ago for airport expansion (map 7.18), none of these options seems to have been tried for noise in Las Vegas. Most land west of Harry Reid is however owned by Clark County, which may reduce development under the flight path.

Noise patterns around airports can be mapped out, showing decibel levels that are experienced at different distances from the runway. Although aircraft landing are slightly quieter than when taking off, they are lower for a longer period and may affect a larger area than when taking off. This is apparent on the noise map, as the noise levels extend farther to the east than west. Seventy-five percent of traffic departing Harry Reid airport is to the west, but many flights are headed to destinations east of the city. In the past these would make a left turn to the south before turning east, visible on the noise map, but to increase the airport's capacity a third of these turns were made to the north beginning in 2007. This led to complaints from residents of Summerlin who suddenly found themselves under these flight paths. Henderson and North Las Vegas airports have much smaller noise footprints than Harry Reid, and they also have curving flight paths to avoid congested airspace by the major airports.

The FAA does not regulate military aircraft, and anyone who has been near one knows they are much louder than civilian planes, especially when afterburners are used to boost power. Las Vegans are fortunate that Nellis Air Force Base has been occupied by small fighter planes rather than B-52 bombers; each takeoff of those would be heard throughout the city. The air force is aware of noise issues, and military planes follow special procedures. Planes using Nellis usually take off to the southwest, toward the city, but then must turn off afterburners and make a sharp turn to the northwest to leave the city on a route passing between Shadow Creek and Craig Ranch Golf Courses. In addition to being louder, they may also be carrying live ordnance. Those that are armed must avoid overflying the city by taking off to the northeast before turning left toward the Nellis range.

Many companies in Las Vegas operate flightseeing services to Lake Mead and the Grand Canyon, sometimes with stops at the Grand Canyon West Airport (map 9.9). While providing spectacular views from above, it creates noise below. Aircraft noise is a growing problem at the Grand Canyon, though not by takeoffs and landings but by low-flying airplanes and helicopters on scenic tours. The National Park Service, committed to preserving the natural sounds and sights of the Grand Canyon, regulates scenic overflights.

A different kind of airplane noise can occasionally be heard in the northeast side of the city when a plane returns to Nellis with live ordnance that must be dropped before landing. An area known as Jettison Hill north of the Las Vegas Motor Speedway is the preferred place to drop unwanted bombs. Although the bombs are not armed before being jettisoned, they sometimes explode, rattling windows and nerves for many miles around.

9 Outside Las Vegas

This chapter examines the land outside the city, almost all of which is in the public domain. Once available for homesteading, these lands have been assigned to the management of the National Park Service, National Forest Service, Bureau of Land Management, US Fish and Wildlife Service, US Department of Energy, and the US Department of Defense.

National forests were organized near Las Vegas soon after the city was founded. Wildlife refuges were established in the 1930s, followed by the country's first National Park Service site dedicated not to scenic beauty or history but outdoor recreation. Southern Nevada was home to Nevada's first state park in one of the region's two outcropping of colorful sandstone.

Since the late 1960s, a new era in protecting the remaining public lands around the city began. This began west of the city in the spectacular Red Rock area, which was protected by the Bureau of Land Management, an agency formerly known for managing lands of no interest to more prestigious federal agencies. This became the state's first national conservation area. The city's spectacular population boom in the 1990s and 2000s swallowed up much of the remaining open land in the valley as well as that in adjacent valleys. More areas of outstanding scenic or historic importance were protected before the city's growth erased them when another national conservation area, three national monuments, and dozens of wilderness areas were created. Many of these lands are well known to Las Vegas as locations for outdoor activities such as boating, hiking, skiing, water-skiing, river rafting, camping, birdwatching, and hunting.

These lands that encircle Las Vegas also provide a growth boundary for the city. Although the city has such a legal boundary in the form of the Las Vegas Valley Disposal Boundary (chapter 8), it can be expanded to allow more development. These public lands provide a more permanent outer boundary, as is the case for many western cities.

More federal lands are off limits to urban growth, but for different reasons. They were reserved during the national emergency of World War II as military training areas and remained (and were expanded) during the Cold War. They include Nellis Air Force Base, the Nellis Test and Training Range, the Nevada National Security Site, and Area 51. They are entirely off-limits to the public (except under rare circumstances) and what goes on within them is usually unknown. They are quite different from wilderness areas or national monuments, but they are still a vital part of the public land heritage of Southern Nevada.

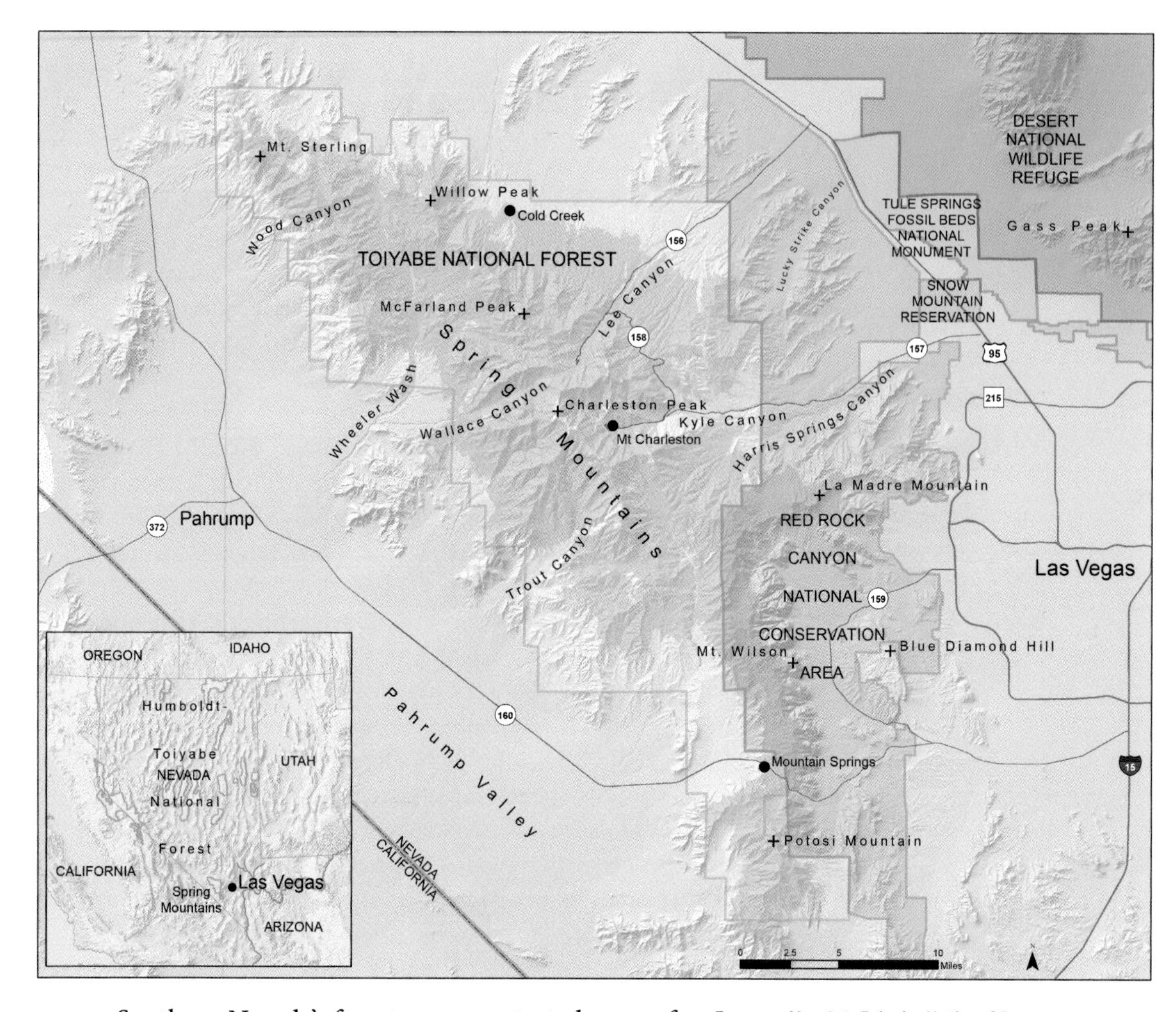

Southern Nevada's forests were protected soon after Las Vegas was founded when President Theodore Roosevelt created the Charleston Forest Reserve in the Spring Mountains and the Vegas National Forest on the Sheep Range (map 4.3). These two forests' names and boundaries changed several times through the years, with the Spring Mountains unit becoming part of the Humboldt-Toiyabe National Forest. This vast forest encompasses fourteen different units across thirteen Nevada and six California counties, and is the largest national forest in the country outside of Alaska. This is because of the basin and range topography of the state, with many narrow north-south mountain ranges throughout the state reaching high enough elevations to be forested (map 2.1).

The importance of recreation to the management of the Spring Mountains was recognized in 1993 when it was designated the Spring Mountains National Recreation Area (SMNRA). Although this is the same designation as Lake Mead, the SMNRA is still administered by the Forest Service and operated largely as before. No commercial logging or grazing is allowed, and the SMNRA is

Map 9.1. Toiyabe National Forest.
Las Vegas's escape from the summer heat is a part of the largest national forest in the country.

managed for hiking, camping, and hunting. It is one of twenty-two such Forest Service–administered national recreation areas.

Most of the recreation development is centered in Lee and Kyle Canyons (map 9.2), but several other canyons have roads that lead up into the mountains. Cold Creek has scattered homes outside the national forest boundary with trail access to the northern end of the mountains. Mountain Springs is a small town on private land within the forest, along the busy Las Vegas–Pahrump highway. Wheeler and Trout Canyons provide trail access to the undeveloped west side of the mountains, and Red Rock Canyon National Conservation Area provides access to the southern end of the range (map 9.6). But the majority of the range has never been developed and remains as it was hundreds of years ago. Aside from agriculture, mining is the oldest legal profession in the Las Vegas area, having been practiced for more than a thousand years at a salt mine along the Virgin River (map 5.1).

Like many recreation areas, the Spring Mountains are threatened by their own popularity. Lee and Kyle Canyons are overwhelmed during holiday weekends, when more than 6,000 vehicles compete for the 1,770 parking spaces. A principal challenge for the US Forest Service is how to accommodate even greater crowds in the future without destroying the natural beauty that draws them.

Another threat is the West's changing climate. As the climate warms, precipitation is forecast to decrease in the summer, though it may increase slightly in the winter. The result, combined with warmer temperatures, would be earlier snowmelts and more intense summer storms. This will put stress on the city's water supplies and flood control system (map 6.8). It will also cause vegetation to move to higher elevations. The tall pines near the top of the mountains may appear only on the higher slopes of Kyle and Lee Canyons, and the tundra above the treeline will shrink or disappear altogether. Like other western forests, wildfires are common in the Spring Mountains. Since 1953, 101 fires have been recorded in the mountains, and their size has been increasing; as the range becomes higher and drier, these forest fires will become more frequent and larger. The challenge for the Forest Service will gradually shift from providing visitor access to preserving the existence of the mountain's forest.

While they were originally valued for their timber, the Spring Mountains more recently became important as a recreation resource, providing relief from the heat in the summer and opportunities to play in the snow in the wintertime. Just as logging required sawmills and roads, recreation also required infrastructure. Many of the recreational facilities we take for granted today were first developed in the 1930s when the Civilian Conservation

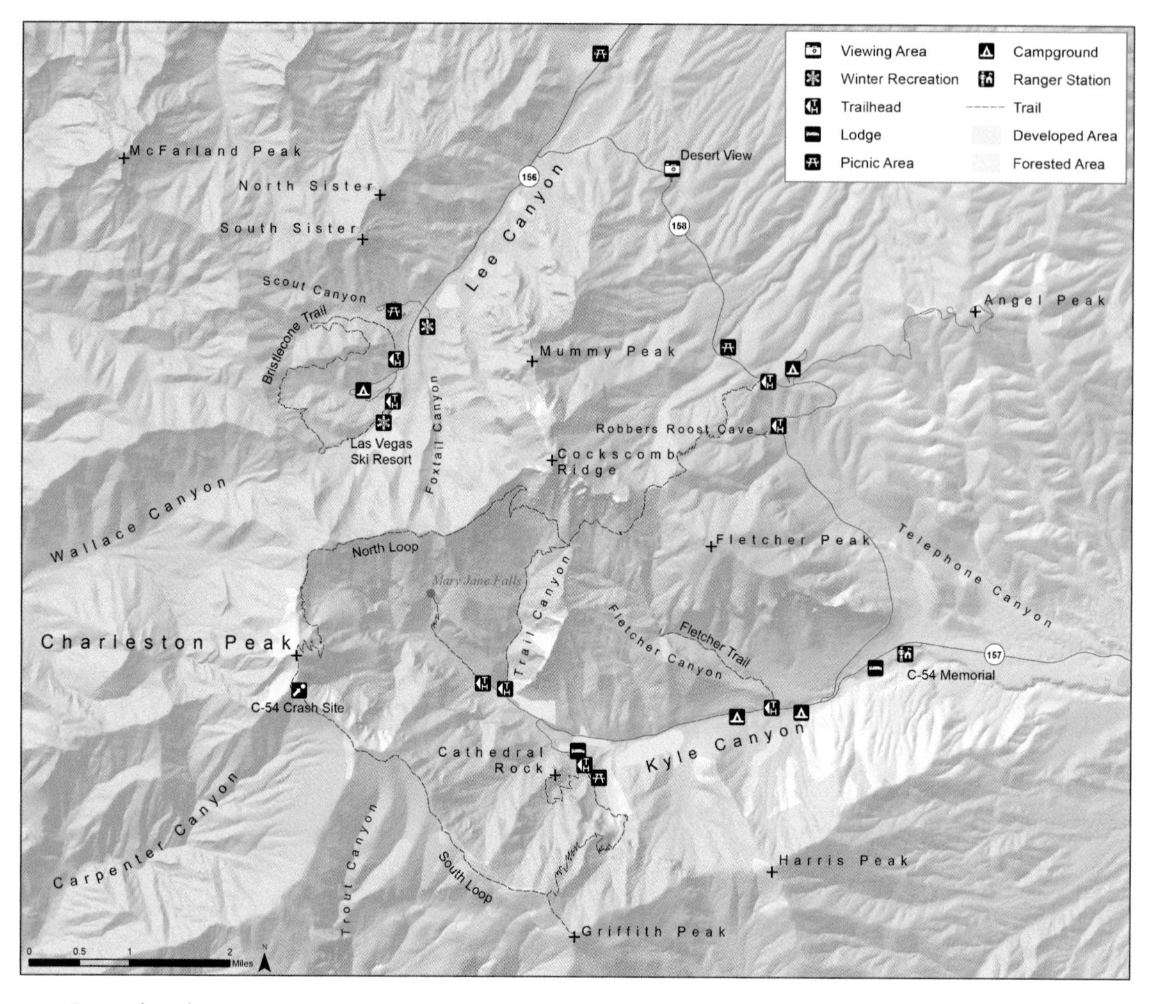

Corps (CCC) was set up to put young men to work on conservation projects (map 4.7).

Map 9.2. Spring Mountains. Kyle and Lee Canyons offer year-round recreation, whether hiking or skiing or just enjoying some peace and quiet.

Several CCC camps were set up in Southern Nevada, including one outside Las Vegas. When the valley floor heated up in the summer, the men were transferred to Camp Charleston Mountain in Kyle Canyon (where the current ranger station is located). This camp operated in summers from 1933 to 1942, and the men assigned here built hiking trails, a ranger station, water systems, and several campgrounds. One of these was where the Cathedral Rock Picnic Area is, another at what is now the Mahogany Grove Group picnic area, and a third is now occupied by the Foxtail snow play area.

The work of these young CCC men one year was almost interrupted by a visit from the president. On September 30, 1935, while President Franklin D. Roosevelt was in Las Vegas, he took the opportunity to visit Camp Charleston Mountain to inspect the work of his CCC. Unfortunately, his group got a late start and had to turn back before reaching the camp. Accounts vary, but it appears that the president's car had to turn around on a narrow section of mountain road and barely avoided tumbling down a steep hillside. Roosevelt fortunately made it back safely to Las Vegas, though he never returned to Southern Nevada.

Recreation in the Spring Mountains got another boost in 1935 when the Forest Service began issuing summer home permits. Several small mountain communities developed out of this construction. Another tenant arrived on the mountain in 1952 when the Las Vegas Air Force Station opened on Angel Peak. This was a radar station that formed part of the country's air defense system. A large number of buildings were necessary to support the radar and its support personnel, including dormitories, mess and recreation halls, and various maintenance buildings. The facility closed in 1969, but the Federal Aviation Administration still operates a radar facility on Angel Peak, and the buildings of the former air force station below it is now the Spring Mountain Youth Camp, a juvenile correctional facility.

Development today is concentrated in Lee and Kyle Canyons on the east side of the range. These long canyons allow access to the interior of the range and end at elevations high enough to support thick forests and heavy snowfalls. Kyle Canyon was named after an early pioneer (who sometimes spelled his name Kiel) that established the first sawmill in the Spring Mountains. His sawmill is now the site of the ranger station; this is also the location of Camp Charleston Mountain. The town of Mount Charleston exists on private land within the forest near the head of the canyon. It has 357 residents, a lodge, library, elementary school, many summer or weekend homes, and is above 7,000 feet above sea level, guaranteeing much cooler temperatures than the valley below. From the town many trailheads reach higher into the mountains, including the trail to Charleston Peak. This is the most heavily used trail to the peak and is a 16.5-mile round trip with a steep climb out of Kyle Canyon. Once on the ridge between Charleston and Griffith Peaks, the trail passes in and out of wooded areas that include ancient bristlecone pines with much of the trail above the treeline. Conditions here are cold and harsh, even in the summertime, but the views are worth it.

Lee Canyon is less developed but includes the Las Vegas Ski and Snowboard Resort, with three chairlifts and eleven runs. This area, developed for skiing in 1940, has continually expanded since. It is the only ski resort near Las Vegas; the next closest are at Flagstaff, Arizona; Brian Head, Utah; and in the mountains near Los Angeles. The end of Lee Canyon is higher than Kyle Canyon, with the high point of the road at 8,736 feet, the highest public road in the Las Vegas area. Lee Canyon also includes picnic areas, trailheads, and private homes. The two canyons are connected by the scenic Nevada Highway 158.

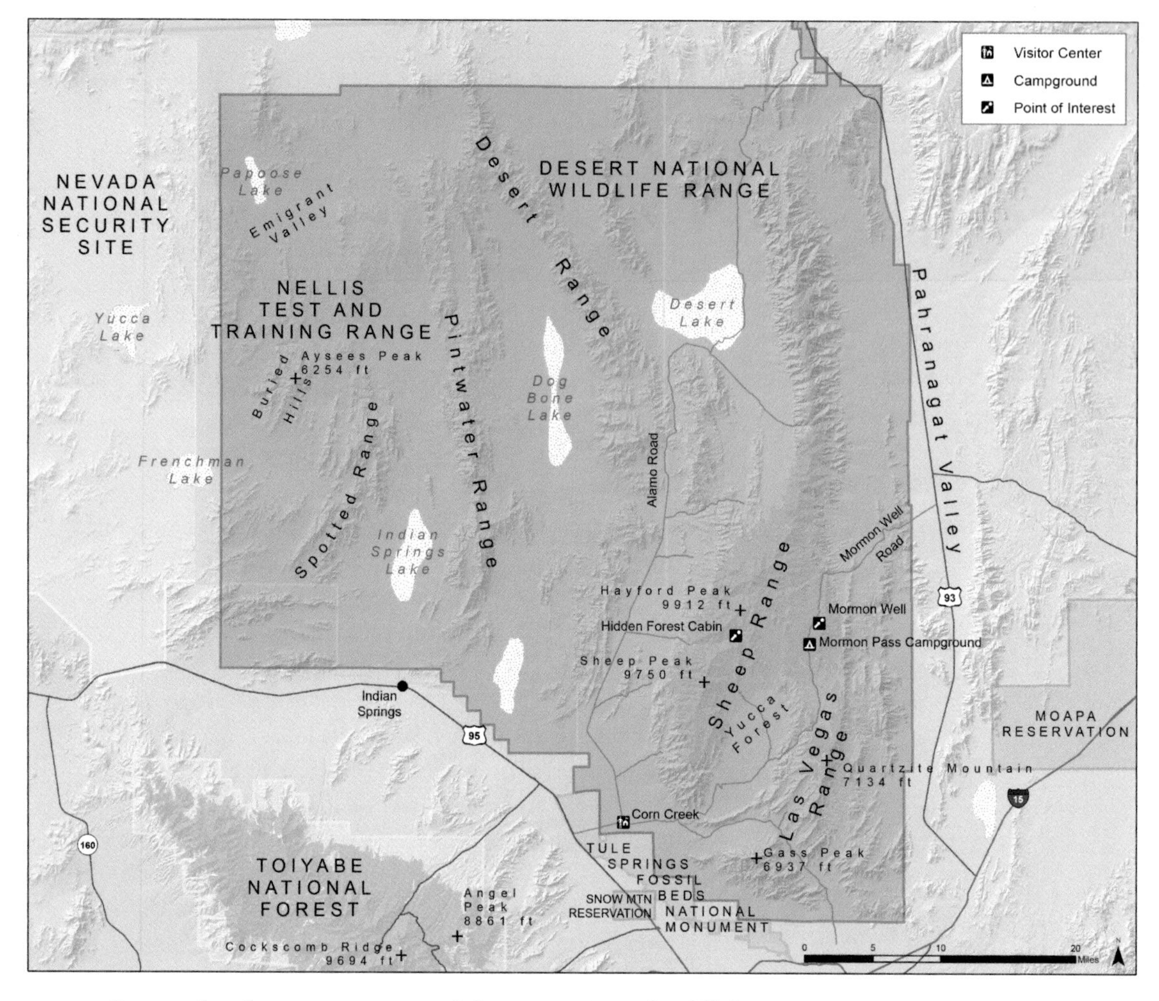

Map 9.3. Desert National Wildlife Refuge. This vast refuge is easy to overlook since most of it is far from roads, and overlaps with the Nellis Test and Training Range. The southeastern portion offers many recreational opportunities.

Conserving forest resources and the protection of wildlife from overhunting were growing concerns in the early twentieth century. President Theodore Roosevelt first acted to protect important habitat in 1903 when he created a wildlife refuge in Florida; many years later, a different President Roosevelt created two wildlife refuges in the Las Vegas area. The first was along the Colorado River and the future Lake Mead. The Boulder Dam Wildlife Refuge was proclaimed by the president in 1933 to protect birds in this area, presumably waterfowl. Its boundaries were nearly the same as the Boulder Dam Reservation created for the construction of the dam, and therefore much the same as the later Lake Mead National Recreation Area. The recreation area absorbed the refuge in 1947.

A second refuge encompassed the forested Spring Mountains and Sheep Ranges north of Las Vegas. These ranges had been protected early in the twentieth century when they became part of national forests (map 9.1), but in addition to timber they were also home to bighorn sheep, a species considered both important and threatened. These mountains, as well as a vast area of

Desert bighorn sheep, 2007. Photo by Andrew Barna.

desert valleys and low mountain ranges around them, became the Desert Game Range on May 20, 1936.

World War II disrupted the solitude of this vast area when the Tonopah Bombing Range was created, overlapping much of the Game Range. The Game Range became the Desert National Wildlife Refuge NWR after the war, and the boundaries were changed in 1966 to include the entire Sheep Range, the Las Vegas range to the east, and many lower desert mountains and valleys to the west. The range today is headquartered at Corn Creek Springs north of Las Vegas, an oasis along the route of the early Las Vegas and Tonopah Railroad. From there several dirt roads enter the east and west sides of the Sheep Range, accessing several forested canyons on the west side as well as the Mormon Pass campground to the east. The Hidden Forest is a popular hiking area, with a trail leading up a forested canyon to a lonely cabin. Visitors here will enjoy far more solitude than in the nearby Spring Mountains.

The Tonopah Bombing Range did not disappear after the end of World War II, and it was instead expanded and is now the Nellis Test and Training Range (map 9.13). The overlap of an air force bombing range on a wildlife preserve is a strange anomaly, but the military restricts its bombing to the valley floors while the sheep prefer the mountain ranges. This area remains closed to the public except for a handful who are able to obtain annual bighorn sheep hunting licenses. The air force also sought to take over much of the Desert NWR permanently for its combat training (map 10.12), but this was rejected in 2020.

Other wildlife refuges were created later around the Desert NWR. The 5,380-acre Pahranagat NWR was added in 1963 to protect migrating birds, especially waterfowl, in the Pahranagat Valley. The Moapa NWR was created in 1979 and is much smaller at just 106 acres. Most of this was once a privately owned retreat and later a resort surrounding hot springs. It now serves to protect the Moapa dace, an endangered fish. The Ash Meadows NWR was created in 1984 to protect 23,000 acres of Ash Meadows, a rare desert wetland northwest of Pahrump. It also surrounds Devils Hole, an underwater cave inhabited by the Devils Hole pupfish. All four of the surviving refuges today are part of the Desert National Wildlife Refuge Complex.

When they were building the dam, Bureau of Reclamation officials realized the new reservoir would have tremendous potential for recreation. They had no interest in providing this themselves so they reached an agreement with the National Park Service to develop the lake with new roads, campgrounds, marinas, and other recreational facilities. This new recreation area was

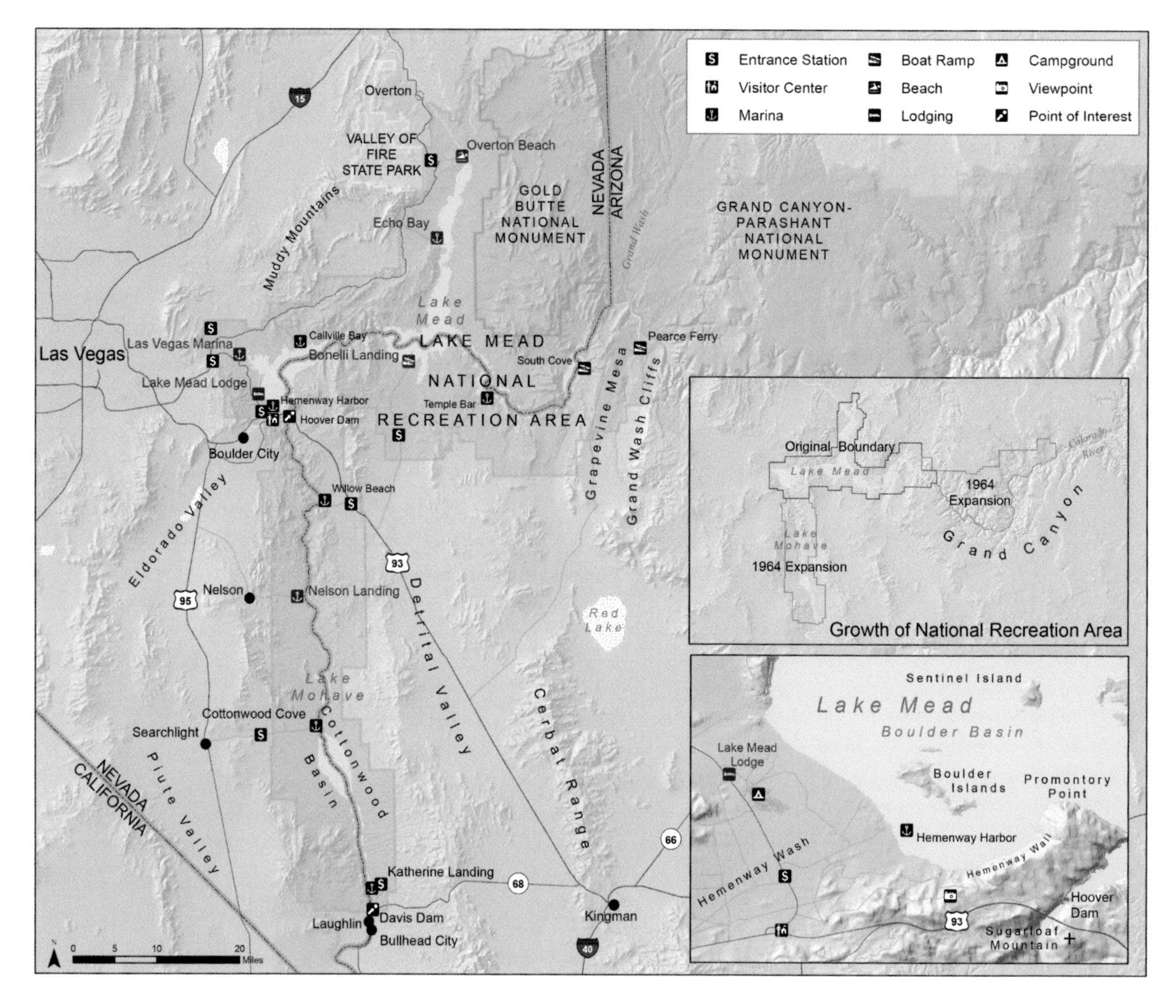

Map 9.4. Lake Mead National Recreation Area. The lake, and Lake Mohave farther downriver, have been heavily developed for recreation. But falling water levels have meant that many are now far from the lakeshore. Symbols in red mark recreation facilities that have been closed. *Lower inset*: Hemenway Wash, where the receding lakeshore has caused a marina to be moved and the Boulder Islands to grow; *upper inset*: the tremendous growth of the national recreation area from the 1930s to today.

initially known as Boulder Dam Recreation Area and in 1947 as Lake Mead National Recreation Area.

The National Park Service put the Civilian Conservation Corps (CCC) beginning in 1936 to work developing recreational facilities in Hemenway Wash (map 4.7). Other projects followed in Callville, Las Vegas, and Echo Washes. A lodge on the lakeshore in Hemenway Wash opened in 1941 but permanently closed in 2009 because of a receding lakeshore and competition from hotels in Las Vegas. The Bureau of Reclamation also leased homes to individuals, making Lake Mead one of the few national park units with a resident population. These homes remain at Stewarts Point and Meadview in Arizona. Airplane landing fields were also built at Echo Bay, Temple Bar, and near Pearce Ferry. Recreation development continued after World War II, including the completion of the Echo Bay and Las Vegas Bay facilities.

A drought since 2000 has substantially lowered the surface level of Lake Mead to record low levels and is likely to continue to lower the lake. This drawdown has already drastically changed the lake's geography. New islands have appeared while others, such as Saddle Island, are now peninsulas. The Overton Arm has shrunk, revealing the site of the former town of St. Thomas and leaving Overton Beach high and dry. The Echo Bay Marina has had to move and extend its boat ramp several times to keep up with the retreating lake. Grand Wash Bay, once an arm of the lake, is now dry. In the Boulder Basin the lake retreated more than a mile from the Las Vegas Marina, causing the NPS to convert it to a boat storage area. Hemenway Harbor marina is still open, but the boat ramp has been extended to keep up with the shrinking lake. Lake Mead Lodge closed in 2009 because of dropping water levels and growing competition from hotels in Las Vegas.

The dropping water levels have created a new attraction—a B-29 bomber on the bottom of the Overton Arm. The plane crashed July 21, 1948, when it hit the lake surface while flying very low on a special mission. The crew survived, but the plane quickly sank and was forgotten; the wreck was not rediscovered until 2001. Diving is allowed at the site, though only under special conditions because of its depth.

Pearce Ferry is the most remote of the lake's developed areas, located on the Arizona shore near the Grand Wash Cliffs. The site has long been a crossing place on the river; it can be reached by washes from the north and south sides, and a ferry boat once operated here. The lake levels dropped many years ago, and the Colorado now runs freely here; the river's edge is now a two-mile drive from the old boat ramp. On the hilltop to the north of the old boat ramp are the remains of a CCC camp and a seismograph building used to monitor earthquakes when the lake filled.

The national recreation area was expanded south in 1964 to include all of Lake Mohave. This lake formed behind Davis Dam on the Colorado River downstream from Las Vegas after it was finished in 1951. This dam is much smaller than Hoover Dam and is made of rock and dirt, with only a small concrete section containing the power plant and spillway, but it provided additional water storage capacity and electricity generation.

The expansion brought all of Black Canyon below Hoover Dam into the national recreation area as well as Cottonwood Basin and Pyramid Canyon. Cottonwood Basin is the largest stretch of open water in the lake and although a relatively featureless basin today was once the site of Cottonwood Island. This was a ten-mile-long and three-mile-wide island, the only sizable island in the Colorado River. It was thickly forested with mesquite and cottonwood and offered grass for grazing. It also accumulated tremendous amounts of driftwood each year, a valuable resource in the desert. The island was important to several Native American groups. The Chemehuevi band of Paiutes and Mohaves both lived along this section of river and farmed on the island (map 3.3), using the annual flood of the Colorado to irrigate their crops.

There are three marina areas on Lake Mohave. Willow Beach is in Arizona in Black Canyon below Hoover Dam. This site is one of the few places in that canyon that can be easily approached, and consequently appeared on the routes of nineteenth-century explorers such as Jedediah Smith and Joseph Ives (map 5.2). The first automobile road between Hoover Dam and Kingman, Arizona, also ran through here, using Jumbo Wash to the south and an unnamed wash to the north (map 7.11). Recreational facilities include a campground, camper parking, marina, fishing, and a restaurant. A fish hatchery keeps the Colorado River stocked with rainbow trout, which thrive in the cold water released by Hoover Dam.

Farther south are more traditional marinas located in large washes. Katherine Landing is the largest marina on the lake, near the south end on the Arizona side. Cottonwood Cove is the only development on the Nevada side. While the lake is more remote than Lake Mead for Las Vegas locals and tourists, it is just north of the twin cities of Laughlin, Nevada, and Bullhead City, Arizona, and also a short drive from Needles, California, and Kingman.

A fourth marina, Nelson Landing, east of the small town of Nelson, existed until 1974. Storms dumped heavy rain in the mountains west of Nelson Landing on September 14 of that year, sending a flash flood down the normally dry washes. The floodwaters swept through the marina and destroyed everything in its path. Nine people were killed, the worst loss of life from any flood in Southern Nevada. The site can still be reached by a paved road, but the few structures that survived were removed and little trace of the flood remains at this peaceful site. A dirt road extends along the side of the hill on the south side of the wash near the shore and once housed several fuel tanks. Debris from the flood, including what may be several cars, remain on the bottom of the lake.

With the 1964 expansion of Lake Mead National Recreation Area, almost the entire Colorado River from Davis Dam to Moab, Utah, is within national parks or recreation areas. The only place not protected is a stretch of the south side of the Grand Canyon that lies within the Hualapai Indian Reservation. This area, Grand Canyon West, has a road to the bottom of the canyon and tourists on helicopter flightseeing trips can even land at the bottom of the Grand Canyon (map 9.9).

Las Vegas's first highway connection to neighboring states was the Arrowhead Trail, which connected Los Angeles to Salt Lake City (map 7.7). Northeast of Las Vegas, the road passed through a valley of red sandstone, originally known as the Red

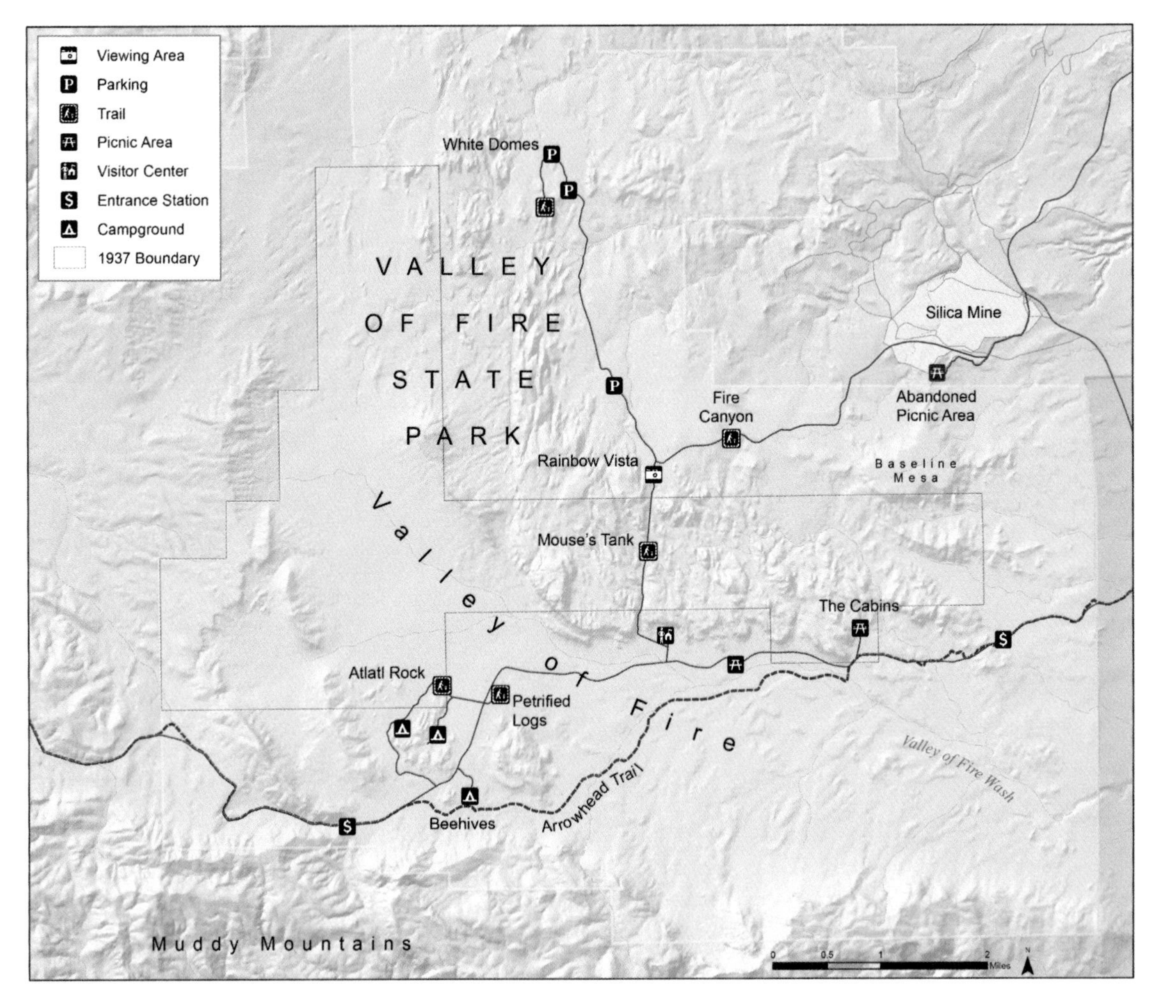

Map 9.5. Valley of Fire State Park. The original boundaries and the recreation facilities built in the 1930s and later in the southern end of the park are depicted. The Arrowhead Trail marks Southern Nevada's first long-distance highway and can be hiked today.

Road region and later named the Valley of Fire by highway promoters. These rocks are Aztec sandstone laid down as dunes in an ancient desert, and are also exposed at Red Rock Canyon west of Las Vegas (map 9.6) as well as much of southeastern Utah, northern Arizona, and southwest Colorado.

The highway was soon rerouted to the north, but interest in this region of striking red sandstone formations remained. The area became Boulder State Park in 1935, one of the first state parks in Nevada. A CCC camp (map 4.7) was established near Overton in the 1930s, and the men stationed there not only took part in archaeological digs and built the Lost City Museum, but also built tourist facilities in this new state park. The stone cabins they built for overnight accommodations (now part of the Cabins picnic area) are among the most picturesque of their projects in Southern Nevada. They also built the roads and trails that still provide the park's main facilities.

After the National Park Service declined to add Valley of Fire to the Lake Mead National Recreation Area in 1947, the park boundaries were greatly expanded and new entrance roads were

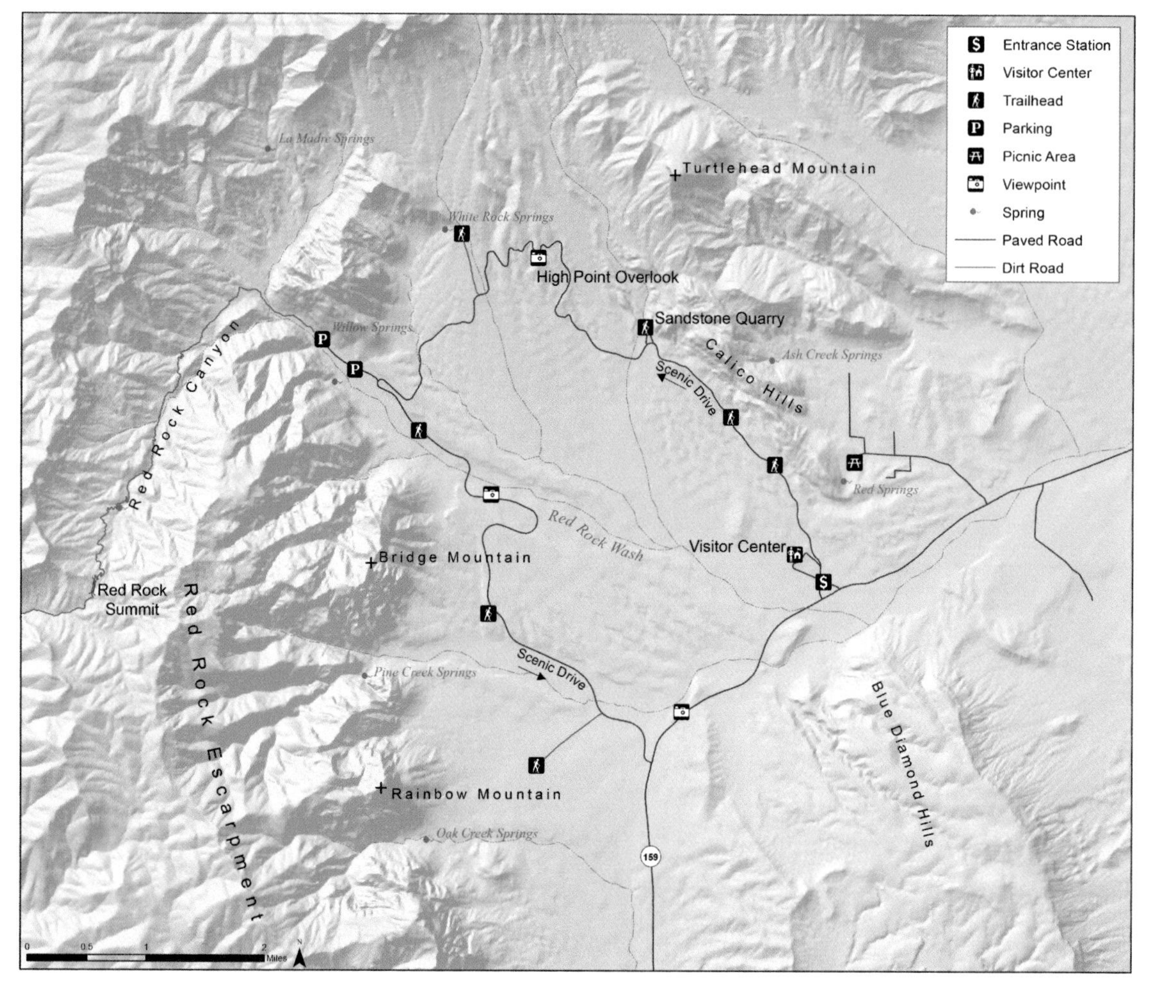

Map 9.6. Red Rock Canyon National Conservation Area. A favorite hiking destination for Las Vegans boasts cool pine-shaded streams and sandstone cliffs.

built in 1952; the old road and a CCC-built picnic area were abandoned but still exist overlooking the Simplot silica sand mine. The park now provides for spectacular scenery as well as petroglyphs, campgrounds, and opportunities for hiking. A portion of the old Arrowhead Trail through the park can be hiked; it is memorialized with a state historic marker near the east entrance.

The park has appeared in many movies, among them 1964's *Viva Las Vegas*, when it forms part of a road race in which Elvis Presley competes (map 11.2).

The Red Rock Canyon National Conservation Area west of Las Vegas includes the spectacular cliffs of the Red Rock Escarpment along the east side of the Spring Mountains, the colorful Calico Hills, and Red Rock Wash between them. To the east are the Blue Diamond Hills, which largely screen view of the cliffs from Las Vegas. Driving the Red Rock loop (Nevada Highway 159) can be an amazing experience for those jaded by views of casinos and billboards. Even better, the hills also block views of the city from many areas, though a large proposed housing development may change this (map 10.2).

In addition to spectacular scenery, the area also has a number of springs that provide it with a reliable water supply. This water also attracted some of the earliest settlers in the Las Vegas area. What is now the Spring Mountain Ranch was established in the 1850s and went through a succession of owners and names. In 1967 it was known as the Krupp Ranch and purchased by Howard Hughes, who had just relocated to Las Vegas (map 11.3). The ranch was to be a gift to his wife, Jean Peters, whom he wished would live in the Las Vegas area. She refused to come, and Hughes sold the ranch in 1972. After another sale, it became Spring Mountain State Park. The park has a 1860s-era blacksmith shop, one of the oldest buildings in Southern Nevada, a picnic area, and trails.

Other ranches were also created at the base of Wilson Cliffs, including Cottonwood Ranch circa 1901, Ash Springs in 1915, and Pine Creek Ranch circa 1922. Little remains at many of these, but a ranch at Bonnie Springs was transformed into the Old Nevada theme park in 1974, with an Old West town, train, zoo, and restaurant.

The most significant of all of these water sources was Cottonwood Springs. This was a major Paiute camp and well known to explorers in the nineteenth century, and along with Las Vegas Springs was one of the landmarks mentioned by early travelers. It was a stopping point on the Old Spanish Trail, the first transport route connecting Las Vegas to the outside world. The town of Blue Diamond was established there in the 1940s, created as a company town to house workers for the Blue Diamond gypsum mines (map 8.19). The nearby Camp Wheeler Springs was named after the Wheeler expedition in 1871. The springs were a rendezvous point for several detachments that separated near Independence, California, before exploring Death Valley. After recuperating at the springs, the expedition continued east: one party headed up the Mormon Trail, while another rowed up the Colorado River into the Grand Canyon (map 5.4).

Calico Hills attracted entrepreneurs who saw value in the colorful rocks. A sandstone quarry was opened in Red Rock Canyon in 1905 but only operated for about a year before shutting down because high shipping costs. A steam traction engine pulled wagons loaded with cut stone to Las Vegas for shipping to San Francisco by train. The quarry reopened for several more years beginning in 1910, and the road built for it became a county road and eventually evolved into Flamingo Road. The quarry site can still be seen along the Scenic Drive in Red Rock Canyon.

The CCC built the first direct highway between Las Vegas and Pahrump through the Red Rock area, perhaps their least-known accomplishment in Southern Nevada. This dirt road branched off of West Charleston Boulevard and went up through Red

Rock Canyon and Wilson Canyon to cross over the mountains into Pahrump Valley. This eventually became NV 85, which first appeared on state road maps in 1947 and remained part of the state highway system until 1964. The current Blue Diamond Highway to Pahrump (NV 160) was not completed to Pahrump until 1955.

Red Rock Canyon. Photo by author.

The spectacular colored rocks and cliffs of the Red Rock area were protected in 1967 when Red Rock Recreation Lands were created under the administration of the Bureau of Land Management (BLM). It had previously been proposed as a national park or monument but was considered lacking in scenic attributes. The first management plan called for an extensive road network, including a Scenic Road running past the Calico Hills and bottom of the Red Rock or Wilson Escarpment and Crest Drive connecting Willow Springs to Mountain Springs to the south (map 10.2). This would have climbed Red Rock Canyon and then run along the top of the escarpment with many overlooks. Although scaled back in later years, the Scenic Road was completed to Willow Springs in 1972 and finished in 1978. The thirteen-mile-long road provides access to the red sandstone and escarpment, with many parking areas and pullouts (usually full on weekends). The park has many hiking trails, ranging from slickrock to deep shaded canyons.

In 1990 the area became the state's first national conservation area, a new kind of protected area that has since become more common. (Sloan Canyon and Beaver Dam Wash are two others in the Las Vegas area.) The region has provided a name for many businesses in Las Vegas, including the Red Rock Casino Resort and the demolished Red Rock Theatre on Charleston Boulevard (map 8.13). Red Rock Wash itself has been removed from the map when it enters the city; a large detention basin was built in the 1980s to catch floodwaters (map 6.8) and below it the wash has been erased by urban growth. The area has also provided a backdrop for countless movies and TV shows. Among the best of these was *The Stalking Moon* with Gregory Peck. This 1968 western involves a siege at a pioneer cabin in a wooded canyon, filmed at the mouth of Pine Creek Canyon.

The National Park Service arrived in Southern Nevada in the 1930s when it began developing the newly created Lake Mead as a recreational facility, now one of the most visited national park units in the country. The National Park Service took on another responsibility in Southern Nevada in 2014 when Tule Springs Fossil Beds National Monument was created just north of Las Vegas.

The monument was created to protect Pleistocene fossils in the upper Las Vegas Wash near Tule Springs. These are found

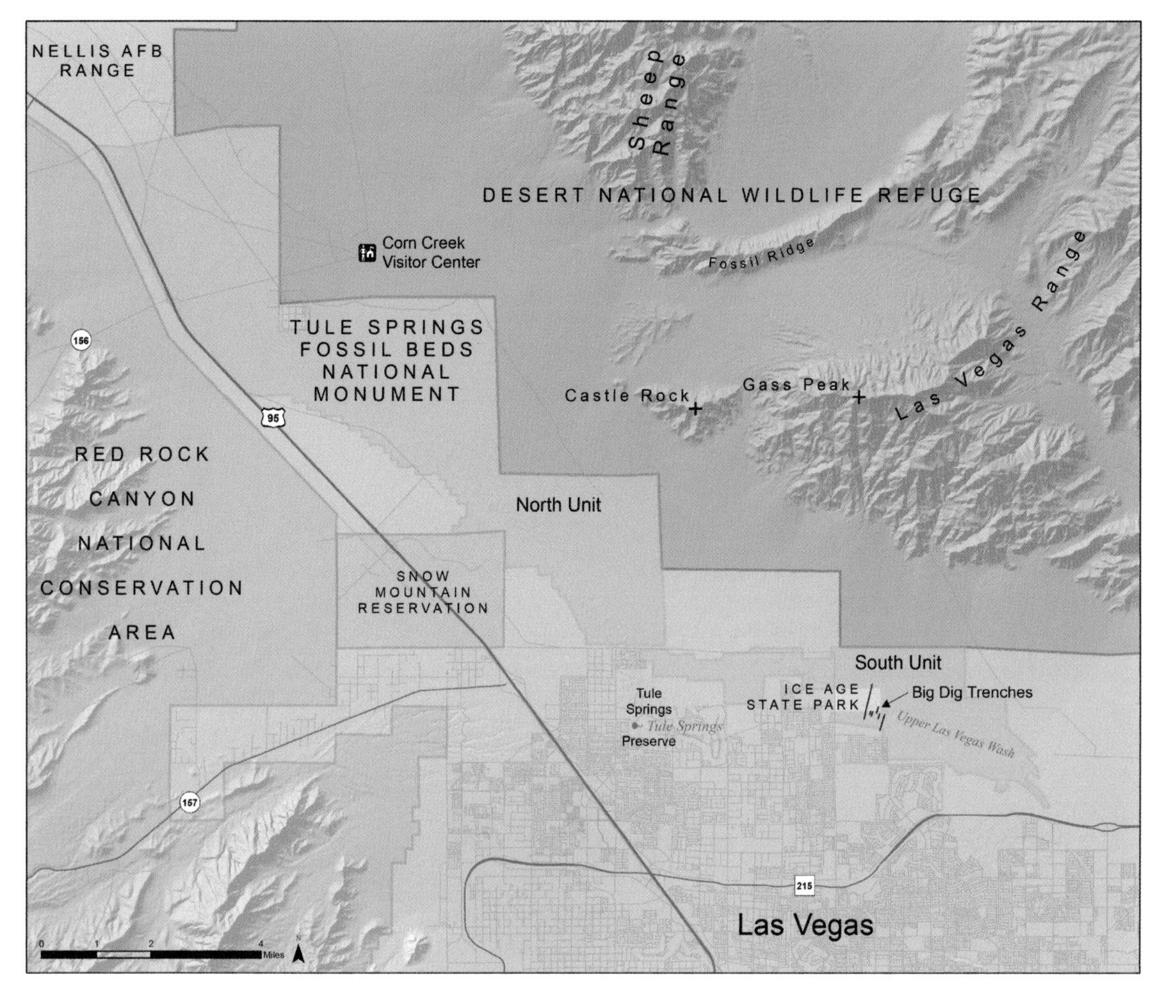

Map 9.7. Tule Springs Fossil Beds National Monument. The monument has no facilities, but the Corn Creek Visitor Center in the Desert National Wildlife refuge can provide information. The Big Dig trenches are the remains of early archaeological excavation done with bulldozers.

in the light-colored, heavily eroded hills seen along the valley bottom. Archaeologists first investigated these deposits in the 1930s (map 3.1); the invention of carbon-14 dating in the 1950s led to the discovery that humans were present in the valley as many as thirty thousand years ago, far longer than previously believed. This attracted more attention to the region and a major excavation project in the winter of 1962–1963. Bulldozers dug twelve trenches, totaling two miles, across Tule Springs Wash. The goal of the "Big Dig" was to create cross sections, up to thirty feet deep, in which Ice Age skeletons and artifacts might be found.

Unfortunately, the Big Dig found no evidence that humans and Ice Age animals coexisted, and later testing found the earlier carbon-14 dates were wrong. Public interest in the area disappeared along with these results, but paleontologists continued to roam the area, finding more and more evidence that the area contains a vast collection of Ice Age animal fossils. It may be one of the most extensive deposits of these fossils, and the remains of mammoths, camels, bison, ground sloths, saber-toothed cats, and many other mammals, reptiles, and mollusks can be found there. These fossils tend to be fragmented and not as visually stunning as those found at the La Brea Tar Pits in Los Angeles and other western sites, but they are still of great value to scientists.

After several years of work, the fossil beds were designated a national monument and assigned to the National Park Service. The monument has not yet been developed for visitors, and lacks a visitor center or any other facilities. It has two sections, a southern one that lies adjacent to the northern edge of Las Vegas, and a much larger northern unit occupying the valley floor east of US 95. The south unit is often within sight of homes and streets and is ringed or crossed by several major power lines. The eroded remains of the Big Dig trenches are in this section, north of the east end of Horse Drive. Ice Age State Park, created in 2017, also preserves parts of several trenches. The northern section is more remote and largely undeveloped. The only paved road into the monument leads from US 95 to the Corn Creek visitor center for the Desert National Wildlife refuge; the road passes through the monument.

One feature not included within the new national monument is Tule Springs. These were a water stop on stagecoach lines between Las Vegas and Rhyolite, and a small ranch operated there for a number of years. It became a divorce ranch in 1941, where couples could wait out their six weeks of Nevada residency before gaining a divorce. It also operated as a dude ranch where tourists could pretend being cowboys. The ranch became a city park in 1964, and from 1977 to 2007 was Floyd Lamb State Park. It then became a city park again. Several lakes and old ranch buildings remain.

Southern Nevada lies within the Basin and Range Province, characterized by hundreds of north-south mountain ranges and valleys (map 2.2), many with playas at their lowest elevations. Most of the valleys and a number of mountains around Las Vegas have been heavily developed, but farther north some basins have changed little from a century ago.

The Basin and Range National Monument was created north of Las Vegas in 2015 to protect some of these undeveloped mountains and desert basins as well as several significant petroglyph sites, though hunting and cattle grazing are allowed within the monument. Like much of Nevada, the land is under the administration of the BLM.

Basin and Range National Monument. Courtesy of the Bureau of Land Management.

Among the mountain ranges found here are the Worthington Mountains, with elevations up to 8,933 feet and the highest in the monument, the Golden Gate Range, Seaman Range, and Mount Irish Range (all with peaks more than 8,000 feet). Big horn, deer, and mountain lions live in these forested mountains, which provide a glimpse of the West before settlement.

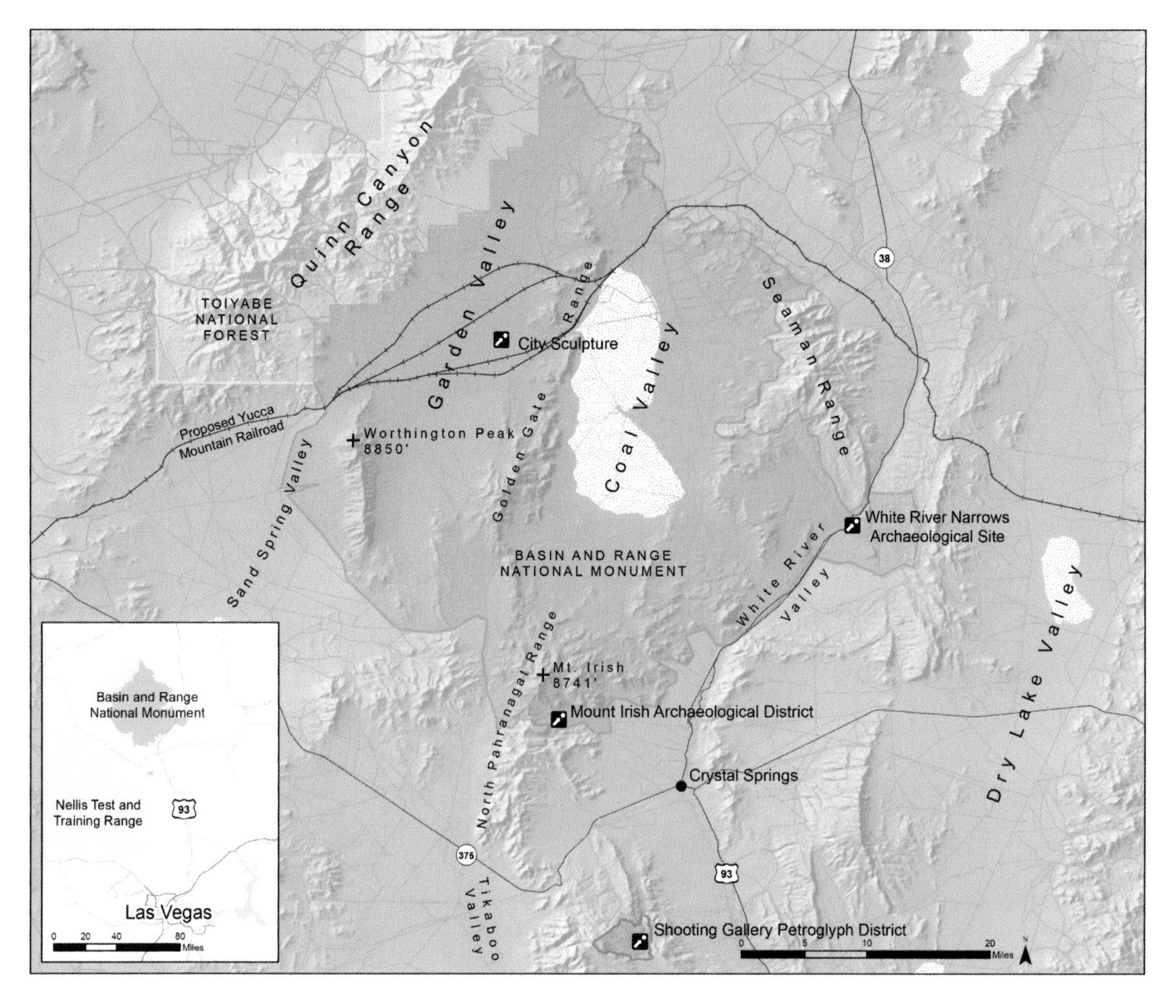

Map 9.8. Basin and Range National Monument. There are no facilities in the monument, but archaeological sites and the *City* sculpture provide points of interest for the adventuresome.

The monument currently has no public services or visitor facilities. The monument does however include *City*, a large piece of artwork being created on private land within the monument by artist Michael Heizer in Garden Valley. This consists of a large outdoor gravel and concrete construction stretching 1.25 miles long and 0.25 miles wide. Heizer started building it in 1972 and it will be under construction for the foreseeable future. It is not open to the public, but it will be when it is eventually finished.

Although nowhere near it, the monument is part of the debate over the Yucca Mountain nuclear waste dump. The proposed Caliente Corridor railroad line to Yucca Mountain (map 10.10) would have passed through what is now the monument. Four possible routes were defined through Garden Valley, two to the north of *City* and two to the south. The existence of the monument closes this area to the railroad, and preventing this rail line may in fact have been the primary reason for the monument's creation. If the Yucca Mountain nuclear waste dump ever proceeds, it will have to do so with a different rail line.

The Grand Canyon is a 277-mile-long canyon created by the

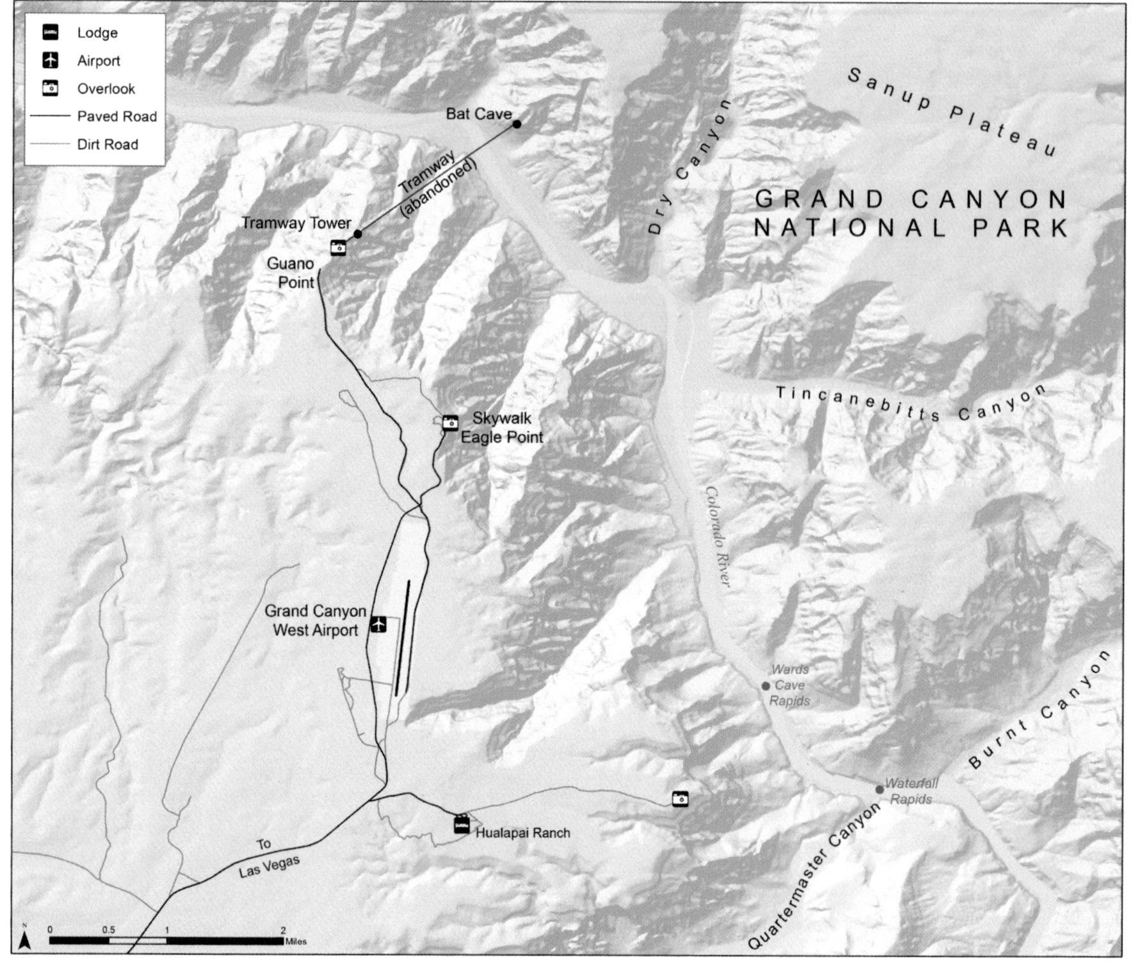

Map 9.9. Grand Canyon West. This development on the Hualapai Indian Reservation is centered on a large airport with roads leading to scenic overlooks and the Grand Canyon Skywalk.

Colorado River east of Las Vegas. Reaching this spectacular feature was a goal of many explorers who passed through the Las Vegas area, including Joseph Ives, John Powell, and George Wheeler (maps 5.2 and 5.4). Reaching the canyon is still a goal of many visitors today, and parts of the canyon have been heavily developed to support millions of tourists.

Most of this development is concentrated at the North and South Rims in Arizona, hundreds of miles to the east of Las Vegas, but from these only a tiny portion of the canyon's length can be seen. Flightseeing tours opened up the entire canyon in 1930s; the first such tours operated from Boulder City Airport and later from Las Vegas.

The western end of the Grand Canyon was opened up to tourists when the Hualapai Indians, neighbors of the Paiutes south of the Colorado River (map 3.3), built what is called Grand Canyon West on their reservation. This runs along the south side of the canyon and before the 1970s was included within Lake Mead National Recreation Area (map 9.4). After a boundary change, the tribe was free to develop their part of the canyon as they saw fit. A large airport opened there in 1988 with an enormous parking area for dozens of airplanes and helicopters for flightseeing tours. The passenger terminal allows easy loading for a fleet of buses that takes visitors to several canyon overlooks, including the transparent Grand Canyon Skywalk. A 122-mile paved route leads to Las Vegas via US 93 south of Hoover Dam for those who wish to drive here. Grand Canyon West is a much closer alternative for Las Vegans than the distant and very crowded South Rim.

Unlike the North and South Rims, visitors here can look directly down into the Colorado River. This stretch was once part of Lake Mead as recently as the 1990s, but since drought lowered the lake it has been a free-flowing river again. The Guano Point overlook takes its name from the deposits within Bat Cave, across the canyon. This cave was considered a good source of fertilizer because it was rich in nitrogen, and a mine operated in the late 1950s. The guano was removed from the canyon by an aerial tramway that crossed the river and climbed to the south rim. Eagle Point includes the Skywalk, the area's best-known attraction, completed in 2007. This extends seventy feet over a side canyon and has a glass floor, allowing the brave to look straight down hundreds of feet. Various plane and helicopter tours are available at Grand Canyon West, and the area is a stop on many flightseeing tours from Las Vegas. Some helicopter tour packages even land at the bottom of the Grand Canyon.

The Sloan Canyon National Conservation Area is perhaps the least-known protected area in the Las Vegas area, though it is within sight of the city and adjacent to new subdivisions in the Anthem master-planned development. It was created in 2002 and encompasses much of the north end of the McCullough Mountains. Its principal feature is the Sloan Canyon Petroglyph Site, which contains more than three hundred rock art panels created as far back as the time of Las Vegas's earliest inhabitants (map 3.1). The carved or pecked imagery usually consists of abstract geometric designs, though there are also depictions of people and animals.

The conservation area has very few facilities but is slowly being developed as money becomes available. A new entrance road and ranger station opened in 2016 south of Anthem, which borders the national conservation area on the northwest. From there a trail heads south into Sloan Canyon to provide access to the petroglyph site. Several additional trailheads around the northern perimeter of the conservation area will include many hiking and equestrian trails. One will reach Black Mountain, at 5,092 feet the high point of the north McCullough Mountains.

The lower Virgin River Valley was the site of early farming, mining, settlement, and transportation routes, and the west side of the valley became known as the Valley of Fire after the

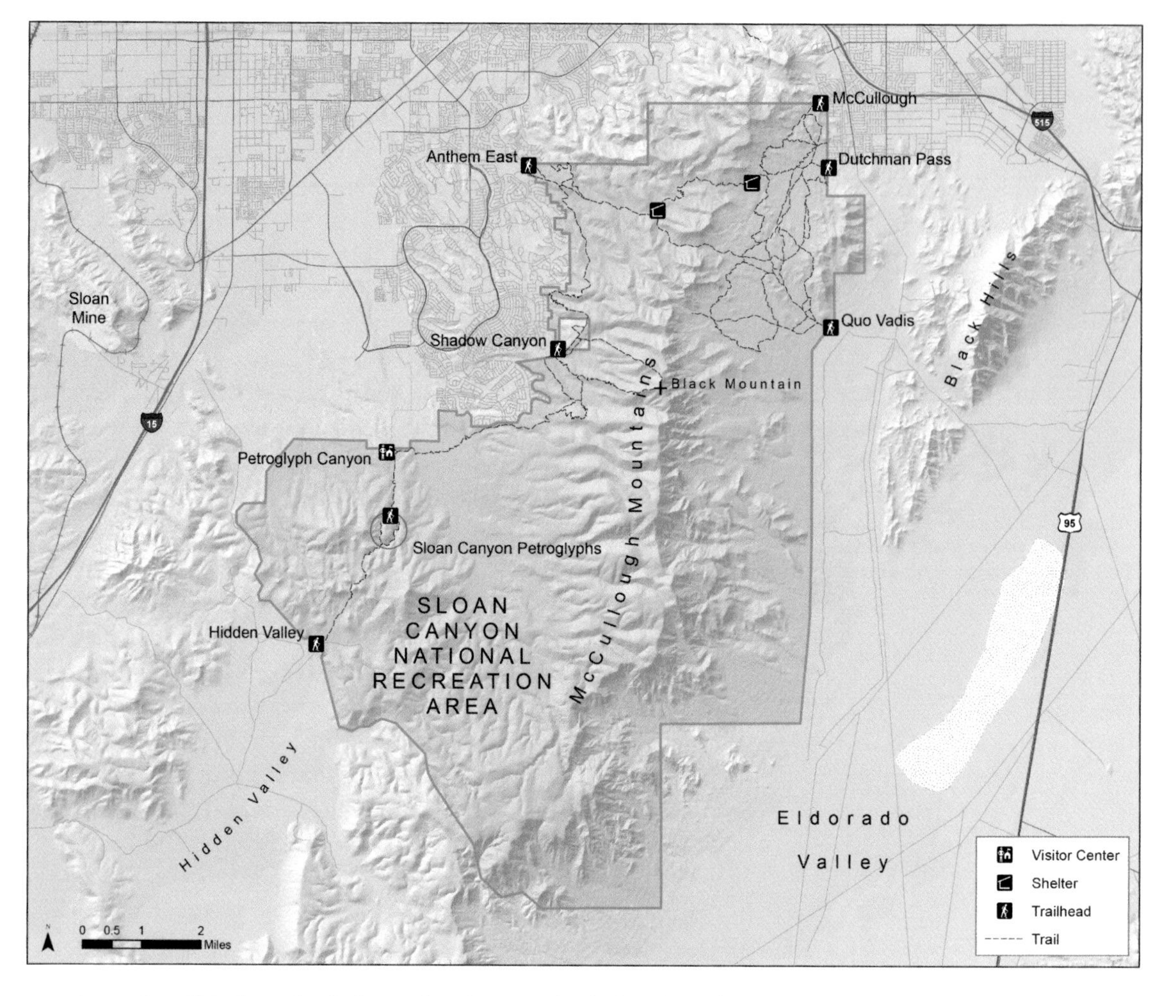

Map 9.10. Sloan Canyon National Conservation Area. One of the newest recreation areas around Las Vegas, with a number of hiking trails through the McCullough Mountains and to the Sloan Canyon Petroglyphs.

Arrowhead Trail was built through it (map 9.5). The lands to the east of the Virgin River remain much less known or visited today.

But this was not the case a hundred years ago. The town of Gold Butte was founded in 1908 during a short-lived mining boom, and some copper mining took place farther east in the 1910s. St. Thomas was the nearest place where supplies could be obtained or ore hauled to the railroad, and the road east from town into the Gold Butte region was even once a state highway. After much of the land surrounding Lake Mead became a government reservation and the lake filled, the Gold Butte region became isolated; no marinas were built along the lake in this area, and a proposed road running east from Overton to Pearce Ferry, with a bridge over the Colorado River, was never pursued. The eastern shore of the lake remained completely undeveloped. But this may change.

Nevada's newest national monument was created December 28, 2016. Gold Butte is surrounded on three sides by Lake Mead National Recreation Area and Grand Canyon-Parashant National Monument and is administered by the BLM. It will

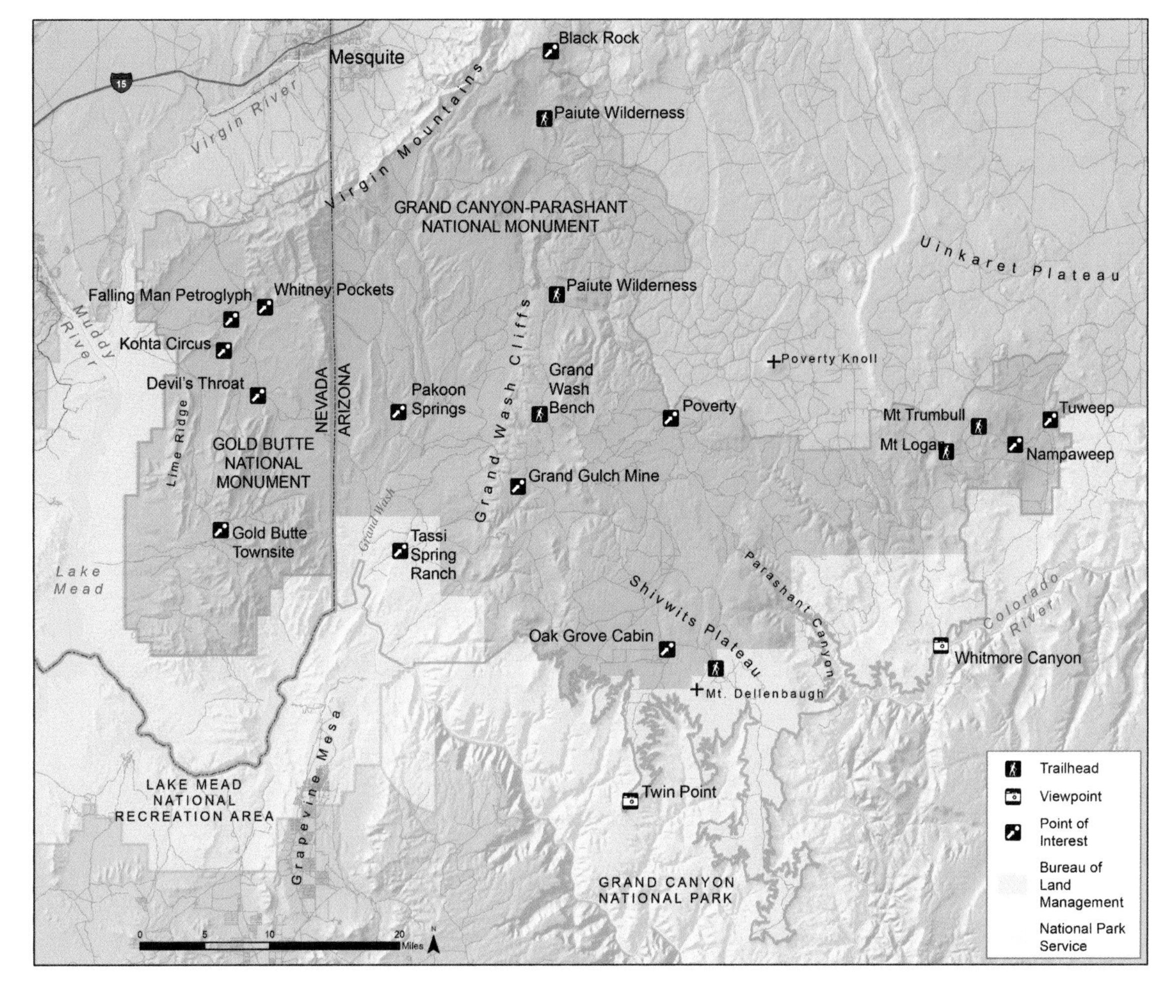

Map 9.11. Gold Butte and Grand Canyon-Parashant National Monuments. These remote monuments reward those interested in rough roads and long trails.

Falling man petroglyph, 2013. Photo by Trylander

never be developed for tourism the way the National Park Service has for its parks and monuments, but there will likely be some facilities added.

The monument can be accessed from the Mesquite area using New Gold Butte Road and a number of often-rough dirt roads. One of these leads down to the Gregg Basin area of Lake Mead using Scanlon Dugway, a spectacular road carved out of the side of a mountain. The monument also contains spectacular petroglyphs, a sinkhole (Devil's Throat), and colorful rock formations.

The Grand Canyon-Parashant National Monument protects a rarely visited area to the northeast of Lake Mead. This was created in 2000 out of lands that were formerly part of Lake Mead National Recreation Area as well as Bureau of Land Management land; the BLM still administers its portion of the monument. The monument has no services, but several rough dirt roads allow access to Grand Canyon overlooks on the edge of the Shivwits Plateau. These can be reached from Mesquite, St. George, Utah, or through the new Gold Butte National Monument.

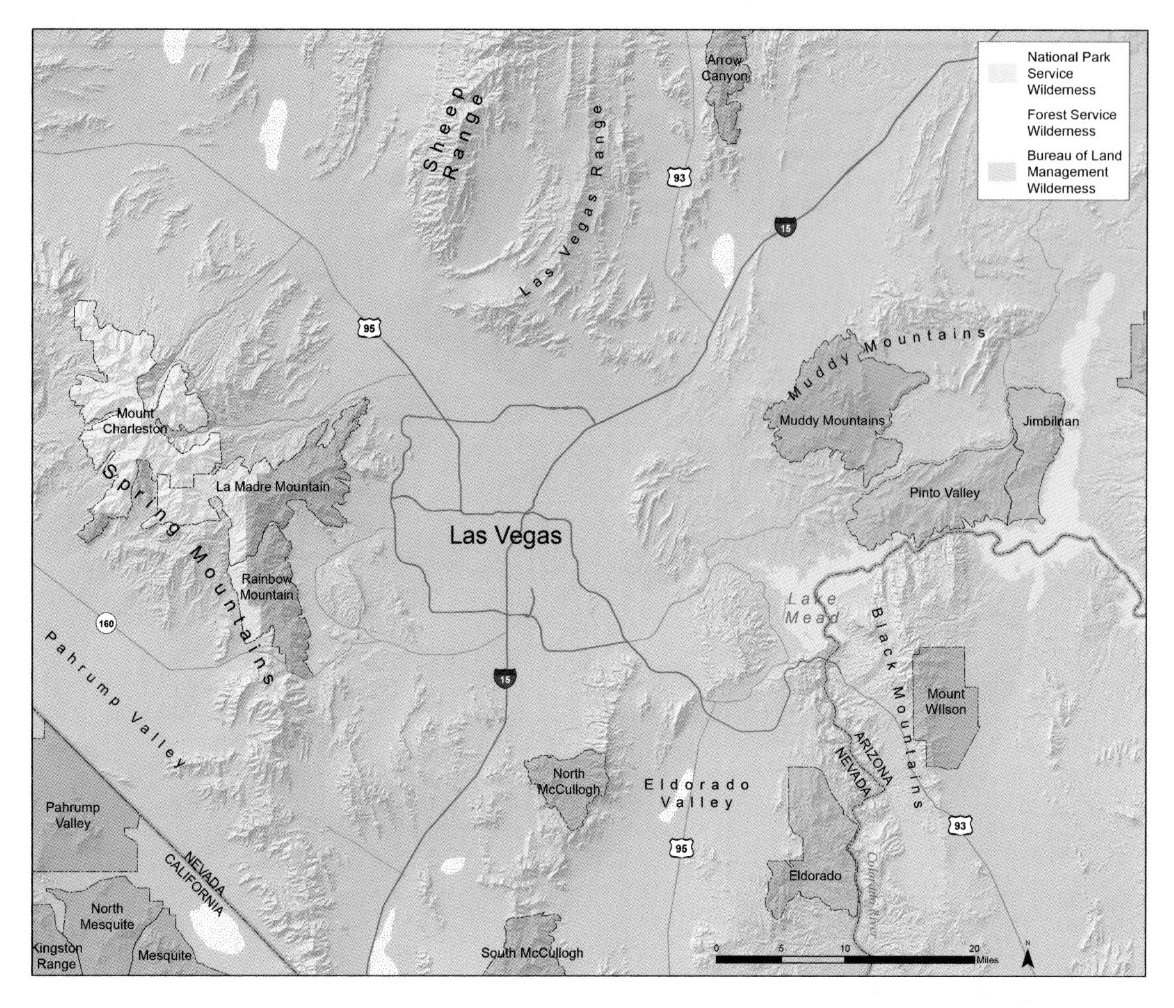

Wilderness in Las Vegas? Yes, there are several wilderness areas within sight of the Strip. These were made possible by the Wilderness Act of 1964, which authorized Congress to create wilderness areas of at least five thousand acres on federally owned lands with little or no development. There are now 765 such areas nationwide, though most are in the West. No mechanical vehicles are allowed in wilderness areas, and mining and logging are prohibited. Natural processes are allowed to operate unhindered. They serve largely as recreational areas, open for hiking and camping.

Map 9.12. Wilderness. There are many designated wilderness areas around Las Vegas, some even offering views of the Strip. Wilderness units are managed by one of three agencies, but all offer solitude and backcountry hiking.

Most of these wilderness areas are in rugged mountainous areas (map 2.1) that were of no interest to farmers in the days of homesteading and difficult to reach or develop. Some areas saw logging or mining early in the twentieth century, but few traces of these remain. Several were of tremendous importance to Paiutes in their annual movements for centuries, with petroglyphs and important archaeological sites remaining within them (map 3.1). All of these areas remained under federal ownership (map 8.1), and several were included within the national forests that have been established in the Spring Mountains (map 9.1) as well as Lake Mead National Recreation Area (map 9.4). However, none are found in the Sheep Range, though most of the Desert NWR was proposed as wilderness in the 1970s (map 10.12).

The 41,180-acre La Madre Mountain Wilderness and the adjacent Rainbow Mountain wilderness include most of the higher elevations of the Red Rock Canyon area as well as the La Madre thrust fault, a geologic landmark where the limestone that makes up the Spring Mountains overlays the sandstone of Red Rock Wash. The Charleston Peak wilderness includes the highest elevations of the Spring Mountains, reachable only by a long and steep trail. The Pinto Valley, Muddy Mountains, Eldorado, and Mount Wilson wildernesses are within Lake Mead National Recreation Area and preserve lower elevation desert ranges and canyons. Others are on land administered by the BLM; the North McCullough Wilderness was established in 2002 and is the closest to the city, only a short distance from Anthem Lakes. It contains the Sloan Canyon petroglyph site, a spectacular collection of prehistoric rock art that is the centerpiece of the Sloan Canyon National Conservation Area (map 9.10).

These wilderness areas are among the last vestiges of Las Vegas before the railroad arrived and the city took root, though from many places the lights of the Strip are clearly visible.

The greatest show in Las Vegas isn't in any casino, and you can't get a ticket for it. Admission is by invitation only, and very few of the most highly skilled men and women from around the world have ever qualified to participate in it. But if you look closely in the skies over the city and along lonely rural highways to the north you may catch a glimpse of it. This show is called Red Flag, and it is has been run several times a year since 1975 by the air force. It involves hundreds of warplanes carrying out a simulated battle in the skies over Southern Nevada, with the Nellis Air Force Range providing bombing ranges and simulated air defenses.

The majority of the pilots who take part are visiting from other air bases. To train them to fight against enemy planes, special units known as Aggressors are used. These use American fighters that resemble the size and performance of foreign planes, and they are often painted in similar schemes. During Operation Constant Peg, Aggressors operated actual Soviet fighter planes secretly imported from unnamed foreign countries; these have since been retired. Aggressor pilots are trained to fly and fight the way enemy countries train their pilots, giving American pilots the closest possible experience to actual combat with them, but without real weapons. Radar and various sensors carefully track the movements of each airplane, allowing each pilot

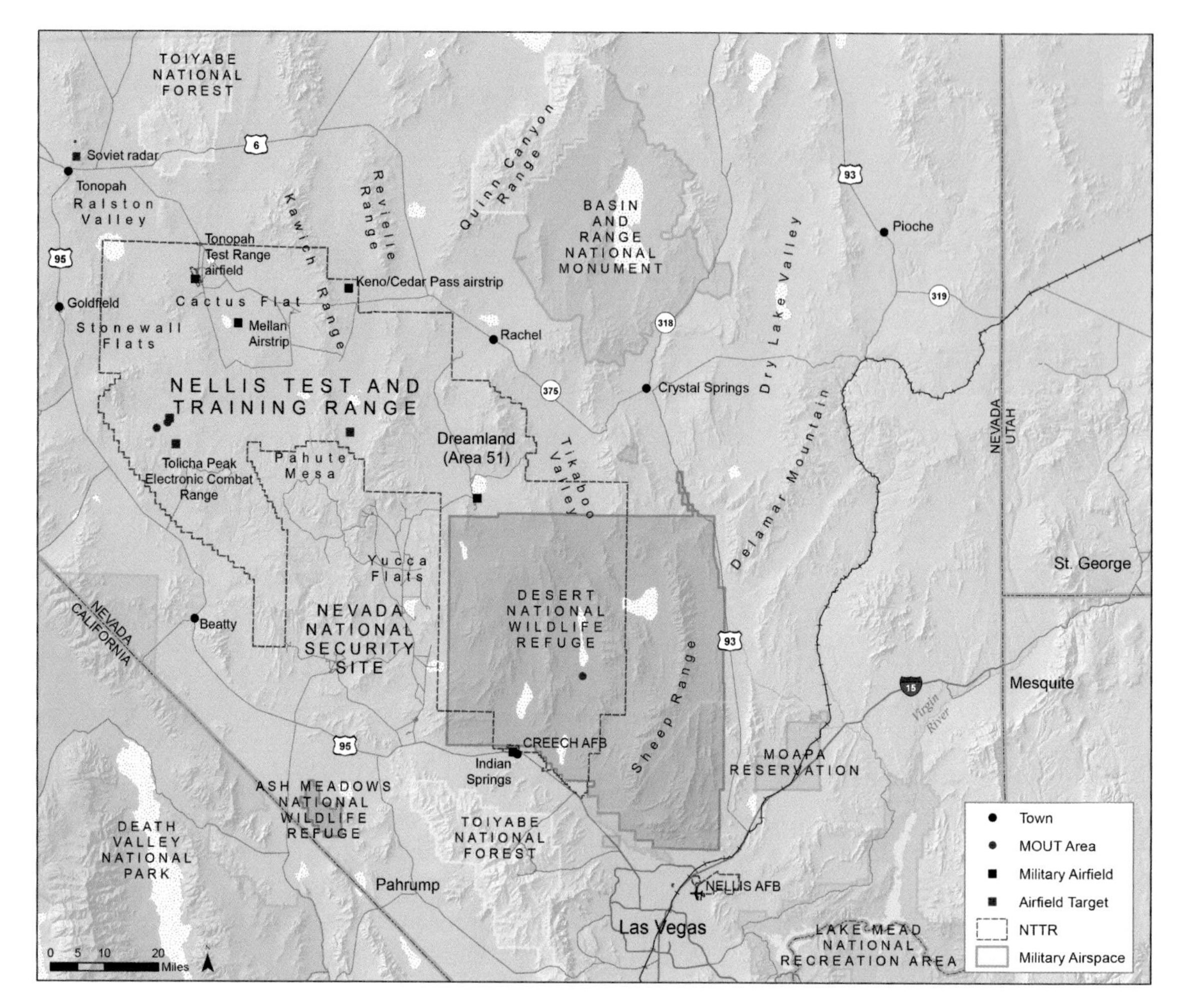

Map 9.13. Nellis Test and Training Range. The military base is shown within a dashed red outline; the pink shading is military airspace used by pilots during Red Flag exercises. The red squares indicate a training area for Military Operations in Urban Terrain (MOUT), where soldiers and pilots practice urban warfare in fake towns built for the purpose.

to see what they did correctly and what needs improvement. As elsewhere in Las Vegas, the odds favor the house and the Aggressors usually emerge victorious, at least during the early part of each Red Flag exercise.

Friendly (blue) forces carry out missions, including ground attack, searching for enemy fighters, electronic warfare, aerial refueling, and search and rescue. Arrayed against them are the Aggressors (the red force), simulated antiaircraft missiles, and guns. If enemy radars track a blue force plane and simulate a launch, a computer will decide if the plane was hit or not. Those that are hit must leave the range and return to base. Similar exercises are carried out at Fallon Naval Air Station in northern Nevada, Alaska, and Canada, but the Nellis operation is the biggest and longest running.

Red Flag isn't just for pilots in the US armed forces. Friendly nations also contribute units, and at least 34 countries have sent aircraft. Pakistani warplanes took to the skies to carry out a bombing mission during the Red Flag in 2016, escorted by Israeli fighters, something unthinkable anywhere but Las Vegas.

Russian-built MIG fighters belonging to India or Poland have also flown over the city.

Those visiting pilots aren't necessarily roughing it during their stay in Las Vegas; in August 2017 the Saudi Air Force booked the entire W Las Vegas hotel (located within what was then known as the SLS hotel, now Sahara Las Vegas), spending at least $1.4 million in the process. To avoid offending the delicate sensibilities of the pilots from that country, female hotel employees had to be reassigned and all artwork in the hotel featuring women had to be removed.

Pakistani fighter planes during a Red Flag exercise over the skies of Southern Nevada, 2010. Photo by Daniel Phelps.

While pilots flying out of Nellis practice for war, an actual war is being fought outside Las Vegas. Anyone driving north from Las Vegas on US 95 passes through the quiet little town of Indian Springs. North of the road is the vast Creech Air Force Base, out of which combat missions are flown around the clock and frequent air strikes are carried out against enemy forces (map 4.19). The aircraft are drones flying out of bases in Iraq and Afghanistan, and the pilots are housed inside windowless buildings in Indian Springs. It is not just American pilots operating out of Creech; in January 2005, the British Royal Air Force's No. 39 squadron was based here and is also flying combat drone missions in Afghanistan. The video footage you see on the news of drone strikes was likely recorded in one of these buildings.

Red Flag isn't the only activity in the Nellis Range. In addition to the secret Area 51 and Tonopah Test Range facilities (map 4.12), Keno Airfield was built in 2003 and is heavily used by the air force and marines to practice parachute operations and transport plane operations on rough dirt airfields. The US Department of Energy continues to carry out research and training activities involving radioactive materials at the Nevada National Security Site (map 9.14).

Although the Nellis Test and Training Range is more than 4,500 square miles, the air force doesn't consider it big enough. Expansions have been proposed along the south and east boundary to provide more operational flexibility in air defense radar locations and attack patterns. This expansion would cut heavily into the Desert NWR (map 10.12).

The first atomic bomb was assembled and tested in New Mexico in July 1945. After the end of World War II, continued atomic testing took place in several south Pacific atolls, but the cost and difficulty of those operations made a domestic testing area necessary. After examining several candidate locations, an area larger than Rhode Island within the Nellis Range was selected in 1951 as a testing ground for atomic weapons. This was the Nevada Test Site, under the control of the Atomic Energy

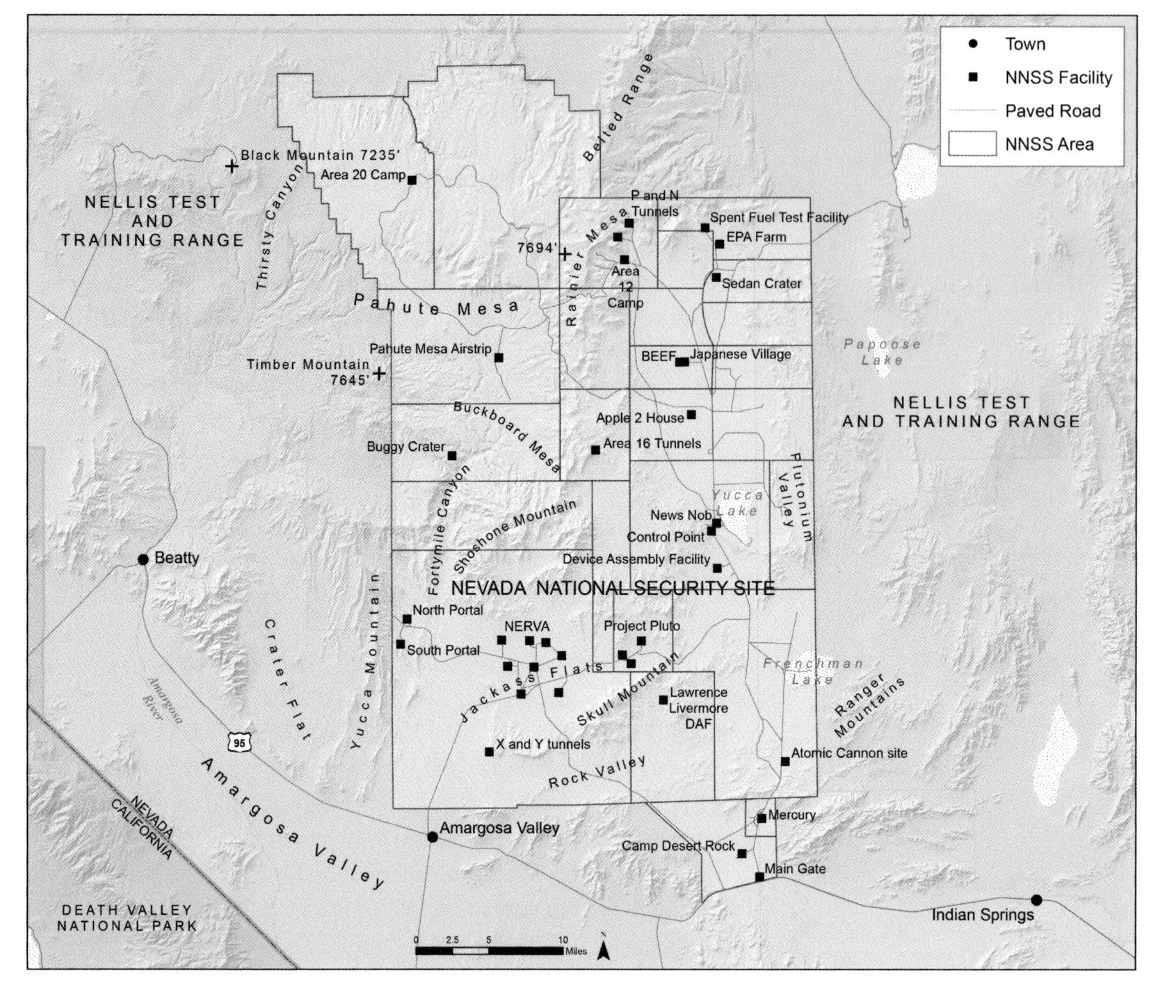

Map 9.14. Nevada National Security Site. Once used for atomic testing, this Department of Energy facility contains many relics of that era.

Commission and later the Department of Energy. After nearly a thousand nuclear tests (map 4.13), testing ended in 1992.

The test site remains active as the Nevada National Security Site and is used for a variety of research and training activities. Much of it is filled with the remains of nuclear testing; these include not just the subsidence craters of Yucca Flats but also a variety of test structures slowly decaying. These were built for experiments on the blast and radiation effects of nuclear bombs. One of the more famous of these was the Japanese Village, designed to understand the radiation exposure experienced by those in Hiroshima and Nagasaki in August 1945.

Because these experiments were not definitive, a more controlled experiment was called for—this time using an unshielded nuclear reactor mounted on a 1,527-foot-tall tower, called the BREN Tower or Bare Reactor Experiment Nevada. The reactor provided radiation similar to a nuclear blast above the houses, but without the blast. Three houses were built along with three front walls, representing common types of houses in Japan. After the testing in 1962, the houses were abandoned; two of the houses and two wall sections collapsed, but one of each remains standing. The BREN Tower was later disassembled and moved to Jackass Flats, where it remained until its demolition in 2012.

From 1964 to 1981, a farm was operated at the north end of Yucca Flats valley. The primary goal was to test the spread of radioactive fallout from the ground through plants to cows and into milk. A small farm for alfalfa was set up and a herd of dairy cows kept on site. Free-range cattle were also kept in the area.

In the southwest portion of the test site, two projects explored the possibilities of nuclear-powered flight. Project Pluto produced a functioning nuclear-powered ramjet engine that could be used on a cruise missile, though the highly radioactive fallout it spewed out might have killed more people than its warhead. The project was canceled in 1964 when it became apparent that ballistic missiles would provide a more effective means of attacking Russia.

Project NERVA (Nuclear Engine for Rocket Vehicle Applications), a joint project of the Atomic Energy Commission and NASA, began in the 1950s in Jackass Flats. The goal was to build a nuclear rocket engine that could be used for a manned mission to Mars. The rocket worked by superheating hydrogen gas inside a powerful but unshielded nuclear reactor and venting the gas to produce thrust. The project succeeded, but the project was shut down in 1973 after the Apollo moon-landing program satisfied the nation's appetite for expensive space voyages.

The project included several buildings where the rockets could be serviced along with a small railroad to transport them to test stands where they were connected to fuel lines for operation. The facilities were abandoned and have mostly been demolished. A railroad engine used to move the reactor around is on display in Boulder City at the Nevada State Railroad Museum.

Project NERVA rocket engine on railroad car in Jackass Flats. Courtesy of the National Nuclear Security Administration.

Along the western border of the test site lies Yucca Mountain, a long north-south ridge. The US government in 1987 designated this ridge as a permanent storage site for high-level radioactive waste. The preferred means of shipping radioactive waste to the facility was by railroad, using a new line branching off from the Union Pacific Railroad at Caliente and running all the way around the Nellis Air Force Range (map 10.10). These trains will likely never be seen as the nuclear dump project stalled and later stripped of its funding.

Unbuilt Las Vegas

10

Although the rapid growth of the city, the construction of Hoover Dam and similar projects, and the creation of massive themed casinos may lead to the impression that in Las Vegas anything is possible, the city's history is also littered with canceled plans, abandoned dreams, and assorted failures. Some of these are well known, others obscure; some appear as reasonable ideas that may in fact eventually come to be while others are bad ideas or just plain bizarre.

All of them—dams not built, railroads that never were, an airport that remains to be built, casinos that never obtained approval, or a water system that may someday be necessary to keep the city alive—are important to show that the city's growth has been anything but inevitable. There were always other possibilities, and many projects that have been completed perhaps only narrowly escaped being included in this chapter.

The chapter also displays how the Las Vegas we know today could have turned out to be very different. It could have been an agricultural center, or located next to the Colorado River, or be walled in by a giant levee rivaling the Great Wall of China, or be served by the fastest passenger trains in America. But none of these schemes survived to become reality.

Many plans, schemes, and dreams have not (yet) come to pass in Las Vegas. Prominent among these are plans for casinos that were never built for one reason or another. The map shows many of these along the Strip and a few off it. The southwest corner of the Strip and Sahara Avenue was the location of the El Rancho Vegas, the first hotel/casino on the strip before it burned down and became an empty lot (map 8.9). At various times this site has been proposed for the Starship Orion hotel and casino (modeled after the starship *Enterprise* from *Star Trek*), Countryland USA (a country music-themed hotel), and the Titanic (a full-size replica of the famous ship). The site is now Las Vegas Festival Grounds. Another popular location farther south was proposed for a World Wrestling Federation-themed hotel and casino and also desired by the DesertXpress high-speed railroad (map 10.7). It was later selected as the site for Allegiant Stadium to house the relocated Oakland Raiders of the NFL.

As the city grew so did the area vulnerable to flooding (map 6.7). The Army Corps of Engineers developed a flood plan in 1959 that involved building a huge levee along the west side of the city to stop flood waters from the Spring Mountains. No growth would be allowed to the west of this barrier,

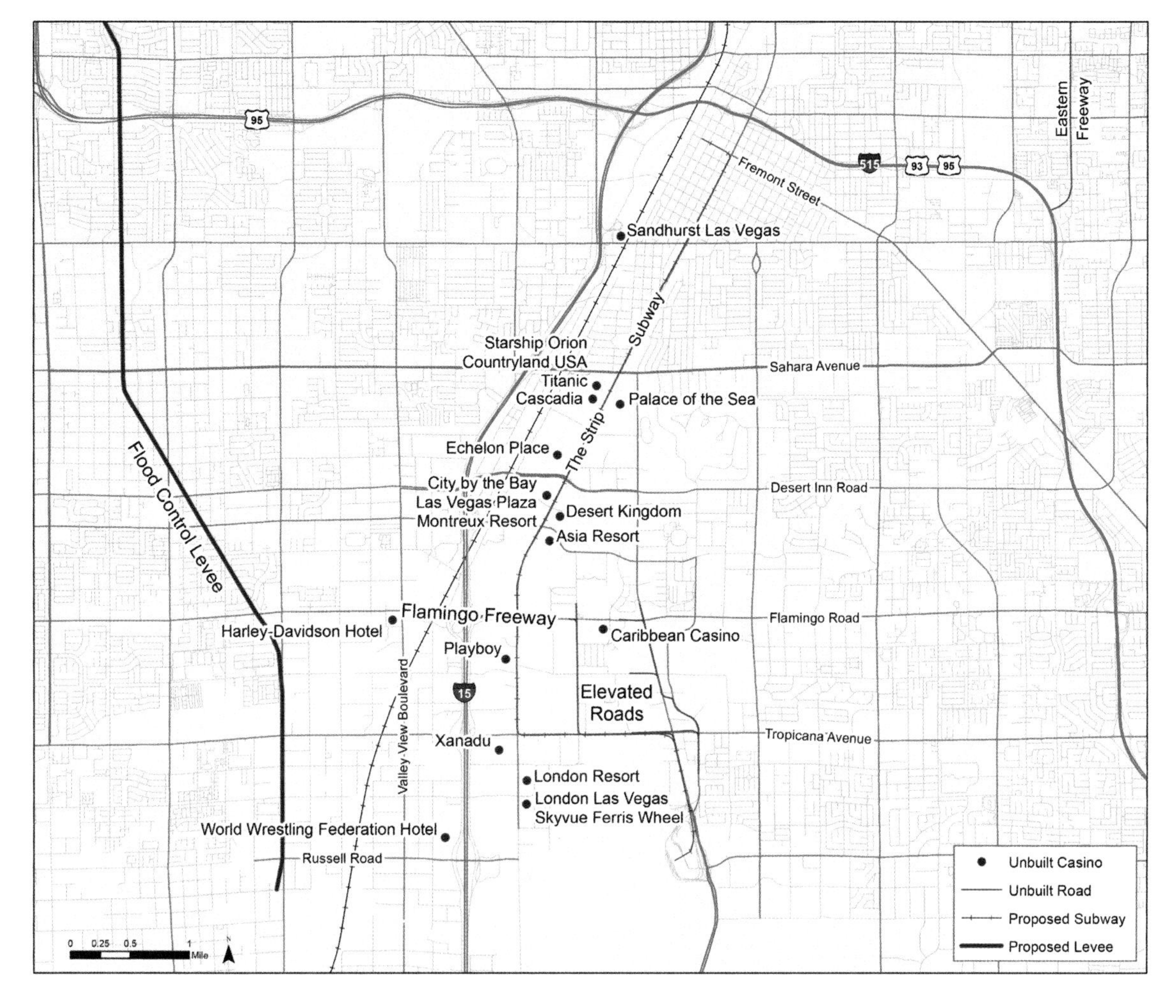

Map 10.1. Unbuilt Las Vegas I. A variety of proposed projects were never built, including casinos, roads, a subway line, and an enormous levee to prevent flash floods from reaching the city.

located along modern-day Jones and Decatur Boulevards. This levee would have prevented the development of the west side of the valley, and with it there would never have been a Summerlin (map 10.2). The plan was rejected, and in its place a series of smaller flood detention structures were built to slow flooding around the periphery of development.

Transportation plans are a rich source of things never built. Many freeways have been built in the city, but not every highway planned was built. A 1963 transportation plan called for converting Flamingo Road into a freeway running from Valley View Boulevard to Paradise Road, while a 2016 plan called for elevated expressways between McCarran International Airport and the Strip. At several times a rail line, perhaps even a subway, has been proposed for the Strip to connect downtown to the airport. The first time this was suggested was in 1963; the most recent occasion was in 2018. Perhaps it will someday happen. These projects would be separate from the monorail connecting a number of casinos east of the Strip. This opened in 1995

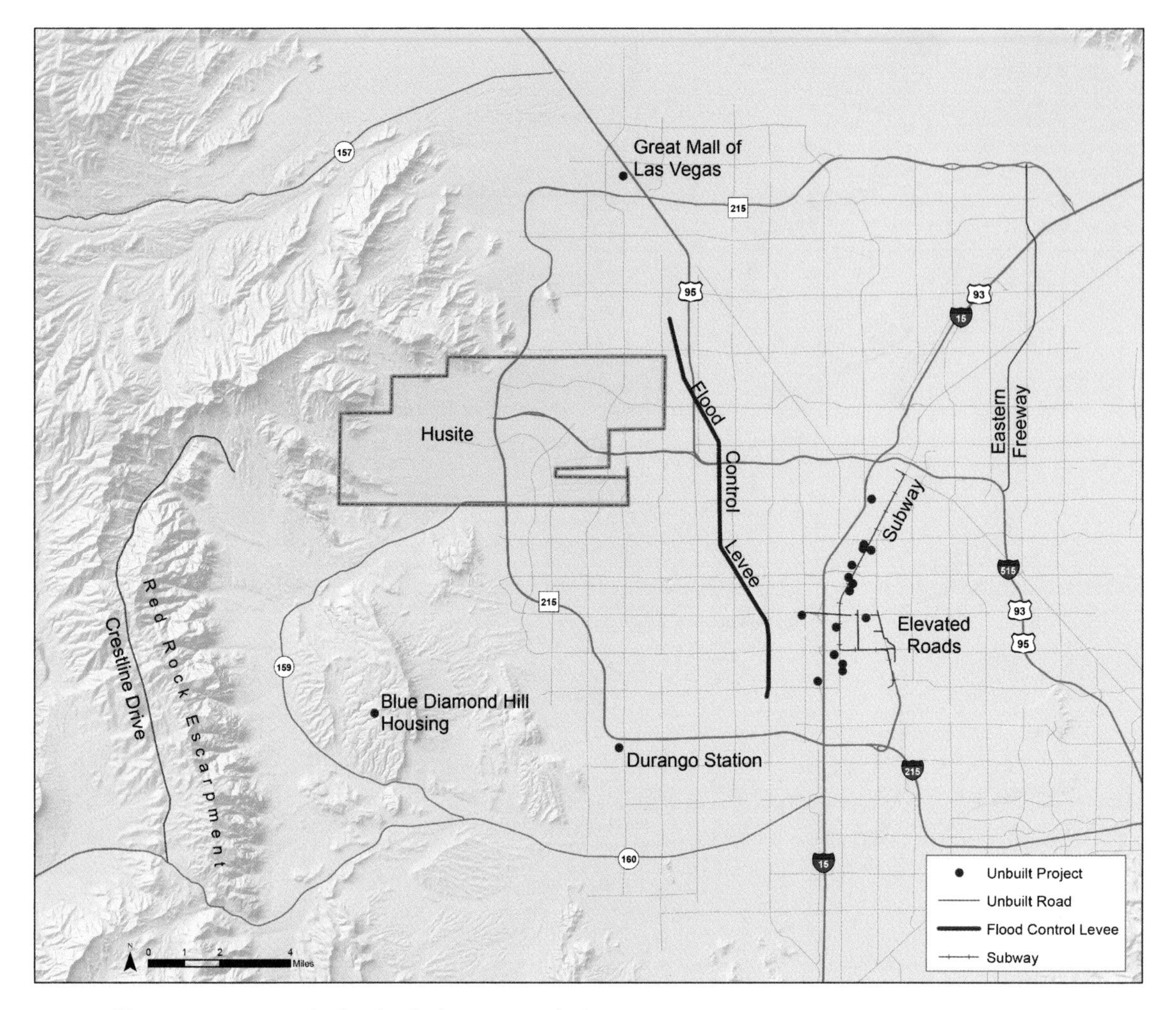

Map 10.2. Unbuilt Las Vegas II. Failed proposals can be found far from the Strip, including a scenic mountain road and Howard Hughes's planned aerospace industrial center at Husite.

and has its own record of unbuilt dreams, including expansions to downtown and the airport.

Many unbuilt schemes exist beyond the Strip, including at least one casino, the Durango Station. The Great Mall of Las Vegas was planned in the 2000s in the northwest along US 95, but the idea was abandoned in 2009. In the 1950s, Howard Hughes (map 11.3) bought twenty-five thousand acres of land west of Las Vegas with the plan to develop it as an electronics and missile factory in an area he called Husite. His workers at the existing plant in Los Angeles rebelled at the thought of relocating to an empty desert, and Hughes eventually backed down. He lost interest in it, and it remained empty until after his death; it was developed as the Summerlin master-planned community in the 1990s. Had the proposed 1959 flood control levee along the west side of the city (map 10.1) been built, Summerlin would not have been developed and the Great Mall or Durango Station would never have been considered for development. The west side of the valley would be empty still today.

Farther west, the spectacular colored rocks and cliffs of the Red Rock area had been proposed as a national park or monument, but nothing came of these dreams (map 9.6). It was finally protected in 1967 when Red Rock Recreation Lands were created under the administration of the BLM. Supporters of the area's preservation were shocked when the first management plan the agency developed called for an extensive road network, including a Crest Drive running up Wilson Canyon, along the top of the escarpment, and connecting to the Blue Diamond Road at Mountain Springs. The Crest Drive plan was soon dropped. Another lost highway originated in a study from 2002 that raised the possibility of an eastern leg of the city's beltway, running from the northeast end of the beltway south to connect to Interstate 515 east of downtown. This would have required the demolition of hundreds of homes and businesses, far more than were required for Interstate 15, and the idea disappeared.

Several major housing developments have been proposed, or even started, that have yet to be completed. Northeast of the city, Coyote Springs was to be a forty-three-thousand-acre residential community around a golf course, located along US 93 near the Clark/Lincoln County line. The land was obtained from the BLM through a land swap; building such a vast development outside the private lands already existing around Las Vegas and a few other areas (map 8.1) is otherwise impossible. In the case of Coyote Springs, land in the Florida Everglades was traded for the Nevada land by an aerospace company looking for a place to build rocket engines. Much like Howard Hughes's plan to build an aerospace plant west of Las Vegas in the 1950s (map 11.3), the factory was not built and instead was developed as a residential property. The golf course opened in 2008 but no homes have been built, a victim of legal problems, the Great Recession, and problems obtaining water for the new community. All previous growth has taken place on private lands clustered around water sources; Coyote Springs is built on what had been public land acquired after most sales were ended in 1976 at a location far removed from any springs or creek. This could have been a model for future growth beyond the Las Vegas Land Disposal Boundary.

When the Mike O'Callaghan–Pat Tillman Memorial Bridge bypassing Hoover Dam was built, several residential developments were proposed near Kingman, Arizona. The new bridge made trips to Las Vegas much quicker, allowing people to live as far away as Kingman and commute to Las Vegas. The Great Recession put an end to these projects before homes were built. Should growth return to earlier levels and Interstate 11 brings

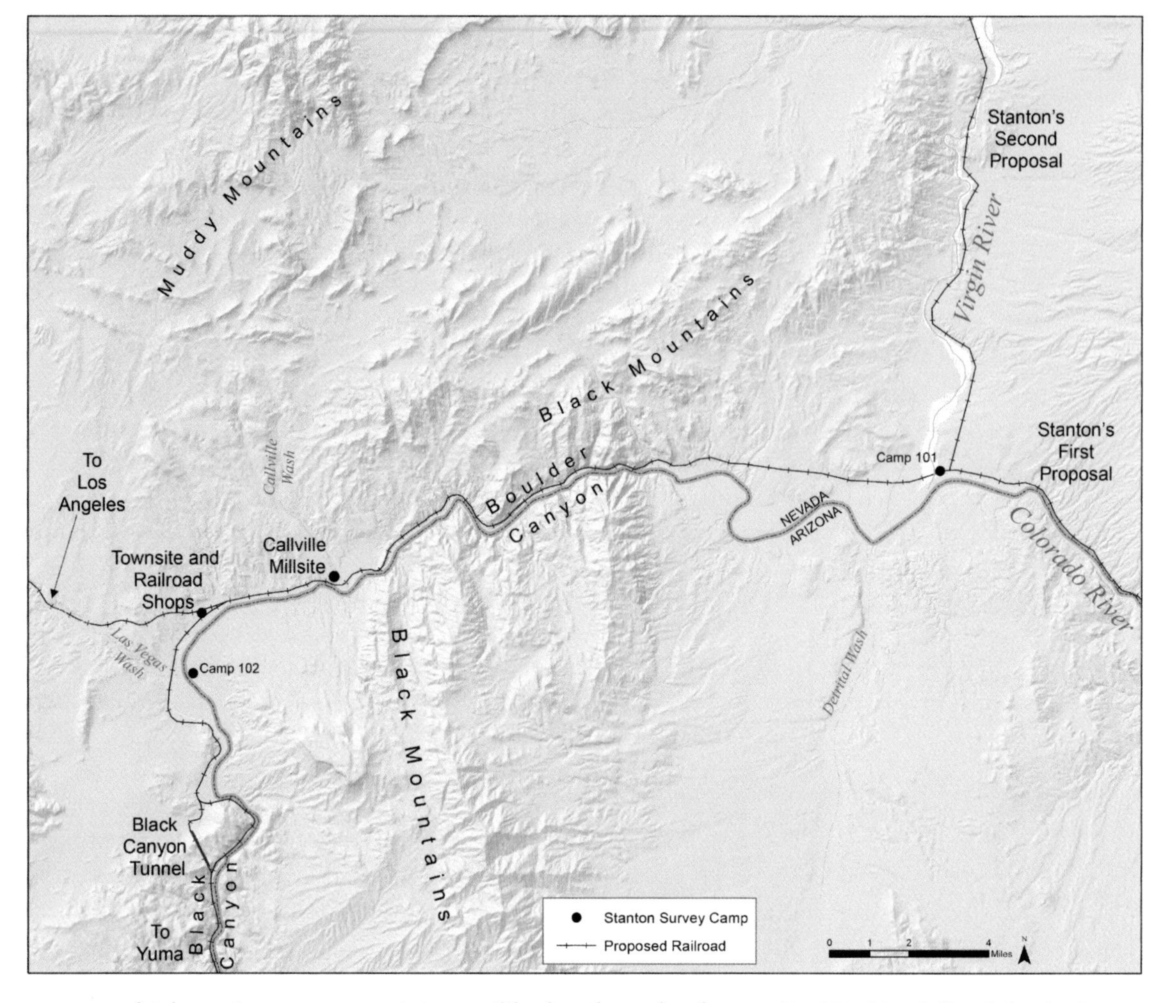

Map 10.3. Colorado River Railroads. The map is based on Robert Stanton's ideas for a railroad to be built along the Colorado River. The red line indicates where a tunnel might have been built to avoid the entrance to Black Canyon.

more highway improvements, it is possible that these developments may eventually be resumed.

A battle over a proposed housing project near the Red Rock area has lingered for years. This is to be built on the site of the former Blue Diamond Gypsum mine (map 8.19). This mine started operations in 1902 and closed in 2004, though the mill and drywall factory nearby are still active. The mining company has been trying to build a large residential area on the site, which would offer spectacular views over the city. A community of 7,250 homes was proposed in 2011, reduced to 5,025 homes in 2016 after strong opposition by the public. The project is still active, and it remains to be seen whether this particular dream will remain unbuilt. This and the McDonald Highlands development in Henderson are among the first attempts at building hilltop homes for the wealthy in Las Vegas, as seen in Los Angeles or Phoenix.

Imagine a railroad running down the bottom of the Grand Canyon and along the Colorado River where Lake Mead now fills the canyons and basins. A seemingly bizarre vision, but it might have been. Engineer Robert Stanton, backed by eastern financiers, led a survey down the Colorado River through the Grand Canyon in 1890 to investigate using the river for a railroad line from Grand Junction, Colorado, to Yuma, Arizona.

The survey found no major obstacles in building a railroad line along the river, especially compared to other western canyons such as Colorado's Royal Gorge. Building here would have avoided the steep grades and high elevations found on railroad lines built through northern Arizona, Utah, and Colorado, and would therefore have been much cheaper to operate.

Where would it have been built? In what would become the Lake Mead area, the proposed route would have stayed on the north side of the Colorado River, usually close to the river at the foot of bluffs with many small bridges over washes. The line would have cut along the base of the Temple, which Stanton called the Citadel. Once past the mouth of the Virgin River, the line would have headed straight west to avoid the bend of the river to the south. Boulder Canyon was not seen as a barrier, though several tunnels were thought likely. A large flat area near the mouth of Las Vegas Wash had room for a railroad yard. A railroad line up Las Vegas wash could have easily been built, allowing a connection to Los Angeles.

This site also had room for a town to house the workers. Railroads commonly created towns, and of course the San Pedro, Los Angeles and Salt Lake Railroad founded Las Vegas itself at the site of its maintenance facilities in 1905. Imagine if Las Vegas had instead been founded in the 1890s where the Las Vegas Wash reached the Colorado River. This would have prevented the construction of Hoover Dam, though a dam could still have been built in Boulder Canyon (map 10.4) if portions of the railroad were moved. This town might have expanded along Hemenway Wash, and US 91 would no doubt have reached the city from this direction. The Strip might even have developed within the confines of Hemenway Wash or in Eldorado Valley.

The entrance to Black Canyon farther south was judged to be one of the few difficult places to build a railroad along the entire route, and a tunnel bypassing this section might have been preferable. Ironically, the entrance to Black Canyon was one of the very few places along Stanton's survey where a railroad was built, as part of the Hoover Dam project (map 5.7). This was a double-track line with two tunnels, carrying a heavy tonnage around the clock on a daily basis. It is now flooded under Lake Mead (the route of a higher elevation railroad line above the canyon still survives as a hiking trail).

Nothing came of Stanton's survey. He later wrote another proposal for a railroad down the Virgin and Colorado Rivers, with

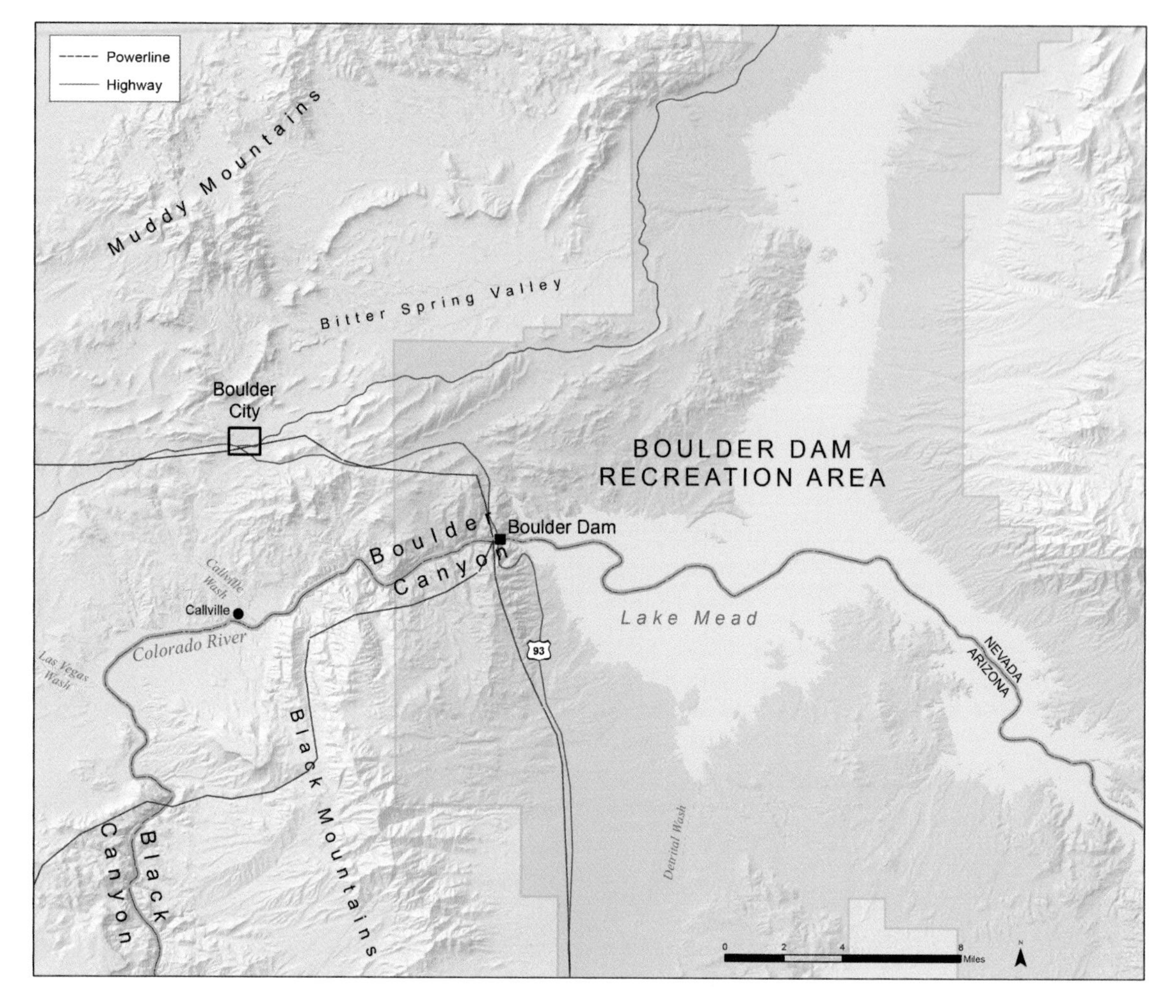

a smelter for Rocky Mountain mines at the site of Callville, but nothing resulted from this scheme either.

More railroad plans were made when the dam project was announced. Union Pacific and Santa Fe Railroad officials expressed interest in building a branch line to the construction site to carry the heavy traffic they knew would be necessary. The Union Pacific projected a line from what is now downtown Las Vegas down the Las Vegas Wash to the Colorado River, while the Santa Fe discussed a line north from Searchlight, which it already served, and down Hemenway Wash to the river. Yet another line was proposed from the southeast, where it would connect to the Arizona and Utah railroad in Chloride, Arizona. None of these lines was ever built; the Union Pacific and other contractors eventually built a line that branched off the mainline well south of Las Vegas, crossed through Railroad Pass, and then down Hemenway Wash instead.

Surely everyone is aware that Hoover Dam was once known as Boulder Dam, though the fact that Boulder Dam was named after Boulder Canyon is less known. This was the original planned

Map 10.4. Boulder Dam. Plans to build a dam in Boulder Canyon were dropped early, but this map speculates on how things might have turned out if it had been built there. The dam location and lake are based on Bureau of Reclamation plans; the rest is speculative.

Drawing of planned Boulder Dam and reservoir, looking east across Black Mountains, 1921. Courtesy of the Los Angeles Times.

location for the huge dam to be built to tame the Colorado River (map 5.6). But once the Bureau of Reclamation began surveying the best location for a dam, Black Canyon emerged as a competing site. Surveys and test drilling in the early 1920s showed that either canyon would make for a sound dam, and in 1928 the Black Canyon site was chosen because the reservoir behind it would be larger and was closer to transportation lines. The result was Hoover Dam.

What if the dam had been built in Boulder Canyon? Bureau of Reclamation documents show a Boulder Canyon dam would have had a design very similar to that of Hoover Dam, and was planned to have the same full-lake elevation as Lake Mead eventually did. This means that upriver of Boulder Canyon the reservoir would have had the same shoreline as the present Lake Mead, but the total storage capacity of Lake Mead would have been greatly reduced. However, since the lake would receive the same runoff with a much smaller capacity it would not have had the same level of drawdown that Lake Mead has had in recent decades.

Other details of Boulder Dam shown on the map are purely speculative. Boulder City would have been built somewhere nearby to house the workers, perhaps to the north where flatter land could be found. US 93 might have run east from Las Vegas to cross over the dam and run south to Kingman. Assuming the lake had become a national recreation area, it would likely have been developed in much the same way. With a smaller lakeshore, perhaps a road along the east side of the Overton Arm would have been built, something that was considered for Lake Mead but never pursued.

The Colorado River would still flow freely through what is now the Boulder Basin, and the Great Bend of the river could still be seen. It is easy to imagine a road running down the Las Vegas Wash to the river with riverside resorts and RV parks similar to that of Arizona's Parker Strip, though the riverside high-rises along the river at Laughlin provide another model for development. Perhaps the stone ruins at Callville would now be a state park, offering a counterpart to the Old Mormon Fort in Las Vegas.

What if neither dam had been built? There would be no Boulder City, of course, but also no Henderson: without water and electricity, the BMI mill would have been built at Needles, California. But other developments would have been unaffected. Nellis Air Force Base would still have been built, and the area would have still made a good site to test nuclear weapons. Hoover Dam provides no power to the Strip, and casinos did not use river water until the 1970s, so the Strip could have still developed without a dam or lake.

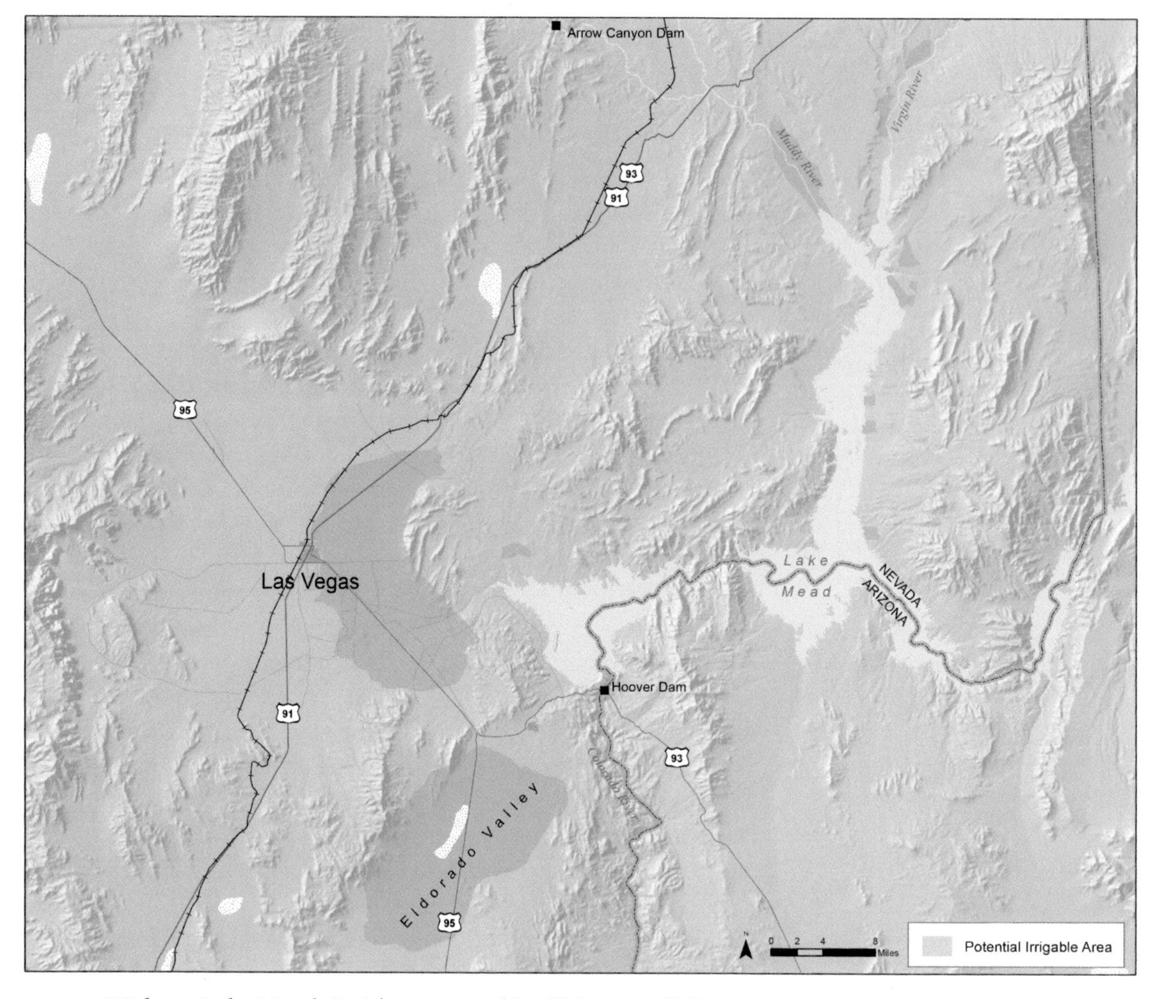

Map 10.5. Agricultural Las Vegas. Hoover Dam was built to regulate the flow of water for downstream farmers. Why not farm in Las Vegas as well? A Nevada state engineer thought irrigated agriculture might be possible in some areas. Only the Muddy and Virgin River valleys were ever farmed.

Without Lake Mead, St. Thomas would still be a small farm town, though given the limited water supplies farming would not likely have spread downriver. There might still be salt mines along the Virgin River, with trucks hauling it up to St. Thomas and the railhead. The Lost City Museum might have been built at the actual Pueblo Grande site, which could be toured. Bonelli's Ferry, once the crossroads of Southern Nevada, might be a small town with a highway bridge over the Colorado. Perhaps it would be a state park as well, much like Yuma Crossing far to the south in Arizona. It might also be a popular takeout point for Grand Canyon rafters and a starting point for those rafting Boulder and Black Canyons. The salt pool would no doubt be a popular local attraction.

Farming is one of the oldest activities of the Las Vegas region. Paiutes farmed around Las Vegas Creek and other springs throughout the valley, and the first white settlement at Las Vegas was an agricultural community. Histories of Las Vegas pay little attention to agriculture after the railroad arrived in 1905, but interest in farming increased as the small town grew. The valley had abundant groundwater, artesian wells, and land available to homestead, leading to an agricultural boom in the town's first decade. Unfortunately, the soil in Las Vegas Valley was not suitable for farming and the boom died. Efforts at farming the Ivanpah Valley south of Las Vegas were abandoned for the same reason.

But the vision of an agricultural Las Vegas persisted. In the 1920s, irrigated agriculture using groundwater was considered the most promising way of developing desert valleys and encouraging population growth; ranching and mining were then in steep decline, and tourism and military bases did not yet exist as significant activities. Only farming would bring people to stay. The US Geological Survey (USGS) investigated these possibilities throughout Southern Nevada and the Mojave Desert, but dreams withered because of limited amounts of suitable groundwater in most valleys. Only in a few areas, such as California's Antelope Valley, were there both good soil and sufficient water close to the surface to permit economical pumping and irrigation.

Another such area was the Pahrump Valley. It had few settlers during this time, but later proved to have both abundant groundwater and good soil. Cotton farming proved a success there in 1948, and it soon became a major crop in the valley. This lasted until the 1970s because of weak prices and the sale of the largest cotton farm, the Pahrump Ranch, to a landholding company in 1970. Without the output of the ranch, the valley's cotton gin could no longer stay in business and other farmers could not afford to ship their cotton to California for ginning. The valley began its transformation from farming into the low-density city it is today (map 4.18).

What about rivers? Agriculture was practiced along the Muddy River for at least a thousand years, leading Mormon pioneers to settle that valley in the 1860s. The water available in this valley quickly became fully allocated, limiting growth. To obtain a larger supply of water, the Civilian Conservation Corps (CCC) began building the Arrow Canyon dam and reservoir in 1935 to help farmers in the Muddy River Valley, but the project was abandoned when the foundation failed.

The Colorado River is an obvious source of irrigation water, but its annual floods made it difficult to harness except on a small scale. A small farm and orchard was run near Greggs Ferry, and Callville was considered a suitable location for agriculture if a canal could be built to divert river water some distance above. The Paiutes had farmed here for generations, using the annual flood to irrigate their crops.

The construction of Lake Mead behind Hoover Dam was justified in part to store irrigation water for downstream farms near Blythe, California; Yuma, Arizona; and in California's Imperial

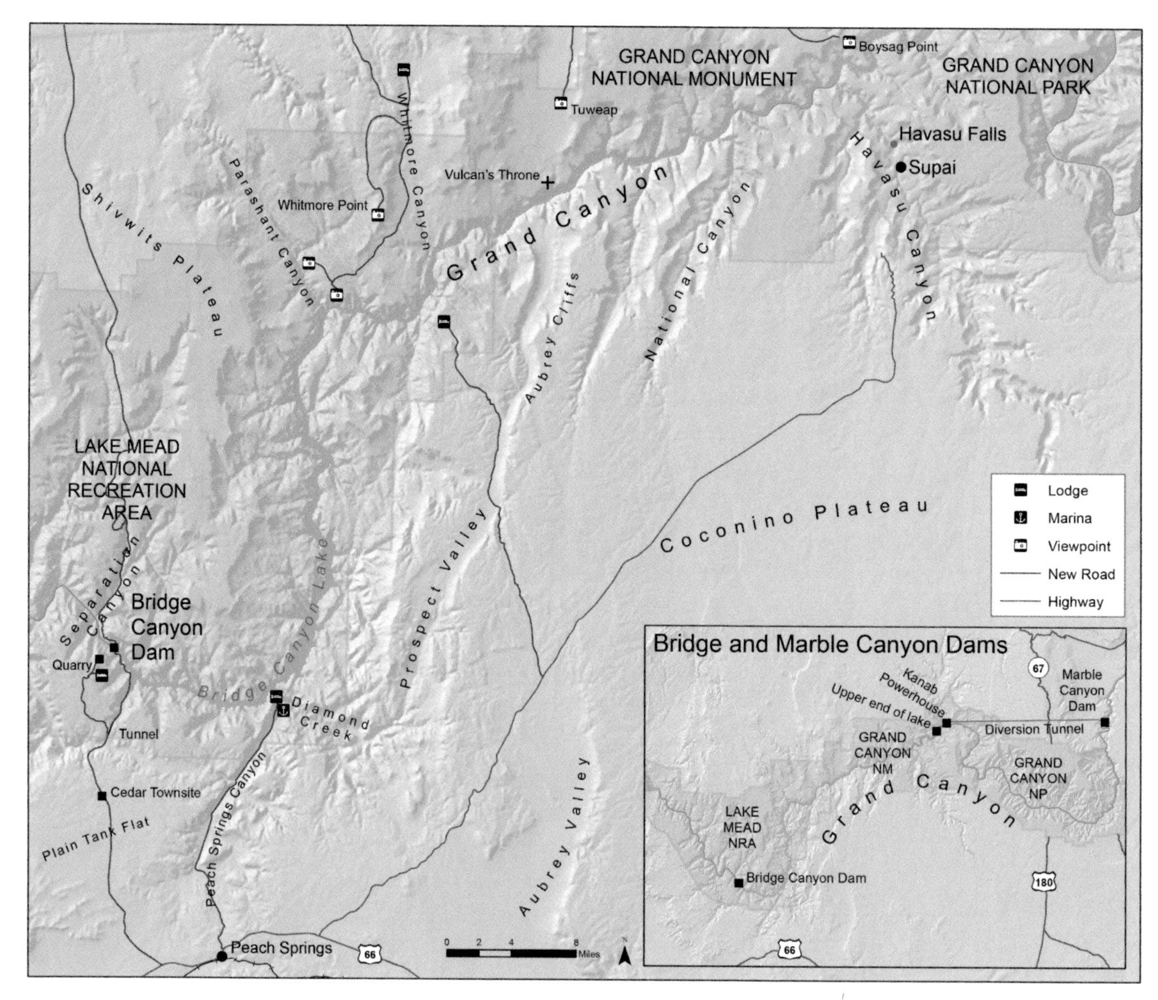

Map 10.6. Bridge Canyon Dam. This dam was never built. Locations are based on plans for new roads and facilities. *Inset*: a diversion tunnel was planned to divert most of the Colorado River away from the Grand Canyon National Park to generate even more electricity.

Valley. Why not use some of that water locally? A map showing potential agricultural areas around Lake Mead was produced. Aside from the Muddy and Virgin River valleys, which were already being farmed, none of these areas were ever developed for agriculture. The map suggests that much of the eastern (and lower) half of Las Vegas Valley could have been developed as farmland. Imagine cotton or alfalfa fields stretching from Nellis Air Force Base south to Henderson, as far west as Maryland Parkway. Another area considered suitable was the Eldorado Valley, including the playa on the valley floor. This might have become a vast agricultural landscape like California's Imperial Valley, with powerlines from Hoover Dam crisscossing the flat fields.

Even if Las Vegas had become an agricultural town, it might have been a brief era. Phoenix was founded as a farm town in 1870, but homes and shopping centers replaced most fields as the city grew. The same might have happened in Las Vegas as well, with only a few curving streets marking the location of former irrigation canals.

After the success of Hoover Dam, other Colorado River

dams were quickly planned, and many built. Parker Dam was completed in 1938, Davis Dam in 1951, and Glen Canyon Dam in 1963. But other potential dams never came to be. One of the most controversial was Bridge Canyon (or Hualapai) Dam, to be built on the Colorado River near the mouth of Bridge Canyon, deep in the Grand Canyon. The Bureau of Reclamation pursued this dam in the 1950s as part of the Pacific Southwest Water Plan but finally abandoned in 1968.

The Bridge Canyon damsite was about twenty air miles upriver from the current Grand Canyon West airport and twenty miles northwest from Peach Springs, Arizona, in an area then and still quite remote. The dam would have filled the inner gorge of the Grand Canyon and stand 740 feet tall, taller than Hoover Dam. The reservoir it created would however have been much smaller than Lake Mead in volume and capacity because it would have been confined within the narrow inner gorge. This small capacity was to have been deliberate; unlike Lake Mead, the dam had not been intended to store water but only to generate electricity. The volume of water behind the dam was considered irrelevant; only the height of the dam was important to maximize the electrical output. The smaller capacity would have been an advantage because it would have made the lake easier to maintain at full capacity (and full electrical output) year round.

If the reservoir had been filled, it would have initially reached ninety-three miles upriver. But like Lake Mead, the reservoir would have retreated downstream as sedimentation began. When full, water would have extended two-and-a-half miles up Peach Springs and Parashant Canyons and a half-mile up Havasu Canyon. At the time the dam was being planned, the reservoir location lay mostly within Lake Mead National Recreation Area. The upper end passed through Grand Canyon National Monument, and when first filled would have included several miles along the boundary of Grand Canyon National Park. The maximum lake extent would not have been visible from any tourist overlook on the north or south rims.

The stretch of Colorado River through the national park was steep and long enough to allow another similar dam and reservoir for power production, but this was impossible because it was within a national park. Instead, to harness the drop in elevation from a Marble Canyon reservoir, upstream from the national park, and a Bridge Canyon reservoir, a forty-five-mile tunnel was planned to connect the two. Most of the river water would have been diverted from Marble Canyon through this tunnel to a powerhouse near Kanab Creek, where it would have fallen through a hydroelectric facility and reenter a river channel. The Colorado River through Grand Canyon National Park would have been reduced to a trickle, making river rafting impossible and likely allowing the spread of plants such as salt cedar throughout the mostly dry river channel. This would have been visible from tourist overlooks.

Following the experience of building Glen Canyon Dam, no railroad was considered necessary for construction. Plans were made that showed the location for the construction road branching off US 66 west of Peach Springs and running to the dam site. The descent on the road into the outer gorge of the Grand Canyon would have been impressive, with a 4,000-foot-long tunnel descending into the canyon through the walls of the outer gorge. Redwall limestone quarries were needed for the construction project, located on the canyon rim to the south of the dam. A town, labeled as Cedar townsite in the plans, would have been required to house workers and was to be on a plateau south of the canyon. This town would likely have been quite different from the model community of Boulder City developed in the 1930s (map 5.8); it might have been laid out more like Page, Arizona, developed in the 1950s, with the curving street pattern popular at the time.

The map shows the reservoir at its maximum extent and the location of planned roads and developments. The project never advanced beyond the early planning stage, so the locations and facilities shown are all speculative. The reservoir was to have been developed for recreation, with lodges and other facilities located on the canyon rim above the dam, in Diamond Creek, overlooking Granite Park, and north of the reservoir in Whitmore Wash. Peach Springs Canyon would have provided the primary access to the lake; the other viewpoints and lodges would have been high on the canyon rim. A highway was planned from the damsite north, climbing out of the canyon and providing the only road crossing between Hoover Dam and the Navajo Bridge near Lees Ferry. About 365,000 visitors a year were expected when the facilities were completed (about 10 percent of the annual visitors that Lake Mead received in the mid-1960s, but similar to the numbers Glen Canyon National Recreation Area received at the time). The long narrow reservoir would have been a very different experience from Lake Mead. The section from Diamond Creek to Parashant Canyon was planned as the heart of the reservoir, and likened to a fjord with cliffs towering spectacularly over the narrow lake.

Opposition to the Bridge and Marble Canyon dams was intense. Both dams were ultimately canceled in 1968. Marble Canyon became a national monument in 1969 and absorbed into Grand Canyon National Park in 1975. Grand Canyon National Monument was absorbed into Grand Canyon National Park that

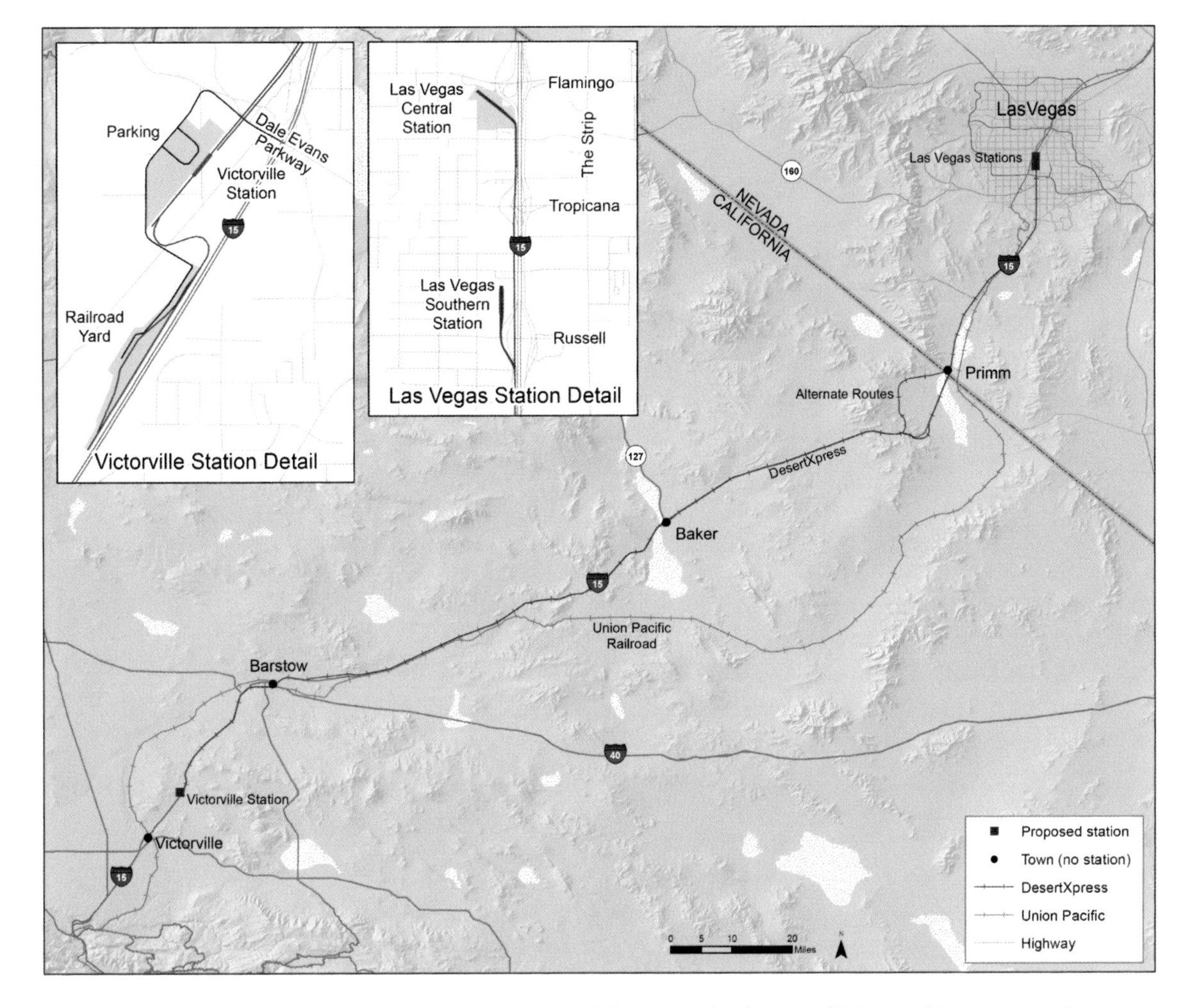

Map 10.7. By Train to Las Vegas. A high-speed passenger train between Victorville and Las Vegas has been proposed. Most of the line would be built within the median of Interstate 15, with alternate routes under study near Primm. *Insets*: details of the stations at each end.

same year, and the part of Lake Mead National Recreation Area that included the Grand Canyon was also transferred to the park.

Portions of the Pacific Southwest Water Plan were later built as the Central Arizona Project, which includes a canal from Lake Havasu to Phoenix and Tucson. The electricity that the Bridge Canyon dam would have generated was to be used to power massive water pumps on this canal. These pumps, the single biggest users of electricity in the state of Arizona, required a new source, furnished by the coal-burning Navajo Generating station near Page (also providing electricity for Las Vegas and other cities [map 8.17]). The coal required for this plant was strip-mined on the Navajo Indian Reservation at enormous environmental cost. The power plant was permanently closed in 2019 because of rising coal prices and the air and water pollution problems it created.

Following the demise of passenger train service to Las Vegas in 1997, several schemes to start new services have been promoted. One of these is the DesertXpress high-speed railroad, which would run between Southern California and Las Vegas, carrying many travelers who would otherwise drive on Interstate 15.

Trains would depart hourly and every twenty minutes Fridays and Sundays, preventing weekend traffic jams.

The line would be a new double-track railroad, completely separated from existing rail lines, and usually far distant from the Union Pacific Railroad. The southern end of the line was to be north of Victorville, near the Dale Evans Parkway exit on Interstate 15. This meant that riders would first have to drive from their homes past Victorville and then leave their vehicles for the train, but building a line down Cajon Pass and through dense urban areas was not considered feasible.

Most of the 190-mile route was to be built alongside Interstate 15, but in a few areas the line would depart from the highway to obtain an easier curve or better grades. Between Mountain Pass and Primm, two routes were considered, one swinging well to the east of Interstate 15 and the other to the west. Two locations were also proposed for the Las Vegas station, one at Interstate 15 south of Flamingo Road and the other west of Interstate 15 at Russell Road. The latter site had the advantage of being an empty lot, but was later selected to become the site of Allegiant Stadium for the Raiders. The train could also have a stop at the proposed Ivanpah airport, though this was not part of the original plan (map 10.9).

The trains would be powered by overhead electric lines, like those in the Northeast or in Europe and Japan. They would run at a maximum speed of 150 mph (though the average speed would be much slower), rivaling the speed of Amtrak's fastest train service in the Northeast. Trains would complete a one-way journey in an hour and twenty minutes with no stops or even railroad crossings along the way. One-way fares were planned at $50. Construction costs were estimated to be as much as $5 billion; it was repeatedly claimed to be a purely private sector project but a $6.5 billion loan from the federal government was sought. Critics have noted that the ridership levels used by the promoters appeared wildly overoptimistic, and why Southern Californians would drive halfway to Las Vegas, park and then ride a train the rest of the way was never explained.

The DesertXpress line was renamed XpressWest in 2012 and became part of a proposal for a high-speed rail network connecting Southern California, Las Vegas, Phoenix, Salt Lake City, and Denver. Later plans also called for the southern end of the line to be located at Palmdale, where it would connect with the California High Speed Rail line under construction between the Bay Area and Los Angeles. Construction dates for the XpressWest line have repeatedly been pushed back, and the project does not appear likely to ever begin. If you want to take a high-speed train to a casino city now, your only option is the Guangzhou-Zhuhai route

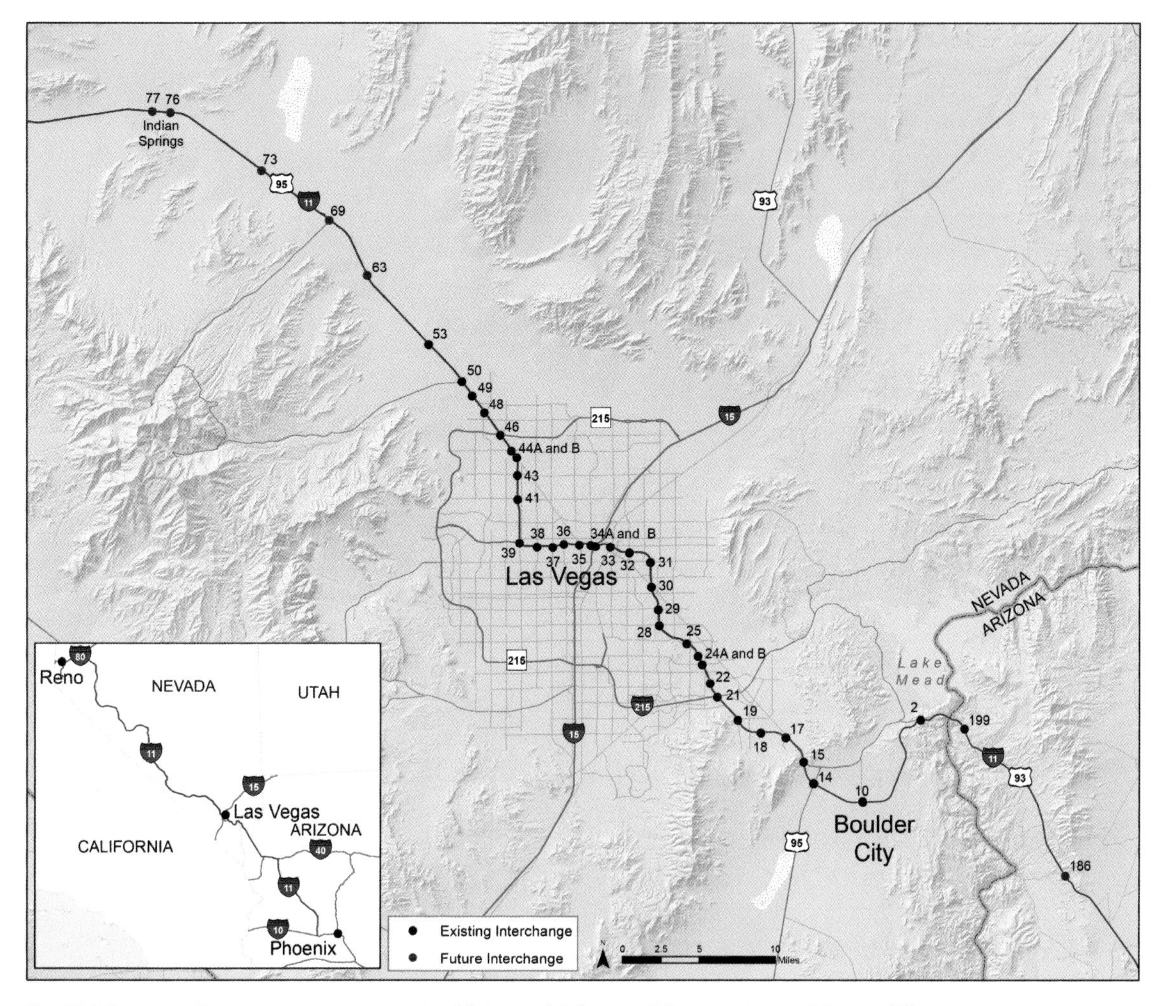

Map 10.8. Interstate 11. Signs for Interstate 11 can already be seen around Boulder City, but the highway may eventually run from Arizona to Reno (and beyond). The map and inset show what this might look like if it is completed, with numbered interchanges based on current highways.

in China, which provides service to Macao. This train connects Las Vegas's biggest competitor to a nearby metropolis of more than fourteen million people and runs at 124 mph.

When the Interstate Highway System began construction in the 1950s, Las Vegas was a small town of about twenty-five thousand people, while Phoenix was a small city with a little more than one hundred thousand people. The main highway between them, US 93, did not even exist then. The two metropolitan areas today have a combined population of six million and heavy traffic between them. Freeways connected other cities of similar size and distance as early as the 1950s, so why not these two?

Federal law authorized Interstate 11 in 2012 to run between these cities, and studies have further examined extending it north to Reno and Canada and south to Tucson and Mexico. Although Interstate 11 signs have been installed in sections in the Las Vegas Valley and south, the highway remains mostly in the planning stage, and there is little or no money available to build it. Interstate routes were built quickly in the 1950s and 1960s because the federal government paid for at least 90 percent of the cost

(93 percent in Nevada). That is no longer the case: highways face increasing competition from other priorities and diminished budgets. The Boulder City Bypass was called Interstate 11 when it was completed in 2018, but drivers will look in vain for the number beyond the immediate region.

Where will Interstate 11 run if it is built? The route has not yet been officially determined, but portions of a freeway between Phoenix and Las Vegas have already been built or could be finished with relatively little effort. As of 2021 two routes are under consideration in Las Vegas, one using Interstate 515 and US 95, and the other running around the west side of the city on the Interstate 215 beltway to US 95. US 93 to Kingman is already a four-lane divided highway that could be upgraded to a freeway, and the Arizona Department of Transportation initiated a study on how to better connect US 93 to Interstate 40 in Kingman. From Kingman, Interstate 11 would presumably use US 93 to Phoenix.

The planned route was extended to Reno in 2015, and it may someday be extended even farther north, though exactly where has not been determined. North of Las Vegas, it will surely follow US 95 at least until close to Reno. The highway is already four lanes to Mercury, and beginning in the 1990s US 95 has crept north when several interchanges were built outside Las Vegas. There would be little barrier to making the remainder of this highway into a freeway. In fact, when US 95 north was rebuilt as a four-lane road in 1963 provisions were made in the highway's right of way for several future interchanges between Indian Springs and Beatty.

Exits along unbuilt sections of Interstate 11 shown on the map are conjectural, though many locations are easy to guess. In Nevada and Arizona, exit numbers are based on mileage from the beginning of the road, which is the southern end for odd-numbered roads. North of Las Vegas, exit 73 might serve Cold Creek Canyon and several prisons west of the highway, exit 69 allows access to Lee Canyon in the Spring Mountains, and exit 63 is for the Corn Creek area in the Desert National Wildlife Refuge. Arizona exit 186 is for the Willow Beach road to the Colorado River, a popular spot within the Lake Mead National Recreation Area.

Interstate 11 will provide Las Vegas with a direct freeway connection to the booming Phoenix metropolitan area as well as smaller towns and cities in northern Nevada. It could allow for spillover growth in the Kingman area, which is only slightly farther than Pahrump or Mesquite. However, it will likely have very little effect within Las Vegas itself because the route will follow existing freeways.

In the early 2000s, passenger traffic to McCarran International Airport (transitioning to be called Harry Reid International

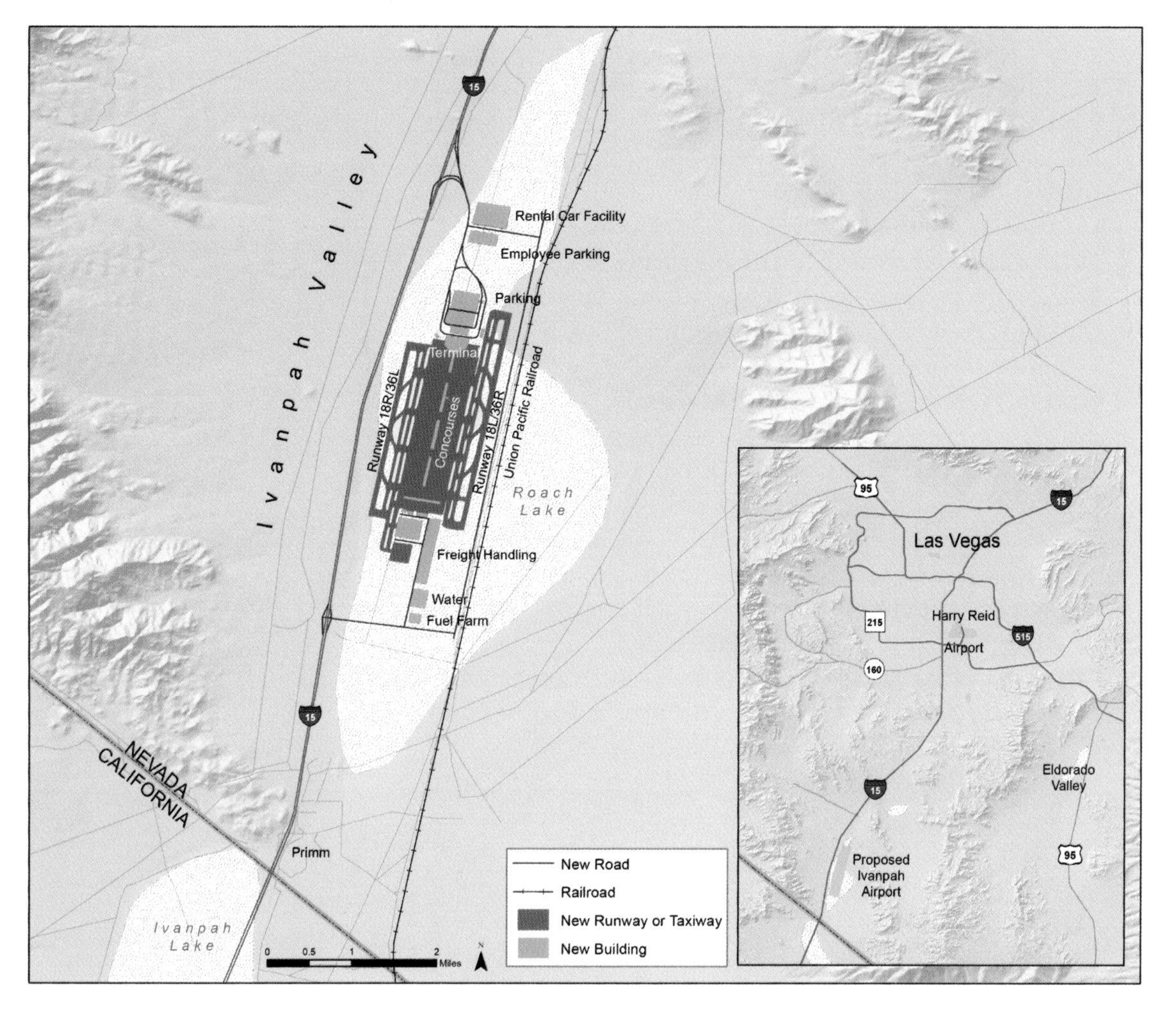

Map 10.9. Ivanpah Airport. Las Vegas will someday need another airport to handle more visitors in Ivanpah Valley north of Primm. Map based on 2006 plans for the airport.

Airport in 2021) was growing at a steady rate, and a series of improvements expanded the airport's capacity (map 7.18). However, with the completion of Terminal 3 and the D concourse, this expansion has come to an end. Flights and passengers will continue to grow, reaching a point where the airport has reached its maximum capacity. The airport is surrounded by built-up land, much of it very valuable, and can no longer be expanded. None of the other airports in the Las Vegas area can be expanded for airline operations even if residents around them supported the increased traffic and noise this would bring. The only solution to future growth is a new airport.

Where should this airport be located? It would have to be built on a vast open valley floor; have no incompatible land uses around it; be distant from restricted military airspace to the north, west, and east of the city (map 7.19); and be convenient to the Las Vegas area, preferably within a thirty-minute drive. Ivanpah Valley, Eldorado Valley, Pahrump Valley, Indian Springs, Apex Valley, and the Moapa Reservation were all examined as sites for a new airport. Only the Ivanpah and Eldorado Valleys met all of these requirements, but most of the Eldorado Valley was within the vast city limits of Boulder City (map 5.8), which was opposed to using it for a new airport. The Ivanpah Valley south of Las Vegas was therefore chosen. This is a long north-south valley stretching from Sloan to California, and is familiar to many as the route of Interstate 15 between Las Vegas and Primm. The preferred location was between Interstate 15 and the Union Pacific Railroad, covering most of Roach Dry Lake.

Most rural land in Southern Nevada, including Roach Lake, remains in the public domain and no longer sold to private owners or even local governments (map 8.1). Clark County had to persuade Congress to pass a law authorizing the BLM to sell 5,800 acres of land to the county. The earliest plans called for the airport to open in 2017, but soon after this announcement the Great Recession began and air traffic to McCarran slowed. The new airport project was shelved, but it will someday return as air traffic picks up. It will eventually be built.

What will the Ivanpah airport look like? Compared to Harry Reid, it will occupy twice the amount of land but will start off modestly. Preliminary plans for the airport call for two runways; the longer runway (18R/36L) would be 15,000 feet long to handle airplanes as large as the double-decker A380 (the largest airliner in the world) during hot summer conditions, which increase takeoff runs. The shorter (18L/36R) would be 12,000 feet long, capable of handling the majority of flights. The longest runway at Harry Reid is 14,512 feet long; a 15,000-foot runway would be the second-longest commercial airport runway in the United States.

The airport would have a midfield terminal between the runways and would open with twelve gates. The number would grow to thirty within five years, and a satellite terminal (or concourse) connected by an underground people-mover would bring the number to forty-two. Space would exist for three additional satellite terminals to the south when demand warranted. An air traffic control tower would stand in the center of the airport, eventually to become part of one of the satellite terminals. This being the Las Vegas airport, the county's plans for the airport include 51,000 square feet of casino space when it opens, which would be expanded to 79,600 square feet when the airport was developed with forty-two gates (in contrast, McCarran had 118,000 square feet of gambling before Terminal 3 opened).

Harry Reid International Airport would remain open after the new airport began operations. Exactly how flights and airlines might be allocated between the two airports was not addressed in the airport plan, except for a statement that the new airport would be more likely to handle long-distance flights. A similar scheme was used when Dulles International Airport was opened outside Washington, DC, with all flights from one thousand or more miles away forced to use this airport rather than the more convenient Washington National Airport (now Ronald Reagan Washington National Airport). An international arrivals area would be included in the terminal building.

Interstate 15 would provide access to the airport, with extra airport-only lanes added in the median and a new interchange built north of the airport. Parking garages and a rental car facility were included in the plans. Maintenance, freight-handling facilities, hangars, and a fuel tank farm were to be built to the south of the airport. Bringing fuel to the airport would not be a problem because the airport was adjacent to the Calnev pipeline, which also supplies Harry Reid with jet fuel (map 8.18). The possibility of providing railroad service to and from the airport, located between the Union Pacific Railroad and the proposed DesertXpress high-speed train (map 10.7), was also not addressed but seems to offer considerable potential. A second Interstate 15 interchange south of the airport would give access to freight facilities, general aviation, and other services. The freight facilities would be modest because Las Vegas is not a major freight destination, and the city generates very little outbound air freight. Express companies such as FedEx and UPS would also prefer to stay at Harry Reid to be closer to their customers. Perhaps the biggest problem to be confronted when building the new airport is that it will be built in the middle of Roach Dry Lake. Dry lakes are not always dry, however, and the airport would require a massive flood control and drainage system to divert water away into nearby lakes, such as Ivanpah Lake south of Primm. The details of such a system remain to be worked out, but Southern Nevada has had tremendous success in building flood control projects (map 6.8).

What will the new airport be called? The airport would be situated in Roach Lake, but that would presumably not be part of the airport's name. The plans refer to it simply as Ivanpah Airport, and the county plans also use the abbreviation IVP as its three-letter airport code (Harry Reid International will keep the familiar LAS code as McCarran).

What will happen in the Ivanpah Valley when the airport is built? To the south are the casinos of Primm at the California stateline, with a resident population of more than 1,100 people. It is likely that these two communities would grow substantially following the opening of the new airport and have to deal with accompanying airplane noise. The Jean airport is about four miles north and will lie under the flight path of the Ivanpah airport. The law that transferred the land for the Ivanpah airport to Clark County requires the county to keep this airport in use even after opening the new one.

This is not the first airport discussed in the Ivanpah area. In September 1967 Howard Hughes announced his intention to build a new airport for supersonic transports, such as the new British/French Concorde and a planned American counterpart. This airport would have been the destination for supersonic flights across North America and the Pacific Ocean, and would have been a hub for passengers to transfer to conventional airplanes to continue their trips to California, Nevada, and Arizona. He suggested the Ivanpah Valley, Pahrump Valley, Mormon Mesa, or even the North Las Vegas airport (which he owned) as possible sites. As was often the case with Hughes's announcements, nothing was ever heard of the plan again, and supersonic transports ultimately proved a commercial failure. Only one new airport was ever started for supersonic transports, in the everglades west of Miami, but only a single runway was ever completed before this project was discontinued.

Yucca Mountain, a low ridge just west of the Nevada National Security Site (chapter 8), has been part of a very long running political battle in Southern Nevada, and it may not even be over. The project originated with the need to safely store highly radioactive waste from nuclear power plants and decommissioned nuclear weapons. These will remain lethally radioactive for thousands of years, and could potentially be used as weapons if they fell into the wrong hands.

In the 1970s, the Atomic Energy Commission planned to build a high-level waste repository in a former salt mine underneath the town of Lyons, Kansas. The opinions of locals were mixed about the idea, but the plan fell apart because of evidence that groundwater might enter the repository and eventually contaminate water supplies with radioactive waste. After this site was canceled the commission began looking for sites across the nation, but the Nuclear Waste Policy Act of 1982 required the agency, now the US Department of Energy, to look at just five sites and recommend three of them for further study by January 1, 1985. The three finalists were a Texas mine, the Hanford Nuclear Reservation in Washington, and Yucca Mountain. But before studies could be carried out on each of them, Congress cut funding and instead passed a bill amending the Nuclear Waste Policy Act that made the Yucca Mountain site the designated waste site. This act was called the "Screw Nevada Bill" and began a long-running saga of scientific tests and political battles over Yucca Mountain. Funding for the project was killed in 2011, thanks in part to the adamant opposition of five-term senator Harry Reid.

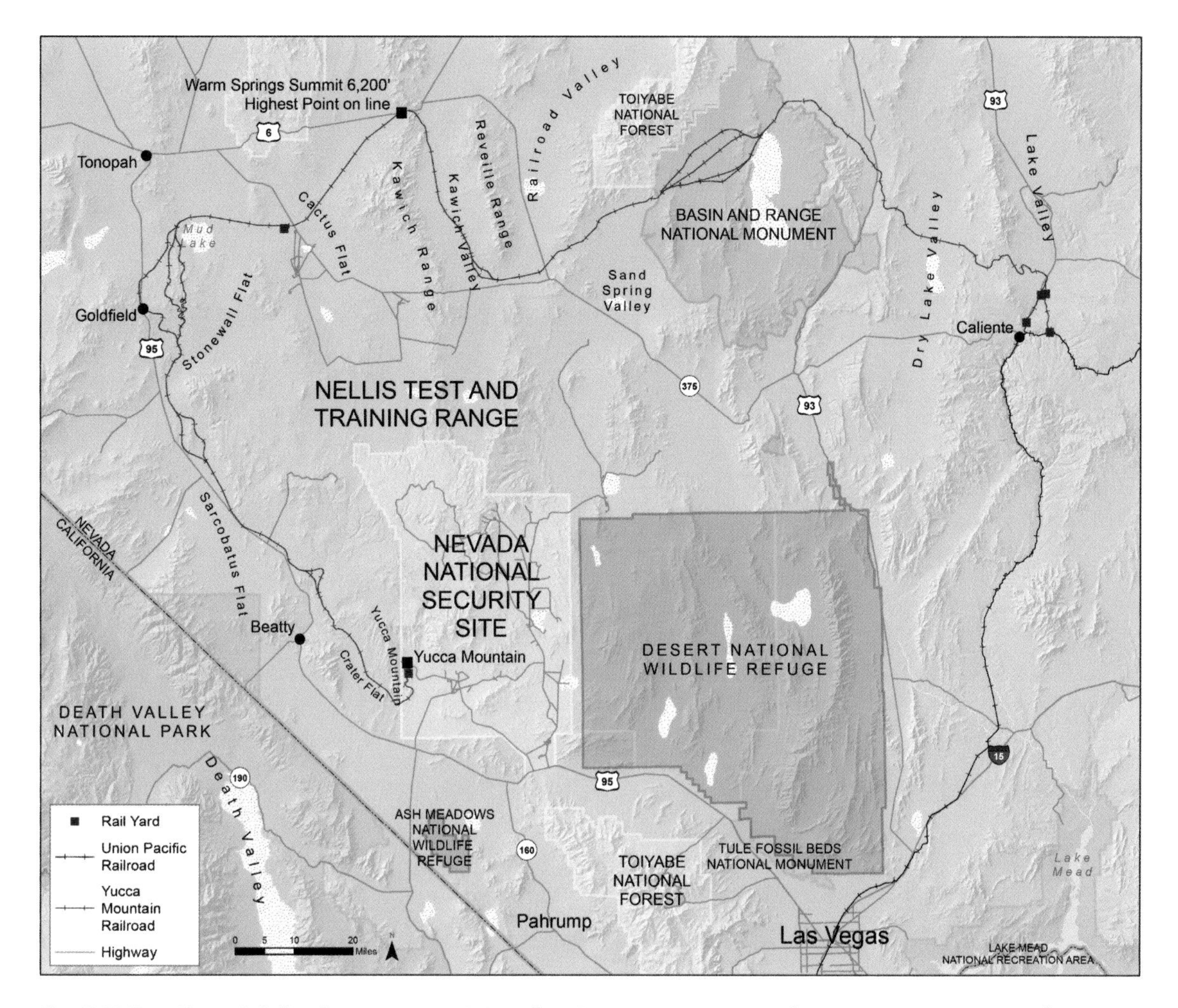

Map 10.10. Yucca Mountain Railroad. The planned route of a railroad that would haul radioactive waste to the Yucca Mountain repository is shown in red. In several locations, there were alternative routes under consideration. The routes for the railroad are from a Department of Energy website devoted to Yucca Mountain transport issues (since taken down).

Were the Yucca Mountain nuclear waste repository to have opened, a railroad was planned to carry the nuclear waste to the site. Several railroads once ran nearby (map 7.2), but the last was torn up in the 1940s and two different possibilities were considered for a new line. The Mina Corridor would have connected to railroad lines at Mina in northern Nevada and run south to Yucca Mountain. The preferred plan, called the Caliente Corridor, would have branched from the Union Pacific Railroad at Caliente, loop around the north end of the Nellis Test and Training Range, pass south of Tonopah (the headquarters for the railroad), then turn south along US 95. Several different routes near Goldfield were discussed, but all would have passed the town to the east. From there the line would run south, passing Beatty to the east before turning north to Yucca Mountain.

The long looping route of the railroad was required to bypass US Air Force bombing ranges and because it would have to go around many north-south mountain ranges. The high point on the line would be about 6,200 feet at Warm Springs Summit, and

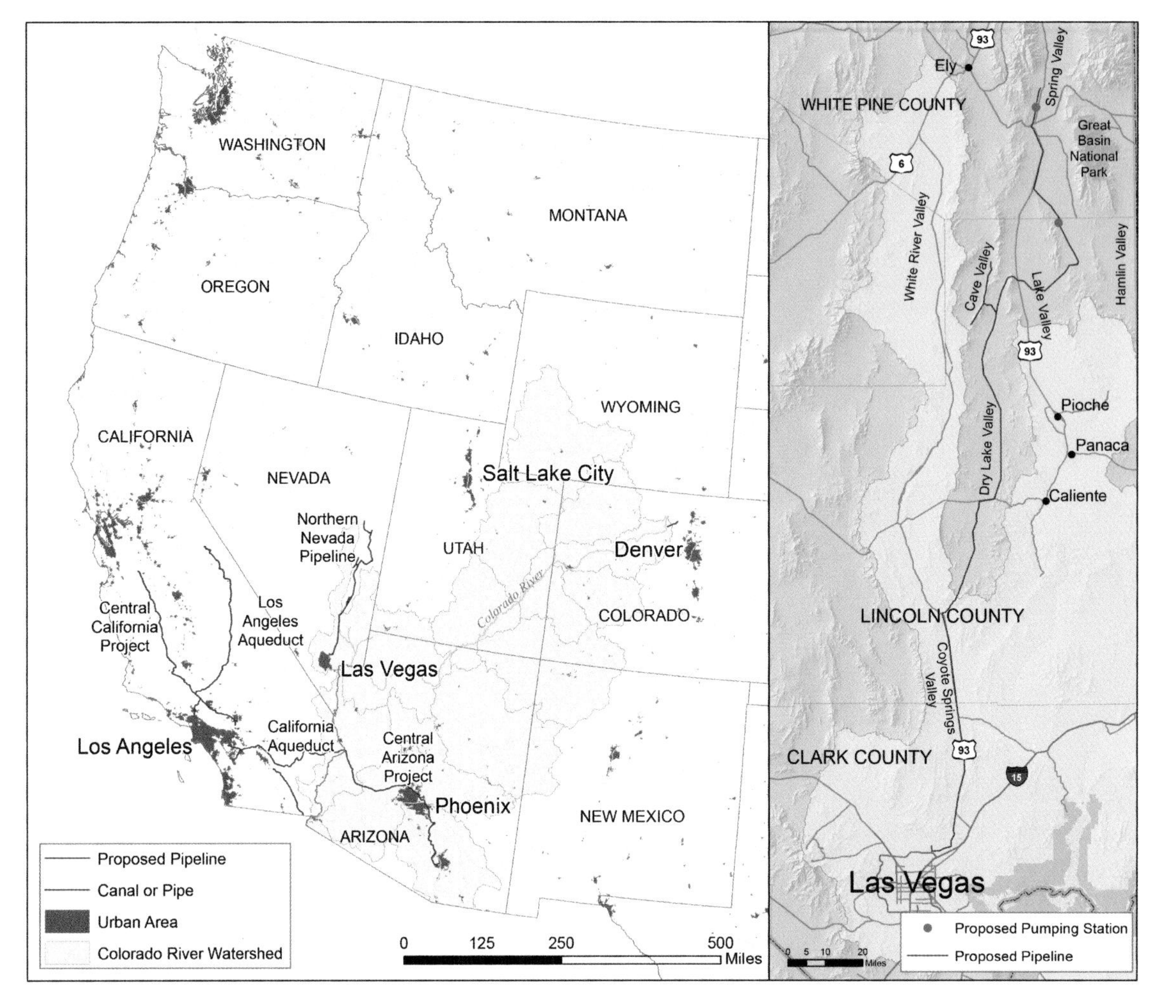

Map 10.11. Northern Nevada Pipeline. Major water projects in the West, including a proposal to bring water from northern Nevada to Las Vegas, are displayed. *Inset*: the detailed route and pumping plants that would be needed. Based on a 2012 plan by the Southern Nevada Water Authority.

several substantial bridges would be required, all of them overpasses of state highways. This new railroad would have been very well engineered and quite different from the quickly built mining railroads of the early twentieth century. The Yucca Mountain railroad would also have differed because it would have hauled valuable (and dangerous) cargo in rather than out.

In 2015, President Barack Obama signed a proclamation creating the Basin and Range National Monument north of Las Vegas (map 9.8). The Caliente Corridor would have passed through what is now the monument, and the existence of the monument closes this area to the route. The monument may even have been created deliberately to block the route. If Yucca Mountain ever proceeds, it will have to do so with a different rail line, or the national monument will have to be eliminated.

For the first sixty-five years of the city's existence almost all water in Las Vegas came from Las Vegas Creek, flowing springs, and a large number of wells. Only in the 1970s, with the completion of the Southern Nevada Water Project, did the city become reliant on water from the Colorado River. A massive system of pumps, pipes, and tunnels was necessary to make this happen. The continuing drought and shrinking lake levels prompted a third water intake to be built, this one a three-mile-long tunnel dug underneath Lake Mead to an intake structure lowered into a dredged hole. This will allow the city to obtain water even if the lake level continues to drop.

Despite this massive infrastructure, there are limits to the use of Colorado River water. In 1922, the Colorado River Compact specified that the river's annual flow in the lower basin (California, Arizona, and Nevada) amounted to 7.5 million acre-feet. In 1964, the US Supreme Court decided that of this amount California was to get 58.7 percent, Nevada 4 percent, and Arizona 37.3 percent. California was quick to divert its share of the river with several aqueducts bringing water to the Imperial Valley and the Los Angeles area. Arizona built the Central Arizona Project much later to bring river water to Phoenix and Tucson.

The distribution of water reflected the population of these states at the time; when the river's water was allocated, Clark County had only 8,532 people, but it now is home to more than two million. The uncertain availability of Colorado River water and hopes for future population growth will require a new source of water. Beginning in the 1980s, plans have been developed to bring more water to Las Vegas by pumping groundwater from several valleys in northern Nevada and then moving it to the city in a pipeline. Like Las Vegas, high mountains that receive much heavier precipitation surround these valleys, supplying the aquifers in the valley bottoms. Even though the pipeline would be built through basin and range terrain (map 2.2), only two pumping stations would be required to move the water over passes; these would also require a parallel power line.

The Las Vegas Valley Water District has purchased groundwater rights throughout much of the area. The Bureau of Land Management approved the pipeline in 2012, but in 2020 the Southern Nevada Water Authority decided not to proceed with it. If it were to be built, it would have been the most expensive construction project in Nevada's history. This project would rival the Los Angeles Aqueduct, which brings water 419 miles from Owens and Mono Valleys to enable Los Angeles to grow beyond its own meager water supplies, and the 336-mile Central Arizona Project.

The environmental consequences of the project have been sharply questioned. Groundwater pumping within Las Vegas dried up the valley's springs by 1957, while pumping within Pahrump Valley dried up several springs by 1975. Extinctions of native fishes in both valleys occurred when these springs went dry. Pumping in Ash Meadows likewise led to sharp declines

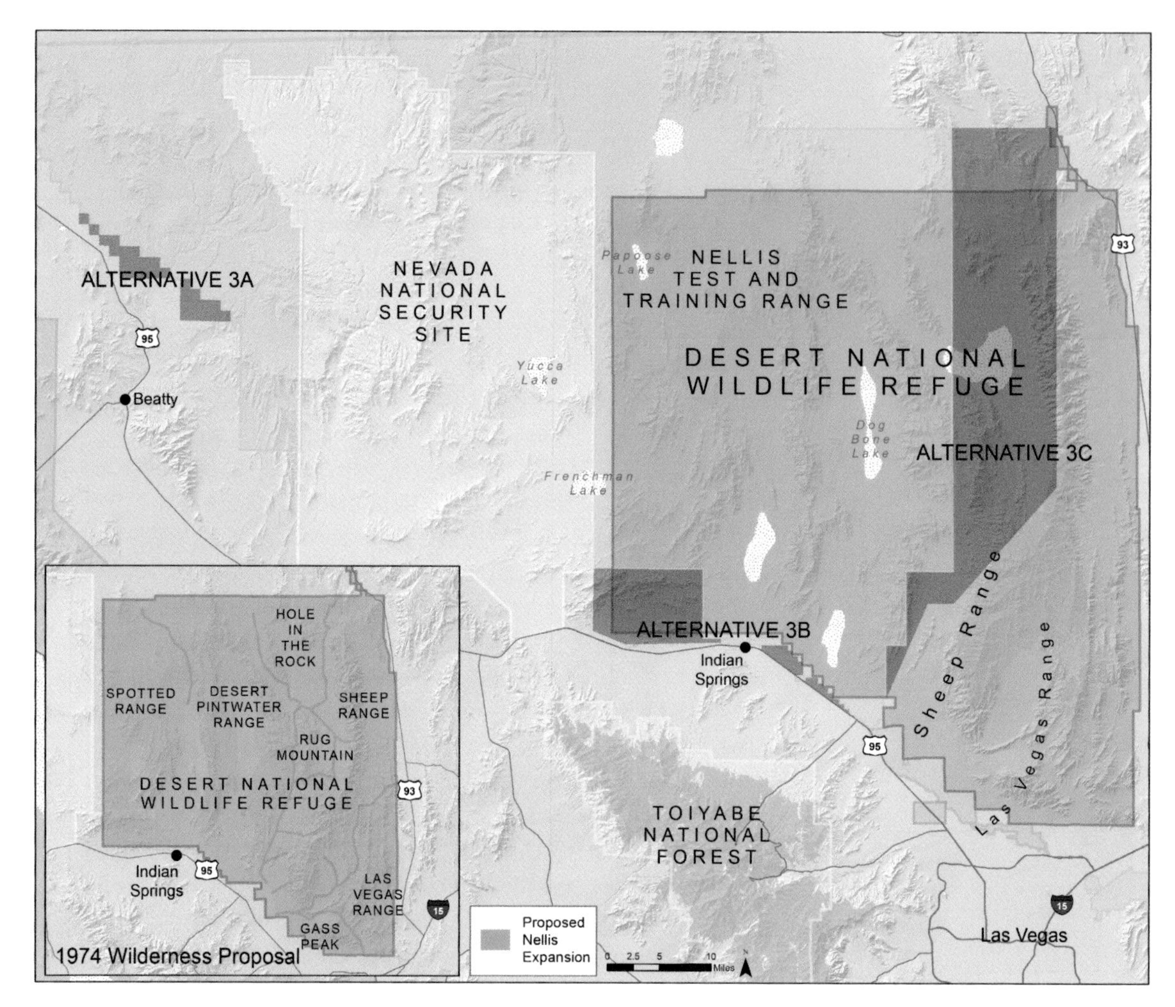

until a US Supreme Court decision in 1976 halted it to protect the Devils Hole Pupfish. Plans by Las Vegas to pump groundwater from northern Nevada will reduce groundwater levels throughout this area, and many springs will go dry.

Map 10.12. Bombing the Bighorn. To expand its reach in 2017, the US Air Force proposed several areas (called Alternatives), two of which were to have taken place in the Desert National Wildlife Refuge.

Scientists have found that groundwater in many Nevada valleys is connected; meaning that heavy pumping in one may eventually reduce water levels in other valleys. These connections likely extend as far west as Death Valley and may include parts of northern Nevada where pumping is planned. The tiny pupfish at Devils Hole, still alive and protected by the Supreme Court decision giving the National Park Service the authority to regulate groundwater pumping in the area, may prevent Las Vegas from pumping in northern Nevada.

The Desert Game Range was created in 1936 (map 9.3), encompassing the Sheep and Spring Mountain Ranges and many desert basins. In World War II, the Tonopah Bombing Range was established and encompassed much of the Game Range. At war's end, the Gunnery Range did not disappear but was instead expanded, with the air force bombing the valley floors and the bighorn roaming the scattered mountains.

The Game Range became the Desert National Wildlife Refuge (NWR) in 1966, and the Fish and Wildlife Service began to create a wilderness plan. This was finalized in 1974 and included most areas more than 3,600 feet in elevation, except for several road corridors and disturbed areas. Almost 88 percent of the NWR would be classified as wilderness, including the portion in the air force range. However, no action was taken, and today the NWR is striking for its absence of wilderness areas (map 9.12).

The temporary overlay of the air force range on the wildlife preserve has been extended several times and is currently set to expire in 2045. The air force wants to not only continue it but also to expand it. For training purposes the training range is operated as a northern and southern section, but the southern section is considered too small to encompass all flight training. It is particularly restricted for bombing missions, as planes may only approach targets from certain directions to avoid sensitive areas. Elevations above 4,000 feet within the Desert NWR overlay are managed as wilderness, and the air force can only effectively use 112,000 acres on the 1.2 million-acre South Range.

The air force has three expansion zones, termed Alternatives 3A, B, and C. Alternative 3A would encompass several square miles outside the current boundary near Beatty, though the state highway department has expressed concern that this not limit the future route of Interstate 11. Alternative 3B expands the border to US 95 east and west of Indian Springs. Alternative 3C is the largest and most contentious. With this the Nellis Range would expand into the Desert NWR, including portions of the Sheep Range, to allow for more flexibility when dropping bombs. With these expansions the air force would also no longer manage any lands as wilderness, removing this protection from bighorn habitat. Fortunately for the bighorn, the air force proposal was rejected in 2020 by the 2021 National Defense Reauthorization Act.

Remembering Las Vegas

11

This chapter begins with the stories of several individuals associated with Las Vegas. These entertainers, singers, fictional secret agents, billionaires, and washed-up cops are an important part of our memory of the city and its image. Their time in Las Vegas is shown on the maps: where did they live and perform, race their cars, hide from the public, fight evil villains or dirty cops, or get married (on- or off-screen)?

The chapter also examines important places in the city that have been lost. Many casinos have burned down, torn down, or imploded over the years to make way for new ones. These were not just obsolete businesses; they were places where important civic events were held, where entertainers became stars, and, to many around the country and the world, they represented the best and worst of Las Vegas.

The last few maps depict what survives from the city's early years. Despite what many people think, Las Vegas has a number of historic buildings and neighborhoods, and many relics can still be found here and there around the city. Only one of these is a casino; they may include water towers, signs and old houses. They are often found in out-of-the-way locations, but may be hiding in plain sight; you may have driven past them and never gave them a second glance. The city's history is all around if we look for it.

Frank Sinatra, Dean Martin, Sammy Davis Jr., Peter Lawford and Joey Bishop are some of the most famous names associated with Las Vegas. Together they were known as the "Rat Pack," a name bestowed on them by actress Lauren Bacall at the Desert Inn in 1955, but they all had separate careers in Las Vegas as well.

Born in 1915, when Las Vegas had fewer than two thousand people, Frank Sinatra was a wildly popular singer before becoming a movie star. He performed in the city for the first time in September 1951 at the Desert Inn and later took up residency at the Sands in 1953. He continued to perform there for fourteen years while staying in its presidential suite. Along with Dean Martin, he later even owned a share of the hotel. The end of his stay at the Sands came abruptly in 1967 after the Sands new owner, Howard Hughes, canceled his line of credit at the casino. In anger, Sinatra moved his show to Caesars Palace, where he performed until 1983. He later starred at the Golden Nugget, and at the MGM Grand until 1994, when he retired. He died in 1998.

Although many sources say Sammy Davis Jr. first performed in Las Vegas in 1944, evidence shows that his first visit to the city was in 1947 as part of the Will Mastin Trio. They performed at

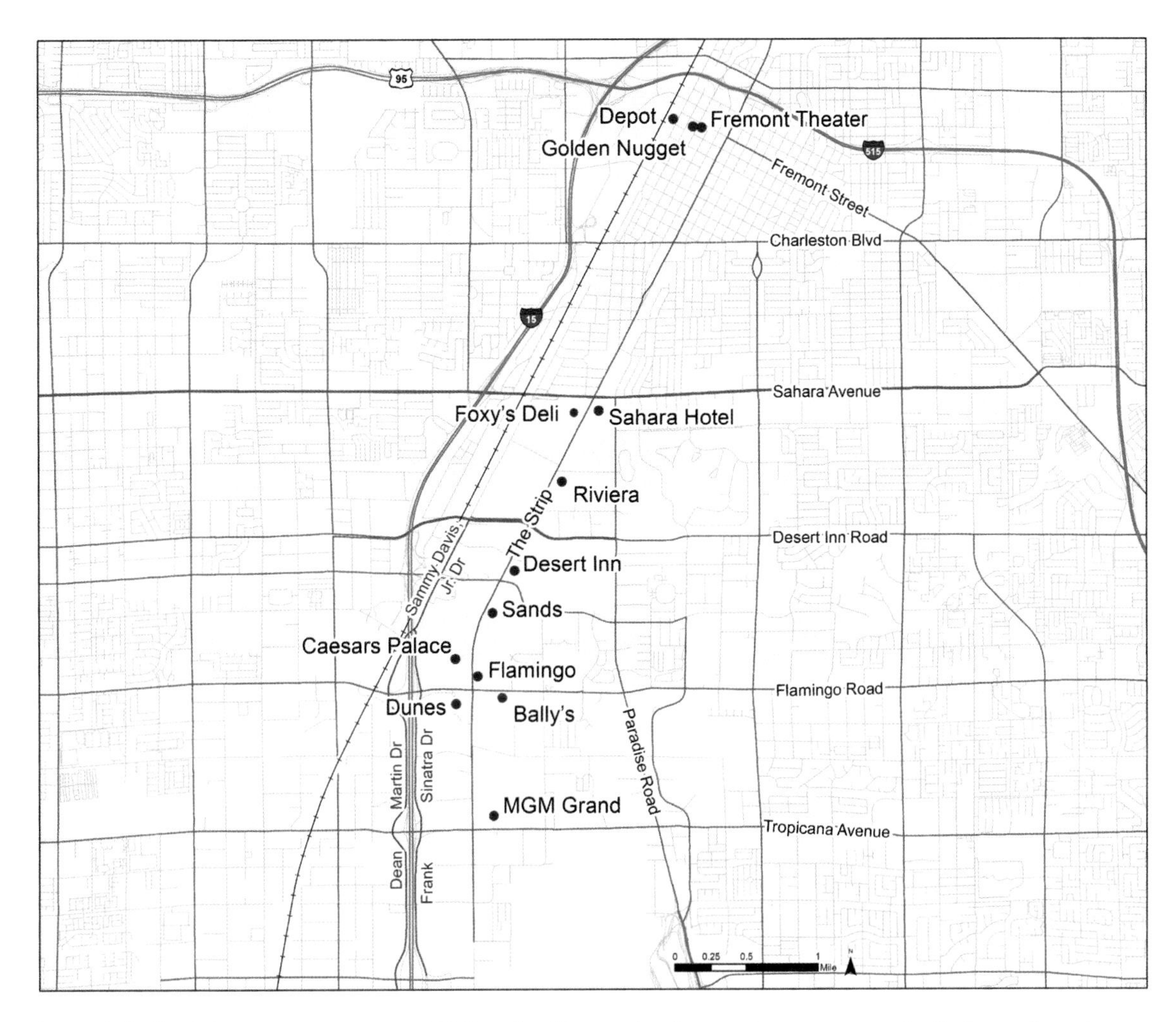

Map 11.1. The Rat Pack. The famous group was associated with casinos, hangouts, and a theater where their movie premiered. Streets renamed for members of the Pack are in red.

El Rancho Vegas, the first major resort on the Strip, and made a repeat visit in 1949. Given the racial restrictions of the time, they were not allowed to stay at the hotel or even watch other performers and had to stay at a boarding house in the Westside (map 4.11). In 1951 and 1952, the Trio played the Flamingo before switching to the Last Frontier in 1954. By this time Davis was a major performer and had enough clout to be allowed to stay at the hotel, one of the few Black entertainers given that privilege in those years. It was while driving back to Los Angeles from one of his Last Frontier performances that he had a car accident in San Bernardino, California, which cost him an eye. In 1957 Davis switched to the Sands, where he continued to perform regularly for the next seventeen years, recording *That's All*, a live album, in 1966. In 1974 he left the Sands for Caesars Palace, switching to the Aladdin in 1983 and finally in 1985 to the Desert Inn. He died in 1990.

Dean Martin became famous as part of a comedy show he developed with Jerry Lewis, and during their time together they made their first appearance in Las Vegas in 1949 at the Flamingo.

The two made *My Friend Irma Goes West* in Las Vegas and attended the premiere downtown the following year. After the duo broke up, Martin continued to perform regularly in Las Vegas at the Sands (1952–1968), the Riviera (1968–1972), and finally at MGM and Bally's before his death in 1995. Joey Bishop (1918–2007) and Peter Lawford (1923–1984) were other members of the Rat Pack, known more for their TV shows than Las Vegas appearances.

The high point of the Rat Pack's existence was in early 1960 during the filming of *Ocean's 11* in Las Vegas. The movie's plot involved the robbing of the Sahara, Riviera, Desert Inn, Sands and the Flamingo, with scenes filmed at each of these over five weeks before the production moved to Hollywood. Filming took up little of the stars' time, as most reported for work after noon and they usually only worked a few hours a day at most. The five stars were on the set together on only one occasion, when they were filmed walking down the Strip for the final scene of the movie. They even found time to make an appearance in another movie, *Pepe*, being filmed in the city at the same time.

During the filming, the Rat Pack stayed at the Sands and appeared each evening in a legendary run of shows referred to by the Pack as the Summit. Fans flocked to the city in the hopes of getting one of the scarce $5.95 tickets. Presidential candidate John F. Kennedy was one of these; he was visiting Las Vegas to raise money for his campaign and was in the audience the night of February 7, 1960. The cast returned to Las Vegas for the film's premiere at the Fremont Theatre downtown on August 3; the theater is long gone (map 8.13), as are most of the casinos associated with the Rat Pack, but their memory endures on the names of several streets commemorating their contribution to the city's history.

Perhaps no celebrity is more closely associated with Las Vegas than Elvis Presley. Born in Tupelo, Mississippi, in 1935, a month before the last bucket of concrete was poured on Hoover Dam, he became famous in Memphis and then nationally in the 1950s. He first set foot in Las Vegas on April 6, 1956, when he flew in from San Diego to have a look around and stayed one night at the Last Frontier. He returned April 23 for a two-week engagement of twice-nightly shows at the Last Frontier. He was not a hit; his performances, which averaged twelve minutes in length, did not go over well with audiences. The first night he was the third of three acts, but after the tepid response to his music he was switched to be the opening act for the remainder of his shows. Only one Saturday matinee show brought in the usual crowd of screaming teenagers. He played his last show May 6 and drove

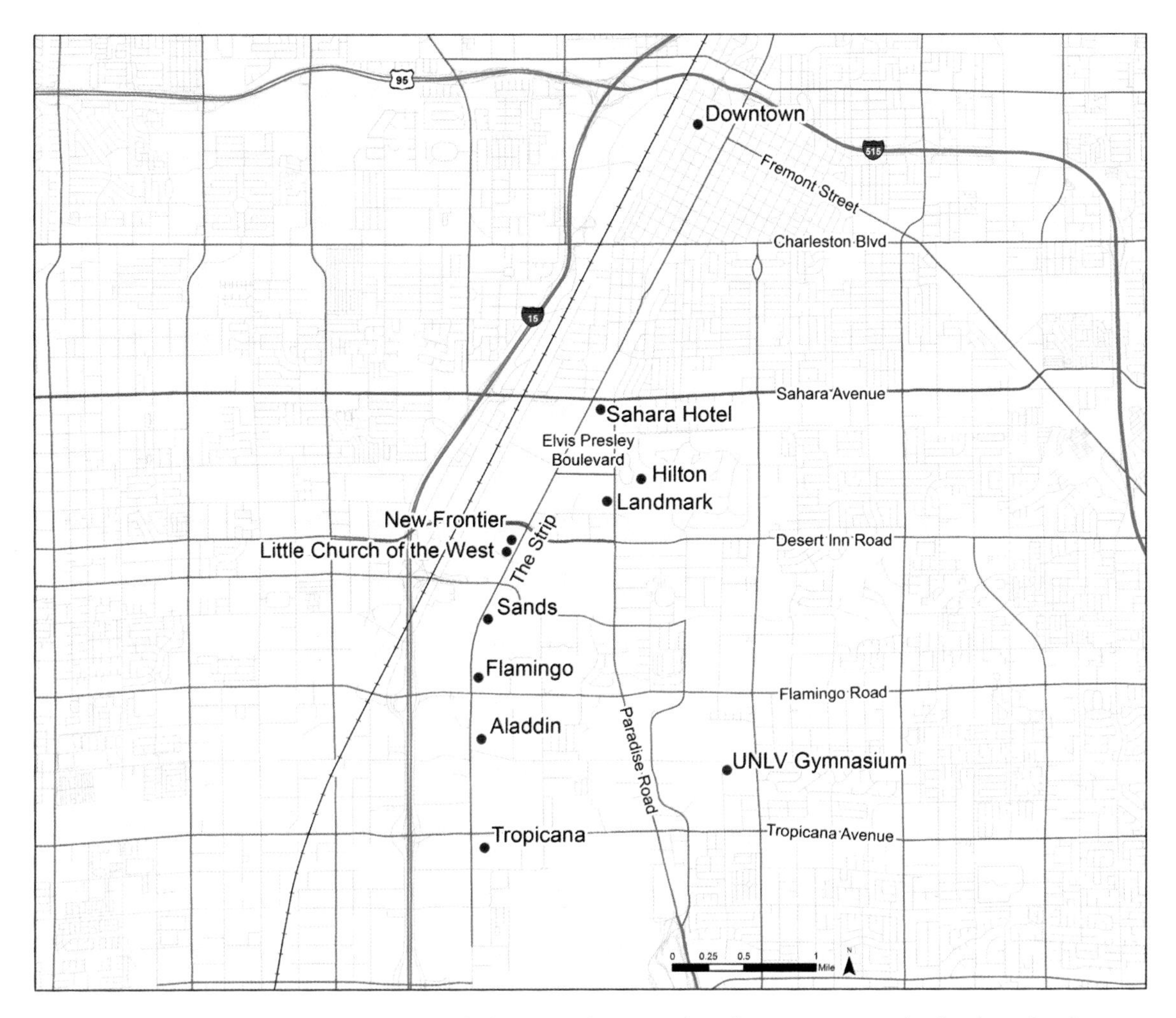

Map 11.2. Viva Las Vegas. The movie was filmed throughout Las Vegas, but several locations on the Strip and UNLV are best remembered. Elvis Presley stayed and performed at several casinos over the years.

back to Memphis two days later. Many no doubt thought this would be the last of Elvis in Las Vegas.

But this brief visit turned out to have momentous consequences for him and all of America. First, while the audience was not interested in him, he loved Las Vegas, the nightlife and music, and the excitement of the place. He was not, however, enthusiastic about gambling; instead he was interested in the other performers, getting to meet Liberace and Paul Anka. But most importantly, he saw a show by the long-forgotten Freddie Bell and the Bellboys at the Sands. As part of their act, this group performed their own version of "Hound Dog," originally recorded by Big Mama Thornton in 1952. The Bellboys' version had different lyrics and was performed much slower than the original; Elvis liked this version and insisted on playing it in his shows. He also noticed how one of the musicians was constantly bending his legs while he performed, and Elvis imitated this by swinging his hips while he sang. The next month he debuted his new song and hip movements on the *The Milton Berle Show*. Not everyone was pleased, but "Hound Dog" went on to become his biggest hit ever and launched him to a new level of stardom. And it all happened because of a visit to the Sands in Las Vegas.

Elvis returned to the city many times throughout the 1960s, vacationing for a weekend or several weeks at a time. The Sahara became his favorite hotel, and he stayed there when he returned to the city to film *Viva Las Vegas* in the summer of 1963. Production for the movie ranged widely over the city; aerial shots of Fremont Street and the Strip were prominent in the movie's opening sequence; casino scenes were filmed at the Sands and Tropicana; and the race car garage was near the Convention Center (the Landmark Hotel is prominent in the background of several scenes). Elvis and Ann-Margret go skeet shooting near McCarran International Airport, dance at the UNLV gymnasium, and board a helicopter at the airport for a sightseeing trip over Hoover Dam and Lake Mead. They even land at a marina and go waterskiing by the dam. Elvis's reason for being in Las Vegas, the Grand Prix race, starts at the Convention Center area, runs down Fremont Street, across Hoover Dam (from the Arizona side), through Valley of Fire, past the junction of US 93 and Lakeshore Drive, up into Mount Charleston and back, before reaching the finish line on Fremont Street. Elvis and Ann-Margret are married at the Little Church of the West in the last scene. This chapel has been moved several times, but when the movie was made in 1963 it was at the south side of the Last Frontier.

Elvis and Liberace at the Riviera, November 15, 1956. Photo by the Associated Press.

Elvis stayed at the Sahara again when he introduced his girlfriend, Priscilla Wagner, to the city, but in later visits he switched to the Aladdin. He chose that hotel when he married Priscilla on May 1, 1967. Following his disappointing 1956 shows, Elvis wisely chose not to perform in Las Vegas again. He did not change his mind until he returned for a show at the International Hotel (later the Hilton and Westgate) on July 31, 1969. The success of that show brought him back for several weeks of shows each year, selling out 636 consecutive appearances. He stayed in room 3000 at the International (now the Tuscany Sky Villa), where legend has it he shot his television in 1974. He last performed in the city on December 12, 1976, and left the city for the last time April 18, 1977, to tour nationally; he was scheduled to return to Las Vegas in October to begin another set of shows. Unfortunately, he died August 16 at his home in Memphis.

The importance of Las Vegas in Elvis's life can be shown by the locations where he performed. Of the 1,684 shows throughout his lifetime, 767 were in Nevada; 641, or 38 percent of all the times Elvis took to the stage, were at the Las Vegas Hilton. To honor him, Riviera Boulevard, connecting the Hilton (now Westgate) with the Strip, was renamed Elvis Presley Way in 2016.

One of the most fascinating people associated with Las Vegas

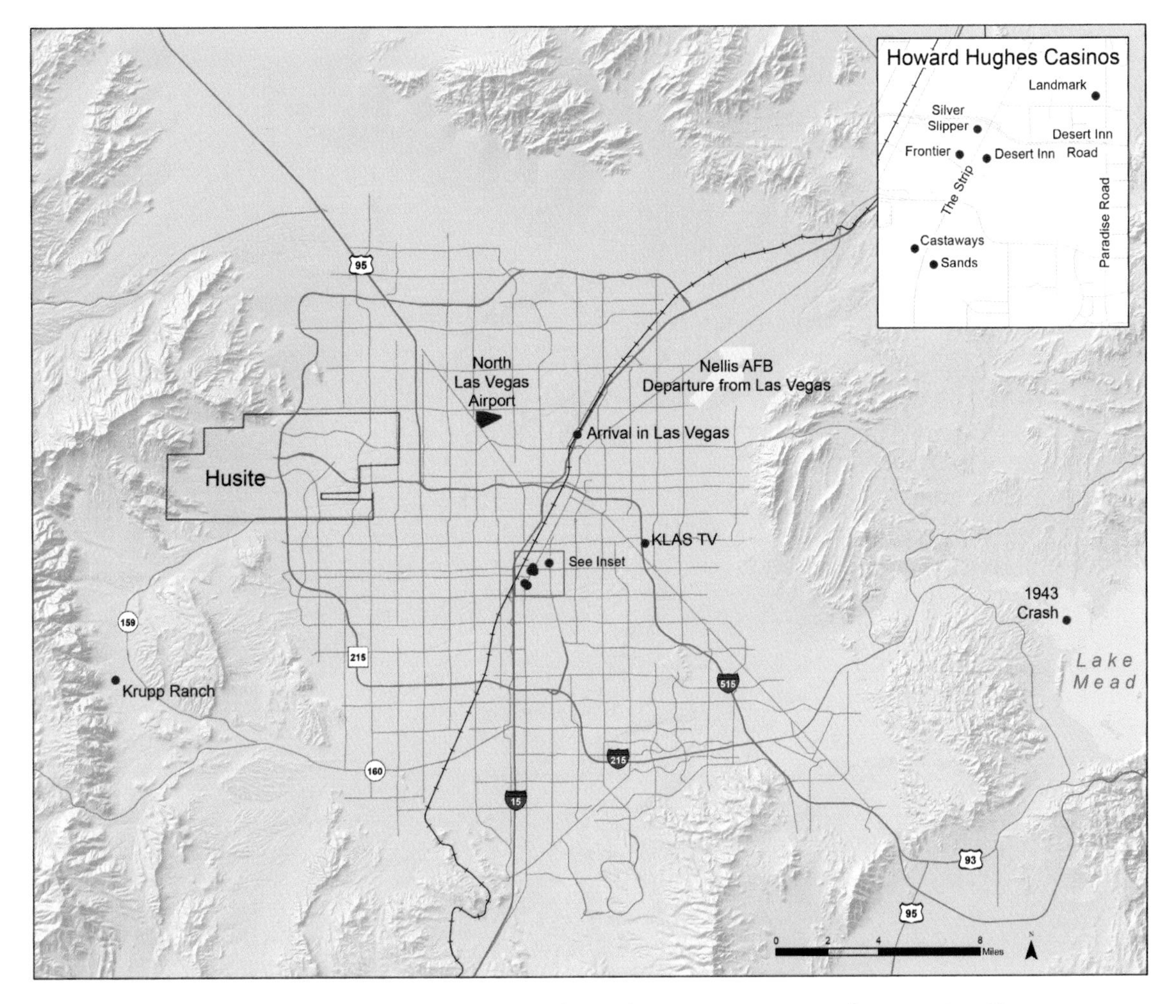

Map 11.3. Howard Hughes. The wealthy recluse bought many casinos (red dots) and lived in the Desert Inn. He also owned an airport, TV station, ranch, and Husite, an empty stretch of desert west of the city that was later developed into Summerlin. One of his earlier visits to the area, when he crashed in Lake Mead, is also shown.

was Howard Hughes. Born in 1905, the year Las Vegas was founded, he became a millionaire at age 19 when he inherited the Hughes Tool Company in Texas. He dropped out of college and used his newfound wealth to dabble in a variety of businesses, though the aviation and movie industries were always two of his favorites. Both of these interests brought him to Las Vegas on several occasions.

On one early visit to Las Vegas to test a new airplane design, he was nearly killed. Flying out of Boulder City Airport on May 17, 1943, he attempted to land his Sikorsky S-43 amphibian on Lake Mead but misjudged the approach and crashed. A crewman and government inspector were killed, the plane sank, but Hughes was uninjured (map 7.20).

In the 1950s he lived briefly in Las Vegas, staying at the Flamingo, Desert Inn, and El Rancho Vegas before buying a small house near the Desert Inn on the grounds of the Sun Villa motel in 1953. During this first stay in the city, *The Las Vegas Story* was filmed. This was made by RKO studios, owned by Hughes, and starred his girlfriend, Jane Russell. Despite her screen appeal, it was not a success. In 1954 he bought twenty-five acres of desert west of Las Vegas, an area he called Husite, to be the new home of his missile and electronics company then located in Culver City, California. He wanted to move out of California to escape the state's income tax, a lifelong obsession for him. However, his engineers refused to make the move into the desert and it was eventually called off. For many years, the property remained vacant.

Hughes arrived in Las Vegas by train in the early morning of November 27, 1966, for his final visit. This was not at the downtown train station but farther north at the Carey Avenue crossing. By this time, he had been living as a recluse and wanted no one outside his inner circle to see him when he arrived. His visit was again motivated by a desire to avoid California income taxes, this time following his sale of TWA airline stock. He rented out the entire top floor of the Desert Inn where he lived in seclusion for three years, seen by only a handful of associates, who occupied the floor below.

The Desert Inn management was not pleased to have two floors taken over by guests who did not gamble and wanted him to leave. Hughes solved their problem by buying the hotel in March 1967. This marked the beginning of a purchasing binge that included the Sands (August 1967), Frontier (September 1967), Silver Slipper (April 1968), Landmark (January 1969), and Castaways (February 1970). He went on to make other purchases off the Strip, including the North Las Vegas airport, Alamo Airways, a residential subdivision near the Desert Inn Country Club, and the KLAS television station (the last so he could watch his favorite movies all night). He also purchased nearly two thousand worthless mining claims in central and northern Nevada.

Although living as a recluse, he was still married to his second wife, Jean Peters, with whom he exchanged vows in a secret ceremony in Tonopah in 1957. The two were not living together when he arrived in Las Vegas, but he tried to entice her to come to the city by buying her a house as well as the Krupp Ranch, west of the city near Red Rock Canyon (map 9.6). She refused to come, and the ranch sat empty until Hughes finally sold it in 1972.

Not all of his attempts at purchases were successful. He tried to buy the Bonanza, Silver Nugget, and Stardust but was blocked out of concerns he would be approaching a monopoly over hotels in Las Vegas. He also was unable to buy the *Las Vegas Review-Journal* to ensure favorable news coverage. He came up with many other schemes that likewise came to nothing, including a plan to build the world's largest hotel on the site of the Sands, building an airport for supersonic passenger jets, and putting radio and TV stations in every small town in the state. He fiercely opposed nuclear testing in Nevada, plans to complete the pipeline bringing water from Lake Mead, and racial integration of casinos, all of which put him at odds with the political leadership in the state. He also fought a losing battle to prevent the construction of the International (later Hilton, now Westgate) Hotel on Paradise Road. To counter it, Hughes purchased the Landmark across the street to compete directly with it. It was a spectacular failure.

He left Las Vegas November 26, 1970, on a private jet departing Nellis Air Force Base on a direct flight to the Bahamas. No public announcement of his leaving or destination was made, and a week passed before many of his closest associates even knew he had left. He never returned to Las Vegas or the United States, and died in Mexico on April 5, 1976.

At the time of his death, he left no known will. But several weeks after he died, one was found in the offices of the Church of Jesus Christ of Latter-day Saints in Salt Lake City. This document, known as the Mormon Will, left a substantial sum of money to the church as well as various employees of Hughes. It also left money to an unknown individual named Melvin Dummar, who was soon tracked down by reporters. He recounted a story of driving between Gabbs and Las Vegas in January 1968 and picking up an old man he found just off the road south of Lida Junction. He took this man to Las Vegas, where he claimed to be Howard Hughes and asked to be dropped off at the Desert Inn. A probate trial was held in Las Vegas in 1978 but Dummar's account and the "Mormon Will" were tossed out. The story is the basis for the Academy Award-winning 1980 film *Melvin and Howard*.

Although popular perception of Hughes is that of an eccentric but successful businessman, he was in fact terrible at business. He spent lavishly, poorly managed what he owned, and bankrupted, ruined, or nearly destroyed many formerly profitable companies such as RKO Pictures and TWA. His time in Las Vegas was no exception. Despite purchasing many profitable casinos, he ignored them and allowed his few confidants to staff them with incompetent relatives, and they all soon began losing money. The Landmark was a particular failure despite its eye-catching Space Needle-like appearance, combining a slender tower with a flying-saucer-shaped section perched on top. He refused to allow his staff to invest in renovations or expansions, and conditions became so bad that the health department condemned ninety rooms at the Desert Inn in 1974. After his death, his holdings were reorganized and began to function like a normal business for the first time, and the casinos he had owned returned to profitability. Hughes made several important contributions to Las Vegas. His corporation spurred passage of laws that allowed public companies to acquire gambling

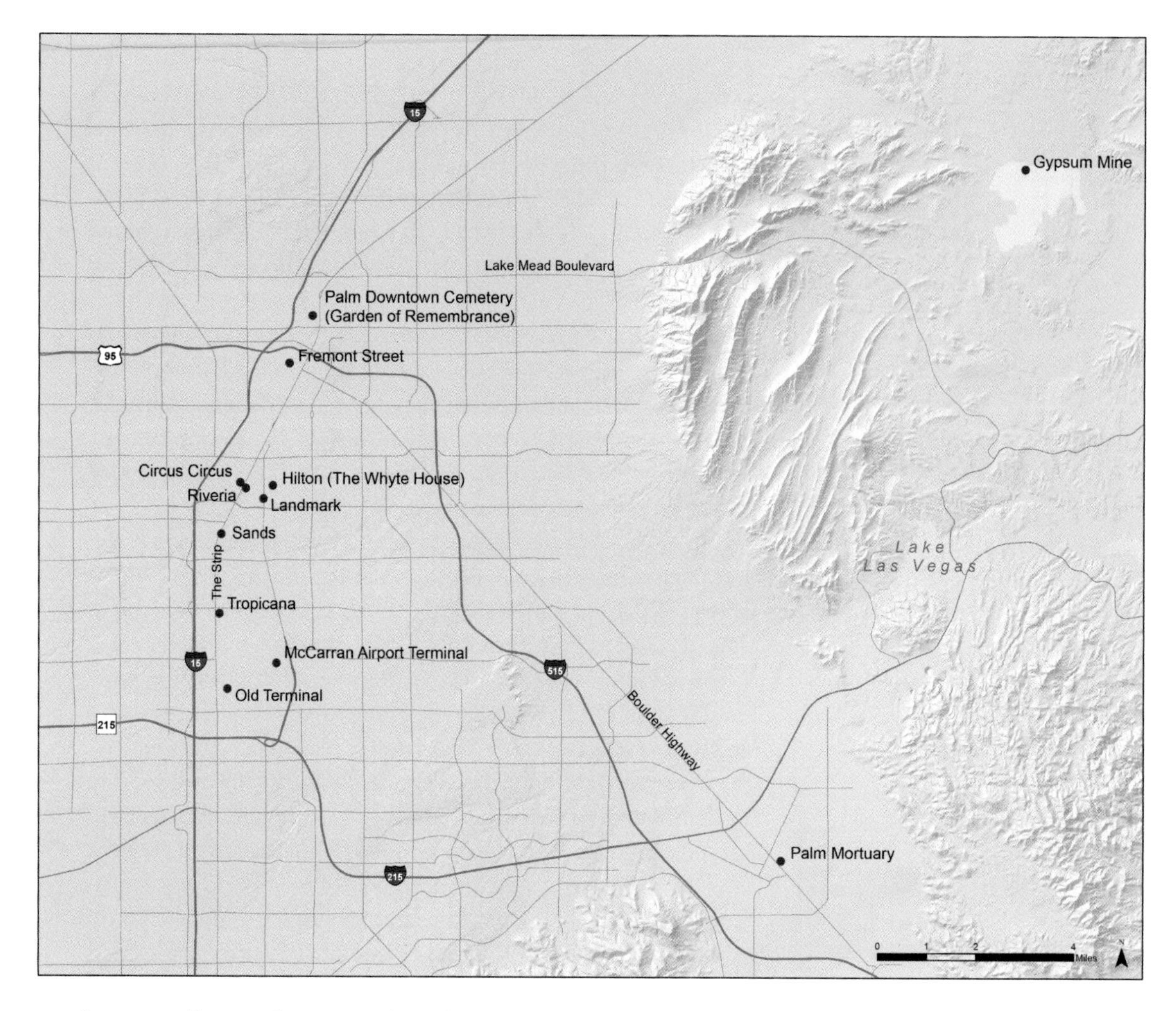

Map 11.4. Bond . . . James Bond. Several businesses provided locations for *Diamonds are Forever.* Except for Fremont Street and the airport terminal, they look much the same today. This map is based on viewing the movie with Andrew Lycett's book, *Ian Fleming*, providing details of Ian Fleming's earlier visit.

licenses, allowing for tremendous future growth of casinos here as well as helping push the Mafia out of town. His company also developed the twenty-five thousand acres that had been purchased for an electronics and missile factory back in the 1950s. This became a master-planned community named Summerlin, after his grandmother's maiden name.

Author and James Bond creator Ian Fleming (1908–1964) visited Las Vegas in 1954 and stayed a single night at the Sands. He managed to visit every casino on the Strip during his brief stay, and the city clearly made an impression on him as he incorporated it into his fourth novel, *Diamonds are Forever*, published in 1956. The plot involves diamond smuggling from Africa, which Bond traces to Las Vegas. He spends time at the fictional Tiara Hotel and a ghost town named Spectreville before escaping on a railroad push car, and the finale takes place on Lake Mead.

The novel was adapted into the seventh James Bond movie in 1971. The spy, played by Sean Connery, pays a memorable visit to Las Vegas, and a number of recognizable locations appeared in the film. The International was the most prominent of these and in the movie was the Whyte House. The top floor was occupied by reclusive billionaire Willard Whyte, obviously modeled after Howard Hughes, a reclusive billionaire then living on the top floor of the Desert Inn (map 11.3). Several other casinos were used in the movie, most of them belonging to Hughes since he was an acquaintance of Bond producer Cubby Broccoli. Filming took place in two funeral homes, the Palm Mortuary on Boulder Highway in Henderson and the Palm Downtown Cemetery (at what's called the Garden of Remembrance in the film) north of downtown. The PABCO gypsum mine (map 8.19), east of town along Lake Mead Boulevard, is shown, though the movie version is a high-tech facility owned by Whyte. Bond also escaped from the police after a memorable chase downtown on Fremont Street.

The movie shows the Las Vegas of the early 1970s very well, including the long-gone original exterior of the 1963 McCarran terminal. (When Fleming flew in to Las Vegas in the 1950s, he would have used the old terminal on the west side of the airport.) However, several scenes set in Las Vegas were shot elsewhere. The interiors of hotel rooms were filmed in Hollywood studios, while other scenes were produced in Palm Springs. The unusual mountaintop house where Bond meets Bambi and Thumper is one of these, at 2175 Southridge Drive and known today as the Elrod House.

Diamonds are Forever was Connery's last film as James Bond (unless 1983's *Never Say Never Again* is counted). The final scene to be filmed was the one in which Bond is almost cremated at Palm Mortuary. This nondescript building at 800 South Boulder Highway still exists, a monument commemorating the end of the Connery era as Bond.

Władziu Valentino Liberace was born in 1919 near Milwaukee, Wisconsin. His parents were Italian and Polish immigrants who encouraged a musical interest; to their delight, he not only showed an early interest in the piano but also enormous talent. In the late 1930s he began to earn a living playing the piano in small town theaters and supper clubs, gradually moving away from his classical background to a more varied mix of old and new. He also shortened his name to Liberace.

He arrived in Las Vegas the first time for a six-week run at the Last Frontier between Thanksgiving and Christmas 1944. He was one of several acts the first night, but his performance got him promoted to the headliner from the second night on. He liked Las Vegas and returned twenty-five times to play the Last Frontier in the 1940s and early 1950s. Bugsy Siegel saw his act and wanted him for his new Flamingo, but Liberace preferred staying where he was; Siegel was murdered before Liberace could find the courage to turn the ruthless mobster down.

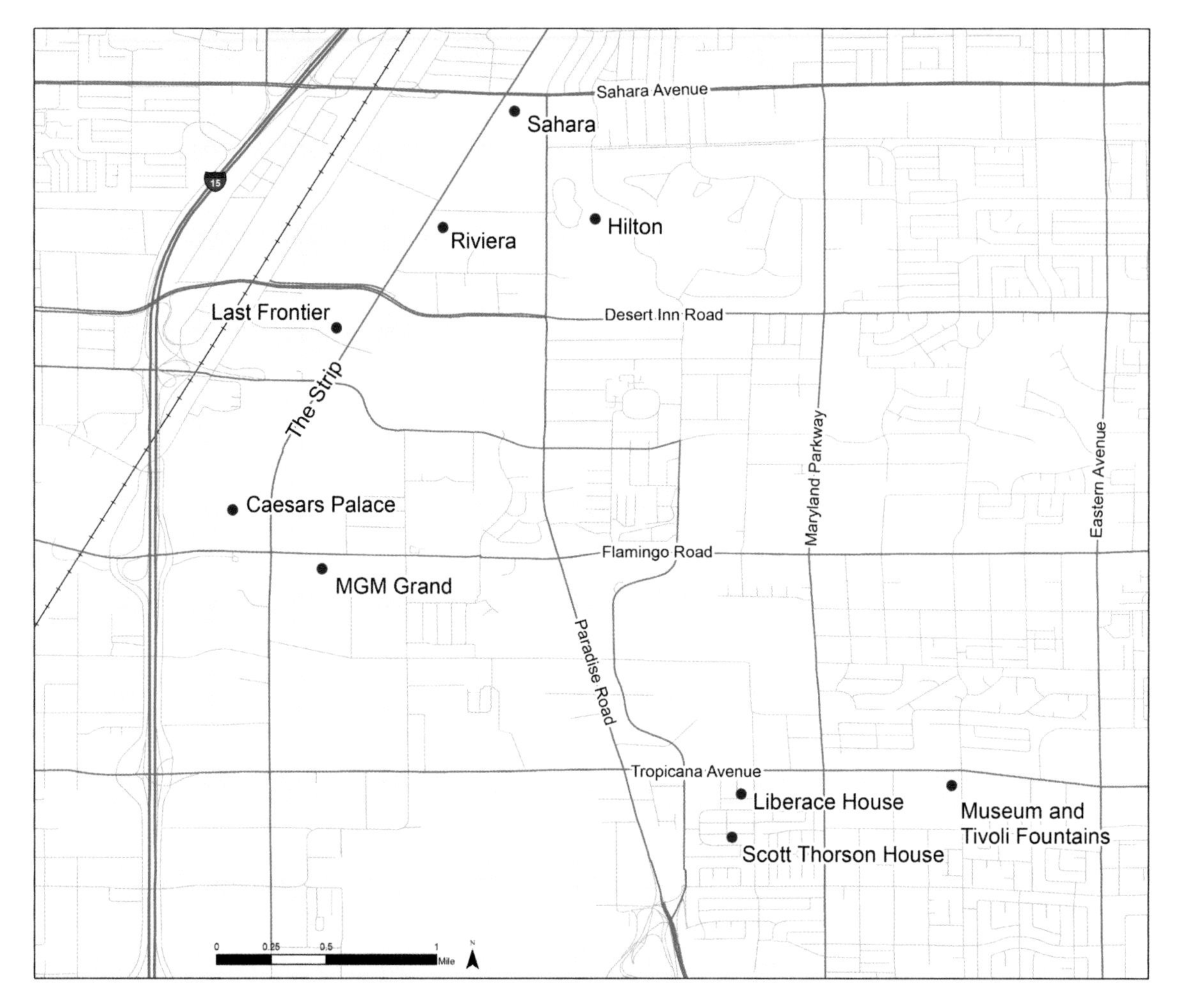

Map 11.5. Mr. Showmanship. Liberace performed in a number of casinos, owned several homes, and even created a museum in las Vegas.

His time at the Last Frontier ended in 1954 when the Riviera opened down the highway; the new casino lured him away with a much larger showroom and salary. The Last Frontier was remodeled and renamed the New Frontier, and sought out new talent. One rising star they found was a young singer named Elvis Presley (map 11.2). Unfortunately, he was not a success, but Liberace liked the young singer and visited him with reporters in tow, telling him he needed more glitz and glamour in his act.

Elvis's failure in Las Vegas must have rubbed off on Liberace, as his act at the Riviera didn't go over well, and his contract was canceled in 1958. He was reduced to returning to the old dinner and nightclub circuit in smaller cities and towns before making a splashy return to Las Vegas and the Riviera on July 2, 1963, calling himself "Mr. Showmanship." He played at the Hilton from 1972 to 1982, then returned to the Riviera, with later shows at the Sahara, Caesars Palace, and the MGM Grand. He established his legal residence in the city in 1963 to take advantage of the state's lack of income tax, though he also kept a palatial home in the Hollywood Hills and another in Palm Springs (he had already moved out of his Hollywood house with the famous piano-shaped swimming pool).

His lavish shows and lifestyle led to many rumors and insinuations about his sexual orientation. Unlike many celebrities, he fought back against the press for these and won several judgments against tabloids. About 1954, he was quoted as saying he didn't mind all these allegations, but his brother and manager, George, was crying all the way to the bank. The tears continued to flow, as Liberace increased his fame and success with ever more extravagant costumes and shows. Despite the glamour of his public appearances and his skillful use of publicity he was an extremely private person, even reclusive near the end of his life.

In 1976 he bought a house on Shirley Street near McCarran International Airport, which he expanded and extensively remodeled into a 10,500-square-foot showpiece of his collections and decorating style. In 1978 he also bought a nearby house at 933 Laramore Drive for his lover Scott Thorson, which also served as a secret retreat that even his closest associates didn't know about. He created the Liberace Museum in 1979 to showcase his costumes and decorations, buying a shopping center at the corner of Tropicana Avenue and Spencer Street (a short distance from his house) to house it. The remodeled shopping center also featured his Tivoli Gardens restaurant.

He died February 4, 1987, at his Palm Springs, California, home. His legacy in Las Vegas has not received the prominence given to many others who performed in the city, and he has yet to have a street named after him. The museum closed in 2010, followed by the restaurant in 2011, and the shopping center has since been redeveloped. However, in 2016, the Liberace Garage opened next to the Hollywood Cars Museum, allowing fans a look at some of his bejeweled automobiles.

Clint Eastwood is not a name strongly associated with Las Vegas, but in 1977 he made *The Gauntlet*, in which he plays a washed-up Phoenix cop sent to Las Vegas to pick up a witness and bring her back for trial. Unfortunately for both of them, the Mafia and high-ranking corrupt police officers want them dead.

After picking up the witness, Gus Mally (Sondra Locke), at the city jail, a shootout ensues on the way to the Las Vegas airport, after which gunfire levels Mally's entire house. They escape and take a local cop prisoner while driving west on Charleston Boulevard toward Red Rock Canyon and the Blue Diamond Loop (Nevada Highway 159). In the movie, this represents the road to Arizona, and the Arizona/Nevada state line was filmed along the Blue Diamond loop, with no sign of Hoover Dam and the Colorado River! After escaping from more crooked cops, Eastwood and his witness spend the night in a gypsum mine on

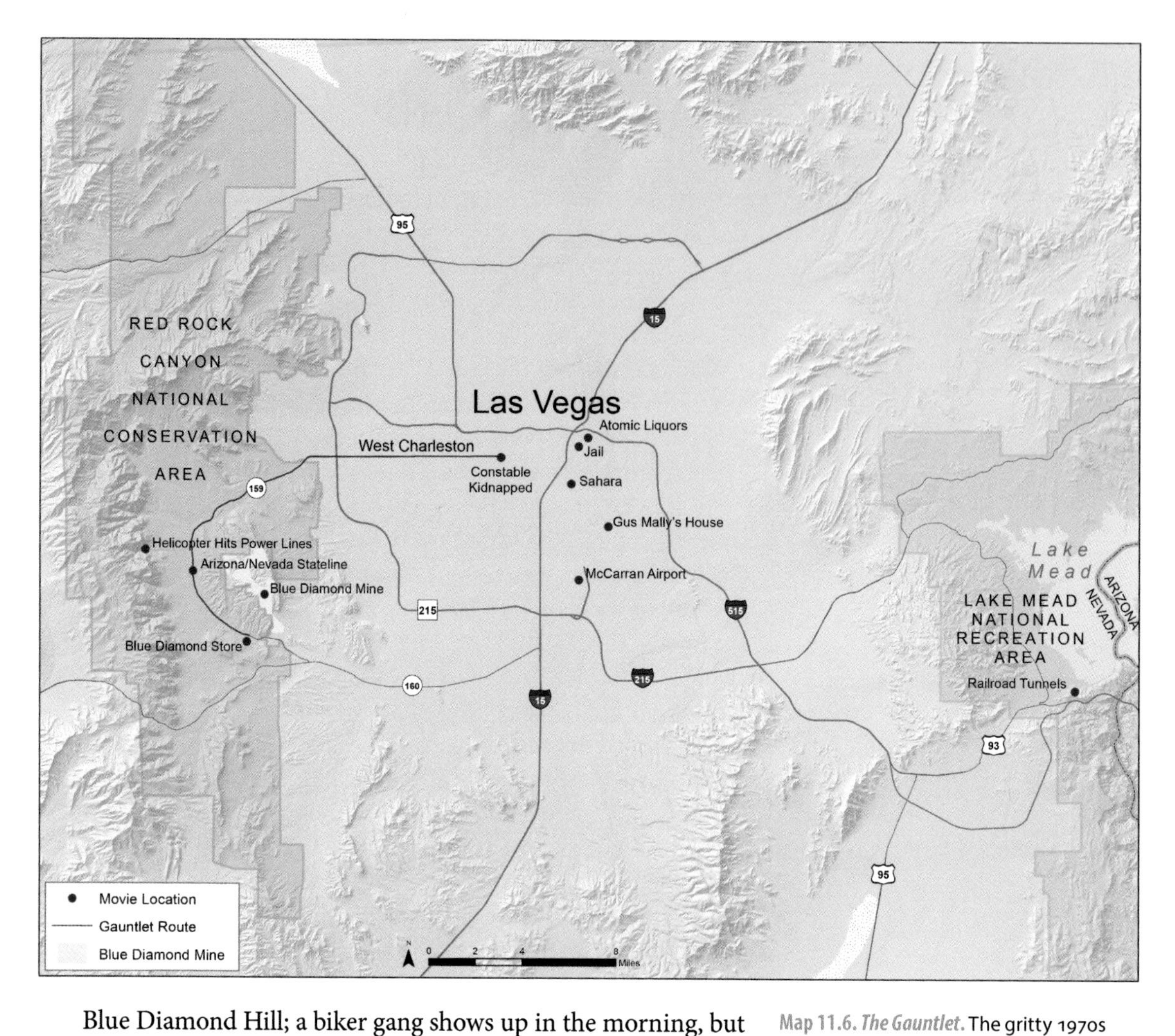

Map 11.6. *The Gauntlet*. The gritty 1970s Clint Eastwood movie had locations throughout the Las Vegas Valley.

Blue Diamond Hill; a biker gang shows up in the morning, but Eastwood's charm and gun clear them out. They set off on one of the biker gang's motorcycles and stop at a store, filmed in the town of Blue Diamond, when a police helicopter arrives and opens fire on them. They escape along Blue Diamond Road, with the chase including scenes filmed in the tunnels on the abandoned Hiline railroad to Hoover Dam (map 5.7) and in Red Rock Canyon (map 9.6). There the helicopter crashes into power lines and is destroyed. The remainder of the movie was filmed in Arizona and concludes with a showdown with the entire Phoenix police department. Despite an enormous amount of gunfire in the movie (a car, bus, and house are destroyed by bullets), Eastwood's cop never shoots anyone.

Although not a highly regarded movie at the time, it captures the look of Las Vegas in the 1970s, especially the enormous amount of vacant land that existed in the city then, even near the Strip. The Cinerama and Red Rock movie theaters (map 8.13) are briefly visible in the background, as is the original MGM Grand Hotel (now Bally's).

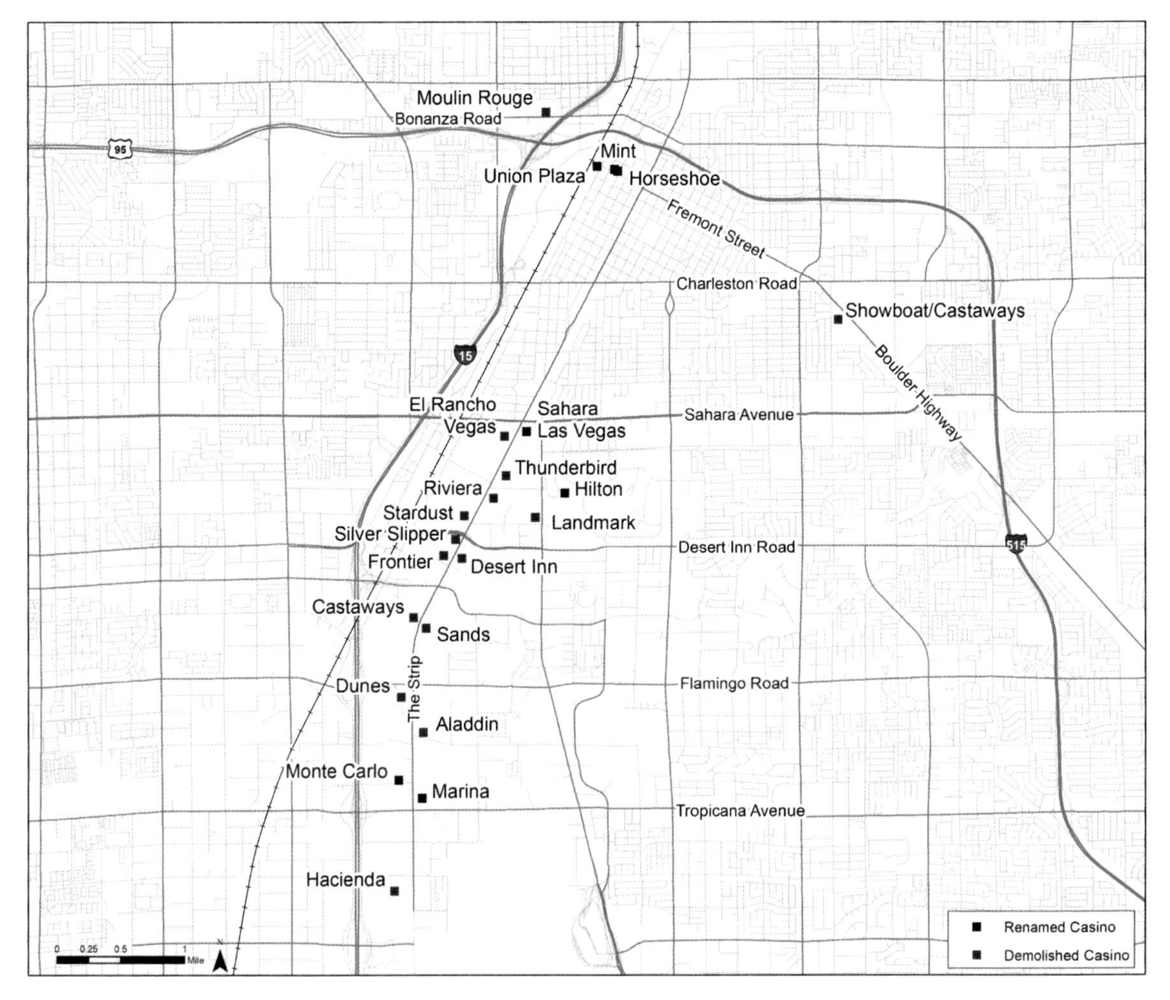

Map 11.7. Lost Vegas. Many casinos have been renamed or demolished over the years.

Much of old Las Vegas, or even the Las Vegas existing before 1990, is gone, having been demolished for new casinos, shopping centers, airport terminals, or the cycle of urban growth. Many of the city's lost landmarks are of course casinos, some of them largely forgotten, but others have become almost legendary. Among those mostly forgotten is the Northern Club, the first legal casino in the town when it opened downtown on March 20, 1931. It went through many name changes before ending up as La Bayou and closing for good in 2012. The nearby Mint opened in 1957 and closed in 1988, after which it was incorporated into Binion's Horseshoe.

The first major casino located on the Boulder Highway was the Showboat Hotel, which opened in 1954 and became the Castaways Hotel in 2000. It was a longtime presence at the north end of Boulder Highway but closed in 2004 and was demolished. The site is currently vacant. Nevada Palace operated farther south along Boulder Highway from 1960 to 2008. It was also torn down, but replaced by the Eastside Cannery. Even farther south two casinos have disappeared from downtown Henderson: The Lucky Star Club (1957–1972) and the Royal (1960–1964).

In the 1950s and 1960s Jackson Street in the Westside neighborhood emerged as the Black Strip, with a number of small gaming halls and a few hotels and rooming houses catering to the city's Black population and visiting entertainers, who were denied entrance to Strip casinos (map 4.11). All of these are long gone, along with the Moulin Rouge, the first racially integrated hotel-casino in Las Vegas in 1955. It remained in business only six months before shutting down, though reopening on several occasions in later years. A series of fires destroyed much of the structure.

The Strip of course has the greatest collection of lost casinos. The Red Rooster was a small casino that opened in 1931 along US 91 where the Mirage is located today, only to burn down later that year. It is considered the first casino along what became the Strip, but El Rancho Vegas is generally considered the first major hotel-casino along US 91. It was located at what is now the southwest corner of the Strip and Sahara Avenue, the current location of the Las Vegas Festival Grounds. The hotel opened in April 1941 and included a casino, swimming pool in front, room wings in an arc behind the parking lot, and tower with a windmill with the hotel's name. At the time it was surrounded by empty desert and was visible from a considerable distance. It burned to the ground in 1960 and was never rebuilt.

Aside from the Flamingo, the oldest surviving on the Strip, most of the casinos built before the 1970s are gone. The building boom that began with the Mirage in 1989 and continues to the present has eliminated twelve hotels, all but a few replaced by newer megaresorts. The Castaways, which opened in 1970, was the first of these when it closed in 1987 and was replaced by the Mirage. The Silver Slipper opened in 1968 and closed in 1988; the site is now a vacant lot. The stunning Landmark tower opened 1969 after being purchased by Howard Hughes in a fit, but it was never a success and finally shut down in 1990. It was imploded in 1995, and its location has become part of the convention center parking lot. The Dunes, with perhaps the most spectacular sign on the entire Strip, opened in 1955 but closed in 1993 and was replaced by the Bellagio. The Sands, which appeared in 1952, was where Elvis learned to swing his hips, the Rat Pack played, and Martin Luther King Jr. visited on his only trip to the city. Closed in 1996, this cultural landmark was torn down and replaced by the Venetian. The Hacienda operated from 1956 to 1996 where Mandalay Bay is now located. The Desert Inn, which opened in 1950 and hosted John F. Kennedy on his second visit to the city before becoming Hughes' home for several years, lasted until

2000. The Wynn and Encore now stand on the site. The Stardust opened in 1958 and was the last casino run by the Mafia and with loans from the Teamsters before it closed in 2006. It was to be replaced by the Echelon Place Hotel, but that project was canceled (map 11.1) and the site turned into Resorts World, which opened in 2021. The Last Frontier, where Elvis received a cold reception to his first performance in the city and Sammy Davis Jr, a warmer one, opened in 1942. It was later known as the New Frontier and finally just the Frontier before closing in 2007. It was imploded in several blasts later that year. A hotel was planned for the site but canceled, and now the site is an empty lot. The Aladdin, where Elvis got married, opened in 1963 and, except for the theater, was demolished in 1998. A new Aladdin hotel opened in 2000 and became the Planet Hollywood Las Vegas in 2007. The most recent demolition was of the Riviera. This opened in 1955 and closed in 2015, and like many others it has not yet been replaced.

The Sands Hotel and Casino, 1967. Curt Teich postcard.

Even surviving casinos have discarded much of their history. The Flamingo is the oldest surviving casino on the Strip, but the last original portion of the building was demolished in 1993. Circus Maximus was the main theater at Caesars Palace and home of many performers over the years, including Celine Dion for her first show in the city, but was torn down in 2000.

Some old casinos are still around but have changed their names. Union Plaza was reduced to the Plaza in 2011. The Sundance became Fitzgeralds in 1987 and transformed into the D Las Vegas in 2012. Lady Luck closed in 2006 but reopened in 2013 as Downtown Grand. Sahara opened its doors in 1952 and closed in 2011, but reopened two years later as SLS and is now Sahara Las Vegas. The International became the Hilton in 1971, the LVH in 2012, and then the Westgate in 2014. The first MGM Grand became Bally's in 1986; the Marina Hotel opened in 1975 but was incorporated into the structure of the second MGM Grand, which opened in 1993 (the western end of the MGM's hotel tower is the old Marina Hotel). Most recently, Monte Carlo was renamed Park MGM in 2018.

The Stardust Hotel, late 1960s or early 1970s. Unknown photographer.

Despite the demolition of so much of Las Vegas's history, much of it survives. Not just in the minds of old-timers and those nostalgic for the old days, but according to the United States government. The National Park Service created the National Register of Historic Places in 1966, a program to record historic buildings and neighborhoods throughout the country. Being listed means a site was judged as being important to the country's history, as determined by each state's historic preservation coordinator. A building may be associated with a famous person or event, have artistic or other value because of its construction, or have the

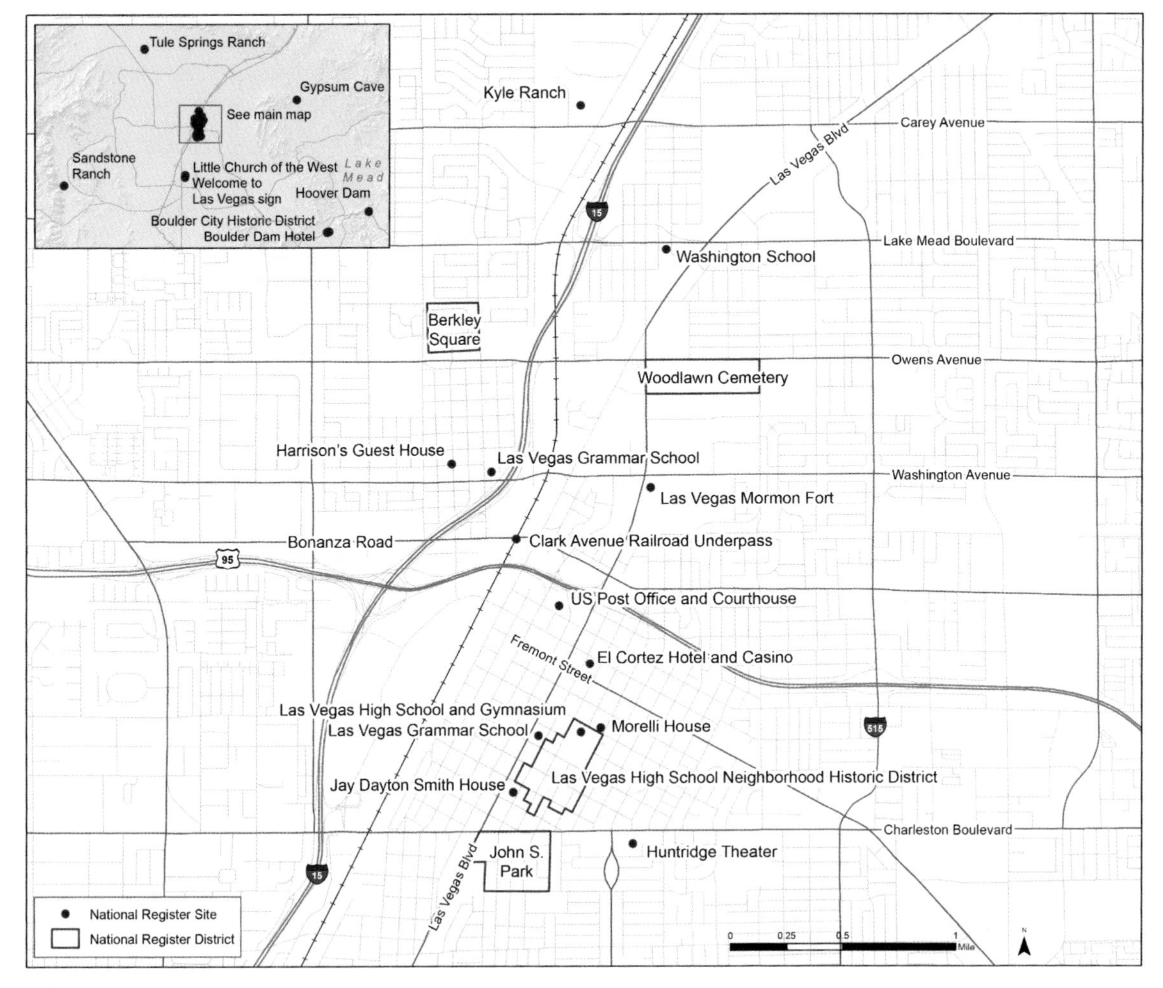

Map 11.8. Historic Las Vegas. The red dots denote properties listed in the National Register of Historic Places, while the red outlines are neighborhoods or historic districts.

potential to provide information about an area's history or prehistory. Buildings (or the events that made them important) must usually be at least fifty years old. Almost all of these sites are privately owned and there are no special protections for them; many have in fact been lost. More than one million properties are now listed nationwide on the register.

Clark County had fifty-nine places listed on the National Register of Historic Places in 2021, about 15 percent of the entire state of Nevada, which had only 382 National Register sites that year. Most of the historic sites in Las Vegas are downtown or just outside this area where the fifty-year age threshold is most likely to be met. The Mormon Fort (map 3.7) is foremost among them, containing the city's oldest building. This was built in 1855 as part of an adobe fort to shelter Mormon pioneers. Most of the fort was demolished years ago; the structure that survives was a storeroom on the interior of the fort. Kyle (or Kiel) Ranch, once the largest Paiute village as well as an early ranch, is also listed.

Aside from these, most of Las Vegas's National Register sites are buildings constructed after the town was founded in 1905.

The Las Vegas High School Neighborhood Historic District contains the original city high school and one of the oldest surviving neighborhoods in the city, with houses built in the 1930s; the nearby Jay Dayton Smith House dates from 1931. Only one of the historic properties is a casino, El Cortez, which opened downtown in 1941 and remains open and little changed on the outside. The Post Office and Courthouse, which opened in 1933 as the city's third courthouse, is now the National Museum of Organized Crime and Law Enforcement (better known simply as the Mob Museum). Woodlawn Cemetery opened in 1914 with land donated by the railroad. Three primary schools are included: one was built in the Westside in 1922 for Indian children, one north of town in 1932, and another east of the tracks in 1936. Historic sites occasionally move to new locations. The Morelli House, an outstanding example of midcentury modern architecture (map 8.7), was built in 1959 in the Desert Inn Country Club but moved to a downtown location in 2001.

From its earliest days, Las Vegas was divided by the railroad into east and west. While Clark's townsite on the east contained the majority of population, the McWilliams Townsite to the west also attracted many settlers in its earliest years. The busy tracks were a barrier to those who lived in what would be known as the Westside. This barrier was finally overcome in 1937 when the Clark Avenue Railroad Underpass was built on Clark Avenue (later Bonanza Road), allowing people to cross the tracks safely without delays. This was a two-lane undercrossing, later expanded to four lanes, and became part of US 95 when that route was established. The prosaic underpass is now one of Las Vegas's official historic sites.

The Westside neighborhood contains several properties, one of them being Harrison's Guest House. This once served Black entertainers such as Sammy Davis Jr., who were not allowed to gamble, eat, or sleep in the Strip casinos in which they performed. Berkley Square was a subdivision developed from 1949 by the African American architect Paul Revere Williams (1894–1980), who also designed the La Concha hotel lobby now at the Neon Museum and the Guardian Angel Cathedral on the Strip. Other Black history landmarks in Westside have disappeared; the Moulin Rouge hotel was listed on the National Register but has largely been destroyed by a series of fires.

The Huntridge Theater and John S. Park Historic District along Charleston Boulevard were part of the city's expansion from the original townsite. The Huntridge was the first movie theater outside downtown when it opened in 1944 and remained in business until 2004 (map 8.13). The building remains intact and may someday be restored to operation.

Other sites are scattered throughout the city. The "Welcome to Fabulous Las Vegas" sign, designed by Betty Willis, was put up in 1959 along US 91 south of the city but just north of the original McCarran airport terminal, ensuring it would be seen by virtually every visitor to the city. Today those arriving by Interstate 15 or at the newer terminals on the east side of the airport must make a special trip to see it. (The current parking lot was installed in 2008 for their convenience.) The Little Church of the West, the city's oldest wedding chapel, opened in 1942 and was featured at the end of the Elvis movie *Viva Las Vegas* (map 11.2). At that time it was located near the New Frontier Hotel, but was relocated in 1979 to the Hacienda Hotel. It was moved again in 1996 to its current location. Among those married here were Redd Foxx, Judy Garland, Betty Grable, Dudley Moore, Telly Savalas, and Zsa Zsa Gabor. An Elvis impersonator is even available to perform or officiate at your wedding ceremony. (The real Elvis got married in Las Vegas with a Nevada Supreme Court judge officiating, but those are not in demand by most couples looking to tie the knot.) The Boulder City Historic District was created in 1983 and contains much of the original part of the city, with the majority of buildings having been constructed by 1942 (map 5.8). The nearby Boulder Dam Hotel is listed separately.

The Historic Welcome to Fabulous Las Vegas Sign, 2012. Photo by Dietmar Rabich.

One National Register site is underwater, at the bottom of Lake Mead. This is a B-29 bomber that crashed in the lake during a low-level flight on July 21, 1948. The crew survived, but the plane quickly sank. The wreck was only rediscovered in 2001; in 2020, the National Park Service allowed two companies to begin guided tours for divers wanting to visit the sunken plane.

The Moulin Rouge is not the only place listed on the National Register that no longer exists. The Green Shack was once the oldest restaurant in the city, having opened in 1929 on Fremont Street. It closed in 1999 and was demolished. In 1909, the San Pedro, Los Angeles, and Salt Lake Railroad built sixty-four houses for employees on several blocks bounded by Second, Third, and Fourth Streets. The Railroad Cottage Historic District was created to include the survivors in 1987, but all but one of the houses have since been demolished or relocated. Several were moved to the Clark County Museum and the Springs Preserve.

Several National Register areas are not shown on the map because the National Park Service has withheld their locations. This is common where there is concern that a site may be damaged or destroyed if its location were to be more widely known. The Brownstone canyon, Corn Creek Campsite, Gold Strike Canyon-Sugarloaf Mountain, Sheep Mountain Range, and Tule Springs sites are all archaeological sites. Sloan, Tim Springs and Grapevine Canyon are petroglyphs. The Eureka railroad

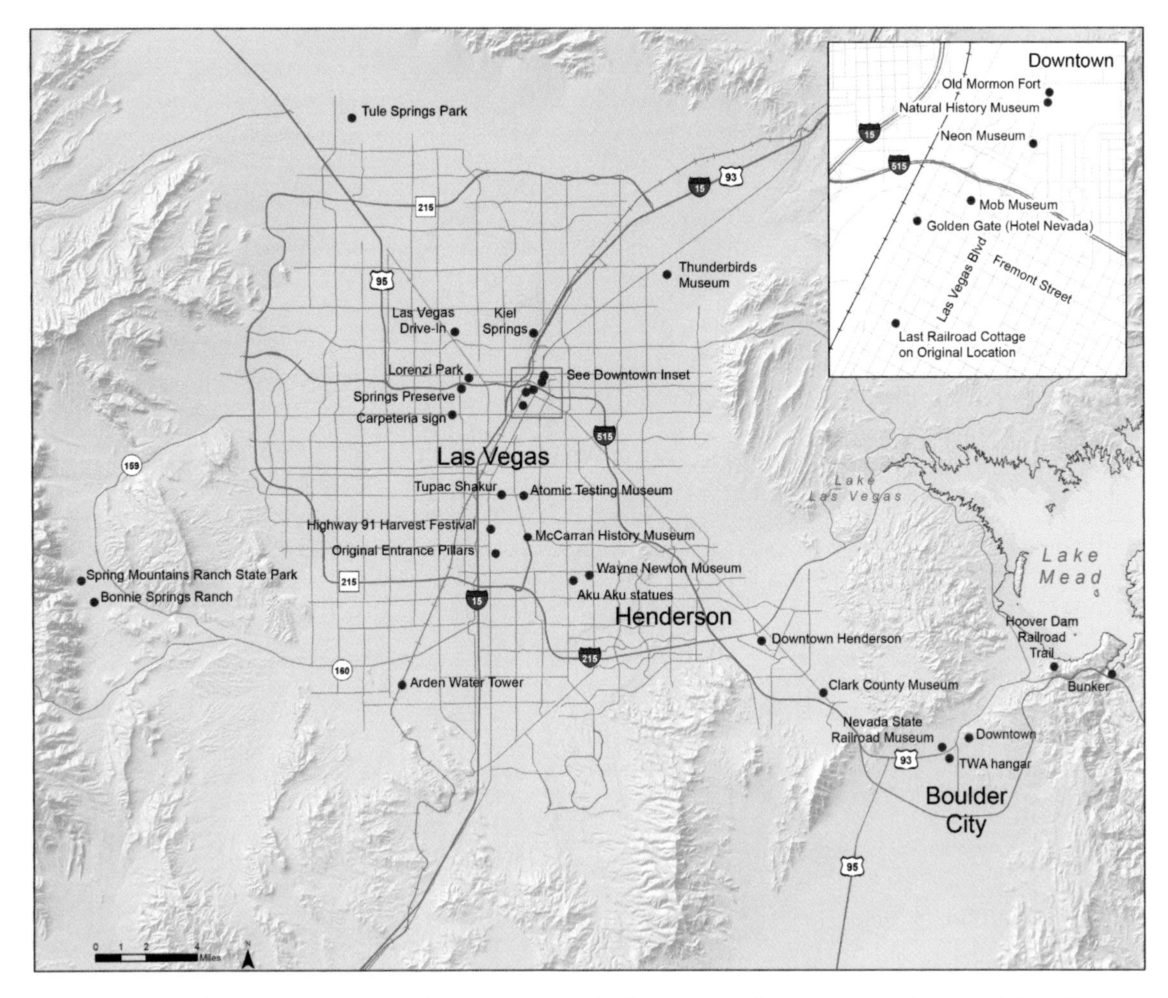

Map 11.9. Las Vegas Memories. Many museums, historic sites, and other places still have traces of old Las Vegas.

locomotive was built in 1875 and is now privately owned; it makes occasional appearances at railroad events throughout the West. The Homestake Mine is an abandoned property near Searchlight, and like many old mines is a hazard to the public.

The future of historic sites in Las Vegas is uncertain. Every year new buildings or locations will meet the conditions for eligibility, and the list of National Register sites will grow. But the city has little appreciation for its past, and each year will likely see more and more of its oldest buildings be demolished to make way for new growth. A positive step was the founding of the Las Vegas Historical Society in 2013 to preserve the city's history. Perhaps preservation will become more of a significant force in the future.

In addition to those places officially recognized as historic sites (map 11.8), there are many places where Las Vegas history can be experienced. Foremost among them are the city's several history museums. The Clark County Museum in Henderson includes a variety of exhibits as well many historic buildings moved there from various locations. Seven of these are relocated homes built from 1910 to 1941. One is a 1910 railroad cottage from downtown Las Vegas, another a 1933 Boulder City home built for dam workers, while a 1941 home from Henderson preserves early workers' housing for that city. The museum also has a reconstructed mining camp, the 1932 Boulder City railroad depot, railroad equipment and other exhibits. Among the attractions at the Nevada State Museum at the Springs Preserve is the Boomtown 1905 exhibit, a reconstructed early Las Vegas street with a replica of the original train station, several businesses, and four railroad cottages moved here from downtown. More recent Las Vegas history is the focus of the National Atomic Testing Museum. This provides an amazing overview of what went on at the Nevada Test Site and why.

Many casinos have come and gone (map 11.7), leaving little trace of their existence. Perhaps the best place to remember them is at the Neon Museum, which has so far saved more than two hundred old casino signs. Other remnants of the Strip survive under new names. The Sands Expo and Convention Center outlived the Sands Hotel, which closed in 1996. The seven-thousand-seat Aladdin Theater for the Performing Arts, once the city's largest venue for musical acts, survived the demolition of the first Aladdin in 1998 and became the AXIS Theater at Planet Hollywood in 2007 (and is now known as the Zappos Theater). Among its claims to fame was the very first American show by Iron Maiden and the first Las Vegas appearance of Judas Priest, together on June 3, 1981.

Lots of history can be experienced away from the Strip. One old-time Las Vegas sign is still in place on West Charleston Boulevard, where the Carpeteria Genie once held the name of the carpet store above his head. He can still be seen advertising the current business at that location. Those seeking out other aspects of the city's history can see sites associated with the Mafia (map 4.15) or visit the intersection of Flamingo Road and Koval Lane where rapper Tupac Shakur was gunned down in 1996. The site of the October 1, 2017, Route 91 Harvest music festival, where a gunman in room 32–135 of Mandalay Bay opened fire and killed sixty, injuring hundreds more, is another place that will not be forgotten.

Although Las Vegas began as a railroad town and depended on the railroad for its first few decades of survival, there are surprisingly few remnants of railroad history. The Nevada State Railroad Museum in Boulder City preserves many engines and rolling stock, including an engine used on a nuclear-powered rocket engine project at the Nevada Test Site (map 9.14). A portion of the nearby Hoover Dam rail Hiline, including several old tunnels, can be hiked within Lake Mead National Recreation Area. Other relics of the town's railroad history are more obscure. A small shed at Corn Creek is built out of old ties from the Las Vegas and Tonopah Railroad, one of the few reminders of that railroad. A steel water tower stands near the tracks at Arden; this is perhaps the only original building left from the San Pedro, Los Angeles, and Salt Lake Railroad, and one of the oldest structures in the valley.

Relics of the early automobile age in Las Vegas are also hard to find. A state historical marker for the Arrowhead Trail can be found near the eastern entrance of Valley of Fire State Park, and a short stretch of the old road can even be hiked within the park. Highway history is sometimes hidden in plain sight; the oldest surviving highway bridge in Southern Nevada is an otherwise unremarkable 1931 bridge on the southbound lanes of Boulder Highway over Flamingo Wash, between Sahara Avenue and Lamb Boulevard (it has been widened many times).

The aviation history of the city is perhaps the most difficult to find. Stone pillars still stand on the east side of the Strip at the southwest corner of Reid (formerly McCarran) International Airport and mark the original entrance to the 1948 airport terminal. The Howard W. Cannon Aviation Museum exists in Harry Reid International Airport's Terminal 2 and has exhibits on the airport's history as well as aviation in Southern Nevada. One of the stories told there is of the longest airplane flight ever made, which began at the airport on December 4, 1958, and returned on February 7, 1959, for a total of sixty-four days, twenty-two hours in the air. The pilots had to refuel twice a day, transferring fuel from a car while flying low at the Blythe, California, airport or a nearby straight highway. The pilots spent most of their time flying over the desert as far south as Yuma, Arizona, but also visited Los Angeles. The airplane used for the flight now hangs from the ceiling in the Harry Reid International Airport terminal.

Other aviation history sites are scattered across the region. Along US 93 in Boulder City, just west of the turnoff to Hoover Dam, the old TWA hangar at the original Boulder City Airport can be found, a legacy of the city's airline service (map 7.15). The Thunderbirds have a museum in Nellis Air Force Base, though only military personnel can enter the base. Outside of Las Vegas, other aviation landmarks include the remains of a concrete airway arrow on Mormon Mesa and the Kyle Canyon memorial to the Charleston Peak C-54 crash (map 7.20).

Other features are left over from the construction of Hoover Dam and the development of Lake Mead. As the lake began to fill, gravel was piled above the future water line in case more cement would be needed. These gravel piles are visible in Hemenway Wash to the west of the road between Highway 93 and Boulder Beach. A World War II bunker overlooking the dam can still be seen on the hilltop above the lower Arizona parking lots. And at the far eastern end of the lake, many remnants of the Pearce Ferry Civilian Conservation Corps (CCC) camp and seismometer stations survive.

Some of Las Vegas history survives elsewhere. An early bridge on the Boulder City railroad line was moved to the Virginia & Truckee Railroad outside Carson City, where it now carries the line over US 50. The Las Vegas and Tonopah (LV&T) Railroad coach #30 survives and is at the Nevada State Railroad Museum in Carson City. Atomic Annie, the cannon used to fire a nuclear warhead at Frenchman Flat at the Nevada Test Site in 1953, can be seen at the US Army Artillery Museum in Fort Sill, Oklahoma. The Smithsonian's National Air and Space Museum in Washington, DC, has a Douglas M-2 biplane painted in Western Air Express markings; a plane like this provided Las Vegas's first airline service in 1926. And finally, what of the overlooked but heroic air force CH-3E helicopters, which rescued hundreds of guests trapped on top of the burning MGM Grand hotel in 1980? They have long since been retired by the air force, but surviving examples can be seen at the Pima Air and Space Museum in Tucson, Yanks Air Museum in Chino, California, the Aerospace Museum of California in Sacramento, and at Edwards Air Force Base, as well as many eastern aviation museums.

Aku Aku moai statue in Sunset Park. Photo by author.

Sunset Park is a good place to go to remember the old Las Vegas. It was once a thriving Paiute encampment, and there are still remnants of the irrigation ditches they used for farming, erased everywhere else in the valley. It is the original Las Vegas (map 6.3), visited and named by Antonio Armijo in 1830 while his men rested in the shade from the hot January sun. It is one of the last places where any of the original valley's vegetation and sand dunes can still be seen; elsewhere they have long since been paved over. The park also has a replica of a moai, or Easter Island statue, originally on display in front of the Polynesian-themed Aku Aku lounge at the Stardust, the last of the Mafia-ran and Teamster-financed casinos (maps 4.15 and 8.12). The end of those eras was soon followed by the destruction of most of the older Strip properties when megaresorts replaced them. The moai managed to survive as a relic of a lost era; it is fitting that it rests here in this quiet refuge from time.

Acronyms

AASHO	American Association of State Highway Officials
BLM	Bureau of Land Management
BMI	Basic Magnesium, Incorporated
CAB	Civil Aeronautics Board
CCC	Civilian Conservation Corps
EPA	Environmental Protection Agency
FAA	Federal Aviation Administration
FBI	Federal Bureau of Investigation
FRA	Federal Railroad Administration
NDOT	Nevada Department of Transportation
NERVA	Nuclear Engine for Rocket Vehicle Application
NTTR	Nevada Test and Training Range
PEPCON	Pacific Engineering and Production Company of Nevada
PLSS	Public Land Survey System
SCOP	Systems Conveyance and Operations Program
UNLV	University of Nevada, Las Vegas
USGS	United States Geological Survey
VEA	Valley Electric Association
WECC	Western Electricity Coordinating Council
WPA	Works Progress Administration

References

The maps in this atlas were made using a range of data sources. Most maps have a shaded relief or hillshaded map as their base, using a US Geological Survey (USGS) ⅓ arc-second (10-meter) resolution Digital Elevation Models (DEMS) created by the USGS and downloaded from www.nationalmap.gov. Several maps covering the entire state or western United States used a 100-meter resolution shaded relief map from the National Map small-scale data set. The website also provided the aerial imagery used in many maps. These images were taken by airplane as part of the National Agriculture Imagery Program and have one-meter resolution.

Many maps show current streets and main highways. These are derived from the US Census Bureau TIGER GIS database (www.census.gov). Other data were obtained from Clark County's Geographic Information Systems GIS data, which can be downloaded free (http://www.clarkcountynv.gov/gis/Services/Pages/FreeGISData.aspx). This includes flood control facilities, casinos, and many other features.

The USGS website (www.usgs.gov) allows the downloading of current and historic topographic maps at a variety of scales. These were used as the source for many roads, springs, vegetation, boundaries, and other features for many time periods (these are discussed in more detail in the notes for chapter 3). The University of Nevada, Las Vegas (UNLV), has many early maps of Las Vegas available online in its digital collections (http://digital.library.unlv.edu/collections/maps), and these were also used to identify early boundaries, vegetation, and other features. In addition to these, many other data sources were required. A wide range of other data were used or created, and these are discussed under individual chapters and maps. In many cases, the map data were created in Google Earth using a variety of current and historic imagery to map out objects such as lakes, railroads, roads, and other features at various points in time.

Most maps use the Nevada State Plane (Nevada East) coordinate system with a transverse Mercator projection, but those covering all of Southern Nevada are in the Universal Transverse Mercator UTM coordinate system (Zone 11). Those showing the United States use the Albers Equal Area projection, and several specialized projections were used for other areas.

CHAPTER 1. INTRODUCTION

In addition to the sources discussed under each following chapter and map, there are many general sources about Las Vegas history. Among the best are:

Anderton, Frances and John Chase. 1997. *Las Vegas: A Guide to Recent Architecture*. London: Ellipsis.

Balboni, Alan. 1996. *Beyond the Mafia: Italian Americans and the Development of Las Vegas*. Reno: University of Las Vegas Press.

Cameron, Robert. 1996. *Above Las Vegas, Its Canyons and Mountains*. San Francisco: Cameron and Company.

Denton, Sally, and Roger Morris. 2001. *The Money and the Power: The Making of Las Vegas and its hold on America*. New York: Alfred A. Knopf.

Fischer, Steve. 2005. *When the Mob Ran Vegas: Stories of Money, Mayhem, and Murder*. New York: MJF Books.

Gottdiener, M., Claudia C. Collins, and David R. Dickens. 1999. *Las Vegas: The Social Production of an All-American City*. Malden, MA: Blackwell.

Land, Barbara, and Myrick Land. 1999. *A Short History of Las Vegas*. Reno: University of Nevada Press.

Littlejohn, David, ed. 1999. *The Real Las Vegas: Life Beyond the Strip*. Oxford: Oxford University Press.

Marschall, John P. 2008. *Jews in Nevada: A Brief History*. Reno: University of Nevada Press.

McCracken, Robert D. 1996. *Las Vegas: The Great American Playground*. Reno: University of Nevada Press.

Miranda, M. L. 1997. *A History of Hispanics in Southern Nevada*. Reno: University of Nevada Press.

Moehring, Eugene P. 2000. *Resort City in the Sunbelt: Las Vegas 1930–2000* (second edition). Reno: University of Nevada Press.

Moehring, Eugene P., and Michael S. Green. 2005. *Las Vegas: A Centennial History*. Reno: University of Nevada Press.

Rothman, Hal K., and Mike Davis, eds. 2002. *The Grit Beneath the Glitter: Tales from the Real Las Vegas*. Berkeley: University of California Press.

Rowley, Rex J. 2013. *Everyday Las Vegas: Local Life in a Tourist Town*. Reno: University of Nevada Press.

Schumacher, Geoff. 2012. *Sun, Sin and Suburbia: The History of Modern Las Vegas, Revised and Expanded*. Las Vegas: Stephens Press.

Simich, Jerry L., and Thomas C. Wright, eds. 2005. *The Peoples of Las Vegas: One City, Many Faces.* Reno: University of Nevada.

Simich, Jerry L., and Thomas C. Wright, eds. 2010. *More Peoples of Las Vegas: One City, Many Faces.* Reno: University of Nevada.

CHAPTER 2. THE NATURAL SETTING

Bell, John W. 1981. *Subsidence in Las Vegas Valley.* Reno: University of Nevada School of Mines.

Houghton, John G., Clarence M. Sakamoto, and Richard O. Gifford. 1975. *Nevada's Weather and Climate.* Reno: Nevada Bureau of Mines and Geology.

Malmberg, Glenn T. 1965. *Available Water Supply of the Las Vegas Ground-Water Basin Nevada.* USGS Water Supply Paper 1780.

Von Till Warren, Elizabeth. 2001. *The History of Las Vegas Springs, a Disappeared Resource.* PhD Dissertation, Washington State University.

Map 2.3. Mojave Desert

Pavlick, Bruce M. 2008. *The California Desert: An Ecological Rediscovery.* Berkeley: University of California Press.

Trimble, Stephen. 1989. *The Sagebrush Ocean: A Natural History of the Great Basin.* Reno: University of Nevada Press.

Map 2.4. Rainfall

Discussion about flooding in the city is based on:

Marlow, Jarvis. 2011. *Taming the waters that taketh from the devil's playground: A history of flood control in Clark County, Nevada, 1955–2010.* Thesis, UNLV.

United States Department of Agriculture. 1977. *History of Flooding, Clark County, Nevada, 1905–1975.* http://nevadafloods.org/docs/clark_flood_hist.pdf

Other sources:

Houghton, John G., Clarence M. Sakamoto, and Richard O. Gifford. 1975. *Nevada's Weather and Climate.* Reno: Nevada Bureau of Mines and Geology.

Orndorff, Richard L., John G. Van Hoesen, and Marvin Saines. Implications of new evidence for late quaternary glaciation in the Spring Mountains, southern Nevada. *Journal of the Arizona-Nevada Academy of Science* 36(1): 37–45.

Map 2.5. Watersheds and the Great Basin

Carpenter, Everett. 1915. *Ground Water in Southeastern Nevada.* USGS Water Supply Paper 365.

National Park Service. 2013. *Lower Las Vegas Wash Flow Regulation Environmental Assessment.* https://parkplanning.nps.gov/document.cfm?parkID=317&projectID=42094&documentID=51964

Plume, Russell W. 1989. *Ground-Water Conditions in Las Vegas Valley, Clark County, Nevada; Part 1 Hydrogeologic Framework.* USGS Water Supply Paper 2320-A

Map 2.6. Ecoregions

Boyer, Diane E., and Robert H. Webb. 2007. *Damming Grand Canyon: The 1923 USGS Colorado River Expedition.* Logan: Utah State University Press.

Carpenter, Everett. 1915. *Ground Water in Southeastern Nevada.* USGS Water Supply Paper 365.

Cole, Kenneth L., Kirsten Ironside, Jon Eischeid, Gregg Garfin, Phillip B. Duffy, and Chris Toney. 2011. Past and ongoing shifts in Joshua Tree distribution support future modeled range contraction. *Ecological Applications* 21(1): 137–49.

Cornett, James W. 2008. The desert fan palm oasis. In *Aridland Springs in North America* (edited by Lawrence E. Stevens and Vicky J. Meretsky). Tucson: University of Arizona Press.

Cornett, James W. 2010. *Desert Palm Oasis.* Palm Springs, CA: Nature Trails Press.

Di Tomaso, Joseph M. 1998. Impact, biology, and ecology of saltcedar (Tamarix spp.) in the Southwestern United States. *Weed Technology* 12:326–36.

Environmental Protection Agency. 2017. *Level III and IV Ecoregions of the Continental United States.* https://www.epa.gov/eco-research/level-iii-and-iv-ecoregions-continental-united-states

Gagnon, Jeffrey W., Chad D. Loberger, Scott C. Sprague, Mike Priest, Kari Ogren, Susan Boe, Estomih Kombe, and Raymond E. Schweinsburg. 2013. Evaluation of desert bighorn sheep overpasses along US Highway 93 in Arizona, USA. *Proceedings of the 2013 International Conference on Ecology and Transportation.*

Hill, James M. 1914. *The Yellow Pine Mining District, Clark County, Nevada.* USGS Bulletin 540-F.

Jones, Florence Lee, and John F. Cahlan. 1975. *Water: A History of Las Vegas, Volume 1.* Las Vegas: Las Vegas Valley Water District.

Lanner, Ronald M. 2007. *The Bristlecone Book: A Natural History of the World's Oldest Trees.* Missoula, MT: Mountain Press Publishing Company.

Schoenwetter, James, and John W. Hohmann. 1997. Landuse reconstruction at the founding settlement of Las Vegas, Nevada. *Historical Archaeology* 31(4): 41–58.

Fowler, Catherine S., and Darla Garey-Sage. 2016. *Isabel T. Kelly's Southern Paiute Ethnographic Field Notes, 1932–1937.* Salt Lake City: University of Utah Press.

Paher, Stanley W. 1971. *Las Vegas: As it Began—As it Grew.* Las Vegas: Nevada Publications.

Von Till Warren, Elizabeth. 2001. *The History of Las Vegas Springs, a Disappeared Resource.* PhD Dissertation, Washington State University.

Map 3.1. Anasazi Las Vegas

Harrington, M. R. 1926. A pre-pueblo site on Colorado River. *Indian Notes* 3(4) 274–84.

Harrington, M. R. 1926. Another ancient salt mine in Nevada. *Indian Notes* 3(4) 221–32.

Harrington, M. R. 1925. Ancient salt mine near St. Thomas, Nevada. *Indian Notes* 2(3) 227–31.

Harry, Karen, and James T. Watson. 2010. The archaeology of Pueblo Grande de Nevada. *Kiva* 74(4): 403–24.

Mcarthur, Aaron James. 2012. *Reclaimed from a Contracting Zion: The Evolving Significance of St. Thomas, Nevada.* PhD Dissertation, UNLV.

Rafferty, Kevin A. 1984. *Cultural Resource Overview of the Las Vegas Valley.* BLM Technical Report No. 13.

Roberts, Heidi, and Richard V. N. Ahlstrom. No date. *Native Americans in Southern Nevada Before 1492.*

Shutler, Dick, Mary Elizabeth Shutler, and James S. Griffith. 1960. *Stuart Rockshelter: A Stratified Site in Southern Nevada.* Nevada State Museum Anthropological Papers No. 3.

Steward, Julian H. 1997 [1938] *Basin-Plateau Aboriginal Sociopolitical Groups.* Salt Lake City: University of Utah Press.

Wheat, Margaret M. 1967. *Survival Arts of the Primitive Paiutes.* Reno: University of Nevada Press.

Wormington, H. M., and Dorothy Ellis, editors. 1967. *Pleistocene Studies in Southern Nevada.* Nevada State Museum Anthropological Papers No. 13.

Map 3.2. Yiwaganti

Baker, Jeffrey L. 2012. *A Paiute Canal in Sunset Park, Clark County, Nevada.* https://sierradeagua.wordpress.com/2009/06/17/paiutecanal/

Fowler, Catherine S., and Darla Garey-Sage. 2016. *Isabel T. Kelly's Southern Paiute Ethnographic Field Notes, 1932–1937.* Salt Lake City: University of Utah Press.

Inter-Tribal Council of Nevada. 1976. *Nuwuvi: A Southern Paiute History.* Reno: Inter-Tribal Council.

Knack, Martha C. 2001. *Boundaries Between: The Southern Paiutes, 1775–1995.* Lincoln: University of Nebraska Press.

Las Vegas Paiute Tribe. 2017. *History.* http://www.lvpaiutetribe.com/history.html

Reeve, W. Paul. 2006. *Making Space on the Western Frontier: Mormons, Miners, and Southern Paiutes.* Urbana: University of Illinois Press.

Wheeler, George M. 1872. *Preliminary Report Concerning Explorations and Surveys Principally in Nevada and Arizona.* Washington, DC: Government Printing Office.

Map 3.3. Paiute Bands

Kelly, Isabel T. 1934. Southern Paiute bands. *American Anthropologist* 36(4): 548–60.

Knack, Martha C. 2001. *Boundaries Between: The Southern Paiutes, 1775–1995.* Lincoln: University of Nebraska Press.

Kroeber, A. L., and C. B. Kroeber. 1973 *A Mohave War Reminiscence, 1854–1880.* New York: Dover.

Inter-Tribal Council of Nevada. 1976. *Nuwuvi: A Southern Paiute History.* Reno: Inter-Tribal Council.

Map 3.4. Mexican Las Vegas

Atlas of Historical County Boundaries. 2017. http://publications.newberry.org/ahcbp/

Bureau of Land Management. 2017. Designated trail alignment of the Old Spanish National Historic Trail. http://nrdata.nps.gov/profiles/NPS_Profile.xml

Denton, Sally. 2004. *American Massacre: The Tragedy at Mountain Meadows, September 1857.* New York: Vintage.

Hafen, LeRoy R., and Antonio Armijo. 1947. Armijo's journal. *Huntington Library Quarterly* 11(1) 87–101.

Map 3.5. The Old Spanish Trail

Bartlett, Richard A. 1962. *Great Surveys of the American West.* Norman: University of Oklahoma Press.

Brooks, George R., ed. 1977. *The Southwest Expedition of Jedediah Smith: His Personal Account of the Journey to California 1826–1827.* Glendale, CA: The Arthur H. Clark Co.

Casebier, Dennis G. 1999. *Mojave Road Guide: An Adventure Through Time.* Goffs: Tales of the Mojave Road Publishing Company.

Crampton, C. Gregory, and Steven K. Madsen. 1994. *In Search of the Old Spanish Trail, Santa Fe to Los Angeles, 1829–1848.* Salt Lake City: Gibbs-Smith Publishing

Dawdy, Doris. 1993. *George Montague Wheeler: The Man and the Myth*. Athens: Ohio University Press.
Hafen, LeRoy R. and Ann W. Hafen. 1993 (originally published 1954). *Old Spanish Trail: Santa Fe to Los Angeles*. Lincoln: University of Nebraska Press.
Inter-Tribal Council of Nevada. 1976. *Nuwuvi: A Southern Paiute History*. Reno: Inter-Tribal Council.
Jackson, Donald, and Mary Lee Spence. 1970. *The Expeditions of John Charles Frémont, Volume 1: Travels from 1838 to 1844*. Urbana: University of Illinois Press.
Lyman, Edward Leo. 2004. *The Overland Journey From Utah to California: Wagon Travel from the City of Saints to the City of Angels*. Reno: University of Nevada Press.
Von Till Warren, Elizabeth, and Ralph J. Roske. *Cultural Resources of the California Desert, 1776–1980: Historic Trail and Wagon Roads*. Riverside, CA: Bureau of Land Management.

Map 3.6. Mormon Las Vegas

Arrington, Leonard J. 1966. Inland to Zion: Mormon trade on the Colorado River 1864–1867. *Arizona and the West* 8: 239–50.
Lingenfelter, Richard E. 1978. *Steamboats on the Colorado River 1852–1916*. Tucson: University of Arizona Press.
Mcarthur, Aaron. 2012. *Reclaimed from a Contracting Zion: The Evolving Significance of St. Thomas, Nevada*. PhD Dissertation, UNLV.
Meinig, D. W. 1965. The Mormon culture region: Strategies and patterns in the geography of the American West, 1847–1964. *Annals of the Association of American Geographers* 55(2): 191–220.
Newberry Library. 2018. *Atlas of Historical County Boundaries*. http://publications.newberry.org/ahcbp/
Smith, Melvin T. 1972. *The Colorado River: Its History in the Lower Canyons Area*. Brigham Young University, PhD Dissertation.

Map 3.7. The Mormon Fort and Potosi

Inter-Tribal Council of Nevada. 1976. *Nuwuvi: A Southern Paiute History*. Reno: Inter-Tribal Council.
Jones, Florence Lee, and John F. Cahlan. 1975. *Water: A History of Las Vegas, Volume 1*. Las Vegas: Las Vegas Valley Water District.
Schoenwetter, James, and John W. Hohmann. 1997. Landuse reconstruction at the founding settlement of Las Vegas, Nevada. *Historical Archaeology* 31(4): 41–58.

Map 3.8. The Las Vegas Rancho

Jones, Florence Lee, and John F. Cahlan. 1975. *Water: A History of Las Vegas, Volume 1*. Las Vegas: Las Vegas Valley Water District.
Roske, Ralph J., and Michael S. Green. 1988. Octavius Decatur Gass: Pau-Ute County Pioneer. *The Journal of Arizona History* 29(4): 371–90.

Map 3.9. Las Vegas, Arizona

Bufkin, Donald. 1964. The lost county of Pah-Ute. Arizoniana 5(2): 1–11.
Comeaux, Malcolm. 1982. Attempts to establish and change a western boundary. *Annals of the Association of American Geographers* 72(2): 254–71.
Hubbard, Bill, Jr. 2009. *American Boundaries: The Nation, the States, the Rectangular Survey*. Chicago: University of Chicago Press.
Newberry Library. 2018. *Atlas of Historical County Boundaries*. http://publications.newberry.org/ahcbp/
Roske, Ralph J., and Michael S. Green. 1988. Octavius Decatur Gass: Pau-Ute County Pioneer. *The Journal of Arizona History* 29(4): 371–90.

Map 3.10. Las Vegas, Lincoln County

Newberry Library. 2018. *Atlas of Historical County Boundaries*. http://publications.newberry.org/ahcbp/

Map 3.11. The Enduring Moapa

Inter-Tribal Council of Nevada. 1976. *Nuwuvi: A Southern Paiute History*. Reno: Inter-Tribal Council.
Powell, J. W., and G. W. Ingalls. 1873. *Report of special commissioners on the condition of the Ute Indians of Utah; the Pai-Utes of Utah, northern Arizona, southern Nevada, and southeastern California; the Go-si Utes of Utah and Nevada; the northwestern Shoshones of Idaho and Utah; and the western Shoshones of Nevada*. Washington, DC: Government Printing Office.
United States Government. 1912. *Executive Orders Relating to Indian Reservations 1855–1922*. Washington, DC: Government Printing Office.

Map 3.12. The Public Land Survey

Bureau of Land Management. 2018. *BLM Navigator*. https://navigator.blm.gov/home
Bureau of Land Management. 2018. *General Land Office Records*. https://glorecords.blm.gov/default.aspx
Hubbard, Bill, Jr. 2009. *American Boundaries: The Nation, the States, the Rectangular Survey*. Chicago: University of Chicago Press.
Powell, John Wesley. 1879. *Lands of the Arid Region of the United States*. Harvard: Harvard Common Press.

Webster, Gerald R., and Jonathan Leib. 2011. Living on the Grid: The U.S. Rectangular Public Land Survey System and the Engineering of the American Landscape. In *Engineering Earth: The Impacts of Mega-Engineering Projects* (edited by Stan Brunn), pp 2123–38. Dordrecht: Springer.

CHAPTER 4. THE FOUNDING OF LAS VEGAS AND ITS GROWTH

Jones, Florence Lee, and John F. Cahlan. 1975. *Water: A History of Las Vegas, Volume 1*. Las Vegas: Las Vegas Valley Water District.

Paher, Stanley W. 1971. *Las Vegas: As It Began—As It Grew*. Las Vegas: Nevada Publications.

Von Till Warren, Elizabeth. 2001. *The History of Las Vegas Springs, a Disappeared Resource*. PhD Dissertation, Washington State University.

Map 4.1. The Las Vegas Townsites

For the outlines of individual blocks, see notes for Map 4.4

Hudson, John. 1982. Towns of the western railroads. *Great Plains Quarterly* 2(1): 41–54.

Reps, John W. 1979. *Cities of the American West: A History of Frontier Urban Planning*. Princeton, NJ: Princeton University Press.

Map 4.2. Early Las Vegas

Myrick, David W. 1963. *Railroad of Nevada and Eastern California, Volume 2: The Southern Roads*. Berkeley, CA: Howell-North Books.

Nevada Department of Highways. 1921. *Second Biennial Report of the Department of Highways, 1919–1920*. Carson City: State Printing Office.

Map 4.3. Logging

Hill, James M. 1914. *The Yellow Pine Mining District, Clark County, Nevada*. USGS Bulletin 540-F.

United States Forest Service. 2012. *Establishment and Modification of National Forest Boundaries and National Grasslands: A Chronological Record 1891–2012*. https://www.fs.fed.us/land/staff/Documents/Establishment%20and%20Modifications%20of%20National%20Forest%20Boundaries%20and%20National%20Grasslands%201891%20to%202012.pdf

Map 4.4. Las Vegas in the 1920s

Krafft, Thomas. 1993. Reconstructing the North American urban landscape: Fire insurance maps—an indispensable source. *Erdkunde* 47: 196–211.

Page, Brian, and Eric Ross. 2015. Envisioning the urban past: GIS reconstruction of a lost Denver district. *Frontiers in Digital Humanities* 2: 1–18.

Sanborn Map Company. 1923. *Sanborn Maps for Las Vegas, Nevada*. https://contentdm.library.unr.edu/cdm/ref/collection/hmaps/id/5051

Map 4.5. Early Additions

Asher, Jack. 1946. *Map of the City of Las Vegas, Nevada, June 1, 1946*. http://d.library.unlv.edu/cdm/singleitem/collection/LV_Maps/id/470/rec/1

Map 4.6. Las Vegas in the 1930s

Hiltzik, Michael. 2010. Colossus: Hoover Dam and the Making of the American Century. New York: Free Press.

McWilliams, J. T. 1927. Map of Clark County, Nevada, 1927. Retrieved from http://digital.library.unlv.edu/objects/LV_Maps/718

Map 4.7. The New Deal in Southern Nevada

Davis, Ren, and Helen Davis. 2011. *Our Mark on this Land: A Guide to the Legacy of the Civilian Conservation Corps in America's Parks*. Granville, OH: McDonald and Woodward Publishing.

Hiltzik, Michael. 2010. *Colossus: Hoover Dam and the Making of the American Century*. New York: Free Press.

Kolvet, Renee Corona and Victoria Ford. 2006. *The Civilian Conservation Corps in Nevada: From Boys to Men*. Reno: University of Nevada Press.

Lowitt, Richard. 1993. *The New Deal and the West*. Norman: University of Oklahoma Press.

Maher, Neil M. 2009. *Nature's New Deal: The Civilian Conservation Corps and the Roots of the American Environmental Movement*. Oxford: Oxford University Press.

Rempel, William C. 2018. *The Gambler: How Penniless Dropout Kirk Kerkorian Became the Greatest Deal Maker in Capitalist History*. New York: HarperCollins.

Rothman, Hal K., and Daniel J. Holder. 2002. *Balancing the Mandates: An Administrative History of Lake Mead National Recreation Area*. Las Vegas: National Park Service

Map 4.8. World War II

Brean, Henry. 2014. World War II left its mark on Las Vegas. *Las Vegas Review-Journal* December 29.

Freeman, Paul. 2007. *Abandoned & Little-Known Airfields*. http://www.airfields-freeman.com/

Green, Michael S. 2004. The Mississippi of the West? *Nevada Law Journal* 5: 57–70.

Mikeshi, Robert C. 1973. *Japan's World War II Balloon Bomb Attacks on North America*. Washington, DC: Smithsonian Institution Press.

Nickel, Robert V. 2004. Dollars, defense, and the desert: Southern Nevada's military economy and World War II. *Nevada Historical Quarterly* 47(4): 303–27.

Webber, Bert. 1992. *Silent Siege III: Japanese Attacks on North America in World War II: Ships Sunk, Air Raids, Bombs Dropped, Civilians Killed.* Medford, OR: Webb Research Group.

Wilman, Catherene J., and James D. Reinhardt. 1997. *A Pictorial History of Nellis Air Force Base, 1941–1996.* Las Vegas: Nellis Air Force Base.

Map 4.9. Segregated Las Vegas

Geran, Trish. 2010. *Beyond the Glimmering Lights: The Pride and Perseverance of African Americans in Las Vegas.* Las Vegas: Stephens Press.

Hershwitzky, Patricia. 2011. *West Las Vegas.* Charleston, SC Arcadia Publishing.

McMillan, James B. 1997. *Fighting Back: A Life in the Struggle for Civil Rights.* Reno: University of Nevada Oral History Program.

Rayle, Greta J., and Helana Ruter. 2015. *World War II Era Residential Housing in Las Vegas, Clark County, Nevada (1940–1945).* http://shpo.nv.gov/uploads/documents/World_War_II_Era_Housing_in_Las_Vegas,_Rayle_and_Ruter,_2015.pdf

Wilman, Catherene J., and James D. Reinhardt. 1997. *A Pictorial History of Nellis Air Force Base, 1941–1996.* Las Vegas: Nellis Air Force Base.

Map 4.10. Postwar Las Vegas

Moehring, Eugene P. 2007. *The University of Nevada, Las Vegas: A History.* Reno: University of Nevada Press.

Nevada Department of Highways. 1933. *Eighth Biennial Report of the Department of Highways, 1931–1932.* Carson City: State Printing Office.

Nevada Department of Highways. 1951. *Seventeenth Biennial Report of the Department of Highways, 1948–1950.* Carson City: State Printing Office.

Smith, John L. 2001. *Running Scared: The Life and Treacherous Times of Las Vegas Casino King Steve Wynn.* New York: Four Walls Eight Windows.

Swanson, Doug. 2015. *Blood Aces: The Wild Ride of Benny Binion, the Texas Gangster Who Created Vegas Poker.* New York: Penguin Books.

Map 4.11. The Mississippi of the West

Bracey, Earnest N. 2009. *The Moulin Rouge and Black Rights in Las Vegas.* Jefferson, NC: McFarland & Company.

Geran, Trish. 2006. *Beyond the Glimmering Lights: The Pride and Perseverance of African Americans in Las Vegas.* Las Vegas: Stephens Press.

Hershwitzky, Patricia. 2011. *West Las Vegas.* Charleston, SC: Arcadia Publishing.

McKee, Robert J. 2014. *Community Action Against Racism in West Las Vegas: The F Street Wall, and the Women who Brought it Down.* Lanham, MD: Lexington Books.

McMillan, James B. 1997. *Fighting Back: A Life in the Struggle for Civil Rights.* Reno: University of Nevada Oral History Program.

Orleck, Annelise. 2005. *Storming Caesars Palace: How Black Mothers Fought Their own War on Poverty.* Boston: Beacon Press.

Sallaz, Jeffrey J. 2004. Civil rights and employment equity in Las Vegas casinos: The failed enforcement of the casino consent decree, 1971–1986. *Nevada Historical Quarterly* 47(4): 283–302.

Swanson, Doug. 2015. *Blood Aces: The Wild Ride of Benny Binion, the Texas Gangster Who Created Vegas Poker.* New York: Penguin Books.

Map 4.12. The Cold War

Davies, Steve. 2008. *Red Eagles: America's Secret MiGs.* Oxford: Osprey.

Frenchu, Bob, and Luis Ramirez. 2018. *Nevada Ghost Towns—Mines as Fallout Shelters.* http://www.forgottennevada.org/sites/fallout.html

Jacobsen, Annie. 2012. *Area 51: An Uncensored History of America's Top Secret Military Base.* Little, Brown and Company.

Peck, Gaillard, Jr. 2012. *America's Secret MiG Squadron: The Red Eagles of Project Constant Peg.* Oxford: Osprey.

Plaskon, Kyril D. 2008. *Silent Heroes of the Cold War.* Stephens Press.

United States Air Force. 2013. *Red Flag Nellis 2013 In-Flight Guide.* http://www.vusaf.us/RFN13/resources/In-Flight-Guide/RFN13IFG.pdf

Wilman, Catherene J., and James D. Reinhardt. 1997. *A Pictorial History of Nellis Air Force Base, 1941–1996.* Las Vegas: Nellis Air Force Base.

Winkler, David F. 1997. *Searching the Skies: The Legacy of the United States Cold War Defense Radar Program.* Langley: United States Air Force Air Combat Command.

Yenne, Bill. 2014. *Area 51—Black Jets: A History of the Aircraft Developed at Groom Lake, America's Secret Aviation Base.* Zenith Press.

Map 4.13. Atomic Las Vegas

Department of Energy. 2000. *United States Nuclear Tests July 1945 through September 1992.* https://web.archive.org/web/20061012160826/http://www.nv.doe.gov/library/publications/historical/DOENV_209_REV15.pdf

Fehner, Terrence R., and F. G. Gosling. 2000. *Origins of the Nevada Test Site*. https://www.nnss.gov/docs/docs_LibraryPublications/DOE_MA0518.pdf

Johnston, William. 2001. *Chronological Listing of Above Ground Nuclear Detonations*. http://www.johnstonsarchive.net/nuclear/atestoo.html

Meier, Charles W. 2006. *Before the Nukes—the Remarkable History of the Nevada Test Site*. Pleasonton, CA: Lansing Publications.

Titus, A. Costandina. 1986. *Bombs in the Backyard: Atomic Testing and American Politics (second edition)*. Reno: University of Nevada Press.

Map 4.14. Downwinders

Department of Energy. 2000. *United States Nuclear Tests July 1945 through September 1992*. https://web.archive.org/web/20061012160826/http://www.nv.doe.gov/library/publications/historical/DOENV_209_REV15.pdf

Fradkin, Philip F. 1989. *Fallout: An American Tragedy*. Tucson: University of Arizona Press.

Fuller, John G. 1984. *The Day We Bombed Utah: America's Most Lethal Secret*. New York: New American Library.

Johnston, William. 2001. *Chronological Listing of Above Ground Nuclear Detonations*. http://www.johnstonsarchive.net/nuclear/atestoo.html

National Cancer Institute. 1997. *I-131 Radiation Exposure from Fallout*. https://www.cancer.gov/about-cancer/causes-prevention/risk/radiation/i-131

Simon, Steven L., André Bouville and Charles E. Land. 2006. Fallout from Nuclear Weapons Tests and Cancer Risks. *American Scientist* 94 (January—February): 48–57.

Map 4.15. The Mafia

Fischer, Steve. 2007. *When the Mob Ran Vegas: Stories of Money, Mayhem and Murder*. New York: MJF Books.

Goodman, Oscar. 2013. *Being Oscar: From Mob Lawyer to Mayor of Las Vegas—Only in America*. New York: Weinstein Books.

Pileggi, Nicholas. 1995. *Casino: Life and Honor in Las Vegas*. New York: Simon & Schuster.

Reid, Ed and Ovid Demaris. 1963. *The Green Felt Jungle*. New York: Trident Press.

Roemer, William F., Jr. 1994. *The Enforcer: Spilotro: The Chicago Mob's Man Over Las Vegas*. New York: Ballantine Books.

Map 4.16. Las Vegas in the 1970s

California State Automobile Association. 1975. *Map of Las Vegas, North Las Vegas, and vicinity*.

Other sources:

Moehring, Eugene P. 2007. *The University of Nevada, Las Vegas: A History*. Reno: University of Nevada Press.

Rempel, William C. 2018. *The Gambler: How Penniless Dropout Kirk Kerkorian Became the Greatest Deal Maker in Capitalist History*. New York: HarperCollins.

Schwartz, David G. 2013. *Grandissimo: The First Emperor of Las Vegas*. Las Vegas: Winchester Books.

Smith, John L. 1997. *No Limit: The Rise and Fall of Bob Stupak and Las Vegas' Stratosphere Tower*. Las Vegas: Huntington Press.

Map 4.17. Las Vegas in 1990

NYU Spatial Data Repository. 2017. *TIGER/Line Files, 1995*. https://geo.nyu.edu/

Other sources:

Cameron, Robert. 1996. *Above Las Vegas, Its Canyons and Mountains*. San Francisco: Cameron and Company.

Smith, John L. 2001. *Running Scared: The Life and Treacherous Times of Las Vegas Casino King Steve Wynn*. New York: Four Walls Eight Windows.

Map 4.18. Pahrump

Brean, Henry. 2017. Nevada bans new residential wells in Pahrump over groundwater decline. *Las Vegas Review-Journal* December 21. https://www.reviewjournal.com/news/politics-and-government/nevada/nevada-bans-new-residential-wells-in-pahrump-over-groundwater-decline/

Harrill, James R. 1986. *Ground-Water Storage Depletion in Pahrump Valley, Nevada-California, 1962–75*. USGS Water Supply Paper 2279

McCracken, Robert. 1992. *Pahrump: A Valley Waiting to Become a City*. Tonopah: Nye County Press.

Public Lands Interpretive Association. 2018. *Nevada*. http://publiclands.org/Get-Books-and-Maps.php?plicstate=NV

Segall, Eli. 2016. Slow comeback from recession leaves Pahrump with lots of wide open spaces. *Las Vegas Sun* February 12. https://lasvegassun.com/news/2016/feb/12/slow-comeback-from-recession-leaves-pahrump-with-l/

Map 4.19. The Drone Wars

Bureau of Investigative Journalism. 2018. *Drone Warfare*. https://www.thebureauinvestigates.com/projects/drone-war

Cockburn, Andrew. 2015. *Kill Chain: The Rise of High-Tech Assassins*. New York: Henry Holt & Company.

Hennigan, W. J. 2015. Drone pilots go to war in the Nevada desert, staring at video screens. *Los Angeles Times* June 17. http://www.latimes.com/nation/la-na-drone-pilots-20150617-story.html

Kaplan, Robert D. 2006. Hunting the Taliban in Las Vegas. *The Atlantic*, September. https://www.theatlantic.com/magazine/archive/2006/09/hunting-the-taliban-in-las-vegas/305116/

Rogers, Keith. 2017. Creech drones having "incredible impact" in war on terror. *Las Vegas Review-Journal* June 11. https://www.reviewjournal.com/news/military/creech-drones-having-incredible-impact-in-war-on-terror/

Scahill, Jeremy. 2016. *The Assassination Complex: Inside the Government's Secret Drone Warfare Program*. New York: Simon and Schuster.

Whittle, Richard. 2014. *Predator: The Secret Origins of the Drone Revolution*. New York: Henry Holt and Company.

Map 4.20. Las Vegas Today

Moehring, Eugene P. 2007. *The University of Nevada, Las Vegas: A History*. Reno: University of Nevada Press.

Tsui, Bonnie. 2009. *American Chinatown: A People's History of Five Neighborhoods*. New York: Free Press.

CHAPTER 5. BUILDING LAKE MEAD

A unique resource available for mapping the area before Lake Mead is GIS data created for a USGS sedimentation study of Lake Mead (Twichell, Cross, and Belew, 2003). This provided a hillshade dataset that combined above water elevations from existing DEMS and underwater hillshade based on sonar data, as well as the lake boundary when full and the extent of sedimentation on the lake bottom. It is used here to represent the area before the lake was built; the filling of the reservoir has changed the topography of the lake bottom in many ways, including sedimentation, landslides above water and slumping below the water level, and wave-cut terraces. Polygons representing current lake levels and dead pool were generated from the USGS DEM. A pre-dam river polygon was created from detailed 1924 Colorado and Virgin River survey maps (USGS, 1924). These were the most detailed representations of the river made before Hoover Dam was built, and remained the basis for river mapping through the Grand Canyon until recently (Boyer and Webb, 2007). While the river changed substantially over time, this map provides the location near the end of the pre-dam area. Because of falling water levels, portions of the Colorado and Virgin Rivers are today flowing streams, but in different locations than they were before the reservoir formed; upriver of Iceberg Canyon it proved difficult to match the 1924 map with current topography because of sedimentation that has occurred.

The USGS operates a website that compiles records from the General Land Office (GLO), allowing the original property boundaries and patented mining claims to be mapped. Many features were mapped in Google Earth using these sources as well as information from narratives from nineteenth- and twentieth-century explorers, surveyors, and other travelers, and a variety of historic sources.

Boyer, Diane E., and Robert H. Webb. 2007. *Damming Grand Canyon: The 1923 USGS Colorado River Expedition*. Logan: Utah State University Press.

Gould, H. R. 1960. Erosion in the reservoir. In *Comprehensive Survey of Sedimentation in Lake Mead, 1948–49*. USGS Professional Paper 295.

Twichell, David C., VeeAnn A. Cross, and Stephen D. Belew. 2003. *Mapping the Floor of Lake Mead (Nevada and Arizona): Preliminary Discussion and GIS Data Release*. US Geological Survey Open File Report 03–320.

USGS 1924. *Map of plan and profile of Colorado River from Lees Ferry, Arizona to Black Canyon, Arizona-Nevada and Virgin River, Nevada*. Washington, DC: USGS.

Other sources:

Smith, Melvin T. 1972. *The Colorado River: Its History in the Lower Canyons Area*. Brigham Young University, PhD Dissertation.

WESTEC Services, Inc. 1980. *Historical and Architectural Resources Within the Lower Colorado River System*. Publisher unknown.

Map 5.1. The Great Bend of the Colorado River

Boyer, Diane E., and Robert H. Webb. 2007. *Damming Grand Canyon: The 1923 USGS Colorado River Expedition*. Logan: Utah State University Press.

Bureau of Reclamation. 1950. *Boulder Canyon Project Final Reports: Part 3-Preparatory Examinations*. Boulder City: Bureau of Reclamation.

Langbein, W. B. 1960. Water budget. In *Comprehensive Survey of Sedimentation in Lake Mead, 1948–49*. USGS Professional Paper 295.

Longwell, Chester R. 1936. Geology of the Boulder reservoir floor, Arizona-Nevada. *Bulletin of the Geological Society of America* 47: 1393–1476.

Longwell, Chester R. 1928. *Geology of the Muddy Mountains, Nevada*. USGS Bulletin 798.

McClellan, Carole, David A. Phillips, and Mike Belshaw. 1980. *The Archaeology of Lake Mead National Recreation Area: An Assessment*. Lake Mead: National Park Service.

Webb, Robert H., Stanley A. Leake, and Raymond M. Turner. 2007. *The Ribbon of Green: Change in Riparian Vegetation in the Southwestern United States*. Tucson: University of Arizona Press.

Map 5.2. Early Explorations

Brooks, George R., ed. 1977. *The Southwest Expedition of Jedediah Smith: His Personal Account of the Journey to California 1826–1827*. Glendale, CA: The Arthur H. Clark Co.

Hafen, LeRoy R., and Antonio Armijo. 1947. Armijo's journal. *Huntington Library Quarterly* 11(1) 87–101.

Ives, Joseph C. 1969 [originally published 1861] *Report Upon the Colorado River of the West*. New York: Da Capo Press.

Jackson, Donald, and Mary Lee Spence. 1970. *The Expeditions of John Charles Frémont, Volume 1: Travels from 1838 to 1844*. Urbana: University of Illinois Press.

Powell, John Wesley. 1961 [originally published 1895] *The Exploration of the Colorado River and its Canyons*. New York: Dover.

Wheeler, George M. 1872. *Preliminary Report Concerning Explorations and Surveys Principally in Nevada and Arizona*. Washington, DC: Government Printing Office.

Wheeler, George M. 1875. *Preliminary Report Upon a Reconnaissance Through Southern and Southeastern Nevada Made in 1869*. Washington, DC: Government Printing Office.

Other sources:

Dawdy, Doris. 1993. *George Montague Wheeler: The Man and the Myth*. Athens: Ohio University Press.

Hafen, LeRoy R. and Ann W. Hafen 1993 (1954). *Old Spanish Trail: Santa Fe to Los Angeles*. Lincoln: University of Nebraska Press.

Huseman, Ben W. 1995. *Wild River, Timeless Canyons: Balduin Mollhausen's Watercolors of the Colorado*. Tucson: University of Arizona Press.

Rowan, Steven. 2012. *The Baron in the Grand Canyon: Friedrich Wilhelm von Egloffstein in the West*. Columbia: University of Missouri Press.

Map 5.3. Early Settlement

Based on early USGS topographic maps and Smith and Westec Services, Inc., and:

Arrington, Leonard J. 1966. Inland to Zion: Mormon trade on the Colorado River 1864–1867. *Arizona and the West* 8: 239–50.

Hill, James M. 1914. *The Grand Gulch Mining Region, Mohave County, Arizona*. USGS Bulletin 580.

Lingenfelter, Richard E. 1978. *Steamboats on the Colorado River 1852–1916*. Tucson: University of Arizona Press.

Mcarthur, Aaron. 2012. *Reclaimed from a Contracting Zion: The Evolving Significance of St. Thomas, Nevada*. PhD Dissertation, UNLV.

Noble, L. F. 1922. *Colemanite in Clark County, Nevada*. USGS Bulletin 735B.

Perkins, Waldo C. 2006. From Switzerland to the Colorado River: Life sketch of the entrepreneurial Daniel Bonelli, the forgotten pioneer. *Utah Historical Quarterly* 74: 4–23.

Boyer, Diane E., and Robert H. Webb. 2007. *Damming Grand Canyon: The 1923 USGS Colorado River Expedition*. Logan: Utah State University Press.

Map 5.4. Colorado River Surveys

Bergland, Eric. 1876. Preliminary Report Upon the Operations of Party No. 3, California Section, Season of 1875–76, With a View to Determine the Feasibility of Diverting the Colorado River for Purposes of Irrigation. In Wheeler, George M., 1876: *Annual Report Upon the Geographical Surveys West of the One Hundredth Meridian, in California, Nevada, Utah, Colorado, Wyoming, New Mexico, Arizona, and Montana, Appendix JJ*. Washington, DC: Government Printing Office.

Boyer, Diane E., and Robert H. Webb. 2007. *Damming Grand Canyon: The 1923 USGS Colorado River Expedition*. Logan: Utah State University Press.

Bureau of Reclamation. 1950. *Boulder Canyon Project Final Reports: Part 3—Preparatory Examinations*. Boulder City: Bureau of Reclamation.

Dawdy, Doris. 1993. *George Montague Wheeler: The Man and the Myth*. Athens: Ohio University Press.

Smith, D. L., and Crampton, C. G. 1987. *The Colorado River Survey*. Salt Lake City: Howe Brothers Books.

Snyder, Joel. 1981. *American Frontiers: The Photographs of Timothy H. O'Sullivan, 1867–1874*. New York: Aperture.

Wheeler, George M. 1872. *Preliminary Report Concerning Explorations and Surveys Principally in Nevada and Arizona*. Washington, DC: Government Printing Office.

Map 5.5. Before Lake Mead
Based on USGS topographic maps, Nevada Highway Department road maps, and Smith and WESTEC Services, Inc.

Bureau of Reclamation. 1950. *Boulder Canyon Project Final Reports: Part 3—Preparatory Examinations*. Boulder City: Bureau of Reclamation.
Hill, James M. 1914. *The Grand Gulch Mining Region, Mohave County, Arizona*. USGS Bulletin 580.
Longwell, Chester R. 1928. *Geology of the Muddy Mountains, Nevada*. USGS Bulletin 798.
Mcarthur, Aaron. 2012. *Reclaimed from a Contracting Zion: The Evolving Significance of St. Thomas, Nevada*. PhD Dissertation, UNLV.
Noble, L. F. 1922. *Colemanite in Clark County, Nevada*. USGS Bulletin 735B.
Weber, Joe. 2018. The Colorado and Virgin Rivers Before Lake Mead. Journal of Maps 14(2): 583–88.

Map 5.6. Taming the Colorado
Based on watersheds obtained from the USGS National Map.gov website and on the sites shown in the Bureau of Reclamation 1948 report.

Boyer, USGS *Colorado River Expedition*. Logan: Utah State University Press.
Bureau of Reclamation. 1922. *Problems of Imperial Valley and Vicinity*. Washington, DC: Government Printing Office.
Bureau of Reclamation. 1948. *Boulder Canyon Project Final Reports: Part 1—Introductory*. Boulder City: Bureau of Reclamation.
Hiltzik, Michael. 2010. *Colossus: Hoover Dam and the Making of the American Century*. New York: Free Press.
Kleinsorge, Paul. 1941. *The Boulder Canyon Project: Historical and Economic Aspects*. Stanford: Stanford University Press.
La Rue, E. C. 1916. *Colorado River and its Utilization*. United States Geological Survey Water Supply Paper 395.
Sutton, Imre. 1968. Geographical aspects of construction planning: Hoover Dam revisited. *Journal of the West* 7(3): 301–44.

Map 5.7. Building Hoover Dam
Bureau of Reclamation. 1941. *Boulder Canyon Project Final Reports: Part 4—Design and Construction, Bulletin 1*. Boulder City: Bureau of Reclamation.
Bureau of Reclamation. 1941. *Boulder Canyon Project Final Reports: Part 4—Design and Construction, Bulletin 2*. Boulder City: Bureau of Reclamation.
Bureau of Reclamation. 1948. *Boulder Canyon Project Final Reports: Part 1—Introductory*. Boulder City: Bureau of Reclamation.
Bureau of Reclamation. 1950. *Boulder Canyon Project Final Reports: Part 3—Preparatory Examinations*. Boulder City: Bureau of Reclamation.
Hiltzik, Michael. 2010. *Colossus: Hoover Dam and the Making of the American Century*. New York: Free Press.
Kleinsorge, Paul. 1941. *The Boulder Canyon Project: Historical and Economic Aspects*. Stanford: Stanford University Press.
Myrick, David W. 1963. *Railroad of Nevada and Eastern California, Volume 2: The Southern Roads*. Berkeley, CA: Howell-North Books.

Map 5.8. Boulder City
Denit, W. Darlington. 1952. Boulder City government town problem. *Public Administration Review* 12(2): 97–105.
Kudialis, Chris. 2017. Boulder City tugged between growth and protecting small-town feel. *Las Vegas Sun* March 6, 2017. https://lasvegassun.com/news/2017/mar/06/boulder-city-tugged-between-growth-and-protecting/
Woodward, James and Cindy Myers. 1983. *Boulder City Historic District: Nomination to the National Register of Historic Places*. https://www.nps.gov/nr/

Map 5.9. Lake Mead's Early Years
Based on a map in Longwell (1960), USGS earthquake data, and locations identified in:

Glendinning, Robert M. 1945. Desert change: A study of the Boulder Dam area. *Scientific Monthly* 61(3): 181–93.
Gould, H. R. 1960A. Character of the accumulated sediment. In *Comprehensive Survey of Sedimentation in Lake Mead, 1948–49*. USGS Professional Paper 295.
Gould, H. R. 1960D. Erosion in the reservoir. In *Comprehensive Survey of Sedimentation in Lake Mead, 1948–49*. USGS Professional Paper 295.
Hiltzik, Michael. 2010. *Colossus: Hoover Dam and the Making of the American Century*. New York: Free Press.
Langbein, W. B. 1960. Water budget. In *Comprehensive Survey of Sedimentation in Lake Mead, 1948–49*. USGS Professional Paper 295.
Longwell, C. R. 1960. Interpretation of the leveling data. In *Comprehensive Survey of Sedimentation in Lake Mead, 1948–49*. USGS Professional Paper 295.

Smith, W. O., C. P. Vetter, G. B. Cummings, and others. 1960. *Comprehensive Survey of Sedimentation in Lake Mead, 1948–49*. USGS Professional Paper 295.

Map 5.10. Lake Mead Developed

Dodd, Douglas W. 2007. Boulder Dam Recreation Area: The Bureau of Reclamation, the National Park Service, and the origins of the National Recreation Area concept at Lake Mead, 1929–1936. *Southern California Quarterly* 88: 431–73.

Glendinning, Robert M. 1945. Desert change: A study of the Boulder Dam area. *Scientific Monthly* 61(3): 181–93.

Rothman, Hal K., and Daniel J. Holder. 2002. *Balancing the Mandates: An Administrative History of Lake Mead National Recreation Area*.

Map 5.11. Sedimentation

Based on data in the Twichell, et al., data set.

Bureau of Reclamation. 2008. *2001 Lake Mead Sedimentation Survey*. http://www.riversimulator.org/Resources/USBR/LakeMeadSedimentation-Survey2001.pdf

Gould, H. R. 1960A. Character of the accumulated sediment. In *Comprehensive Survey of Sedimentation in Lake Mead, 1948–49*. USGS Professional Paper 295.

Gould, H. R. 1960B. Amount of sediment. In *Comprehensive Survey of Sedimentation in Lake Mead, 1948–49*. USGS Professional Paper 295.

Gould, H. R. 1960C. Sedimentation in relation to reservoir utilization. In *Comprehensive Survey of Sedimentation in Lake Mead, 1948–49*. USGS Professional Paper 295.

Gould, H. R. 1960D. Erosion in the reservoir. In *Comprehensive Survey of Sedimentation in Lake Mead, 1948–49*. USGS Professional Paper 295.

Smith, W. O., C. P. Vetter, G. B. Cummings, and others. 1960. *Comprehensive Survey of Sedimentation in Lake Mead, 1948–49*. USGS Professional Paper 295.

Thomas, H. E., H. R. Gould, and W. B. Langbein. 1960. Life of the Reservoir. In *Comprehensive Survey of Sedimentation in Lake Mead, 1948–49*. USGS Professional Paper 295.

Map 5.12. The Dead Pool

Based on elevation data from the Twichell study.

Crampton, C. Gregory. 2009. *Ghosts of Glen Canyon: History Beneath Lake Powell*. Salt Lake City: Bonneville Books.

Fowler, Don D. 2011. *The Glen Canyon Country: A Personal Memoir*. Salt Lake City: University of Utah Press.

McGivney, Annette. 2009. *Resurrection: Glen Canyon and a New Vision for the American West*. Seattle: Braided River.

Map 5.13. The Return of the Lost City

Based on reservoir zones discussed by Lenihan and others, using the elevation data from the Twichell study.

Belshaw, Mike, and Ed Peplow. 1980 *Historic Resources Study, Lake Mead National Recreation Area, Nevada*. National Park Service. https://www.nps.gov/parkhistory/online_books/lake/hrs.pdf

Bureau of Reclamation. 2008. 2001 *Lake Mead Sedimentation Survey*. http://www.riversimulator.org/Resources/USBR/LakeMeadSedimentationSurvey2001.pdf

Glendinning, Robert M. 1945. Desert change: A study of the Boulder Dam area. *Scientific Monthly* 61(3): 181–93.

Lenihan, Daniel J., Toni L. Carrell, Stephen Fosberg, Larry Murphy, Sandra L Rayl, and John A. Ware. 1981. *The Final Report of the National Reservoir Inundation Study, Volume I*. Santa Fe, NM: National Park Service Southwest Cultural Resources Center.

Longwell, Chester R. 1936. Geology of the Boulder reservoir floor, Arizona-Nevada. *Bulletin of the Geological Society of America* 47: 1393–1476.

Mcarthur, Aaron. 2012. *Reclaimed from a Contracting Zion: The Evolving Significance of St. Thomas, Nevada*. PhD Dissertation, UNLV.

McClellan, Carole, David A. Phillips, and Mike Belshaw. 1980. *The Archaeology of Lake Mead National Recreation Area: An Assessment*. http://core.tdar.org/document/3962/the-archeology-of-lake-mead-national-recreation-area-an-assessment.

National Park Service. 2016. *St. Thomas, Nevada, Visual Field Guide*. https://www.nps.gov/lake/learn/nature/upload/St-Thomas-Field-Guide-FINAL.pdf

Reno, Ron, Charles Zeier, David Choate, David Conlin, and Daniel Lenihan. 2009. *Hoover Dam Aggregate Classification Plant*. Historic American Engineering Record. http://lcweb2.loc.gov/master/pnp/habshaer/nv/nv0400/nv0431/data/nv0431data.pdf

Reno, Ron, Charles Zeier, David Choate, David Conlin, and Daniel Lenihan. 2009. *Arizona Gravel Pit Road*. Historic American Engineering Record. http://cdn.loc.gov/master/pnp/habshaer/nv/nv0400/nv0430/data/nv0430data.pdf

Reno, Ron, Charles Zeier, David Choate, David Conlin, and Daniel Lenihan. 2009. *Six Companies Railroad*. Historic America Engineering Record. http://cdn.loc.gov/master/pnp/habshaer/nv/nv0400/nv0432/data/nv0432data.pdf

Weber, Joe, and Matthew LaFevor. 2021. Exploring Remnant Landscapes of Nevada's Arrowhead Trail. Geographical Review 111(3): 437–57

Woodbury, Angus. 1960. Protecting Rainbow Bridge. *Science* 132(34) 519–28.

CHAPTER 6. QUENCHING LAS VEGAS'S THIRST

Jones, Florence Lee, and John F. Cahlan. 1975. *Water: a History of Las Vegas, Volume 1*. Las Vegas: Las Vegas Valley Water District.

Jones, Florence Lee. 1975. *Water: A History of Las Vegas, Volume 2*. Las Vegas: Las Vegas Valley Water District.

Von Till Warren, Elizabeth. 2001. *The History of Las Vegas Springs, a Disappeared Resource*. PhD Dissertation, Washington State University.

Map 6.1. The Meadows

Carpenter, Everett. 1915. *Ground Water in Southeastern Nevada*. USGS Water Supply Paper 365.

Schoenwetter, James, and John W. Hohmann. 1997. Landuse reconstruction at the founding settlement of Las Vegas, Nevada. *Historical Archaeology* 31(4): 41–58.

Waring, Gerald A. 1921. *Ground Water in Pahrump, Mesquite, and Ivanpah Valleys, Nevada and California*. Water Supply Paper 450-C.

Map 6.2. Wells

Armelung, Falk, Devin L. Galloway, John W. Bell, Howard A. Zebker, and Randell J. Laczniak. 1999. Sensing the ups and downs of Las Vegas: InSAR reveals structural control of land subsidence and aquifer-system deformation. *Geology* 27(6): 483–86.

Bell, John W. 1981. *Subsidence in Las Vegas Valley*. Reno: University of Nevada School of Mines.

Malmberg, Glenn T. 1965. *Available Water Supply of the Las Vegas Ground-Water Basin Nevada*. USGS Water Supply Paper 1780.

Nevada Division of Water Resources. 2017. *Well Log Search*. http://water.nv.gov/welllogquery.aspx

Plume, Russell W. 1989. *Ground-Water Conditions in Las Vegas Valley, Clark County, Nevada; Part 1 Hydrogeologic Framework*. USGS Water Supply Paper 2320-A

Map 6.3. The Original Las Vegas

Baker, Jeffrey L. 2012. *A Paiute Canal in Sunset Park, Clark County, Nevada*. https://sierradeagua.wordpress.com/2009/06/17/paiutecanal/

Carpenter, Everett. 1915. *Ground Water in Southeastern Nevada*. USGS Water Supply Paper 365.

Fowler, Catherine S., and Darla Garey-Sage. 2016. *Isabel T. Kelly's Southern Paiute Ethnographic Field Notes, 1932–1937*. Salt Lake City: University of Utah Press.

McWilliams, J. T. 1918. *Map of Las Vegas Valley Showing Artesian Wells*. http://digital.library.unlv.edu/objects/LV_Maps/478

Map 6.4. Water Supply

Boulder City. 2017. Utilities. http://www.bcnv.org/321/Utilities

Bureau of Reclamation. 1941. *Boulder Canyon Project Final Reports: Part 4—Design and Construction, Bulletin 1*. Boulder City: Bureau of Reclamation.

Bureau of Reclamation. 1941. *Boulder Canyon Project Final Reports: Part 4—Design and Construction, Bulletin 2*. Boulder City: Bureau of Reclamation.

Carpenter, Everett. 1915. *Ground Water in Southeastern Nevada*. USGS Water Supply Paper 365.

City of Henderson. 2017. Utility Services. http://www.cityofhenderson.com/utility-services/customer-care-center/water-and-sewer-services

City of North Las Vegas. 2017. Utilities Department. http://www.cityofnorthlasvegas.com/departments/utilities/index.php

Lake Mead National Recreation Area. 1991. *Lakeshore Road Reconstruction Draft Environmental Impact Statement*. Washington, DC: National Park Service.

Las Vegas Valley Water District. 2017. Home Page. https://www.lvvwd.com/

Waring, Gerald A. 1921. *Ground Water in Pahrump, Mesquite, and Ivanpah Valleys, Nevada and California*. USGS Water Supply Paper 450.

Map 6.5. Overdrafting and Subsidence

Amelung, Falk, Devin L. Galloway, John W. Bell, Howard A. Zebker, and Randell J. Laczniak. 1999. Sensing the ups and downs of Las Vegas: InSAR reveals structural control of land subsidence and aquifer-system deformation. *Geology* 27(6): 483–86.

Bell, John W. 1981. *Subsidence in Las Vegas Valley*. Reno: University of Nevada School of Mines.

Map 6.6. Lakes of Las Vegas

Lakes were mapped out in Google Earth.

Abbott, J. K. and H. A. Klaiber. 2013. The value of water as an urban club good: A matching approach to HOA-provided lakes. *Journal of Environmental Economics and Management*. 65 (2): 208–24.

Arizona State University. 2017. *Phoenix: Land of 1400 (artificial) lakes*. https://sustainability.asu.edu/caplter/research-highlights/research-highlight-13/
Deacon, James E, and Cynthia Deacon Williams. 1991. Ash Meadows and the legacy of the Desert Hole pupfish. In *Battle Against Extinction: Native Fish Management in the American West*, W. L. Minckley and James E. Deacon. Tucson: University of Arizona Press.
Jones, Florence Lee, and John F. Cahlan. 1975. *Water: A History of Las Vegas, Volume 1*. Las Vegas: Las Vegas Valley Water District.
Larson, E. K., and N. B. Grimm. 2012. Small-scale and extensive hydrogeomorphic modification and water redistribution in a desert city and implications for regional nitrogen removal. *Urban Ecosystems* 15: 71–85.
Roach, W. J., and N. B. Grimm. 2011. Denitrification mitigates N flux through the stream–floodplain complex of a desert city. *Ecological Applications* 21: 2618–36.
Roach, W. J., J. B. Heffernan, N. B. Grimm, J. R. Arrowsmith, C. Eisinger and T. Rychener. 2008. Unintended consequences of urbanization for aquatic ecosystems: A case study from the Arizona desert. *BioScience* 58: 715–27.

Map 6.7. Sewage

Boulder City. 2017. Utilities. http://www.bcnv.org/321/Utilities
City of Henderson. 2017. Utility Services. http://www.cityofhenderson.com/utility-services/customer-care-center/water-and-sewer-services
City of North Las Vegas. 2017. Utilities Department. http://www.cityofnorthlasvegas.com/departments/utilities/index.php
Clark County Water Reclamation District. 2017. Home Page. https://www.cleanwaterteam.com/Pages/default.aspx
National Park Service. 2008. *Systems Conveyance and Operations Program* SCOP *Environmental Assessment for Activities within the National Park Service Lake Mead National Recreation Area*. https://www.nps.gov/lake/learn/management/upload/SCOP%20Ancillary%20Facilities%20EA-2.pdf

Map 6.8. Flood Control

Marlow, Jarvis. 2011. *Taming the waters that taketh from the devil's playground: A history of flood control in Clark County, Nevada, 1955–2010*. Thesis, UNLV.
O'Brien, Matthew. 2007. *Beneath the Neon: Life and Death in the Tunnels of Las Vegas*. Las Vegas: Huntington Press.
United States Department of Agriculture. 1977. *History of Flooding, Clark County, Nevada, 1905–1975*. http://nevadafloods.org/docs/clark_flood_hist.pdf

Map 6.9. Las Vegas Wash

Carpenter, Everett. 1915. *Ground Water in Southeastern Nevada*. USGS Water Supply Paper 365.
Malmberg, Glenn T. 1965. *Available Water Supply of the Las Vegas Ground-Water Basin Nevada*. USGS Water Supply Paper 1780.
National Park Service. 2013. *Lower Las Vegas Wash Flow Regulation Environmental Assessment*. https://parkplanning.nps.gov/document.cfm?parkID=317&projectID=42094&documentID=51964
National Park Service. 2008. *Systems Conveyance and Operations Program (SCOP) Environmental Assessment for Activities within the National Park Service Lake Mead National Recreation Area*. https://www.nps.gov/lake/learn/management/upload/SCOP%20Ancillary%20Facilities%20EA-2.pdf
National Park Service. 2002. *Lake Plan Management Final Environmental Statement Impact*. https://www.nps.gov/lake/learn/management/park-management-plans.htm
Plume, Russell W. 1989. *Ground-Water Conditions in Las Vegas Valley, Clark County, Nevada; Part 1 Hydrogeologic Framework*. USGS Water Supply Paper 2320-A

CHAPTER 7. TRANSPORTING LAS VEGAS

Map 7.1. The Salt Lake Route; Map 7.2. The Goldfield Route; Map 7.3. Mining Branches; Map 7.4. Northern Branches; Map 7.5. Industrial Spurs

Google Earth was used to locate current and abandoned lines identified in Myrick (1963). Details of the history of the lines were found in Myrick (1963) and Signor (1988). The Union Pacific Railroad Timetable provided details of current operations.

Myrick, David W. 1963. *Railroad of Nevada and Eastern California, Volume 2: The Southern Roads*. Berkeley, CA: Howell-North Books.
Signor, John R. 1988. *The Los Angeles and Salt Lake Railroad Company: Union Pacific's Historic Salt Lake Route*. San Marino, CA: Golden West Books.
Union Pacific Railroad. 2008. *Salt Lake City Area Timetable*. http://denversrailroads.com/Denver/Timetables/UP_Salt_Lake_City_TT3_06-16-08.pdf

United States Department of Agriculture. 1977. *History of Flooding, Clark County, Nevada, 1905–1975*. http://nevadafloods.org/docs/clark_flood_hist.pdf

Map 7.6. Crossing the Tracks

Based on road maps of Las Vegas in the UNLV digital collections (http://digital.library.unlv.edu/collections/maps).

Federal Railroad Administration. 2016. FRA *Releases List of Railroad Crossings with Most Incidents over Last Decade*. https://www.fra.dot.gov/eLib/details/L17404

Mooney, Courtney (2003). *Clark Avenue Railroad Underpass*. National Register of Historic Places-Registration Form. National Park Service.

Nevada Department of Highways, 1963. *Las Vegas Valley Urban Transportation Study*. Reno: Nevada Department of Highways.

Snyder, John W. 2008. *Historic Bridges of Nevada, Volume 2*. Carson City: Nevada Department of Transportation.

United States Department of Agriculture. 1977. *History of Flooding, Clark County, Nevada, 1905–1975*. http://nevadafloods.org/docs/clark_flood_hist.pdf

Map 7.7. The Arrowhead Trail

Based on USGS topographic maps and articles by Lyman. Other sources provide additional background on the Trail and other early roads:

Lyman, Edward Leo. 2004. *The Overland Journey from Utah to California: Wagon Travel from the City of Saints to the City of Angels*. Reno: University of Nevada Press.

Lyman, Edward Leo. 1999. The Arrowhead Trails Highway: The beginning of Utah's other route to the Pacific Coast. *Utah Historical Quarterly* 67: 242–64.

Lyman, Edward Leo. 1999. The Arrowhead Trails Highway: Southern Nevada's first automobile link to the outside world. *Nevada Historical Society Quarterly* 42: 256–78.

Lyman, Edward Leo. 1999. The Arrowhead Trails Highway: California's predecessor to Interstate 15. *Southern California Quarterly* 81: 315–40.

Mcarthur, Aaron. 2012. *Reclaimed from a Contracting Zion: The Evolving Significance of St. Thomas, Nevada*. PhD Dissertation, UNLV.

Thompson, David G. 1921. *Routes to Desert Watering Places in the Mohave Desert Region, California*. USGS Water Supply Paper 490-B.

Thompson, David G. 1929. *The Mohave Desert Region: A Geographic, Geologic, and Hydrologic Reconnaissance*. USGS Water Supply Paper 578.

Waddell, A. G. 1917. Easy Grades on Arrowhead Trail. *Motor Age* 31(4): 102–5.

Weber, Joe and Matthew LaFevor. 2021. Exploring Remnant Landscapes of Nevada's Arrowhead Trail. Geographical Review 111(3): 437–57

Weingroff, Richard F. 2013. The Pikes Peak Ocean to Ocean Highway. http://www.fhwa.dot.gov/infrastructure/pikes.cfm

Whiteley, Lee, and Jane Whiteley, 2003. *The Playground Trail: The National Park-to-Park Highway to and Through the National Parks of the West in 1920*. Boulder: Johnson Printing.

Map 7.8. Nevada's State Highways

Based on the Biennial Reports of the Nevada State Highway Commission, Nevada Highway Department road maps (https://www.nevadadot.com/travel-info/maps/historical-maps), and:

Knight, Kenneth C, and T. Hal Turner. 1988. *An Inventory of Nevada's Historic Bridges*. Reno: Nevada Department of Transportation.

Nevada Department of Transportation. 2014. *2014 Facts and Figures*. https://www.nevadadot.com/uploadedFiles/NDOT/About_NDOT/NDOT_Divisions/Planning/Performance_Analysis/2014%20fact%20book%20web2.pdf

Map 7.9. Numbering Las Vegas Highways

Based on Nevada Highway Department road maps (https://www.nevadadot.com/travel-info/maps/historical-maps), with background on the numbered highway system found in:

American Association of State Highway and Transportation Officials. 2017. www.transportation.org.

James, E. W. 1931. Marking our highway system. *American Highways* October 18–20.

James, E. W. 1933. Making and unmaking a national system of marked routes. *American Highways* October 16–18, 27.

Weingroff, Richard F. 2015. From Names to Numbers: The Origins of the U.S. Numbered Highway System. http://www.fhwa.dot.gov/infrastructure/numbers.cfm

Map 7.10. US 95; Map 7.11. Boulder Highway and US 93; Map 7.12. Spring Mountain Roads

Based on the first to 25th biennial reports of the Nevada State Highway Commission, 1919–1967, Nevada Highway Department road maps (https://www.nevadadot.com/travel-info/maps/historical-maps), and:

Knight, Kenneth C, and T. Hal Turner. 1988. *An Inventory of Nevada's Historic Bridges*. Reno: Nevada Department of Transportation.

McCracken, Robert. 1990. *Pahrump: A Valley Waiting to Become a City.* Tonopah: Nye County Press.

Titus, A. Costandina. 1986. *Bombs in the Backyard: Atomic Testing and American Politics, Second Edition.* Reno: University of Nevada Press.

Map 7.13. Las Vegas Freeways

Based on Nevada Highway Department road maps (https://www.nevadadot.com/travel-info/maps/historical-maps) and online news stories about the progress of highway construction. Additional information from:

Bureau of Public Roads, 1939. *Toll Roads and Free Roads.* Washington, DC: Government Printing Office.

Bureau of Public Roads, 1946. *Annual Report.* Washington, DC: Bureau of Public Roads.

Bureau of Public Roads, 1947. *Annual Report.* Washington, DC: Bureau of Public Roads.

Cron, Frederick W. 1968. Touring by numbers—how and why. *Public Works* 99: 80–82, 140, 142.

Federal Highway Administration. 2005. *Final List of Nationally and Exceptionally Significant Features of the Federal Interstate Highway System.* https://www.environment.fhwa.dot.gov/histpres/highways_list.asp

Lewis, T., 1997. *Divided Highways: Building the Interstate Highways, Transforming American Life.* New York: Viking.

Map 7.14. CAM-4

Based on United States Air Navigation Maps no. 132 and 133, produced by the Aeronautics Branch of the Department of Commerce in 1929, at http://magic.lib.uconn.edu/mash_up/air_navigation_index.html, and:

Bednarek, Janet R. Daly. 2001. *America's Airports: Airfield Development, 1918–1947.* College Station: Texas A&M University Press.

Bubb, Daniel. 2009. Transforming the desert: Commercial aviation as agent of change, Las Vegas, 1926–1945. *Nevada Historical Society Quarterly* 52: 198–212.

Serling, Robert J. 1978. *The Only Way to Fly: The Story of Western Airlines, America's Senior Air Carrier.* New York: Doubleday.

Wright, Frank. 1993. *Desert Airways: A Short History of Clark County Aviation, 1920–1948.* Clark County Heritage Museum Occasional Paper #1.

Map 7.15. Early Aviation in Las Vegas

Based on Abandoned & Little Known Airfields at http://www.airfields-freeman.com/ and Airline Timetable Images at http://www.timetableimages.com/

Other sources:

Bednarek, Janet R. Daly. 2001. *America's Airports: Airfield Development, 1918–1947.* College Station: Texas A&M University Press.

Bubb, Daniel. 2009. Transforming the desert: Commercial aviation as agent of change, Las Vegas, 1926–1945. *Nevada Historical Society Quarterly* 52: 198–212.

Freeman, Paul. 2016. Wright, Frank. 1993. *Desert Airways: A Short History of Clark County Aviation, 1920–1948.* Clark County Heritage Museum Occasional Paper #1.

Serling, Robert J. 1978. *The Only Way to Fly: The Story of Western Airlines, America's Senior Air Carrier.* New York: Doubleday.

Map 7.16. The Jet Age

Based on information from Daniel Bubb (1996) and Airline Timetable Images at http://www.timetableimages.com/

Bubb, Daniel. 2001. *Thunder in the Desert: Commercial Air Travel and Tourism in Las Vegas, 1959–2001.* Master's thesis, UNLV.

Serling, Robert J. 1978. *The Only Way to Fly: The Story of Western Airlines, America's Senior Air Carrier.* New York: Doubleday.

Map 7.17. Nonstop Service

Based on information from McCarran International Airport at https://www.mccarran.com/

Airport statistics from Clark County Department of Aviation at https://www.mccarran.com/Business/Statistics?id=62911

Map 7.18. Harry Reid International Airport

Based on current and historic imagery in Google Earth and USGS topographic maps from the 1950s.

Other sources:

Harry Reid airport at https://www.mccarran.com/

The FAA airport diagram for Harry Reid International Airport at https://www.faa.gov/airports/runway_safety/diagrams/

The *Red Flag Nellis 2013 In-Flight Guide* at http://www.vusaf.us/RFN13/resources/In-Flight-Guide/RFN13IFG.pdf

Map 7.19. Airspace

Based on the 2016 *Las Vegas Terminal Area Chart,* from the Federal Aviation Administration (https://www.faa.gov/air_traffic/flight_info/aeronav/digital_products/vfr/). These maps are replaced at regular intervals. Other details about airspace controls over Las Vegas and the Grand Canyon found in:

Ambrose, Skip, and Chris Florian. 2008. *Sound Levels and Audibility of Common Sounds in Frontcountry and Transitional Areas in Grand Canyon National Park, 2007–2008*. National Park Service. https://www.nps.gov/grca/learn/nature/upload/GRCAFrontcountryRep20081112.pdf

Federal Aviation Administration. 2015. *Aeronautical Information Manual*. https://www.faa.gov/air_traffic/publications/media/aim.pdf

Federal Aviation Administration. 2001. *Grand Canyon Aeronautical Chart*. https://www.faa.gov/air_traffic/flight_info/aeronav/digital_products/vfr/

United States Air Force. 2013. *Red Flag Nellis 2013 In-Flight Guide*. http://www.vusaf.us/RFN13/resources/In-Flight-Guide/RFN13IFG.pdf

Map 7.20. Airplane Crashes

There is no single database of airplane crash sites. Based on the Plane Crash Sites in Nevada at https://www.google.com/maps/d/viewer?mid=1AxDpW-wFseoXHv6f9d7aej1IFp64&hl=en_US, and supplemented by a search of the National Transportation Safety Board records and accounts of specific crashes from these sources:

Barlett, Donald L., and James B. Steele. 1979. *Howard Hughes: His Life and Madness*. New York: W. W. Norton & Company.

Chenowith, Bob, Rosie Pepito, Gary Warshefski, and Dave Conlin. 2006. *Lake Mead's Cold War legacy: The Overton B-29*. Harmon, David, ed. 2006. People, Places, and Parks: Proceedings of the 2005 George Wright Society Conference on Parks, Protected Areas, and Cultural Sites. Hancock, Michigan: The George Wright Society.

Davies, Steve. 2008. *Red Eagles: America's Secret MiGs*. Oxford: Osprey.

Hawley, Tom. 2016. Air force pilot makes ultimate sacrifice to save schoolkids. *Video Vault*. https://news3lv.com/features/video-vault/video-vault-air-force-pilot-makes-ultimate-sacrifice-to-save-schoolkids

Matzen, Robert. 2017. *Fireball: Carole Lombard and the Mystery of Flight 3*. Pittsburgh: GoodKnight Books

National Transportation Safety Board. 2016. *Accident Reports*. https://www.ntsb.gov/investigations/accidentreports/pages/accidentreports.aspx

Plaskon, Kyril D. 2008. *Silent Heroes of the Cold War*. Las Vegas: Stephens Press.

CHAPTER 8. CONTEMPORARY GEOGRAPHIES OF LAS VEGAS

Map 8.1. Land Ownership

Brean, Henry. 2018. Clark County wants input on expanding Las Vegas metro area. *Las Vegas Review-Journal* June 4.

Brean, Henry. 2018. Clark County unveils land proposal, draws ire from groups. *Las Vegas Review-Journal* June 5.

Inter-Tribal Council of Nevada. 1976. *Nuwuvi: A Southern Paiute History*. Reno: Inter-Tribal Council.

Kelly, Isabel T. 1934. Southern Paiute bands. *American Anthropologist* 36(4): 548–60.

Knack, Martha C. 2001. *Boundaries Between: The Southern Paiutes, 1775–1995*. Lincoln: University of Nebraska Press.

Von Till Warren, Elizabeth. 2001. *The History of Las Vegas Springs, a Disappeared Resource*. PhD Dissertation, Washington State University.

Bureau of Land Management. 2004. *Las Vegas Disposal Draft Boundary Environmental Impact Statement*. https://books.google.com/books?id=svkxAQAAMAAJ&printsec=frontcover&dq=las+vegas+valley+disposal+area&hl=en&sa=X&ved=0ahUKEwjPqPHmqf_UAhVDQSYKHVD2Av8Q6AEIJjAA

Public Lands Interpretive Association. 2018. *Nevada*. http://publiclands.org/Get-Books-and-Maps.php?plicstate=NV

Map 8.2. City and Town Boundaries

Elliott, Russell R. 1987. *History of Nevada, Second Edition*. Lincoln: University of Nebraska Press.

Fischer, Steve. 2005. *When the Mob Ran Vegas: Stories of Money, Mayhem, and Murder*. New York: MJF Books.

Griffin, Dennis N. 2005. *Policing Las Vegas: A History of Law Enforcement in Southern Nevada*. Las Vegas: Huntington Press.

Legislative Counsel Bureau. 1997. Reconfiguring the Structure of School Districts. https://www.leg.state.nv.us/Division/Research/Publications/InterimReports/1997/Bulletin97-04.pdf

McCracken, Robert D. 1992 *A History of Pahrump, Nevada*. Tonopah: Nye County Press.

Nevada Department of Taxation. 2021. Certified Population Estimates of Nevada's Counties, Cities and Towns 2000 to 2020. https://tax.nv.gov/uploadedFiles/taxnvgov/Content/TaxLibrary/Final%20Popul%20of%20Nevada%27s%20Counties%20and%20Incorp%20Cities%20and%20Unincorporated%20Towns%202020%20for%20Certification.pdf

United States Census Bureau. 2021. City and Town Population Totals: 2010–2020. https://www.census.gov/programs-surveys/popest/technical-documentation/research/evaluation-estimates/2020-evaluation-estimates/2010s-cities-and-towns-total.html

Map 8.3. People

Simich, Jerry L., and Thomas C. Wright, 2005. *The Peoples of Las Vegas: One City, Many Faces.* Reno: University of Nevada Press.

Simich, Jerry L., and Thomas C. Wright, Editors. 2010. *More Peoples of Las Vegas: One City, Many Faces.* Reno: University of Nevada Press.

Tsui, Bonnie. 2009. *American Chinatown: A People's History of Five Neighborhoods.* New York: Free Press.

United States Census Bureau. 2021. County Population Totals: 2010–2020. https://www.census.gov/programs-surveys/popest/technical-documentation/research/evaluation-estimates/2020-evaluation-estimates/2010s-counties-total.html

United States Census Bureau. 2018. *American FactFinder.* https://factfinder.census.gov/faces/nav/jsf/pages/index.xhtml

Map 8.4. Living in Las Vegas

Allen, Marshall, and Alex Richards. 2010. "Do no Harm: Hospital Care in Las Vegas". *Las Vegas Sun* https://lasvegassun.com/hospital-care/

American Library Association. 2018. *The Nation's Largest Public Libraries.* https://libguides.ala.org/libraryfacts

Coughenour, Courtney, and Jennifer R. Pharr. 2014. Community Health Indicators in Southern Nevada. *Nevada Journal of Public Health* 11: 92–112.

Robert Wood Johnson Foundation. 2021. *County Health Rankings & Roadmaps.* http://www.countyhealthrankings.org/

Rowley, Rex J. 2013. *Everyday Las Vegas: Local Life in a Tourist Town.* Reno: University of Nevada Press.

Silverstein, Lauren. 2018. Collaborative effort needed to combat doctor shortage in Nevada. *Las Vegas Sun* July 19. https://lasvegassun.com/news/2018/may/21/collaborative-effort-needed-to-combat-doctor-short/

The Trust for Public Land. 2021. *Park Score.* https://www.tpl.org/city/las-vegas-nevada

Map 8.5. Section Line Roads

Bureau of Land Management. 2017. BLM *Navigator.* https://navigator.blm.gov/home

Map 8.6. Street Names

Baldwin, Lawrence M., and Michel Grimaud. 1989. Washington, Montana, the Dakotas—and Massachusetts: A comparative approach to street names. *Names* 37(2) 115–38.

Hall-Patton, Mark P. 2009. *Asphalt Memories: Origins of Some Street Names of Clark County.* Henderson: Clark County Museum.

Rose-Redwood, Reuben, and Lisa Kadonaga. 2016. The corner of Avenue A and Twenty-Third Street: Geographies of street numbering in the United States. *Professional Geographer* 68(1) 39–52.

Map 8.7. Mid Mod Las Vegas

Corbin, April. 2017. Where the past resides: Touring local neighborhoods with historical patina. *Las Vegas Sun* June 25.

Henle, Mike. 2014. Rancho Circle boasts rich history, possible comeback. *Las Vegas Review-Journal* December 6. https://www.reviewjournal.com/life/rancho-circle-boasts-rich-history-possible-comeback/

Levine, Jack. 2018. *Uncle Jack's Very Vintage Las Vegas—Mid Century Modern Homes, Historic Las Vegas Neighborhoods, Historic Las Vegas and Urban Living.* https://veryvintagevegas.com/

PRW Project. 2021. La Concha Motel–Las Vegas–Paul Revere Williams. https://www.paulrwilliamsproject.org/gallery/la-concha-motel-the-neon-museum-las-vegas-nv/

Sisson, Patrick. 2017. *Vegas Modernism off the Strip, Mapped.* www.curbed.com

View, Andrew Taylor. 2016. East valley neighborhood designated by Las Vegas city council as historic. *Las Vegas Review-Journal* November 17.

Map 8.8. Homeless in Las Vegas

City of Las Vegas. 2018. *Homeless Services.* https://www.lasvegasnevada.gov

Hubert, Cynthia, Philip Reese, and Jim Sanders. 2013. Nevada buses hundreds of mentally ill patients to cities around country. *Sacramento Bee* April 14 https://www.sacbee.com/news/investigations/nevada-patient-busing/article2577189.html

National Alliance to End homelessness. 2018. *The State of Homelessness in America.* https://endhomelessness.org/homelessness-in-america/homelessness-statistics/state-of-homelessness-report/

O'Brien, Matthew. 2007. *Beneath the Neon: Life and Death in the Tunnels of Las Vegas.* Las Vegas: Huntington Press.

Map 8.9. Casinos

Al, Stefan. 2017. *The Strip: Las Vegas and the Architecture of the American Dream*. Cambridge: MIT Press.

Bracey, Earnest N. 2009. *The Moulin Rouge and Black Rights in Las Vegas*. Jefferson, NC: McFarland & Company.

Fowler, Catherine S., and Darla Garey-Sage. 2016. *Isabel T. Kelly's Southern Paiute Ethnographic Field Notes, 1932–1937*. Salt Lake City: University of Utah Press.

Hess, Alan. 1993. *Viva Las Vegas: After-Hours Architecture*. San Francisco: Chronicle Books.

Loi, Kim-Ieng, and Woo Gon Kim. 2010. Macao's casino industry: Reinventing Las Vegas in Asia. *Cornell Hospitality Quarterly* 51(2): 268–83.

Moody, Eric N. 1997. *The Early Years of Casino Gambling in Nevada, 1931–1945*. PhD Dissertation, University of Nevada, Reno.

Reid, Ed, and Ovid Demaris. 1963. *The Green Felt Jungle*. New York: Trident Press.

Rempel, William C. 2018. *The Gambler: How Penniless Dropout Kirk Kerkorian Became the Greatest Deal Maker in Capitalist History*. New York: HarperCollins.

Rowley, Rex J. 2011. Where locals play: Neighborhood casino landscapes in Las Vegas. UNLV Center for Gaming Research, Occasional Paper 9.

Schwartz, David G. 2013. *Grandissimo: The First Emperor of Las Vegas*. Las Vegas: Winchester Books.

Smith, John L. 2001. *Running Scared: The Life and Treacherous Times of Las Vegas Casino King Steve Wynn*. New York: Four Walls Eight Windows.

Swanson, Doug J. 2014. *Blood Aces: The Wild Ride of Benny Binion, the Texas Gangster Who Created Las Vegas Poker*. New York: Viking.

Venturi, Robert, Denise Scott Brown, and Steven Izenour. 1977. *Learning From Las Vegas*. Cambridge: MIT Press.

Map 8.10. Visiting Las Vegas

Las Vegas Convention and Visitors Authority. 2018. Las Vegas Statistics and Frequently Asked Questions. https://www.lvcva.com/

Map 8.11. Entertainment Capital of the World

Las Vegas Convention and Visitors Authority. 2018. Las Vegas Statistics and Frequently Asked Questions. https://www.lvcva.com/

Loi, Kim-Ieng, and Woo Gon Kim. 2010. Macao's casino industry: Reinventing Las Vegas in Asia. *Cornell Hospitality Quarterly* 51(2): 268–83.

Map 8.12. Union Las Vegas

Hopkins, A. D. 1999. Al Bramlet. *Las Vegas Review-Journal*. February 7. https://www.reviewjournal.com/news/al-bramlet/

Kraft, James P. 2010. *Vegas at Odds: Labor Conflict in a Leisure Economy, 1960–1985*. Baltimore: Johns Hopkins University.

Waddoups, C. Jeffrey. 2001. Wages in Las Vegas and Reno: How much difference do unions make in the hotel, gaming, and recreation industry? UNLV *Gaming Research & Review Journal* 6(1): 7–21.

Map 8.13. The Silver Screen

BYU Law Review. 1992. The Syufy Rosetta Stone. BYU *Law Review* 1992(2): 457–77.

Cinema Treasures. 2017. www.cinematreasures.org

DuVal, Gary. 2002. *The Nevada Filmography: Nearly 600 Works Made in the State, 1897 Through 2000*. Jefferson, NC: McFarland.

Las Vegas Sun. 1999. Red Rock closing further thins ranks of discount movie houses. *Las Vegas Sun*. https://lasvegassun.com/news/1999/oct/15/red-rock-closing-further-thins-ranks-of-discount-m/

Lenz, Richard. 1993. *Huntridge Theater*. National Register of Historic Places Nomination Form. http://focus.nps.gov/pdfhost/docs/NRHP/Text/93000686.pdf

Rivest, Mike. 2017. *Nevada Movie Theaters and Drive-ins*. www.movie-theatre.org

Rowley, Rex J. 2011. *Where locals play: Neighborhood casino landscapes in Las Vegas*. UNLV Center for Gaming Research, Occasional Paper 9.

Sanders, Don, and Susan Sanders. 2013. *The American Drive-In Movie Theatre*. London: Crestline.

Segrave, Kerry. 2006. *Drive-in Theaters: A History from Their Inception in 1933*. New York: McFarland & Company

Map 8.14. The Skyline

Al, Stefan. 2017. *The Strip: Las Vegas and the Architecture of the American Dream*. Cambridge: MIT Press.

Center for Land Use Interpretation. 1996. *The Nevada Test Site: A Guide to America's Nuclear Proving Ground*. Los Angeles: Center for Land Use Interpretation.

Garfield, Leanna. 2016. The "death ray hotel" burning Las Vegas visitors came up with a simple fix. *Business Insider*. http://www.businessinsider.com/the-vdara-death-ray-hotel-is-still-burning-people-in-las-vegas-2016-6

Hess, Alan. 1993. *Viva Las Vegas: After-Hours Architecture. San Francisco*: Chronicle Books.

Skyscraper Center. 2017. http://www.skyscrapercenter.com/

Stutz, Howard. 2014. Coming down: MGM begins dismantling Harmon Hotel. *Las Vegas Review-Journal*. June 20. https://www.reviewjournal.com/business/casinos-gaming/coming-down-mgm-begins-dismantling-harmon-hotel/

Venturi, Robert, Denise Scott Brown, and Steven Izenour. 1977. *Learning From Las Vegas*. Cambridge: MIT Press.

Map 8.15. Fire

Clark County Fire Department. 1980. *MGM Fire Investigation Report*. http://fire.co.clark.nv.us/(S(2eiykhxvo2ijka3z4gwh2v5p))/MGM.aspx

Coakley, Deirdre, Hank Greenspun, and Gary C. Gerard. 1982. *The Day the MGM Grand Hotel Burned*. Secaucus, NJ: Lyle Stuart, Inc.

Reed, Jack W. 1988. *Analysis of the Accidental Explosion at Pepcon, Henderson, Nevada, May 4, 1988*. Albuquerque, NM: Sandia National Laboratories.

Rempel, William C. 2018. *The Gambler: How Penniless Dropout Kirk Kerkorian Became the Greatest Deal Maker in Capitalist History*. New York: HarperCollins.

Routley, J. Gordon. 1988. *Fire and Explosions at Rocket Fuel Plant Henderson Nevada, May 4, 1988*. Washington, DC: Federal Emergency Management Agency.

Schwartz, David G. 2020. Could the Strip become a model of safety in a post-pandemic world? The response to the 1980 MGM fire might offer some lessons. https://knpr.org/desert-companion/2020-08/history-it-took-fire-change-us

Smith, John L. 1997. *No Limit: The Rise and Fall of Bob Stupak and Las Vegas' Stratosphere Tower*. Las Vegas: Huntington Press.

Map 8.16. Electricity

Nevada Energy. 2018. Home Page. https://www.nvenergy.com/

Map 8.17. Hoover Dam and Valley Electric

Arizona Power Authority. 2018. http://www.powerauthority.org/

Colorado River Commission of Nevada. 2018. http://crc.nv.gov/

McCracken, Robert D. 1992 *A History of Pahrump, Nevada*. Tonopah: Nye County Press.

McElroy, Sheila. 2010. *Boulder Dam-San Bernardino (115kV) Transmission Line*. Historic American Engineering Record. https://www.loc.gov/pictures/item/nv0441/.

Nevada Energy. 2018. https://www.nvenergy.com/

Scattergood, E. F. 1935. Engineering Features of the Boulder Dam–Los Angeles Lines. *Electrical Engineering* (May) 494–512.

Valley Electric Association. 2018. http://www.vea.coop/

Western Area Power Administration. 2018. https://www.wapa.gov/Pages/Western.aspx

Western Area Power Administration. 2014. *Desert Southwest Region's Facilities Historic Context Statement*. https://www.google.com/url?sa=t&rct=j&q=&esrc=s&source=web&cd=1&cad=rja&uact=8&ved=0ahUKEwjThPvzs9rYAhXLzVMKHU-_DmIQFggnMAA&url=https%3A%2F%2Fwww.wapa.gov%2Fregions%2FDSW%2FEnvironment%2FDocuments%2FDSWHistoricContextFinalSept2014.pdf&usg=AOvVaw3dNEsSv3xnINE4pYLJsBIA

Map 8.18. Pipelines

Associated Press. 1980. Jet fuel leak sparks Vegas fire. *Beaver County Times* December 23. https://news.google.com/newspapers?id=98YuAAAAIBAJ&sjid=GNsFAAAAIBAJ&pg=2134,5016662&dq=pipeline+rupture&hl=en

Bureau of Land Management. 2008. *Draft Environmental Impact Statement for the UNEV Pipeline*. Washington, DC: BLM.

Bureau of Land Management. 2012. *Calnev Pipeline Expansion Project Draft Environmental Impact Statement/Environmental Impact Report*. Washington, DC: BLM.

California State Lands Commission. 1987. *Mojave-Kern River–El Dorado Natural Gas Pipeline Projects Final Environmental Impact Report/Statement*. Sacramento: State Lands Commission.

National Transportation Safety Board. 1990. *Accident Report, Derailment of Southern Pacific Transportation Company Freight Train on May 12, 1989, and Subsequent Rupture of Calnev Petroleum Pipeline on May 25, 1989, San Bernardino, California*. http://www.pipelinesafetytrust.com/docs/ntsb_doc26.pdf

National Transportation Safety Board. 2011. *Pacific Gas and Electric Company Natural Gas Transmission Pipeline Rupture and Fire*. https://www.ntsb.gov/investigations/AccidentReports/Pages/PAR1101.aspx

Pipeline and Hazardous Materials Safety Administration. 2017. *National Pipeline Mapping System*. https://www.npms.phmsa.dot.gov/

United States Energy Information Administration. 2017. https://www.eia.gov/

Map 8.19. Mining

Adams, George. 1904. *Gypsum Deposits in the United States*. USGS Bulletin No. 223.

Castor, Stephen B., Steve Ludington, Brett T. McLaurin and Kathryn S. Flynn. 2006. *Mineral Resource Potential of the Rainbow Gardens Area of Critical Environmental Concern, Clark County, Nevada*. US Geological Survey Scientific Investigations Report 2006–5197.

Deiss, Charles. 1951. Dolomite Deposit near Sloan, Nevada. *United States Geological Survey Bulletin* 973C.

Hildebrand, G. H. 1982. *Borax Pioneer: Francis Marion Smith*. San Diego: Howell-North Books.

Hill, James M. 1914. *The Yellow Pine Mining District, Clark County, Nevada*. USGS Bulletin 540-F.

Hill, James M. 1916. *Notes on Some Mining Districts in Eastern Nevada*. USGS Bulletin 648.

Inter-Tribal Council of Nevada. 1976. *Nuwuvi: A Southern Paiute History*. Reno: Inter-Tribal Council.

Lincoln, Francis Church. 1923. *Mining Districts and Mineral Resources of Nevada*. Reno: Nevada Newsletter Publishing Company.

Moyer, Lorre A. 2004. *Mining Activity in Nevada, 1851–1995*. US Geological Survey Open-File Report 2004–1244

Perry, Rich and Mike Visher 2015. *Major Mines of Nevada 2014: Mineral Industries in Nevada's Economy*. Nevada Bureau of Mines and Geology Special Publication P-21.

Shine, Conor. 2013. Neighbors fighting to block mining in Sloan Hills emerge victorious. *Las Vegas Sun*. April 26.

Tingley, Joseph V. 1998. *Mining Districts of Nevada*. Nevada Bureau of Mines and Geology Report 47.

Tingley, Joseph V., Becky W. Purkey, Ernest M. Duebendorfer, Eugene I. Smith, Jonathan G. Price, and Stephen B. Castor. 2008. *Geologic Tours in the Las Vegas Area*. Reno: Nevada Bureau of Mines and Geology Special Publication 16.

Williams, Chuck, Linda McCollum, Dan Wray, Cam Camburn, Crystalaura Jackson, Norm Kresge, and Sharon Schaaf. 2015. *Seekers, Saints and Scoundrels: The Colorful Characters of Red Rock Canyon*. Salt Lake City: Paragon Press.

Map 8.20. Garbage and Pollution

Bortle, John E. 2001. Introducing the Bortle Dark-Sky Scale. *Sky and Telescope*. February: 126–29.

Cinzano, P., F. Falchi and C. D. Elvidge. 2001. The first world atlas of the artificial night sky brightness. *Monthly Notices of the Royal Astronomical Society* 328(3): 689–707.

Environmental Protection Agency. 2017. *Toxic Release Inventory* https://www.epa.gov/toxics-release-inventory-tri-program

Nevada Division of Environmental Protection. 2017. *Solid Waste*. https://ndep.nv.gov/land/waste/solid-waste

Robison, Jennifer. 2013. Apex landfill: There's no place like home for Las Vegas garbage. *Las Vegas Review-Journal* April 21. https://www.reviewjournal.com/local/local-las-vegas/apex-landfill-theres-no-place-like-home-for-las-vegas-garbage/

Map 8.21. Airport Noise

Ambrose, Skip, and Chris Florian. 2008. *Sound Levels and Audibility of Common Sounds in Frontcountry and Transitional Areas in Grand Canyon National Park, 2007–2008*. National Park Service. https://www.nps.gov/grca/learn/nature/upload/GRCAFrontcountryRep20081112.pdf.

Oldham, Jennifer. 2006. Air war in Vegas won't stay in Vegas. *Los Angeles Times*. November 19. http://articles.latimes.com/2006/nov/19/local/me-vegasflights19

Rogers, Keith. 2011. Emergency bomb drop rattles Nellis neighbors. *Las Vegas Review-Journal*. February 11. https://www.reviewjournal.com/local/local-las-vegas/emergency-bomb-drop-rattles-nellis-neighbors/.

United States Air Force. 2013. *Red Flag Nellis 2013 In-Flight Guide*. http://www.vusaf.us/RFN13/resources/In-Flight-Guide/RFN13IFG.pdf

CHAPTER 9. OUTSIDE LAS VEGAS

Map 9.1. Toiyabe National Forest

United States Forest Service. 2018. *Humboldt-Toiyabe National Forest*. https://www.fs.usda.gov/htnf

Map 9.2. Spring Mountains

Davis, Ren, and Helen Davis. 2011. *Our Mark on this Land: A Guide to the Legacy of the Civilian Conservation Corps in America's Parks*. Granville, Ohio: McDonald and Woodward Publishing.

Kolvet, Renee Corona, and Victoria Ford. 2006. *The Civilian Conservation Corps in Nevada: From Boys to Men*. Reno: University of Nevada Press.

Map 9.3. Desert National Wildlife Refuge

Executive Order 6065. 1933. *Boulder Canyon Wild Life Refuge*.

Fish and Wildlife Service. 2017. Desert National Wildlife Range. https://www.fws.gov/refuge/desert/

Map 9.4. Lake Mead National Recreation Area

Dodd, Douglas W. 2007. Boulder Dam Recreation Area: The Bureau of Reclamation, the National Park Service, and the origins of the National Recreation Area concept at Lake Mead, 1929–1936. *Southern California Quarterly* 88: 431–73.

Rothman, Hal K., and Daniel J. Holder. 2002. *Balancing the Mandates: an Administrative History of Lake Mead National Recreation Area.* Lake Mead: National Park Service.

Map 9.5. Valley of Fire State Park

Cox, Thomas R. 1993. Before the casino: James G. Scrugham, state parks, and Nevada's quest for tourism. *Western Historical Quarterly* 24 (3): 333–50.

DuVal, Gary. 2002. *The Nevada Filmography: Nearly 600 Works Made in the State, 1897 through 2000.* Jefferson, NC: McFarland.

Morrow, David K., et al. 2010. *General Management Plan, Valley of Fire State Park.* http://parks.nv.gov/wp-content/uploads/2012/ValleyofFire/VF_2010_GMP.pdf

Map 9.6. Red Rock Canyon National Conservation Area

DuVal, Gary. 2002. *The Nevada Filmography: Nearly 600 Works Made in the State, 1897 Through 2000.* Jefferson, NC: McFarland.

Moulin, Tom. 2013. *Red Rock Canyon Visitor Guide.* Las Vegas: Snell Press.

Tingley, Joseph V., Becky W. Purkey, Ernest M. Duebendorfer, Eugene I. Smith, Jonathan G. Price, and Stephen B. Castor. 2008. *Geologic Tours in the Las Vegas Area.* Reno: Nevada Bureau of Mines and Geology Special Publication 16.

Williams, Chuck, Linda McCollum, Dan Wray, Cam Camburn, Crystalaura Jackson, Norm Kresge, and Sharon Schaaf. 2015. *Seekers, Saints and Scoundrels: The Colorful Characters of Red Rock Canyon.* Salt Lake City: Paragon Press.

Map 9.7. Tule Springs Fossil Beds National Monument

National Park Service. 2015A. *Geology and Paleontology Explorations and Resources at Tule Springs Fossil Beds National Monument.* https://www.nps.gov/tusk/learn/historyculture/upload/Hardy_Bonde_Final-Report.pdf

National Park Service. 2015B. *Geologic Resources Inventory Scoping Summary, Tule Springs Fossil Beds National Monument, Nevada.* https://www.nps.gov/tusk/learn/historyculture/upload/TUSK_GRI_scoping_summary_2015-1016.pdf

Rafferty, Kevin A. 1984. *Cultural Resource Overview of the Las Vegas Valley.* BLM Technical Report No. 13.

Wormington, H. M., and Dorothy Ellis, eds. 1967. *Pleistocene Studies in Southern Nevada.* Nevada State Museum Anthropological Papers 13.

Map 9.8. Basin and Range National Monument

Houghton, John G., Clarence M. Sakamoto, and Richard O. Gifford. 1975. *Nevada's Weather and Climate.* Reno: Nevada Bureau of Mines and Geology.

Tetreault, Steve. 2015. New national monument blocks rail to Yucca. *Las Vegas Review-Journal,* July 13. http://www.reviewjournal.com/news/new-national-monument-blocks-rail-route-yucca.

Wilderness Connect. 2017. Download Wilderness Boundary Data. http://www.wilderness.net/NWPS/geography

Map 9.9. Grand Canyon West

Boyer, Diane E., and Robert H. Webb. 2007. *Damming Grand Canyon: The 1923 USGS Colorado River Expedition.* Logan: Utah State University Press.

Gould, H. R. 1960A. Character of the accumulated sediment. In *Comprehensive Survey of Sedimentation in Lake Mead, 1948–49.* USGS Professional Paper 295.

Gould, H. R. 1960D. Erosion in the reservoir. In *Comprehensive Survey of Sedimentation in Lake Mead, 1948–49.* USGS Professional Paper 295.

Powell, John Wesley. 1961 [originally published 1895] *The Exploration of the Colorado River and Its Canyons.* New York: Dover.

Smith, D. L., and Crampton, C. G. 1987. *The Colorado River Survey.* Salt Lake City: Howe Brothers Books.

WESTEC Services, Inc. 1980. *Historical and Architectural Resources within the Lower Colorado River System.*

Map 9.10. Sloan Canyon National Conservation Area

Bureau of Land Management. 2018. *Sloan Canyon National Conservation Area.* https://www.blm.gov/programs/national-conservation-lands/nevada/sloan-canyon-nca

Map 9.11. Gold Butte and Grand Canyon-Parashant National Monuments

Hill, James M. 1914. *The Grand Gulch Mining Region, Mohave County, Arizona.* USGS Bulletin 580.

WESTEC Services, Inc. 1980. *Historical and Architectural Resources within the Lower Colorado River System.*

Boyer, Diane E., and Robert H. Webb. 2007. *Damming Grand Canyon: The 1923 WESTEC Colorado River Expedition.* Logan: Utah State University Press.

Map 9.12. Wilderness

Wilderness Connect. 2017. Download Wilderness Boundary Data. http://www.wilderness.net/NWPS/geography

Map 9.13. Nellis Test and Training Range

Davies, Steve. 2008. *Red Eagles: America's Secret MiGs*. Oxford: Osprey.

Dreamland Resort. 2018. *Red Flag Air Exercises*. http://www.dreamlandresort.com/info/flags.html

Prince, Todd. 2017. Middle East air force books entire hotel tower on Las Vegas Strip. *Las Vegas Review-Journal* August 4.

Rininger, Tyson. 2006. *Red Flag: Air Combat for the 21st Century*. St. Paul, MN: Zenith Press.

United States Air Force. 2013. *Red Flag Nellis 2013 In-Flight Guide*. http://www.vusaf.us/RFN13/resources/In-Flight-Guide/RFN13IFG.pdf

United States Air Force. 2018. Flying Operations. http://www.nellis.af.mil/Home/Flying-Operations/

Map 9.14. Nevada National Security Site

Center for Land Use Interpretation. 1996. *The Nevada Test Site: A Guide to America's Nuclear Proving Ground*. Los Angeles: Center for Land Use Interpretation.

Department of Energy. 2000. *United States Nuclear Tests July 1945 through September 1992*. https://web.archive.org/web/20061012160826/http://www.nv.doe.gov/library/publications/historical/DOENV_209_REV15.pdf

Dewar, James A. 2007. *To the End of the Solar System: The Story of the Nuclear Rocket, Second Edition*. Burlington, Ontario: Apogee Books

Fehner, Terrence R., and F. G. Gosling. 2000. *Origins of the Nevada Test Site*. https://www.nnss.gov/docs/docs_LibraryPublications/DOE_MA0518.pdf

Historic American Engineering Record. n.d. *Nevada Test Site, Test Cell A Facility*. https://www.loc.gov/

Historic American Engineering Record. n.d. *Nevada Test Site, Test Cell C Facility, Building No. 3210*. https://www.loc.gov/

Historic American Engineering Record. n.d. *Nevada Test Site, Reactor Maintenance and Disassembly Facility, Building No. 3110*. https://www.loc.gov/

Historic American Engineering Record. n.d. *Nevada Test Site, Engine Maintenance and Disassembly Facility*. https://www.loc.gov/

Historic American Engineering Record. n.d. *Nevada Test Site, Pluto Facility, Disassembly Building*. https://www.loc.gov/

Historic American Building Survey. n.d. *Environmental Protection Agency Farm*. http://cdn.loc.gov/master/pnp/habshaer/nv/nv0200/nv0225/data/nv0225data.pdf

Historic American Building Survey. n.d. *Japanese Village*. https://cdn.loc.gov/master/pnp/habshaer/nv/nv0100/nv0166/data/nv0166data.pdf

Johnston, William. 2001. *Chronological Listing of Above Ground Nuclear Detonations*. http://www.johnstonsarchive.net/nuclear/atestoo.html

Meier, Charles W. 2006. *Before the Nukes—The Remarkable History of the Nevada Test Site*. Pleasonton, CA: Lansing Publications.

Titus, A. Costandina. 1986. *Bombs in the Backyard: Atomic Testing and American Politics, Second Edition*. Reno: University of Nevada Press.

CHAPTER 10. UNBUILT LAS VEGAS

Map 10.1. Unbuilt Las Vegas I

Adams, Mark. 2012. *Failed Vegas Projects and Old Dreams*. http://www.vegastodayandtomorrow.com/page4dreams.htm

Army Corps of Engineers. 1959. *Report on Survey for Flood Control Las Vegas Wash and its Tributaries: Las Vegas and Vicinity, Nevada*. Los Angeles: US Army Corps of Engineers.

Center for Gaming Research. 2016. *Welcome to Paradise, Misplaced*. http://gaming.unlv.edu/Xanadu/index.html

Marlow, Jarvis. 2011. *Taming the waters that taketh from the devil's playground: A history of flood control in Clark County, Nevada, 1955–2010*. Thesis, UNLV.

Nevada Department of Highways, 1963. *Las Vegas Valley Urban Transportation Study*. Reno: Nevada Department of Highways.

Vellota, Richard. 2016. Elevated McCarran airport expressway plan moves forward. *Las Vegas Review-Journal*. April 14. https://www.reviewjournal.com/traffic/elevated-mccarran-airport-expressway-plan-moves-forward/

Map 10.2. Unbuilt Las Vegas II

Adams, Mark. 2012. *Failed Vegas Projects and Old Dreams*. http://www.vegastodayandtomorrow.com/page4dreams.htm

Army Corps of Engineers. 1959. *Report on Survey for Flood Control Las Vegas Wash and its Tributaries: Las Vegas and Vicinity, Nevada*. Los Angeles: US Army Corps of Engineers.

Friess, Steve. 2008. Arizona Developers Welcome Spillover From Las Vegas. *New York Times* August 9. http://www.nytimes.com/2008/08/10/us/10kingman.html

Garrison, Omar. 1970. *Howard Hughes in Las Vegas*. New York: Lyle Stuart.

Marlow, Jarvis. 2011. *Taming the waters that taketh from the devil's playground: A history of flood control in Clark County, Nevada, 1955–2010.* Thesis, UNLV.

Nevada Department of Transportation. 2002. *I-15 Northeast Corridor Study.* Carson City: NDOT.

Phelan, James R., and Lewis Chester. 1997. *The Money: The Battle for Howard Hughes's billions.* New York: Random House.

Rothberg, Daniel. 2017. The battle to build near Red Rock Canyon is coming to a head–again. *Las Vegas Sun.* February 20. https://lasvegassun.com/news/2017/feb/20/save-red-rock-clark-county-commission/https://lasvegassun.com/news/2017/feb/20/save-red-rock-clark-county-commission/

Segall, Eli. 2015. Developers still see potential in Coyote Springs, plan to revive the housing development. *Las Vegas Sun* August 14. https://vegasinc.lasvegassun.com/business/real-estate/2015/aug/14/coyote-springs-development-golf-community/

Wargo, Buck. 2011. Developer buys Great Mall of Las Vegas parcel. *Las Vegas Sun* January 4. https://lasvegassun.com/news/2011/jan/04/developer-buys-great-mall-las-vegas-parcel/

Map 10.3. Colorado River Railroads

Jones, Florence Lee, and John F. Cahlan. 1975. *Water: A History of Las Vegas, Volume 1.* Las Vegas: Las Vegas Valley Water District.

Myrick, David F. 2010. *Railroads of Arizona, Volume 6: Jerome and the Northern Roads.* Berkeley, CA: Signature Press.

Smith, D. L., and Crampton, C. G. 1987. *The Colorado River Survey.* Salt Lake City: Howe Brothers Books.

Map 10.4. Boulder Dam

Boyer, Diane E., and Robert H. Webb. 2007. *Damming Grand Canyon: The 1923* USGS *Colorado River Expedition.* Logan: Utah State University Press.

Bureau of Reclamation. 1948. *Boulder Canyon Project Final Reports: Part 1-Introductory.* Boulder City: Bureau of Reclamation.

Map 10.5. Agricultural Las Vegas

Carpenter, Everett. 1915. *Ground Water in Southeastern Nevada.* USGS Water Supply Paper 365.

Fowler, Catherine S., and Darla Garey-Sage. 2016. *Isabel T. Kelly's Southern Paiute Ethnographic Field Notes, 1932–1937.* Salt Lake City: University of Utah Press.

McWilliams, J. T. 1918. *Map of Las Vegas Valley Showing Artesian Wells.* http://digital.library.unlv.edu/objects/LV_Maps/478

Nevada State Engineer. 1935. *Irrigable Areas Near Boulder Dam.* http://digital.library.unlv.edu/objects/LV_Maps/730

Thompson, David G. 1921. *Routes to Desert Watering Places in the Mohave Desert Region, California.* USGS Water Supply Paper 490-B.

Waring, Gerald A. 1921. *Ground Water in Pahrump, Mesquite, and Ivanpah Valleys, Nevada and California.* USGS Water Supply Paper 450.

Map 10.6. Bridge Canyon Dam

Bureau of Reclamation. 1964. *Pacific Southwest Water Plan.* Washington, DC: Bureau of Reclamation. https://www.usbr.gov/lc/region/programs/crbstudy/PSWPRptJan64.pdf

Dean, Robert. 1997. Dam building still had some magic then: Stewart Udall, the Central Arizona Project, and the evolution of the Pacific Southwest Water Plan, 1963–1968. *Pacific Historical Review* 66(1): 81–98.

National Park Service. 1950. *A Survey of the Recreational Resources of the Colorado River Basin.* Washington, DC: Government Printing Office.

Pearson, Byron. 1994. Salvation for Grand Canyon: Congress, the Sierra Club, and the Dam Controversy of 1966–1968. *Journal of the Southwest* 36(2): 159–75.

Map 10.7. By Train to Las Vegas

Federal Railroad Administration. 2011. *Final Environmental Impact Statement and Final Section 4(f) Evaluation for the Proposed DesertXpress High-Speed Passenger Train Victorville, California to Las Vegas, Nevada.* https://www.fra.dot.gov/eLib/Details/L01361

Map 10.8. Interstate 11

Arizona Department of Transportation. 2009. *I-40/US 93 West Kingman TI Final Feasibility Report.* Phoenix: Arizona DOT.

Arizona Department of Transportation and Nevada Department of Transportation. 2014. *Corridor Concept Report I-11 and Intermountain West Corridor Study* http://i11study.com/IWC-Study/PDF/2014/I-11CCR_Report_2014-12_sm.pdf

Nevada Department of Transportation and Arizona Department of Transportation. 2014. I-11 and Intermountain West Corridor Study. http://i11study.com/IWC-Study/index.asp

Map 10.9. Ivanpah Airport

Clark County Department of Aviation. 2006. *Project Definition and Justification: Proposal to Construct and Operate a New Supplemental Commercial Service Airport in the Ivanpah Valley.*

Garrison, Omar. 1970. *Howard Hughes in Las Vegas.* New York: Lyle Stuart.

Map 10.10. Yucca Mountain Railroad

Department of Energy. 2008. *Final Supplemental Environmental Impact Statement for a Geologic Repository for the Disposal of Spent Nuclear Fuel and High-Level Radioactive Waste at Yucca Mountain, Nye County, Nevada–Nevada Rail Transportation Corridor and Final Environmental Impact Statement for a Rail Alignment for the Construction and Operation of a Railroad in Nevada to a Geologic Repository at Yucca Mountain, Nye County, Nevada.* https://energy.gov/downloads/final-supplemental-environmental-impact-statement-geologic-repository-disposal-spent-2

Tetreault, Steve. 2015. New national monument blocks rail to Yucca. *Las Vegas Review-Journal,* July 13. http://www.reviewjournal.com/news/new-national-monument-blocks-rail-route-yucca.

Walker, J. Samuel. 2009. *The Road to Yucca Mountain: The Development of Radioactive Waste Policy in the United States.* Berkeley: University of California Press.

Map 10.11. Northern Nevada Pipeline

Bureau of Land Management. 2012. *Clark, Lincoln, and White Pine Counties Groundwater Development Project Record of Decision.* Washington, DC: Bureau of Land Management.

Deacon, James E., Austin E. Williams, Cindy Deacon Williams, and Jack E. Williams. 2007. Fueling population growth in Las Vegas: How large-scale groundwater withdrawal could burn regional biodiversity. *BioScience* 57(8): 688–98.

Southern Nevada Water Authority. 2012. *Topographic Maps: Conceptual Plan of Development.* Las Vegas: Southern Nevada Water Authority.

Map 10.11. Bombing the Bighorn

United States Air Force. 2017. *NTTR Military Land Withdrawal Legislative EIS.* http://www.nttrleis.com/

United States Fish and Wildlife Service. 2009. *Desert National Wildlife Refuge Complex: Ash Meadows, Desert, Moapa Valley, and Pahranagat National Wildlife Refuges Final Comprehensive Conservation Plan and Environmental Impact Statement—Volume 2.* https://www.fws.gov/uploadedFiles/CCP%20Vol%202.pdf

CHAPTER 11. REMEMBERING LAS VEGAS

Map 11.1. The Rat Pack

Fischer, Steve. 2005. *When the Mob Ran Vegas: Stories of Money, Mayhem, and Murder.* New York: MJF Books.

Fishgall, Gary. 2003. *Gonna do Great Things: The Life of Sammy Davis, Jr.* New York: Scribner.

Levy, Shawn. 1998. *Rat Pack Confidential.* New York: Anchor Books.

Summers, Anthony and Robyn Swan. 2005. *Sinatra: The Life.* New York: Alfred A. Knopf.

Taraborrelli, J. Randy. 1997. *Sinatra: A Complete Life.* Secaucus, NJ: Birch Cone Press.

Map 11.2. Viva Las Vegas

Cotton, Lee. 1998. *All Shook up: Elvis Day-by-Day 1954–1977, Second Edition.* Ann Arbor, MI: Popular Culture.

Guralnick, Peter, and Ernst Jorgensen. 1999. *Elvis Day by Day.* New York: Ballantine Books.

Guralnick, Peter. 1999. *Careless Love: The Unmaking of Elvis Presley.* Boston: Little, Brown and Company.

Guralnick, Peter. 1994. *Last Train to Memphis: The Rise of Elvis Presley.* Boston: Little, Brown and Company.

Map 11.3. Howard Hughes

Barlett, Donald L., and James B. Steele. 1979. *Howard Hughes: His Life and Madness.* New York: W. W. Norton and Company.

Garrison, Omar. 1970. *Howard Hughes in Las Vegas.* New York: Lyle Stuart.

Magnesen, Gary. 2005. *The Investigation: A Former FBI Agent Uncovers the Truth Behind Howard Hughes, Melvin Dummar, and the Most Contested Will in American History.* Fort Lee, NJ: Barricade Books

Phelan, James R., and Lewis Chester. 1997. *The Money: The Battle for Howard Hughes's Billions.* New York: Random House.

Rempel, William C. 2018. *The Gambler: How Penniless Dropout Kirk Kerkorian Became the Greatest Deal Maker in Capitalist History.* New York: HarperCollins.

Map 11.4. Bond . . . James Bond

Lycett, Andrew. 1995. *Ian Fleming.* London: Weidenfeld & Nicolson.

Raento, Paulina. 2017. All in—and More! Gambling in the James Bond Movies. *UNLV Gaming Research & Review Journal* 21(1): 49–67.

Map 11.5. Mr. Showmanship

Pyron, Darden Asbury. 2000. *Liberace: An American Life.* Chicago: University of Chicago Press.

Thomas, Bob. 1987. *Liberace: The True Story*. New York: St. Martin's Press.

Map 11.6. The Gauntlet

Papa, Paul W. 2014. *Discovering Vintage Las Vegas: A Guide to the City's Timeless Shops, Restaurants, Casinos, & More*. Guilford, CT: Globe Pequot Press.

Map 11.7. Lost Vegas

Al, Stefan. 2017. *The Strip: Las Vegas and the Architecture of the American Dream*. Cambridge: MIT Press.

Bracey, Earnest N. 2009. *The Moulin Rouge and Black Rights in Las Vegas*. Jefferson, NC: McFarland & Company.

Hess, Alan. 1993. *Viva Las Vegas: After-Hours Architecture*. San Francisco: Chronicle Books.

Map 11.8. Historic Las Vegas

National Park Service. 2017. *National Register of Historic Places Official Website*. https://www.nps.gov/nr/

Woodward, James and Cindy Myers. 1983. *Boulder City Historic District: Nomination to the National Register of Historic Places*.

Wright, Dorothy. 1990. *Las Vegas High School Neighborhood Historic District*. National Register of Historic Places Registration Form.

Little Church of the West. 2017. https://littlechurchlv.com/

Map 11.9. Las Vegas Memories

Knight, Kenneth C., PhD and T. H. Turner. 1988. *An Inventory of Nevada's Historic Bridges*. Carson City: Nevada Department of Transportation.

Kudialis, Chris. 2017. Last rounds on the Strip: Golf course made big impressions, but succumbs to economics. *Las Vegas Sun* December 3.

Las Vegas Historical Society. 2017. http://lasvegashistoricalsociety.org/

Papa, Paul W. 2014. *Discovering Vintage Las Vegas: A Guide to the City's Timeless Shops, Restaurants, Casinos, & More*. Guilford, CT: Globe Pequot Press

Roman, James. 2011. *Chronicles of Old Las Vegas: Exposing Sin City's High-Stakes History*. New York: Museyon.

Snyder, John W. 2008. *Historic Bridges of Nevada, Volume 2*. Nevada Department of Transportation.

Index

About the Author

Joe Weber grew up in the Las Vegas area. His fondest memories of the city during the 1970s and 1980s include spending time at the Meadows Mall, dining at the Food Factory restaurant, and driving by the Aku Aku statues on the Strip. Although he left the area to attend the University of Arizona, he often returns to Las Vegas. He is now a professor of geography at the University of Alabama. His academic work focuses on the changing geography of national parks and the American highway system. His favorite activities include searching out abandoned roads, bridges, and other remnants of past landscapes on foot and in Google Earth.